Ku-Band Satellite

Theory Installation & Repair

Fourth Edition - International

Frank Baylin

contributions by
Brent Gale and John McCormac

Published by Baylin Publications

ACKNOWLEDGEMENTS

A number of people helped in the production of this book. Richard Zlotky, president of Earthbound, Inc. provided calculations on antenna wind loading. Jane Maier and Bee Ferrigno proofed our manuscript during its production stages. Amy Lockard created some of the professional line drawings and Jennifer Stewart-Laing produced all the maps.

We are grateful to the hard working staff at Johnson Publishing Company in Boulder, Colorado who have never missed a deadline.

Steven Birkill, a TVRO pioneer, kindly provided many of the European satellite footprint maps that appear in the Appendix. He has 15 years experience in the field and is now based in Gloucestershire, England where he runs his own design and engineering consultancy business Real-World Technology Ltd. His recent design credits included the tuner and signal circuits of the Amstrad SRX and SRD receivers.

For speaking engagements or technical consulting services contact:

Frank Baylin
1905 Mariposa
Boulder, CO 80302
Tel: (303) 449-4551
FAX: (303) 939-8720

Brent Gale
2748 Winding Trail Drive
Boulder, CO 80302
(303) 449-0122

John McCormac
22 Viewmount
Waterford, Ireland
353-51-73640

Copyright 1991 by Baylin Publications

All rights reserved. Reproduction or publication of the content in any manner, without express written permission of the publisher, is prohibited. No liability is assumed with respect to the use of this information herein.

Artist: Amy Lockard and Rod Schubert
Maps: Jennifer Stewart-Laing
Cover Illustration: Peter Stallard

First Edition - September 1986
Second Edition - September 1987
Third Edition - April 1990
Fourth Edition - July 1991

ISBN: 0-917893-14-X

DEDICATION

"The best thing for being sad is to learn something. That is the only thing that never fails. You may grow old and trembling in your anatomies, you may lie awake at night listening to the disorder of your veins, you may miss your only love, you may see the world about you devastated by evil lunatics, or know your honour trampled in the sewers of baser minds.

"There is only one thing for it then - to learn. Learn why the world wags and what wags it. That is the only thing that the mind can never exhaust, never alienate, never be tortured by, never fear or distrust, and never dream of regretting. Learning is the thing for you."

from: TC White's "The Once and Future King" - Merlin's Advice to Arthur

Brief Table of Contents

I. The Basics of Satellite Communication 3
II. Component Theory and Operation 41
III. Designing and Testing Ku-Band Systems 147
IV. Selecting Equipment for Ku-Band Earth Stations 167
V. Installing Ku-Band Satellite TV Systems 189
VI. Retrofitting C-Band Systems for Ku-Band Reception 277
VII. Multiple Receiver Satellite TV and Distribution Systems 287
VIII. Worldwide Ku-Band Satellite Television 327
IX. Troubleshooting and Repair 335

APPENDICES

A. The Decibel Notation 359
B. Satellite TV Equations 361
C. Satellite Footprints and Az-El Aiming Charts 367
D. Glossary of Common Broadcast and Satellite Television Terms 393
E. Channels, Wavelengths & Bandwidths 402
F. Equipment Manufacturers 403
G. Reference Publications 409
H. Order Forms 411

Table of Contents

I. THE BASICS OF SATELLITE COMMUNICATION ... 3
   A. INTRODUCTION ... 3
   B. THE SATELLITE CIRCUIT ... 4
      The Uplink ... 4
      The Communication Satellite ... 6
      The Receiving Station ... 7
   C. MICROWAVES ... 7
      Signal Frequency ... 7
      Phase ... 8
      Transmission Power ... 8
      Signal Polarity ... 9
   D. COMMUNICATION FUNDAMENTALS ... 10
      Coding the Message ... 10
      Modulation ... 11
          Analogue Modulation ... 11
          Digital Modulation ... 12
      Bandwidth ... 12
      Amplification and Attenuation ... 12
      Noise ... 13
   E. MICROWAVES AND SATELLITE COMMUNICATION ... 14
   F. UPLINK OPERATION ... 16
      Portable Uplinks, Satellite News Gathering and Educational TV ... 15
   G. WORLDWIDE FREQUENCY ALLOCATIONS ... 18
      European DBS Satellite TV - TV-SAT/TDF/BSB ... 20
   H. SATELLITE DESIGN AND OPERATION ... 22
      The Downlink Path and Satellite Footprints ... 22
      Understanding Footprint Maps ... 22
      Satellite Operation - Audio and Video Channels ... 25
          Video Transponder Formats ... 25
          Audio Transponder Formats ... 29
   I. SATELLITE LAUNCHING AND MAINTENANCE ... 30
      Launch Sequence ... 30
          Transfer Orbit ... 32
          Plane Change ... 32
          Drift Orbit ... 32
          Geostationary Orbit ... 33
          Stationkeeping ... 33
      Future Trends in Satellite Design and Operation ... 35
          Low and Inclined Orbit Satellites ... 36
          High Power Satellites ... 37
          Frequency Reuse and Multiple Frequency Satellites ... 38
          Digital Compression and Other Improvements in Signal Processing ... 38
          Intersatellite Communication ... 38

II. COMPONENT THEORY AND OPERATION ... 41
   A. INTRODUCTION ... 41
   B. ANTENNAS ... 41
      Single-Focus Antennas ... 42
          Prime Focus Reflector ... 42
          Cassegrain Reflector ... 42
          Offset Fed Reflector ... 43
          Horn Reflector ... 44
      The Flat Plate Antenna ... 45
          Active Flat Plate Antenna ... 45
              The Squarial ... 47
      Passive Flat Plate Antennas ... 48
      Multi-focus Antennas ... 49

TABLE OF CONTENTS

- Judging Antenna Performance ... 50
  - Antenna Gain ... 51
  - Antenna Efficiency ... 51
  - Beamwidth and Side Lobe Patterns ... 52
  - Antenna Noise ... 54
  - f/D - Antenna Depth ... 55
- Antenna Construction ... 56
  - Spinning or Hydroforming ... 57
  - Stamping ... 58
  - Fiberglass ... 58
  - Wire Mesh or Expanded Metal ... 60
  - Molded Plastic ... 60
  - Selecting an Antenna ... 60
- Wind Loading ... 61

C. ANTENNA MOUNTS ... 62
- Types of Mounts ... 63
  - Polar Mounts ... 64
  - The Modified Polar Mount ... 64
  - Az-El Mounts ... 65

D. ACTUATORS ... 66
- Monitoring Antenna Position ... 66
- Feedback Element Types ... 67
  - Resistive Feedback ... 67
  - Reed Switch Feedback ... 68
  - Optical Feedback ... 68
  - Hall Effect Feedback ... 68
- Mechanical Construction ... 69
  - Linear Actuators ... 69
  - Horizon-to-Horizon Actuators ... 70
- Electrical Connections ... 70
- User and Receiver Interfaces ... 70
- Improving Actuator Designs ... 71

E. FEEDHORNS ... 72
- Illumination Patterns and f/D Ratios ... 72
  - Flat Plate Antennas and Feed Illumination ... 73
  - Horn Antennas and Feed Illumination ... 73
- Waveguides ... 74
- The Feedhorn/LNB Probe ... 74
- Varieties of Feedhorns ... 74
  - Pyramidal Feedhorn ... 74
  - Circular Feedhorn ... 75
  - Conical Feedhorn ... 76
  - Scalar Feedhorn ... 76
  - Polyrod Lens Feed ... 77
  - Feed Configurations ... 78
  - Prime Focus Feed ... 78
  - Cassegrain Feed System ... 78
  - Splash Plate Feed ... 78
  - Offset Feed ... 79
- Voltage Standing Wave Ratio (VSWR) ... 80
- Polarity Selection ... 80
  - Mechanical Rotors ... 80
  - Ferrite or Magnetic Devices ... 82
  - Pin Diode Polarity Switching ... 83
- Orthomode Transducers (OMTs) ... 84
- Detection of Circularly Polarised Signals ... 85
- Feedhorn/Customer Interfaces ... 86

F. LOW NOISE BLOCK DOWNCONVERTERS (LNBs) ... 86
- LNAs, LNBs and LNCs ... 87
- Sources of Noise ... 88
  - Thermal Noise ... 88
  - Transistor Noise ... 88

TABLE OF CONTENTS

    Noise Temperature and Noise Figure 89
    Judging LNB Performance 89
        Noise Temperature 89
        Gain 90
        VSWR 91
    Electrical Connections 91
    Mechanical Packages 91
    The Future of LNBs 91
    LNB Performance Calculations 91
        Importance of the LNA 91
        Transition from LNAs to LNA/BDCs to LNBs 92
        LNB Design/Performance Considerations 92
            C-Band LNB 93
            Ku-band LNB 94
G. DOWNCONVERSION METHODS 94
    Downconversion and Satellite Receiver Design 94
        Single Downconversion 95
        Dual Downconversion 95
        Block Downconversion 96
    Downconversion in Ku-band Systems 97
    Downconversion Technical Background 97
        Microwave Integrated Circuits (MICs) 98
        3 dB Hybrid Mixer 98
        The Image Rejection Mixer 98
        The Antiparallel Diode Mixer 99
        Single Conversion 100
H. LNB DESIGN 100
  Isolator 102
  Waveguide to Microstrip Transition 102
  Low Noise Amplifier 102
  GaAsFET Amplifier Power Supply 103
  Bandpass Filter 103
  The Mixer 104
  Dielectric Resonator Oscillator 104
  IF Amplifier Block 105
I. COAXIAL CABLE AND CONNECTORS 105
    Types of Coax 106
        Multi-Run Cables 106
    Judging Cable Performance and Use 107
        Characteristic Impedance 107
        Signal Losses 108
    Coax Connectors 109
    Water and Aging 110
    Summary 110
    The SCART Connector 110
J. SATELLITE RECEIVERS 112
    Receiver Design 112
        Power Supply 112
        Downconverter/Tuner 113
            Choice of IF 113
        Tuner - Selecting Channels 114
            Rotary Tuning 114
            Preset and Detent Tuning .114
            Synthesised Tuning 114
        The Block Tuner 115
            The Mitsumi Tuner 116
            The Mitsumi IF Downconverter 116
        Final IF Stage 116
        Bandpass Filter 116
        Limiter Circuits 117
        Detector/Demodulator 117

TABLE OF CONTENTS

        PLL Demodulators 117
        Quadrature Demodulators 118
        Balanced Demodulators 119
        Delay Line Discriminator 119
        The Mitsumi and Other Demodulator Modules 120
      Baseband Processing 120
        Video Filtering 120
        De-Emphasis Circuit 120
        Clamp Circuit 120
        Filtering 122
        Audio Processor 122
        Audio Demodulators 122
      Modulator 122
      Judging Receiver Performance 123
        Signal-to-Noise Ratio and Picture Quality 123
          Signal-to-Noise Ratio 123
          Picture Quality 123
        Video Bandwidth 124
        Receiver Threshold 124
          Threshold Extension Techniques 125
        Electrical Connections 126
      User Interfaces and Adjustments 127
  K. MODULATORS 127
      Commercial Grade Modulators 128
  L. TELEVISION RECEIVERS AND MONITORS 129
      Understanding Television Operation 129
        Scanning 129
        Vertical Resolution 130
        Trace Rate and Horizontal Resolution 130
      The Black and White Television Signal 130
      The Colour Television Signal 131
  M. WORLDWIDE BROADCAST FORMATS 134
      PAL, SECAM and NSTC 134
      The MAC System 140
  N. MONOPHONIC AND STEREO AUDIO RECEPTION 141
      Transmitting High Fidelity Sound 141
        Pre-Emphasis and De-Emphasis 141
        Amplitude Companding 142
        Spectral Companding 143
      Satellite Audio Reception 143
        Single Channel per Carrier Audio 144
      TVRO Stereo Reception 144
        Multiplex Stereo 144
        Warner Amex Stereo 145
        Discrete Stereo 145
        Processed Narrow Deviation Stereo 145

III. DESIGNING AND TESTING KU-BAND SYSTEMS 147
  A. OVERALL DESIGN CONSIDERATIONS 147
      The Link Equation 147
      EIRP 148
      Path Loss 149
        Effects of Rain and Atmospheric Attenuation 149
        Typical Ku-band Path Loss 152
      System Noise Temperature 152
      How Small Can an Antenna Be? 154
        Determining C/R 154
        Determining G/Tsys 154
        Recommended Antenna Sizes 155
  B. BEAMWIDTH AND SATELLITE SPACING 156
      The Importance of Beamwidth 156
      The Role of Antenna Quality 157

TABLE OF CONTENTS

- Is One Degree Spacing Really One Degree? 157
- Satellite Spacing and Regulation 158
- C. INTERPRETING TELEVISION TEST SIGNALS 158
  - Test Parameters and Picture Distortion 158
  - The Basic Test Signals 159

IV. SELECTING EQUIPMENT FOR KU-BAND EARTH STATIONS 167
- A. EVALUATING EQUIPMENT FOR RECEIVING KU-BAND BROADCASTS 168
  - Mounts 168
    - Actuators 169
    - Antennas 169
      - Solid Antennas 171
      - Mesh Antennas 172
    - Feed Supports 173
    - Feedhorns 174
    - LNBs 175
    - Satellite Receivers 176
    - Stereo Processors 178
    - Television Monitors 178
    - Warranties 178
    - Multistandard Televisions 181
- B. KU-BAND TERRESTRIAL INTERFERENCE 181
- C. SCRAMBLING TECHNOLOGIES 182
  - Scrambling Methods 182
    - Basic Parameters 182
    - Video Scrambling 183
    - Audio Scrambling 183
  - A Review of Scrambling Systems 183
    - Oak Orion and VideoCipher II 183
    - The Filmnet System 184
    - Teleclub Payview III 185
    - VideoCrypt 186
    - Multiplexed Analog Components (MAC) 186
      - B-MAC 187
      - EuroCypher 187
      - D2-MAC 188

V. INSTALLING KU-BAND SATELLITE TV SYSTEMS 189
- An Overview 189
- A. THE SITE SURVEY 191
  - Ensuring a Clear View of the Arc 191
    - Determining a True North/South Bearing 192
    - Determining the Azimuth and Elevation Look Angles 199
    - Choosing the Best Antenna Site 199
  - Planning the Installation 201
- B. ANTENNA SUPPORT STRUCTURES 205
  - Pole and Pad Mounts at Ground Level 205
    - Preparing for the Job 205
    - Pre-Trenching for Conduit Placement 205
    - The Fundamentals of Concrete 205
    - Freezing, Underground Water and Stability 207
    - Pole or Post Supports 207
    - Concrete Pads 210
    - Three Point Pads and Pier Foundations 211
    - Combinations of Pole and Pad Supports 214
    - Ground Supports Not Requiring Concrete 214
  - Above-Ground Mounts 215
    - Roof, Wall and Eave Mounts 215
      - Large Antennas 216
      - Small Antennas 219
    - Towers and Long Pole Supports 221

TABLE OF CONTENTS

- C. CABLE RUNS AND TRENCHING 223
  - Digging the Trench for Ground Mounts 223
  - The Use of Conduit 224
  - Types of Cables for Satellite TV 226
  - The Entry into the Home 228
- D. ASSEMBLING THE ANTENNA, FEEDHORN & LNB 229
  - Assembling the Antenna 229
    - Small Antennas 229
    - Large Antennas 229
  - Mounting an Antenna on the Pole 231
  - Assembling the Feedhorn/LNB Structure 231
  - Correctly Aligning Mechanical Polarisers 231
    - Polarisation Offset 232
  - Retrofitting for Detection of Circularly Polarised Signals 232
  - Setting the Focal Distance and Centring the Feed 233
- E. INSTALLING ACTUATORS 235
  - Mechanical Assembly 235
    - Preventing Water Damage 236
  - Wiring the Actuator 236
    - Safety Considerations 237
- F. ELECTRICAL CONNECTIONS 237
  - Connector Types and Techniques 238
    - F-Connectors 238
    - N-Connectors 239
    - Phono Connectors or RCA Jacks 239
    - Nylon Block Plugs 239
    - Scotch Loks 239
    - The Solderless Lug 240
    - SCART Connectors 240
    - Other Connectors - DIN, BNC, PL259 240
    - Additional Pointers on Installing Connectors 240
  - The Final Wiring 241
    - Small Dishes 241
    - Large Dishes 241
  - Safety Considerations 243
  - Protection Against Power Surges and Voltage Fluctuations 243
    - Ground Rods 243
    - Surge Protectors 244
    - Constant Voltage Transformers 245
  - Protection Against Overheating 245
- G. DECODER INTERFACING 246
  - Clamped Or Unclamped Input? 246
  - Baseband Or Composite Video? 246
  - Modulators and VCRs 246
  - Documentation 247
- H. ALIGNING THE ANTENNA ONTO THE SATELLITE ARC 249
  - Fixed Antennas 249
  - Testing Components Before Setting Angles 250
  - Aiming a Tracking Antenna 251
    - North/South Orientation 251
    - The Polar Axis Angle 252
    - Declination Offset Angle 252
  - Powering On and Aligning Onto the Arc 252
  - Aligning a Polar Mount Without a Compass 254
    - An Example 259
  - Final Feedhorn Centring Adjustments 259
  - Alignment with Test Instruments 260
    - RF Heads – Simple Wideband Detectors 261
    - Narrow Band Carrier Meters 262
    - Spectrum Analysers 262
    - More Practical Meters 263

TABLE OF CONTENTS

- I. FINE TUNING THE SATELLITE RECEIVER 264
- J. PROGRAMMING THE RECEIVER AND ACTUATOR 265
  - Making the System "User Friendly" 265
- K. WATERPROOFING 266
- L. CONNECTING STEREO PROCESSORS AND OTHER ACCESSORIES 267
  - Extra Television Sets 267
  - Stereo Processors 267
  - Modulators 267
- M. TOOLS REQUIRED FOR SATELLITE TV INSTALLATIONS 268
  - Heavy Machinery 268
  - Major Tools 268
  - Occasionally Needed Tools 268
  - Connectors, Cables and Other Accessories 268
- N. COLD WEATHER INSTALLATIONS 270
- O. CUSTOMER RELATIONS 273
  - Contracts and Warranties 273
- P. DOCUMENTATION 274

VI. RETROFITTING C-BAND SYSTEMS FOR KU-BAND RECEPTION 277
- A. PERFORMANCE REQUIREMENTS FOR KU-BAND COMPONENTS 277
  - Antennas 277
  - Mounts and Actuators 278
  - Feeds and LNBs 279
    - Offset Feeds 279
    - Prime-Prime or Coaxial Feeds 281
    - Feed Supports 281
    - Linear versus Circular Polariity 281
  - Satellite Receivers 287
- B. INSTALLATION PROCEDURES 282
  - Mechanical Requirements 282
  - Wiring 283
  - Tracking the Arc 284
  - The Effect of a Dual-Band Retrofit on C-band Reception 284
  - Tuning onto Ku-Band Transponders 284

VII. MULTIPLE RECEIVER SATELLITE TV AND DISTRIBUTION SYSTEMS 287
- A. BASIC COMPONENTS OF DISTRIBUTION SYSTEMS 287
  - Television Signal Requirements 287
  - A/B Switches and Combiners 288
  - Cables, Connectors and Splitters 288
  - Line Amplifiers and Tilt 290
  - Attenuation Pads 291
  - Terminators 291
  - dc Power Blocks 292
  - Input Selectors 292
  - Taps 293
  - Modulators, Processors and Bandpass Filters 294
  - Multi-purpose Components 295
- B. HEADENDS AND DISTRIBUTION NETWORKS 296
  - The Headend 296
    - Signal Sources 296
  - Television Channel Layout and Balancing Signals 297
  - Combining and Distributing the SIgnal 300
- C. HEADEND AND DISTRIBUTION SYSTEM DESIGNS 302
  - The Basic Satellite TV Headend 302
  - Single Receiver, Two TVs and Extra Remote Control Headend 304
  - Dual Receiver, Single Polarity Headend 305
  - Multiple Receiver, Dual Polarity Headend 309
    - M/A COM Switching Matrix 309
    - DX Switching Matrix 309
  - Dual Frequency Band Headend 311
  - 4, 7, and 12 Channel Headends 319

TABLE OF CONTENTS

VIII. WORLDWIDE KU-BAND SATELLITE TELEVISION	327
A. THE AMERICAS	328
B. EUROPE	328
C. AUSTRALIA, NEW ZEALAND AND NEW GUINEA	330
D. JAPAN	331
E. THE SOVIET UNION	332
F. INTELSAT KU-BAND SATELLITES	333
IX. TROUBLESHOOTING AND REPAIR	335
A. INTRODUCTION	335
B. A SYSTEMATIC APPROACH TO TROUBLESHOOTING	336
The Subsystems Approach	336
The Diagnostic Interview Identifying the Problem	336
C. VISUAL INSPECTION	337
RF Subsystem	338
Electromechanical Subsystem	339
Mechanical Subsystem	339
D. TECHNICAL TROUBLESHOOTING	340
Test Instruments	340
Voltmeters and Multimeters	340
Spectrum Analysers	341
Portable Signal Simulators	344
Portable Receivers	345
Switching Components	346
Power Supplies	346
Voltage Checking	347
Grounding Problems and Hum Bars	349
Solar Outages	349
E. UNDERSTANDING COMPONENT FAILURES	350
Satellite Receiver	350
Stereo Processor	350
Cable and Connectors	351
LNB	351
Feedhorn	352
Antenna Actuators	352
Antenna and Mount	354
Television	354
F. SPARE PARTS KIT	354
G. A SYMPTOM/CAUSE MAP	355
APPENDIX A. THE DECIBEL NOTATION	359
APPENDIX B. SATELLITE TV EQUATIONS	361
Link Equations	361
Antenna Gain	362
Loss of Gain with Surfac e Irregularities	362
Antenna Beamwidth	362
Noise Temperature and Figure	362
The Effect of Bandwidth on System Noise Power	363
Declination Angle	363
Azimuth and Elevation Angles	363
Voltage Standing Wave Ratio	364
Antenna Parabolic Geometry	364
Wind Loading	364
APPENDIX C. SATELLITE FOOTPRINTS & AIMING CHARTS	367
APPENDIX D. GLOSSARY OF COMMON BROADCAST AND SATELLITE TELEVISION TERMS	393
APPENDIX E. CHANNELS, WAVELENGTHS & BANDWIDTHS	402
APPENDIX F. LISTS OF MANUFACTURERS OF SATELLITE EQUIPMENT	403
APPENDIX G. REFERENCE PUBLICATIONS	409

INTRODUCTION

Communication satellites, that broadcast radio, television, telephone, computer and other messages in the Ku-band frequency range are becoming commonplace in many countries around the world. While the development of this technology lagged far behind the introduction of C-band systems a few short years ago, Ku-band satellite broadcasting is now leading the way in many countries.

The 10.95 to 14.50 GHz region of the electromagnetic spectrum has been allocated solely to satellite transmissions. There is little concern for interference with other communicators. Therefore, only technical restrictions limit the powers that can be broadcast. Satellites achieving 100 watts per transponder are now possible. So dishes with diameters as small as 50 centimetres are commonplace in certain regions. One installer in Europe is rumoured to have received satellite TV pictures by simply aiming an LNB at the sky.

The purpose of this manual is to provide a comprehensive exploration of the theory and practice underlying Ku-band satellite television. In this third edition, we explore numerous additional areas such as flat plate antennas and receiver design and also clarify other topics previously discussed. We have every effort to make the material clear and understandable to anyone having a curiosity but not necessarily a technical background. There is no reason why most laymen should not be able to participate in this exciting field in one fashion or another.

I. THE BASICS OF SATELLITE COMMUNICATION

A. INTRODUCTION

The concepts underlying satellite broadcasting, first stated in Arthur C. Clarke's ground-breaking article in the October 1945 edition of *Wireless World*, are straightforward: signals are beamed into space by an "uplink" antenna, received by an orbiting satellite, electronically processed, broadcast back to earth by a "downlink" antenna and detected by an earth station consisting of antenna and the associated electronics. The earth station or TVRO (television receive-only) can be situated anywhere within the satellite's "footprint" (as shown in Figure 1-1).

Nearly all communication satellites destined for commercial use are positioned or "parked" in the "Clarke" orbit, also known as the "geostationary" arc. The Clarke orbit lies in the equatorial plane of the Earth at 35,786 kilometres (22,247 miles) above the Earth's surface at its equator. This circle around the earth is unique because in this orbit the velocity of a spacecraft matches that of the surface of the earth below. Therefore each satellite appears in one fixed orbital slot in the sky above. This allow a fixed antenna to be permanently aimed towards a targeted geostationary satellite (see Figure 1-2).

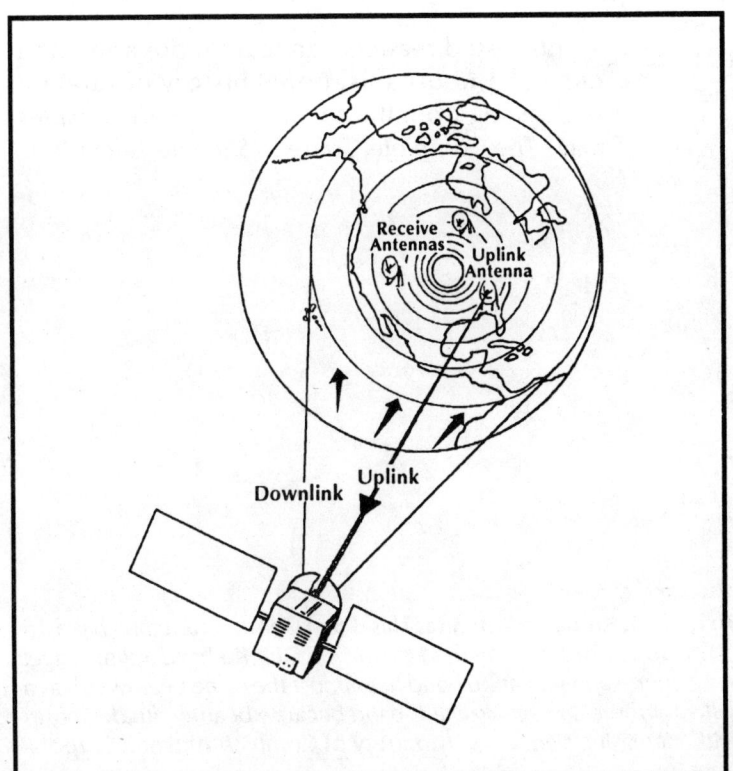

Figure 1-1. The Satellite Communication Circuit.
Signals are relayed to a satellite by one uplink antenna and then rebroadcast to an unlimited number of receiving stations that fall within the beam of the downlink antenna.

BASICS

Figure 1-2. The Geosynchronous Arc. *Satellites in higher or lower positions than the geostationary orbit of 35,803 kilometres (22,247 miles) over the equator will rotate more slowly or quickly, respectively, than the speed of the earth below. Only satellites in the geostationary orbit appear to stay in a fixed position relative to an observer on the earth below.*

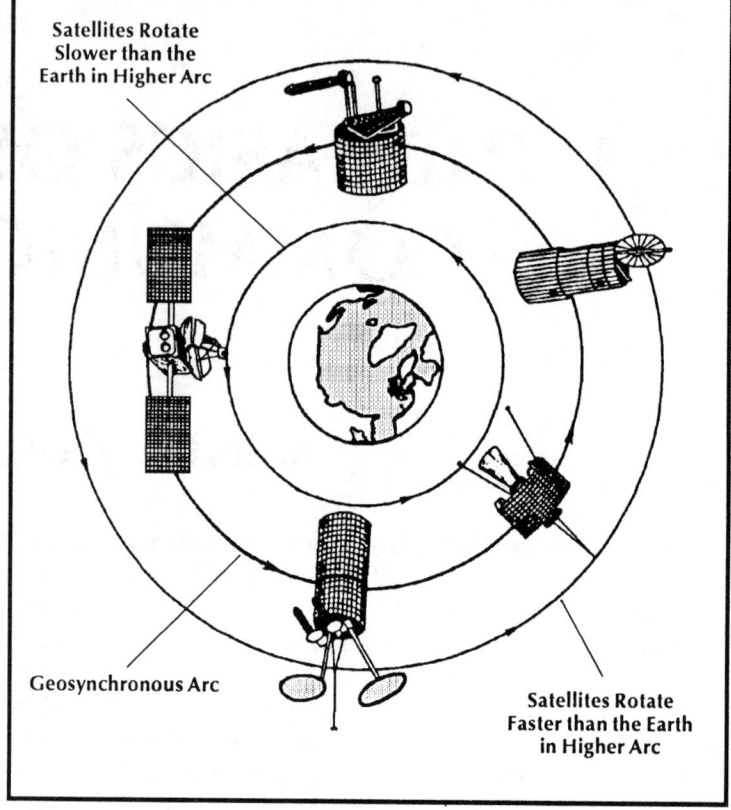

Although the terms geostationary and geosynchronous are used interchangeably in this book, the former is really more accurate. A geosynchronous orbit is defined as an orbit in which a satellite completes one orbit in a period of time that is a multiple of the time in which the earth makes one complete revolution.

Pioneer communication satellites were placed into lower, more complex elliptical orbits because sufficiently powerful launching vehicles were not available to lift them into the distant geostationary arc. Therefore, both transmitting and receiving antennas had to track the spacecraft. Telstar, one of the first communication satellites launched by the United States in 1962, had to be targeted by very costly and bulky mobile equipment that was mounted on rails. It relayed communications between Europe and the United States during a 36 minute window while it was visible from both sides of the Atlantic. Today some communication satellites are still boosted into various types of elliptical orbits. These non-geosynchronous systems include some military and experimental spacecraft as well as the Soviet Molniya television broadcast satellites.

Interested readers can learn more about the technical and organisational history of satellite broadcasting from the second edition of *Satellites Today - The Complete Guide to Satellite Television*.

Figure 1-3. Ku-Band Antenna. *This 3.5 m Comtech antenna has a 15 dB beamwidth of 0.87° in the 14.0 to 14.5 GHz Ku-band uplink range. A 3.5 metre antenna at Ku-band has almost the same beamwidth as a reflector three times its size at C-band because beamwidth decreases with increasing frequency. (Courtesy of Comtech Antenna Corporation)*

INTRODUCTION

B. THE SATELLITE CIRCUIT

A satellite communication circuit consists of an uplink, a communication satellite and an unlimited number of earth-based receiving stations (see Figure 1-3). The overwhelming strength of satellite broadcasting lies in its ability to reach any number of customers regardless of their geographic location, without the need for any physical connection.

The Uplink

Uplinks are complex devices designed to send a beam of microwaves to a pinpoint target in space. The operation of an uplink antenna is similar in principle to that of a car headlight which has a small but powerful light bulb at its centre and a parabolic reflector to direct the illumination to the road in front. Microwaves having powers of hundreds of watts are reflected from the antenna surface and beamed into outer space (see Figures 1-4 and 1-5).

Operators of uplinks are carefully regulated in most nations of the world. Their signals have relatively high powers and can therefore potentially interfere with other communicators or harm people and other living organisms. In addition, each uplink antenna relays messages which can be detected at millions of sites. So an improperly aimed uplink could easily disturb communications relayed by an adjacent satellite and could interfere with thousands or millions of receiving stations.

Figure 1-4. A C-Band Uplink Antenna. *Typical C-band uplink antennas are 10 to 13 metres in diameter. Such a large surface area is required because relatively high signal powers must reach a satellite and because large antennas have smaller "beamwidths." The beamwidth must be less than 1° to keep from activating adjacent satellites in the same frequency range. The 15 dB beamwidth at the mid-band transmission frequency of 6.2 GHz on a 10 metre C-band uplink antenna having a gain of 54.4 dBi is 0.70°. In contrast, antenna beamwidths at Ku-band are more narrow than those operating at the lower C-band frequency range. (Courtesy of Scientific Atlanta)*

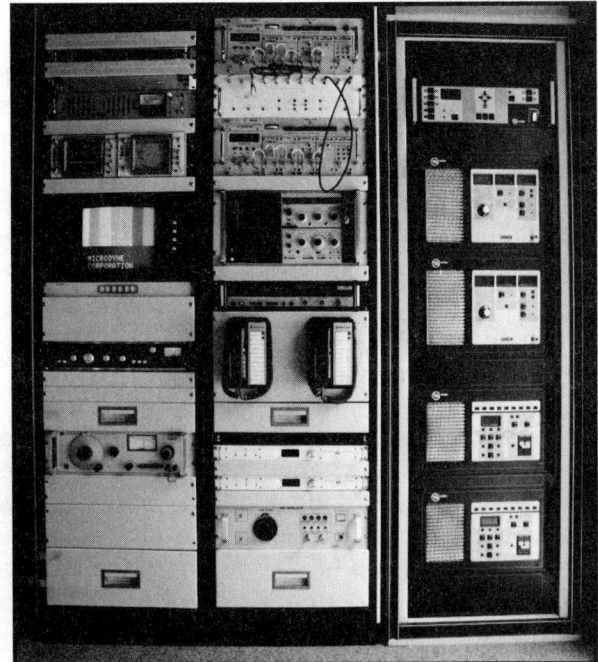

Figure 1-5. Uplink Equipment Racks. *These equipment racks house all the necessary processing equipment used to convert the video, audio, data or voice baseband signals into microwaves to be fed to the uplink antenna. (Courtesy of Scientific Atlanta)*

BASICS

Uplinks are increasingly being used in many segments of the business community. Today, many television and radio stations, telephone systems and data networks are becoming reliant on satellite links. Numerous businesses with multiple locations have established privately owned satellite networks for managing internal communications. This transformation of our communication system has been speeded with the introduction of Ku-band systems because smaller antennas can be used compared to those employed in conjunction with the more mature C-band technology.

A fixed uplink can be fed by fibre optic or coaxial cable links from broadcast studios or from other sources such as telephone switching centres or mainframe computers. Many television stations also relay signals by conventional, over-the-air methods to distant uplinks and subsequently to communication satellites for rebroadcast. Organisations needing uplink facilities on an occasional basis for teleconferencing, one-time events such as concerts or sporting matches or for news gathering activities have another option, that of "taking Mohammed to the mountain," i.e. of bringing the uplink equipment directly to the action. This is a viable economical alternative to over-the-air broadcasting to a fixed uplink site, particularly over Ku-band satellite links. Portable uplinks complete with all equipment and operating crews have become an exciting cost-saving alternative technology.

The Communication Satellite

Geosynchronous broadcast satellites receive the uplinked signal, change the frequency of the message and then broadcast it to any chosen geographic area below (see Figure 1-6). Downlink antennas can target up to 40 percent of the earth's surface with global beams, can broadcast to selected countries or continents via zone beams, or can pinpoint smaller areas with spot beams. To illustrate, many American C-band broadcast satellites employ one downlink antenna which blankets the continental United States and a second smaller one which directs a more localised beam to the Hawaiian islands. Ku-band spacecraft have the attractive capability of easily downlinking more localised spot beams than their C-band counterparts. European satellites such as Astra or the Eutelsat series transmit three or more beams to various regions.

Figure 1-6. Hughes Aussat Ku-Band Satellite. *The Hughes Aussat 1 is a HS-376 spin-stabilised spacecraft. It has a complex antenna array that features three pairs of cross-polarised reflectors in conjunction with a beam-forming network of numerous individual feedhorns. These are used to provide multiple transmission beams throughout Australia, and the Pacific shelf ocean areas. (Courtesy of Hughes Communications)*

THE SATELLITE CIRCUIT

BASICS

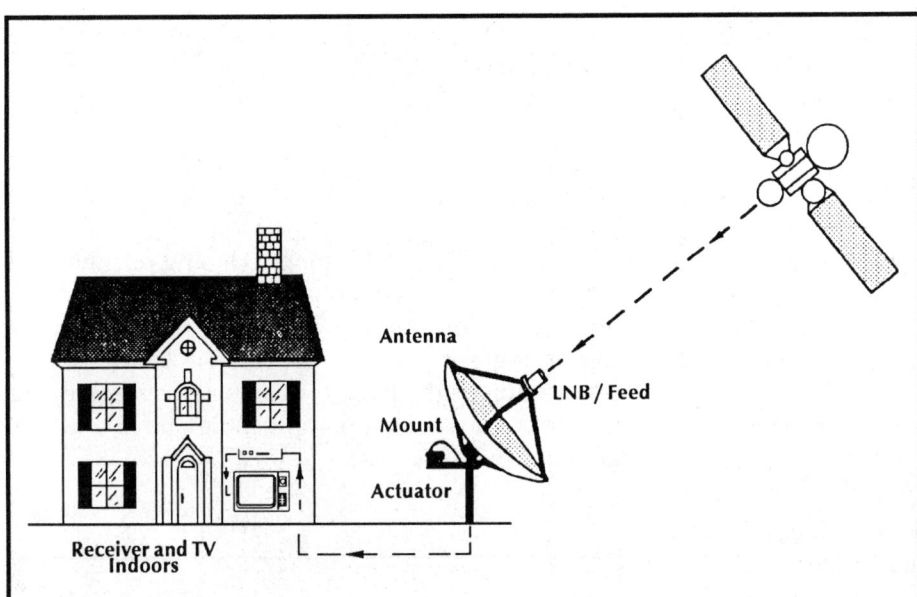

Figure 1-7. Home Satellite TV Receiving System. *Microwave signals are intercepted by the antenna where they are magnified approximately 10,000 times, focused into the feedhorn where polarity is selected, and fed into the LNB where they are amplified another 100,000 times before being lowered in frequency to typically 950-1450 MHz. The signal is then fed into RG-6 or RG-11, 75 ohm coaxial cable and transmitted to the satellite receiver. The receiver selects one channel and converts this satellite FM signal to an AM one for input to a conventional television receiver.*

The Receiving Station

A system for receiving satellite signals uses an antenna designed to collect and concentrate as much of the very weak downlinked signal as possible by reflection to its focus. The feedhorn or sub-reflector in the case of a Cassegrain design, located precisely at this focus, channels microwave signals reflected and concentrated by the antenna into the first electronic component, the low noise block converter or LNB. The signal is then cabled indoors to the satellite receiver and processed into a form understandable by either a television or a stereo. An earth receiving station is essentially an uplink operating in reverse (see Figure 1-7).

C. MICROWAVES

The transmission of extremely low power microwaves underlies the entertainment and information relayed by satellite broadcasting. Microwaves and radio waves, the agents by which radio, conventional television and other man-made devices work, are part of a larger phenomena known as electromagnetic waves. These can be compared to the disturbances that travel outward in concentric circles when a pebble is tossed into a pond or to the waves of vibrating air molecules that we know as sound. While sound travels at a relatively slow 1223 kilometres per hour (760 miles per hour), radio waves and all other electromagnetic waves travel 15,000 times faster at the speed of light, equal to 299,339 kilometres per second (186,000 miles per second). At this rate, a signal travels from an uplink antenna to a satellite and back again to earth in about 4 tenths of a second, or over-the-air from a television station or along a cable network to a customer virtually instantaneously.

Signal Frequency

The frequency of an electromagnetic wave or electrical signal transmitted by wire is the number of vibrations that occur every second as the energy passes any point in space (see Figure 1-8). Just as the frequency of sound vibrations determines whether a musical note is either a soprano or a bass, so the frequency of electromagnetic waves determines whether they are used to transmit regular AM radio broadcasts or satellite television broadcasts. Microwaves have frequencies in excess of one billion cycles per second (known as one gigaHertz and abbreviated 1 GHz). By comparison, the electricity from a wall outlet has a frequency of 60 Hertz, or 60 cycles per second in North America

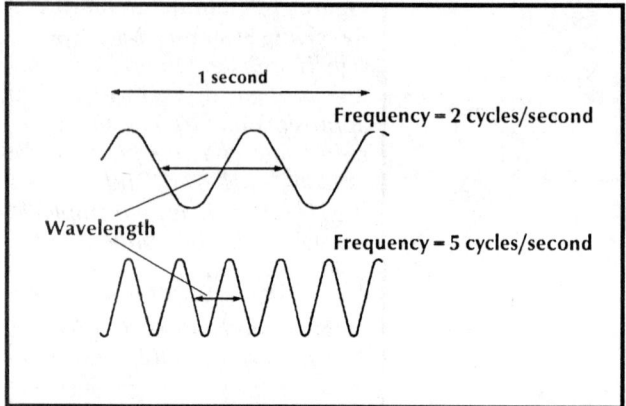

Figure 1-8. Frequency. *All electromagnetic waves are defined by how many vibrations occur each second, the frequency. Higher frequency waves have shorter wavelengths and vice versa. For example, a 12 GHz microwave signal has a wavelength of just under 2.5 cm..*

and 50 Hertz in Europe. In other words, the voltage changes from positive to negative 60 or 50 times each second.

Many seemingly different phenomena encountered in nature including light, X-rays, infrared heat rays, microwaves used for both cooking and communicating and gamma rays from the cosmos are electromagnetic waves. Surprisingly, the only difference among them is their frequency. Since all electromagnetic waves travel at the same speed, the speed of light, as the frequency increases the wavelength must decrease. For example, the wavelength of visible light is comparable to the dimensions of atoms and molecules, and their frequencies are many billions of billions of cycles per second. In contrast, microwaves have lower frequencies of one to fifty gigaHertz and wavelengths ranging from 30 centimetres to a few millimetres. Radio waves have wavelengths which can be kilometres long and frequencies in the millions of cycles per second (megaHertz or MHz) range.

Phase

The simplest form of electromagnetic or electrical wave is a pure sine wave. For example, the line power from a wall outlet is either a 60-cycle or 50-cycle per second sine wave depending upon the convention adapted in a particular country. The phase of a sine wave is always measured relative to another reference wave or starting point (see Figure 1-9). Not knowing the reference would be comparable to attempting to time a 100 metre dash without knowing when it began.

Each full vibration encompassing positive and negative voltages around the baseline is considered to occupy 360 degrees which equals the number of degrees in one complete circle. Therefore, for example, if the peaks of two sine waves are one quarter of a wavelength apart, they have a 90 degree phase difference.

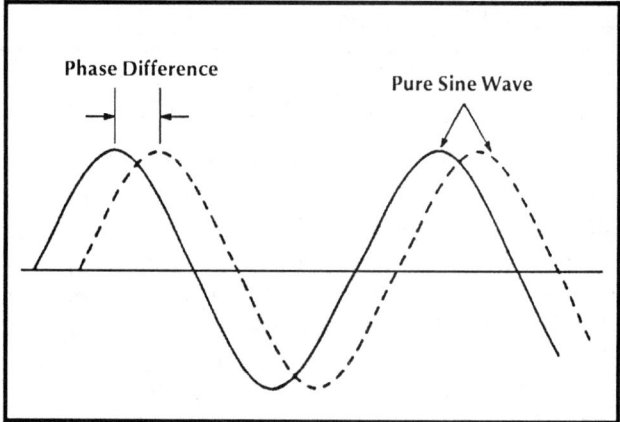

Figure 1-9. Phase. *Phase defines the relationship between two signals. The signals shown here are approximately one eighth of a wavelength or 45° (one eighth of 360°) out-of-phase with each other. One wave is said to lead and the other to lag by 45°.*

Transmission Power

The power of electromagnetic waves is commonly measured in watts or watts per square metre. For example, 10 watts per square metre means that the power passing through each square metre is ten watts. Satellite TV broadcasts are usually received by an antenna at powers of less than 10^{-17} watts per square metre!

The power of a signal relayed over a conductor is commonly measured in watts. To illustrate, an electronic circuit might transform a 1 milliwatt signal into a more powerful one of 10 watts.

BASICS

Signal Polarity

Electromagnetic waves are capable of being polarised. Two standard formats used in C and Ku-band satellite communication links are linear and circular polarity.

In the linear form where two polarisations, vertical and horizontal, are employed, the electric and magnetic fields of the signal remain in the same planes in which they were originally transmitted. To understand the idea of linear polarisation, imagine a car driving along a highway. It can travel to the same destination by following a curving road along flat ground or by following a straight road over rolling hills. Horizontally polarised waves vibrate in a horizontal plane like the car traveling along a curving road. Vertically polarised waves vibrate in a vertical plane (see Figure 1-10).

The dominant linear polarisation format is primarily employed with C and Ku-band domestic domestic satellite television broadcasts even though the original specifications for "Direct Broadcast via Satellite" in Europe specified circular polarisation. Circular polarisation is used primarily on Intelsat telephony links. (Intelsat spacecraft are owned and operated by the International Satellite Consortium). In addition, Soviet C-Band transmissions as well as the French TDF satellite and the proposed BSB satellite use circular polarisation.

In the circular form of signal transmission, the electrical and magnetic fields rotate in a circular motion as they travel (see Figure 1-11). This motion is somewhat analogous to a spiral. The direction of the rotation determines the type of circular polarisation. A signal rotating in a right-hand direction is termed right- hand circular polarisation (RHCP) and a signal rotating in the left-hand direction is termed left-hand circular polarisation (LHCP).

Circular polarisation was used on the early satellites because this eliminated the need for skew adjustment. These spacecraft were generally not in geostationary orbit so that both the receiving and transmitting antennas had to track the satellite. If a linearly polarised signal had been transmitted both

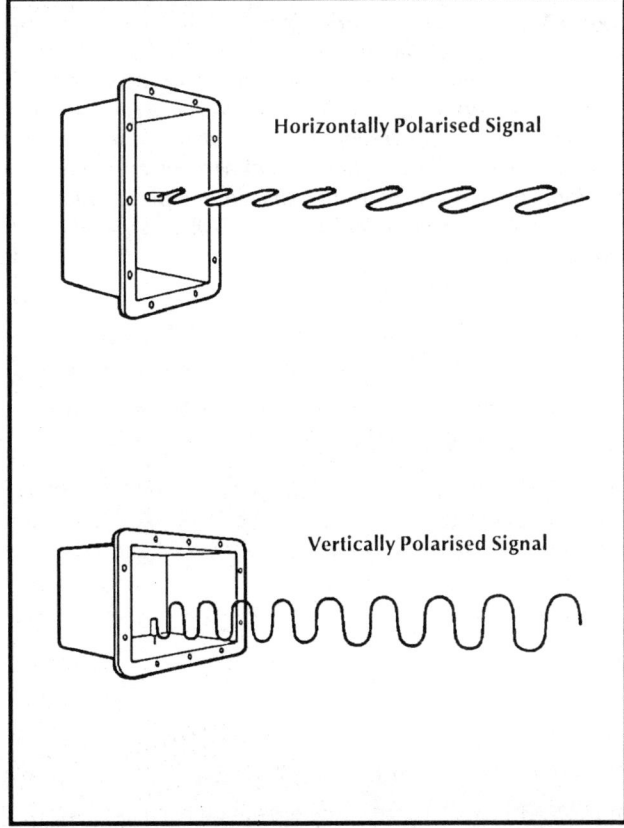

Figure 1-10. Vertically and Horizontally Polarised Waves. *Polarisation is the orientation of the electric field radiated from the transmitting antenna. When the orientation is parallel to the ground, it is defined as horizontal polarisation. When the direction of the radiated electric field is perpendicular to the ground it is defined as vertical polarisation. The reason for using horizontal and vertical polarisation on the same satellite is known as frequency re-use. This allows twice the number of channels to be transmitted in a given bandwidth.*

the transmitting and the receiving antenna would have required a continuous skewing feed. In addition, with the transient problem of Faraday Rotation, some form of skew control would theoretically also be required on the satellite. Faraday Rotation is the rotation of polarity caused by the Earth's magnetic field and/or magnetic storms when a signal travels through the atmosphere. This resistance to Faraday rotation makes circular polarisation eminently suitable for voice and data telephony via satellite.

Figure 1-11. Circular Polarisation. *Circular polarisation can also be used for satellite transmissions. Instead of positioning the microwave energy in either a vertical or horizontal plane, circularly polarised signals are transmitted in a spiral pattern like a coiled spring. The direction of rotation of the electric field vibrations can follow either a clockwise or counterclockwise motion. The two senses, right hand circular polarisation (RHCP) and left hand circular polarisation (LHCP), are typically found on Intelsat spacecraft.*

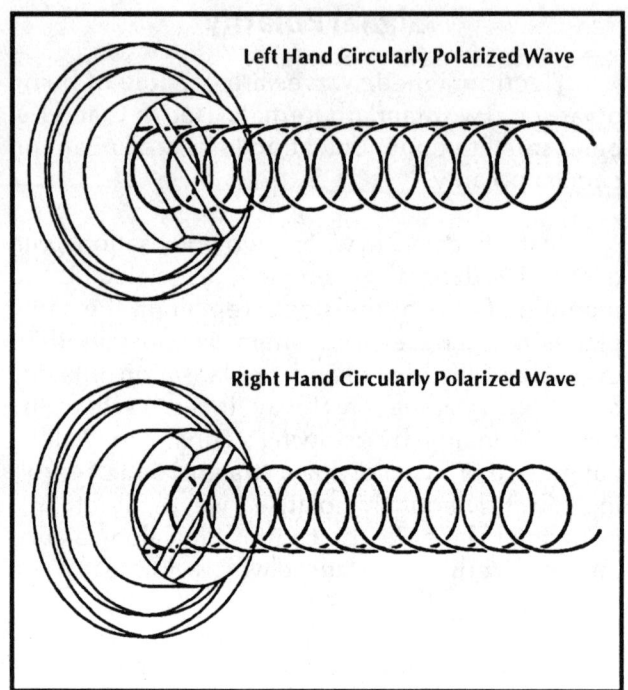

D. COMMUNICATION FUNDAMENTALS

The same principles underlie all forms of man-made communication. The first step is creating and coding the message. Next, this information must be modulated or added onto the medium designed to carry the signal. At the receiving end, the signal is demodulated and the original information is extracted. The amount of information that can be relayed is determined by its "bandwidth." The signal can be amplified or attenuated so that its power is increased or decreased. In addition, communication engineers must always account for unwanted signals or noise which are always present and hinder perfect communication.

Coding the Message

Any message, whether it be the image and voice of an entertainer or details of stock-market transactions, must first be changed into a form that can be transmitted by electromagnetic waves. Analogue coding methods mimic the pattern of a message by changes in electrical voltages. For example, a voice can be changed into an analogue signal by a microphone that creates a voltage pattern determined by the loudness and frequency of the sound. The louder the sound, the higher the voltage. The higher the sound frequency, the more rapid are the changes that occur in this voltage.

In contrast, a digital coding method uses only the binary numbers 0 and 1, represented by any two different voltage levels, to convey all information about these signal powers and frequencies. For example, a voice could be expressed in digital form if the loudness at each point in time was expressed by numbers, or patterns of 0's and 1's. A photograph can also be described by a long series of 1's and 0's that are coded so that some impart information about the location of the dots composing the picture and others communicate the brightness and color of these dots. Computers use digital coded messages exclusively.

Uplinks can relay either digital or analogue forms of the same message. Analogue-to-digital (A/D) or digital-to-analogue (D/A) converters can translate between these two languages or forms of representation. For example, compact disc players convert digital streams detected by a laser from the surface of a revolving plastic disc via a D/A converter into analogue waveforms that can drive loudspeakers. Today, all TV broadcasts are expressed in analogue form for transmission over satellite circuits. However, digital transmissions are becoming more commonplace as satellites are packed with more sophisticated electronics which manage greater amounts of information.

Modulation

Analogue or digital signals are added onto or impressed upon radio or microwaves by a process called "modulation." Once the message is modulated onto the carrier wave of a predetermined frequency it can be relayed from a sending to a receiving location whether it is via a satellite link, over-the-air or along a cable route. The term carrier wave describes precisely what the modulated wave does, it carries the information. Radios, televisions and other communications equipment demodulate or extract the original message from the carrier wave. The demodulation process is also known as detection.

Analogue Modulation

The traditional modulation techniques, developed in the early days of radio communications, are amplitude modulation (AM) and frequency modulation (FM). These are used extensively, for example, in AM and FM radio broadcasts. In amplitude modulation, the power of a carrier wave is varied in accordance with the voltage level of the message being relayed, while in frequency modulation the frequency of the carrier wave is varied (see Figure 1-12).

Each of these types of modulation has advantages and disadvantages. On one hand, AM messages must have relatively high powers to be capable of traveling long distances without being weakened too severely by atmospheric disturbances and other forms of noise. They are also more prone to picking up static than are FM messages. On the other hand, FM signals generally need relatively lower power for successful, long distance transmission, but must use a substantially wider range of frequencies than AM messages would to carry the same amount of information.

Satellite messages are nearly always frequency modulated. As signals cover the long distances between uplinks, satellites and earth receiving stations, their power becomes so low that standard AM transmissions would be unusable. (However, wideband AM satellite audio relays are sometimes used). Also, because extremely high frequencies are used in satellite communication, the large bandwidth required by FM broadcasts is available. Over-the-air television broadcasts as well as cable network signals, in contrast, use an amplitude modulated video signal in order to conserve space in their limited frequency spectrum.

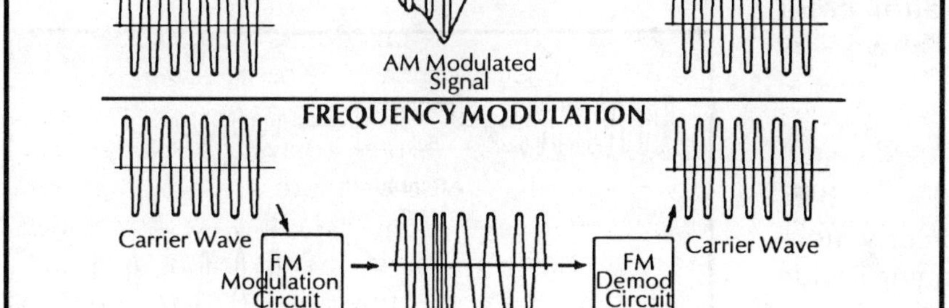

Figure 1-12. Modulation. *Two formats used to impress an audio, video or data signal onto a carrier wave are amplitude and frequency modulation. The modulated signal, the carrier wave, can then be relayed by cables, over-the-air, by satellite or any other transmission media to its final destination.*

BASICS

Digital Modulation

Digital modulation techniques, derived from the original Morse code, are geared to the language used by computers., wherein a pattern of 0s and is relay all information. Digital transmissions are quite tolerant to noise because a receiver must only determine whether the signal is either "on" or "off."

The simplest method to modulate a carrier wave is to switch it on and off. For example, Morse code can be relayed as a series of dots and dashes by turning the carrier wave on and off. In the most common modern form of digital modulation, phase shift keyed (PSK), the position of the 0s and 1s are varied relative to each other.

Digital satellite relays are characterised by the number of bits transmitted per second. For example, a typical high quality television broadcast requires about 32 to 68 million bits per second or megabits. Telephone channels have traditionally required circuits that transmit 64 kilobits per second.

Digital communications have a number of distinct advantages in addition to their noise tolerance. Computers can easily manipulate the data streams to create highly secure encoded transmissions or they can use compression techniques to squeeze more information into each satellite channel. The techniques to compress a digital television signal are based on a simple method. If any given area of the picture does not change from frame to frame, it is not re-transmitted. Today methods have been developed to use as low as 3 megabits per second to transmit a full motion color picture or 16 kilobits per second for a two-way voice circuit.

Bandwidth

Just as a large-diameter pipe can carry more water than a small one, so a signal spanning a wide band of frequencies can carry more information than can one using a narrow band. This range of frequencies is termed the "bandwidth." For example, TV channel 4 in North America is broadcasted in the frequency range from 66 to 72 MHz and has a bandwidth of 6 MHz. When transmitted over satellite, this signal is spread onto a band spanning up to 36 MHz.

Each type of communication medium requires a characteristic bandwidth. Television links need a substantially wider bandwidth than do radio or telephone because much more information is necessary to recreate a picture than music or a voice. To illustrate, television signals broadcast by satellites typically require a video bandwidth of from 6 to 8 MHz, depending upon the television system in use. However, voice channels normally require a bandwidth of only 3,000 to 4,000 cycles per second (3 to 4 kHz) for quality sound reproduction. The bandwidth requirements in MHz or kHz are directly related to the number of bits per second required for transmissions as discussed in the previous section on digital modulation.

Amplification and Attenuation

Man-made communications must often be amplified during their voyage from sender to receiver to keep the information intact because their power is usually weakened or attenuated. In the same fashion that a photograph is enlarged but not changed, correct amplification retains the original message (see Figure 1-13). All receiving devices amplify a signal before demodulation occurs. For example, messages beamed into space from an uplink antenna are weakened on their voyage to a satellite as the signal becomes spread out and is absorbed by water vapour, clouds and other materials. The purpose of a satellite receiving antenna is to collect and concentrate these weak signals much as a magnifying glass focuses light.

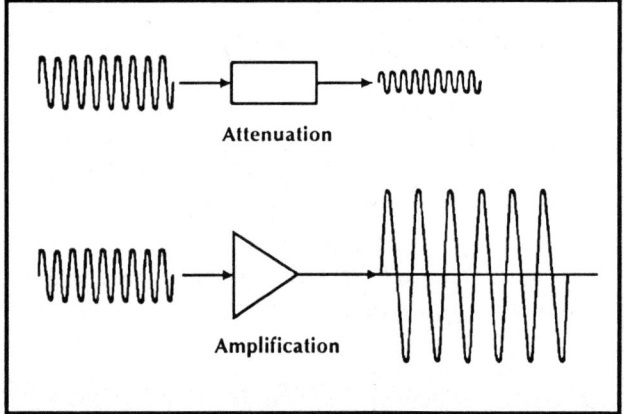

Figure 1-13. Amplification and Attenuation. *Amplification increases a signal's peak-to-peak voltage, while attenuation reduces its strength.*

COMMUNICATION FUNDAMENTALS

Occasionally, a signal will be intentionally attenuated. For example, a cable TV headend may deliver an excessively powerful signal to a feeder line which could over-drive connected TVs and cause distortion. Pads or line attenuators are then inserted to reduce the power. Some satellite receivers when installed very close to the receiving antenna can also be over-driven. Power-passing attenuators are then required to avoid distortion.

Noise

In an ideal communication system transmitted signals would be free of interference or noise. However, television or radio broadcasts are occasionally of poor quality or "noisy." Noise is present in all matter at temperatures above absolute zero, 0°K, the temperature at which all molecular motion ceases. There is no temperature colder than absolute zero! Kelvin degrees have the same magnitude as degrees Centigrade, they are just measured from a different starting point. Absolute zero, zero degrees Kelvin, equals minus 273.16 °C or minus 459.69 °F.

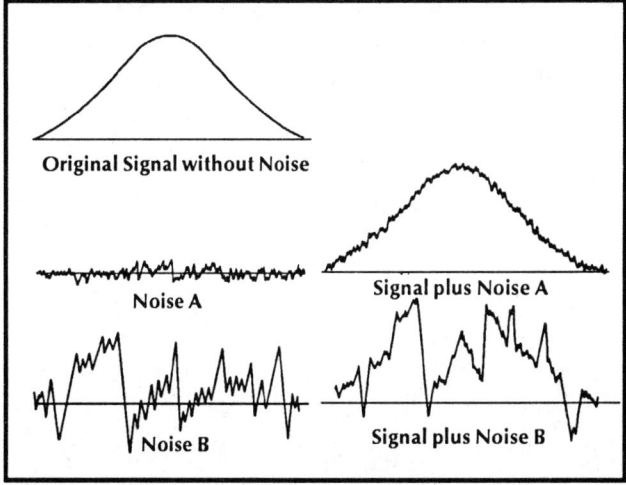

Figure 1-14. Noise. *The ratio of signal power to noise power, the signal-to-noise ratio (SNR), defines the quality of a communication link. Excessive amounts of noise can completely garble an audio or video transmission so as to make it unintelligible.*

Noise is caused by the endless motion of the molecules that compose all matter. These small, vibrating charged particles generate electromagnetic waves that can mask the organised signal sent by man-made devices. Noise from the environment becomes stronger as the temperature increases.

Satellite antennas detect noise from the warm ground. The temperature on an average day is 17 °C (62 °F) or 290°K. At the frequencies employed at Ku-band, outer space also generates noise at about 6°K, as a result of the "big bang." So an antenna pointed straight towards earth like one on a satellite, will pick up an additional 290°K; one on earth pointed straight into space will see approximately 10°K of noise, about 6°K from space and the rest leaking into the antenna from the ground. Receiving antennas detect more of this environmental noise as the signal bandwidth increases. Additional noise is generated by internal heat in power supplies, amplifiers, receivers and other electronic equipment.

Man-made signals not intended for the receiving station are also considered a type of noise called interference. Terrestrial interference, known as TI, results when earth-based communications other than those from satellite broadcasters are inadvertently received by an earth station. This type of interference, while of annoying familiarity in receiving C-band broadcasts, is uncommon, although possible, in Ku-band transmissions.

Random noise is always present in satellite communication systems (see Figure 1-14). The quality of a communication link is determined by the ratio of signal-to-noise power. This is an indicator of how well the signal can be transmitted so that it is clearly stronger than the noise power. For example, if a signal of 1 volt is received along with 1 millivolt of noise, equal to an S/N of 1000, the picture quality could be poorer than if a signal of 2 volts is received with 4 millivolts of noise, an S/N of 500. Typically, televisions must detect a signal having power greater than 63,000 times the accompanying noise in order for a "high-quality" picture to be reconstructed.

E. MICROWAVES AND SATELLITE COMMUNICATION

There are five persuasive reasons why microwaves have been used in satellite communication circuits. First, higher frequency electromagnetic waves have the potential for relaying larger quantities of information because, as the frequency increases, any given bandwidth becomes a smaller fraction of the operating frequency. To illustrate, a 1 MHz wide band located in the 10 MHz region of the spectrum occupies a relatively larger percentage of the frequency space than the same band in the 10 GHz region. Since more bandwidth is available, wider bands with higher information capacities can be used at microwave frequencies. This allows designers to transmit as much information as possible per satellite so that the expensive investment in satellite launching, operation and maintenance can be recouped more quickly.

A second reason for using microwaves stems from the requirement for uplink antennas to aim a highly directional beam towards an extremely small target in space. Physics dictates that electromagnetic waves can be better focused by an antenna that is substantially larger than the wavelength of radiation it is managing. For example, sending a directional beam of AM radio signals having 100-metre long wavelengths would require an extremely large, cumbersome and expensive antenna over hundreds of metres in diameter. By contrast, Ku-band uplinks use 14 GHz signals having wavelengths of approximately 2 centimetres, so that a typical 3 or 4 metre uplink antenna can aim most of its radiation into a very narrow beam, and relatively low power can be used.

Third, microwave transmissions to and from satellites are not as susceptible to noise from atmospheric disturbances as are lower frequency transmissions. To illustrate, several times each year, for periods as long as two or three days, short-wave radio is useless for long-distance communication because sun-spot activity disturbs the required reflection of these relatively low frequency radio waves by the upper atmosphere.

Fourth, the most important property of microwaves that determines their use in satellite communication is their ability to pass through the upper atmosphere into outer space. At frequencies lower than approximately 30 MHz, a radio wave will be reflected back towards earth from the ionosphere layer in the atmosphere. Since microwave frequencies are far above the 30 MHz range, they easily pass through the ionosphere shield.

Fifth, the microwave region of the electromagnetic spectrum was a relatively untouched and available territory during the late 1950s and 60s when frequency spectrum was being allocated by the International Telecommunications Union (ITU). Lower frequencies were already being used by many different communication media and users.

As geosynchronous orbital space has become increasingly populated, progressively higher microwave frequencies have been allocated to satellite communications. Until the early 1980s most satellite broadcasters used C-band frequencies. Today, Ku-band transmissions are commonplace and the higher frequency Ka-band is being eyed by numerous potential users.

Using higher frequency satellite broadcasts is a double-edged sword. Microwaves are depolarised and more strongly absorbed by water vapor in the atmosphere as frequencies are elevated. However, transmitting and receiving antennas have higher gain or concentrating power as signal frequency increases thus compensating for the increased atmospheric absorption.

Ku-band broadcasts have the advantage of using potentially higher transmission powers than those operating at C-band (see Table 1-1). Satellite downlink power levels cannot be increased past a certain point in the C-band because the same range

TABLE 1-1. MICROWAVE FREQUENCY BANDS	
Band Name	Bandwidth (GHz)
P-band	10.20 - 0.40
L-band	1.53 - 2.70
S-band	1.50 - 2.70
C-band	3.40 - 4.20
	4.40 - 4.70
	5.725 - 6.425
X-band	7.25 - 7.75
	7.90 - 8.40
Ku-band	10.95 - 14.50
Ka-band	17.7 - 21.2
K-band	27.5 - 31.0

of frequencies is shared in many countries by land-based microwave towers which operate as part of the telephone communication system. For example, North American C-band satellite broadcasts must usually be relayed at less than 10 watts per circuit, or transponder. Ku-band satellites are not limited in this way and can potentially operate at one hundred watts or more per transponder.

When the lower frequency S-band is employed, lower cost, more readily available electronic components can be used. Although losses caused by signal absorption in the atmosphere are lower than those at higher frequencies, this attenuation is offset because antenna gain is also reduced by nearly the same factor. However, like Ku-band spacecraft, S-band transponders can also be equipped with more powerful transponders in order to reduce the size and lower the cost of earth receiving stations. Some spacecraft like India's Insat 1B and 1C use the 2.50 to 2.70 GHz S-band.

F. UPLINK OPERATION

Uplinks must carefully aim a signal towards a specified communication satellite. This is accomplished as follows. The raw audio, video or data signal is modulated onto an intermediate frequency carrier, often centred on 70 MHz. The carrier is then "upconverted" or raised in frequency to that of a selected satellite channel. The uplink frequency is typically centred near 14 GHz. The signal is then passed into a high power amplifier (HPA) in which it is boosted in power. This microwave band is then transmitted to a "waveguide" at the antenna focus from which it is fed towards the reflective surface. The end result is that a very narrow beam of microwaves is relayed into space to a targeted spacecraft (see Figures 1-15, 1-16 and 1-17).

Ku-band uplinks have one notable advantage over the lower frequency C-band systems which employ extremely large antennas, occasionally in excess of ten metres in diameter. Such larger reflector areas are necessary to ensure that a sufficiently narrow beam which does not interfere with adjacent satellites is transmitted. The larger the antenna diameter, the more narrow the "beamwidth" (this concept is discussed in more detail in Chapter II). In contrast, Ku-band uplinks can use antennas with diameters as small as one third the diameter of their C-band counterparts.

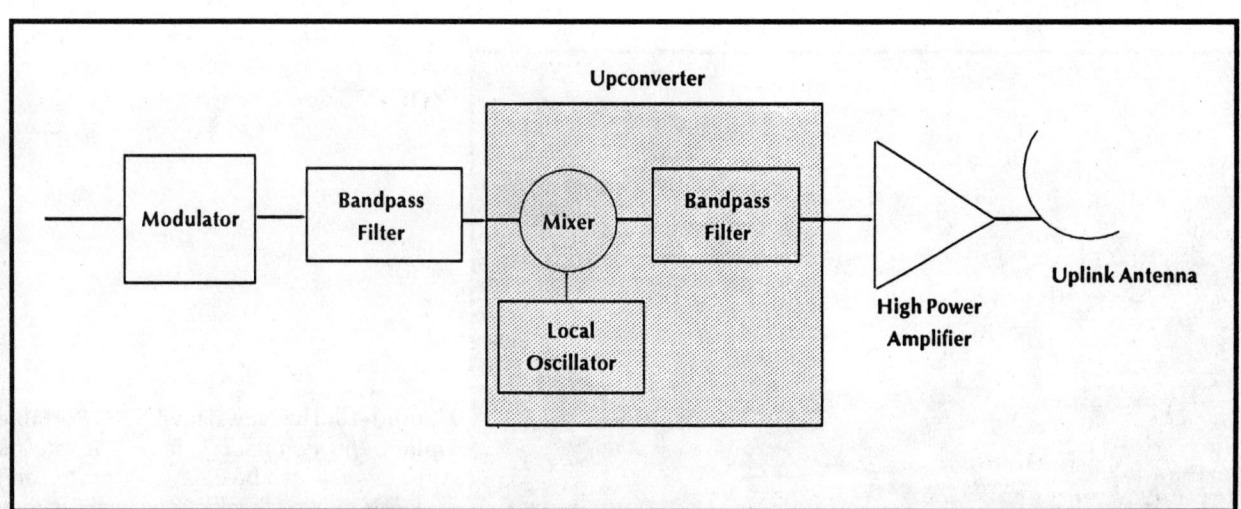

Figure 1-15. Uplink Circuit Diagram. *In an uplink, the baseband signal is fed into a modulator, the unwanted sidebands are removed by a bandpass filter and the signal is then upconverted to the desired transponder frequency. Additional bandpass filtering is accomplished in the upconverter. The signal is than amplified by a high power amplifier (HPA), fed into a feedhorn and reflected by an antenna towards a communication satellite. A typical uplink transmits hundreds of watts of power.*

BASICS

Figure 1-16. SES/ASTRA Betzdorf Uplink Control Centre. *(Courtesy of WV Publications, London)*

Figure 1-17. British Telecom Uplink Site. *The British Telecom International control centre has been operational in the Docklands since 1984. (Courtesy of WV Publications, London)*

Portable Uplinks, Satellite News Gathering and Educational TV

Ku-band satellite technology has led to the development of inexpensive, lightweight, self-contained, transportable uplinks (see Figures 1-18 to 1-21). Some are small enough to be packed into suitcases and carried onto commercial airplanes. For example, the American NBC television network used a "7-suitcase" Ku-band uplink to cover the transfer of power in the Philippines in early 1986. Many local television stations now own one or more uplinks which can be fitted into the confines of small, inexpensive vans.

The burgeoning fields of satellite news gathering (SNG) and classrooms-by-satellite are made possible by Ku-band technology. Inexpensive, grab-and-go setups may one day even be strapped onto the back of an avid reporter who parachutes into a remote jungle, are revolutionary tools allowing even small organisations to compete with entrenched network broadcasters.

Portable SNG uplinks have another important advantage over the earlier C-band technology in that the licensing procedure is much simpler. Ku-band systems, unlike their C-band counterparts, do not interfere with common carriers sharing a common frequency range.

Figure 1-18. The NewsHawk SNG Portable Uplink. *This compact, light-weight satellite earth terminal can be easily transported and rapidly deployed by electronic news gathering teams. The entire system consists of three components contained in three packages: a 2 by 1 metre elliptical antenna, a high power amplifier and all other electronics. (Courtesy of Marconi Communications Systems)*

UPLINK OPERATION

BASICS

Figure 1-19. A Satellite News Gathering Van. *(Courtesy of Centro Corporation)*

Figure 1-20. Satellite News Gathering Truck. *The SNG Uplink Van has many advantages over existing C-band transportable uplinks. A news feed can be broadcast back to the main television studio in a fraction of the time it takes to set up a C-band uplink, without any concern about interfering with terrestrial point-to-point microwave links. (Courtesy of Harris Corporation)*

Figure 1-21. Satellite New Gathering Portable Uplink. *This Harris Fly-a-Way Video Uplink consist of a 1.8 metre prime focus Ku-band satellite antenna, a 160 watt traveling wave tube amplifier (TWTA), an amplifier known as an LNB, a synthesizer-tuned upconverter/downconverter, and a video exciter and video receiver. All these components can be packed in seven suitcases for fast, portable satellite news gathering and uplinking from virtually any location within view of a communication satellite. (Courtesy of Harris Corporation)*

FREQUENCY ALLOCATIONS

G. WORLDWIDE FREQUENCY ALLOCATIONS

Nations around the world have worked together in allocating portions of the electromagnetic frequency spectrum to the various communication media and user groups. The International Telecommunication Union (ITU), an agency of the United Nations, coordinates meetings every four years, known as World Administrative Radio Conferences (WARC), to plot the paths for future technologies and to develop standards for the worldwide communication system. National organisations such as the Federal Communications Commission (FCC) in the United States, the Japanese Ministry of Post and Telecommunications, the Independent Broadcast Authority (IBA) in Great Britain and the Department of Communications (DOC) in Australia have sent representatives to these WARCs. Their decisions affect other international organisations such as INTELSAT, the International Telecommunications Satellite Consortium and INMARSAT, the International Maritime Satellite Organisation. The frequency spectrum was allocated up to only 30 MHz at the first World Administrative Radio Conference (WARC) in 1927. Today, worldwide frequency allocations range up to 300 GHz.

The history of communication is unfolded by an exploration of these frequency assignments. As technology progressed, man has become capable of utilising ever higher frequencies (see Table 1-2). Wire transmissions were first relayed on relatively low frequency carriers because the pioneers were limited by the rather primitive electronics available at that time. Ironically, when radio waves above 1.5 megahertz frequency were first produced, they were relegated to the "hams" because regulators could not see a use for this region of the spectrum. Ham radio operators have contributed numerous innovations since then. For example, the first home satellite receiving stations were built by hams. As technology developed, coaxial cable transmission, microwave relays and then satellite communication were allocated successively higher frequencies.

It is important to understand that identical audio or video messages can be modulated onto and transmitted over any portion of the allocated spectrum. The frequency of the carrier wave is cho-

TABLE 1-2. NORTH AMERICAN ASSIGNMENT OF SOME RADIO FREQUENCIES

Frequency (MHz)	FCC User Assignment
3-30	Wire Service, CB, Ham, etc (HF)
30-54	Mobile Radio
54-72	TV Channels 2-4 (VHF)
72-76	Radio Services
76-88	TV Channels 5 & 6 (VHF)
88-108	FM Radio
108-120	Aeronautical
120-136	Aeronautical
136-144	Government
144-148	Amateur Radio
148-151	Radio Navigation
151-174	Land, Mobile, Maritime
174-216	TV Channels 7-13 (VHF)
216-329	Government
329-420	TV Channels 7-13 (VHF)
420-450	Ham
450-470	Mobile Radio
470-806	TV Channels 14-69 (UHF)

sen according to the type of communication as well as the available spectrum.

In spite of the rapid development in man-made forms of communication, a relatively small portion of the frequency spectrum is presently being used. However, in those ranges where frequency space is available, it is in heavy use and competition for allocated space can be fierce. As a result, innovative methods have been developed to "re-use" or to have more than one user simultaneously share the same portion of this spectrum.

The ITU frequency assignments for satellite communications developed at WARC 1979 are listed in Table 1-3. ITU assignments of microwave frequencies for Ku-band broadcasts are outlined in Table 1-4. The fixed satellite services band in the United States, while also used to transmit non-broadcast services, presently carries television signals. These are lower in power than those envisaged for transmission in the formally relegated DBS band. In region 3, Australian DBS satellites operate in the 12.5 to 12.75 GHz band while Japanese spacecraft have been assigned to the 11.7 to 12.5 GHz range. In region 1, the business band service

BASICS

TABLE 1-3. WARC 1979
FREQUENCY ALLOCATIONS
for SATELLITE COMMUNICATIONS

Downlink Frequency (GHz)	Uplink Frequency (GHz)
Fixed Satellite Services (FS):	
2.500 - 2.655 (Region 2)	
2.500 - 2.535 (Region 3)	
2.655 - 2.690 (Region 2/3)	
3.400 - 4.200	5.725 - 5.850 (Region 1)
4.500 - 1.800	5.850 - 7.075
7.250 - 7.750	7.900 - 8.400
10.70 - 11.70	12.75 - 13.25
12.50 - 12.75 (Region 1/3)	14.00 - 14.80
11.70 - 12.30 (Region 2)	17.30 - 18.10
17.70 - 21.20	27.00 - 27.50 (Region 2/3)
	27.50 - 31.00
Maritime Mobile Satellite	
1.530 - 1.544	1.6265 - 1.645
Aeronautical Mobile Satellite	
1.545 - 1.559	1.6465 - 1.6605
Mobile Satellite	
1.544 - 1.545	1.6455 - 1.6465
19.70 - 21.20	29.50 - 31.00
Broadcast Satellite	
2.500 - 2.690	
11.70 - 12.50 (Region 1)	
12.10 - 12.70 (Region 2)	
11.70 - 12.20 (Region 3)	
22.50 - 23.00 (Region 2/3)	

TABLE 1-4. ITU SATELLITE KU-BAND
FREQUENCY ASSIGNMENTS
(Frequencies in GHz)

REGION 1: EUROPE, MIDDLE EAST and AFRICA
($35°E$ to $56°W$)

Fixed Satellite Service (FSS)	10.95 to 11.20
	11.20 to 11.45
	11.45 to 11.70
Direct Broadcast Service (DBS)	11.70 to 12.50
Business Band Service (BBS)	12.50 to 12.75

REGION 2: NORTH, CENTRAL and SOUTH AMERICA
($57°W$ to $146°W$)

Fixed Satellite Service (FSS)	11.70 to 12.20
Direct Broadcast Service (DBS)	12.20 to 12.70

REGION 3: INDIA, ASIA and the PACIFIC
($170°W$ to $40°E$)

Fixed and/or Broadcast Service	11.70 to 12.75

BASICS

(BBS), originally intended for high-speed digital business communications, will also be used for European cable TV broadcasts.

European DBS Satellite TV - TV-SAT/TDF/BSB

Direct broadcasting via satellite was proposed at the 1977 WARC. The committee developed technical specifications based on extrapolations of equipment available at that time. However, the WARC bid to predict and direct the future was somewhat off target.

Under the WARC 77 proposal, each European country was allocated a set of frequencies and a shared orbital slot (see Tables 1-5 and 1-6). The predicted beam centre EIRP was 64 dBw and the front end noise figure was 7 dB (these concepts are explained in more detail in Chapter II). However, in 1977, silicon was still the prevalent technology used in microwave transistors. Gallium arsenide technology, the engine that drove prices for satellite technology to affordable levels, still in its infancy, was extremely expensive.

Although the available frequency block for DBS was a full 800 MHz wide, the entire range was not allocated. Instead, the channel assignments for each country were organised so that, all together, they occupied just 400 MHz. This bandwidth, incidentally, could have been easily managed by UHF television tuners of the day.

The evolution of the satellite marketplace was more complex in Europe than in North America where C-band home satellite systems developed in a "grass-roots" fashion. In Europe, after a brief experience with C- band reception, the market focused on Ku-band satellite delivery. Due to a mixture of politics, inappropriate technical expectations and some inefficiencies, Ku-band DBS as specified by the WARC 77 committee did not develop as expected.

TABLE 1-5. CHANNEL & FREQUENCY ALLOCATIONS FOR EUROPEAN Ku-BAND DBS
(Frequencies in GHz)

Channel	Frequency	Channel	Frequency
1	11.72748	2	11.74666
3	11.76584	4	11.78502
5	11.80420	6	11.82338
7	11.84256	8	11.86174
9	11.88092	10	11.90010
11	11.91928	12	11.93846
13	11.95764	14	11.97682
15	11.99600	16	12.01518
17	12.03436	18	12.05354
19	12.07272	20	12.09190
21	12.11108	22	12.13026
23	12.14944	24	12.16862
25	12.18780	26	12.20698
27	12.22616	28	12.24534
29	12.26452	30	12.28370
31	12.30288	32	12.32206
33	12.34124	34	12.36042
35	12.37960	36	12.39878
37	12.41796	38	12.43714
39	12.45632	40	12.47550

BASICS

Earlier satellites such at the Eutelsat series and those Intelsat communication vehicles that transmitted signals to the European continent used the 10.90 to 11.20 GHz band. ASTRA, relaying television broadcasts in the 11.20 to 11.50 GHz band, had already beaten the BSB's DBS satellite, the quaintly named "Marco Polo 1", into orbit and operation. The other European DBS satellites are TDF-1, Tele-X and the ill-fated TV Sat-1 which had to be abandoned after it failed to deploy one of its transmitting antennas in orbit. TV Sat-1 is soon to be replaced by TV-Sat 2. Some industry authorities believed that these DBS satellites have a better chance of success than BSB.

The USSR has been allocated two orbital slots, at 23 and 44°E and a large number of channels, primarily due to its large geographical area. This nation possesses a C-band satellite network that rivals Intelsat and has tested a Ku-band transmission system via its Gorizont at 14°W. The "Lutch" test transmitted a cross-strapped Moscow Channel 1 at 11.67 GHz. The USSR channels are 1, 3-5, 7-9, 11-13, 15- 20, 22- 28, 29, 30-32, 33, 34-36, 37, 38-40 to be relayed with right and left hand circular polarity.

The five Nordic countries, (Norway, Denmark, Sweden, Iceland and Finland) have been permitted to use some of their allocated channels to provide a joint regional service. Channels 22, 24, 26, 28, 30, 32, 36 and 40 will be transmitted with left hand circular polarity.

TABLE 1-6. DBS ORBITAL SLOT AND CHANNEL ALLOCATIONS

Orbital Slot: 5°East

COUNTRY	CHANNELS	POLARISATION
Denmark	12, 16, 20, 24, 36	Left Hand
Finland	02, 06, 10, 22, 26	Left Hand
Sweden	04, 08, 30, 34, 40	Left Hand
Norway	14, 18, 28, 32, 38	Left Hand
Greece	03, 07, 11, 15, 19	Right Hand
Cyprus	21, 25, 29, 33, 37	Right Hand
Turkey	01, 05, 09, 13, 17	Right Hand
Iceland	23, 27L, 31, 35L, 39	Right Hand

Orbital Slot: 1°West

COUNTRY	CHANNELS	POLARISATION
Hungary	22, 26, 30, 34, 38	Right Hand
Poland	01, 05, 09, 13, 17	Left Hand
East Germany	21, 25, 29, 33, 37	Left Hand
Czechoslovakia	03, 07, 11, 15, 19	Left Hand
Bulgaria	04, 08, 12, 16, 20	Right Hand
Rumania	02, 06, 10, 14, 18	Right Hand

Orbital Slot: 7°West

COUNTRY	CHANNELS	POLARISATION
Albania	22, 26, 30, 34, 38	Right Hand
Yugoslavia	21, 25, 29, 33, 37	Right Hand

Orbital Slot: 19°West

COUNTRY	CHANNELS	POLARISATION
France	01, 05, 09, 13, 17	Right Hand
West Germany	02, 06, 10, 14, 18	Left Hand
Italy	24, 28, 32, 36, 40	Left Hand
Luxembourg	03, 07, 11, 15, 19	Right Hand
Belgium	21, 25, 29, 33, 37	Right Hand
Netherlands	23, 27, 31, 35, 39	Right Hand
Austria	04, 08, 12, 16, 20	Left Hand
Switzerland	22, 26, 30, 34, 38	Left Hand

Orbital Slot: 31°West

COUNTRY	CHANNELS	POLARISATION
Ireland	02, 06, 10, 14, 18	Right Hand
Portugal	03, 07, 11, 15, 19	Left Hand
UK	04, 08, 12, 16, 20	Right Hand
Spain	23, 27, 31, 35, 39	Left Hand

Orbital Slot: 37°West

COUNTRY	CHANNELS	POLARISATION
Liechtenstein	03, 07, 11, 15, 19	Right Hand
Andorra	04, 08, 12, 16, 20	Left Hand
Monaco	21, 25, 29, 33, 37	Right Hand
San Marino	01, 05, 09, 13, 17	Right Hand
Vatican City	23, 27, 31, 35, 39	Right Hand

H. SATELLITE DESIGN AND OPERATION

Satellites are the central component in the telecommunication revolution. The capacity and reliability of the worldwide communication network is rapidly improving in part because any location within a satellite's "view" can be linked without the use of land-based cables or line-of-sight relay towers. A single communication satellite can simultaneously serve vast areas of our globe.

The Downlink Path and Satellite Footprints

Downlink antennas are capable of broadcasting microwave signals to any chosen geographic region within view of a communication satellite. The satellite footprint, which characterises this geographic coverage, is determined by precise details of the downlink antenna design as well as by the level of microwave power generated by each onboard channel. Although each communication channel, known as a transponder, is a physically separate circuit, signals from one or more transponders can be downlinked by just one antenna. As would be expected, power levels are higher in regions targeted along the main downlink antenna axis and weaker in off-target areas. A larger antenna is therefore required to receive off-axis signals than those at the centre of the footprint, directly along the downlink antenna "boresight." Extremely large diameter antennas would be necessary for reception of broadcasts in regions far off the downlink boresight.

Footprint maps are calculated for every communication satellite. In practice, there is some variation in output powers among transponders on any particular spacecraft. In addition, as a spacecraft ages, transponder power outputs generally weaken. Such variations are not reflected in the calculated footprint maps. But in those cases where footprint maps have actually been measured on-site, differences from the calculated values have been observed.

Understanding Footprint Maps

Footprint maps provide valuable information in sizing TVRO components (see Figures 1-22 and 1-23). Power levels on these maps are measured by the "effective isotropic radiated power," the EIRP, which is expressed in units of decibels relative to one watt (dBw). A footprint map is constructed by joining all those points on the map having an equal

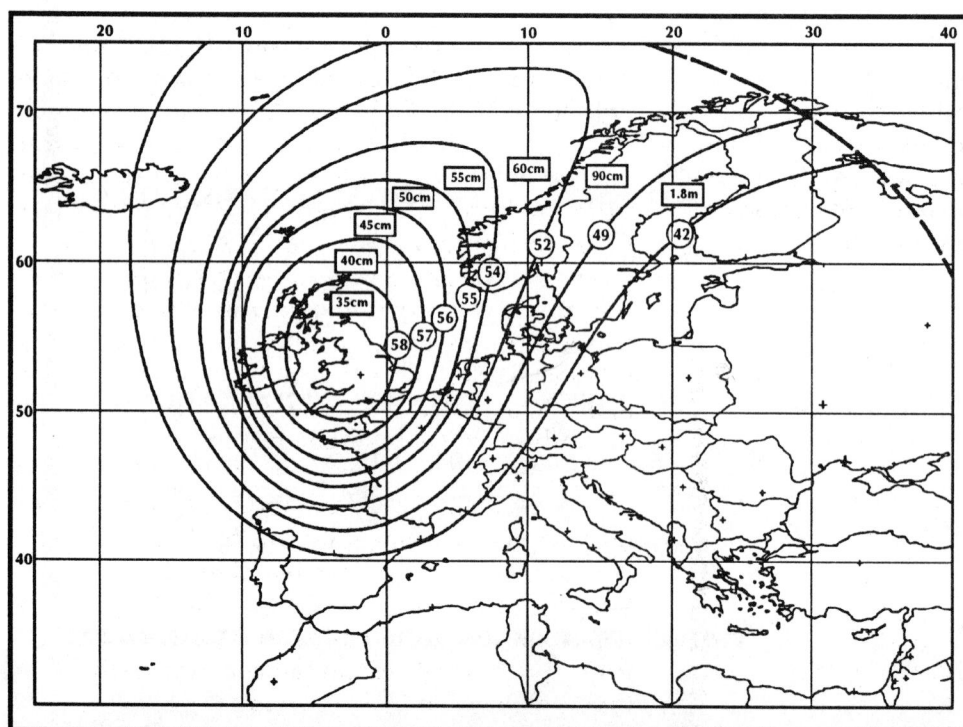

Figure 1-22. BSB Footprint Map. *Every footprint map has a series of equi-power EIRP lines. The intensity of the downlinked signal decreases from those values at the beam centre or boresight. Each contour has a defined signal level that is expressed in dBw, decibels referenced to one watt of power. Footprint maps for each satellite are calculated and drawn for the satellite owners by the satellite manufactures. The BSB Marco Polo satellite is a high powered spacecraft. In this case, a the numbers in the rectangles are suggested antenna sizes for adequate signal reception. (Courtesy of Steve Birkill)*

BASICS

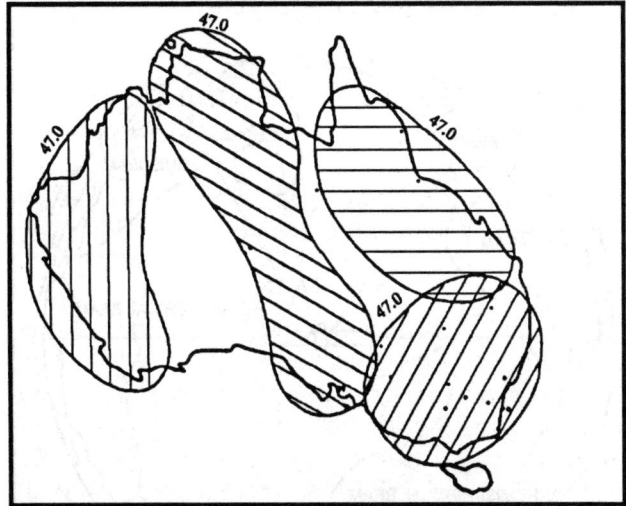

Figure 1-23. Footprints of the Aussat Satellite. *This calculated EIRP contour map of the Aussat 1 satellite shows coverage of the more heavily populated areas of Australia. Each beam has na EIRP of 47.0 dB at its edge. The central spot beam actually resembles a footprint.*

EIRP by continuous lines. Thus a distinctive "footprint" is published for every orbiting satellite as a series of contour lines superimposed upon the map of the region served.

The decibel scale, named in honour of Alexander Graham Bell, was devised to describe the potentially enormous changes in power at various stages in an electronic communication chain by relatively small numbers. This was necessary because components like amplifiers and antennas could increase power levels many hundreds of thousands of times. Constantly writing out such numbers could have become very cumbersome. Very large changes in power or voltage levels are therefore translated into manageable values of decibels. Thus, 3 dB means a doubling of power; a

TABLE 1-7. THE DECIBEL NOTATION	
Number of Decibels	Relative Increase in Power
0	1
1	1.26
3	2
10	10
20	100
30	1,000
50	100,000
100	10,000,000,000

relative change of 30 dB means an increase of power by a factor of 1000. Note that dBw means decibels relative to one watt; dBm means decibels relative to one milliwatt. Therefore, by comparison with Table 1-7, 3 dBw means 2 watts and 0 dBm means 1 milliwatt. Please see Appendix A for more details on the decibel scale.

The apparently small variation in EIRPs on footprint maps can be deceptive as they translate into large differences in signal power. For example, the American C-band satellite Satcom I relayed 33 dBw to Anchorage, Alaska, roughly half that received along the downlink antenna boresight in Denver, Colorado at 36 dBw because of the 3 dB difference in EIRP.

Satellites which broadcast to more limited geographical regions have footprints covering a smaller area. These "spot beams" have proportionately higher EIRPs than those of a satellite targeting a much larger area with either "hemispherical," "zone" or "global" beams because the same power is more concentrated. Global beams cover 42.4% of the earth's surface, the maximum view of any one geosynchronous satellite. Hemispheric and zone antennas target 20% and 10%, respectively, of the globe. Spot beams are focused on even smaller areas (see Figure 1-24). Clearly both the shape and orientation of the downlink antenna as well as the power generated by each transponder determine geographic coverage and EIRPs.

There are characteristic variations in footprint maps among satellites. European Ku-band spacecraft have generally had footprint patterns which were circularly symmetric and decreased in power linearly in regions removed progressively further from the boresight. American C-band satellites have typically had rather "flat" powers over the region of coverage and a fairly sharp drop-off in power past the edges. Recently launched satellites have employed "beam shaping" downlink antennas that permit designers to mold footprints to reach just the targeted region. Such footprints do not "waste" power in non-targeted areas.

In general, most Ku-band footprints do not blanket entire continental areas but have more limited geographic coverage than their C-band counterparts. Therefore, a knowledge of the local EIRP is important when attempting to receive broadcasts from a particular Ku- band satellite. Appendix B lists footprint maps for many Ku-band satellites.

SATELLITE DESIGN AND OPERATION

BASICS

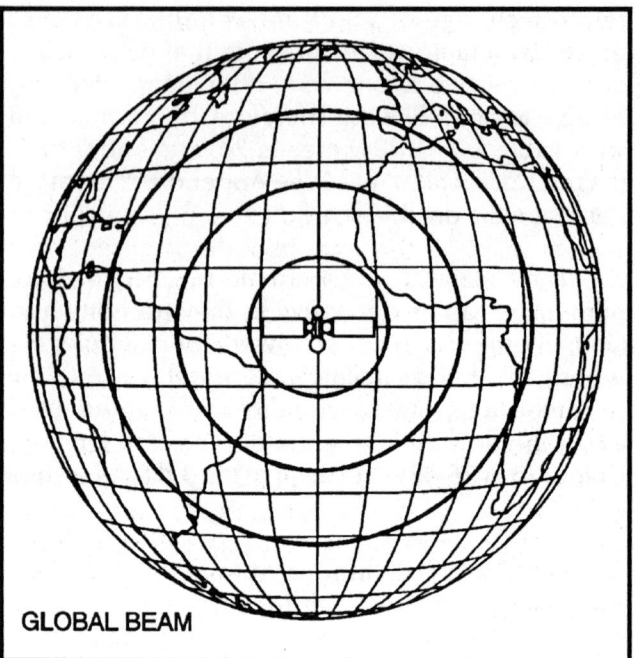 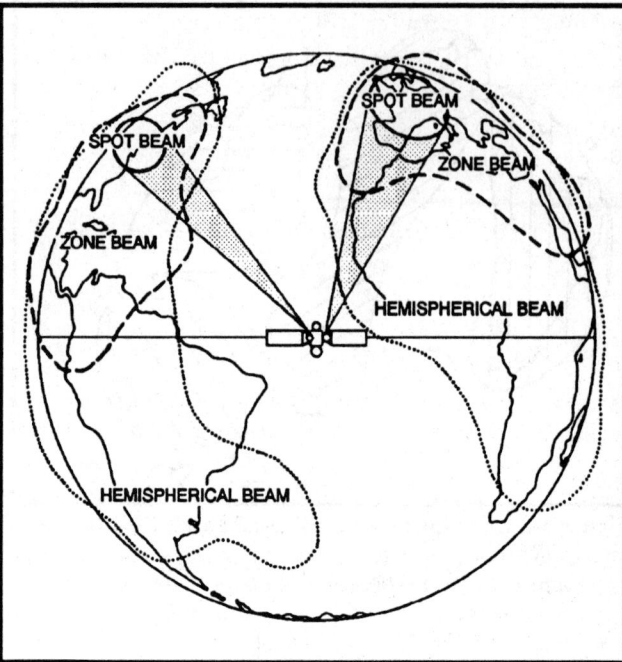

Figure 1-24. Geographic Coverage of Downlink Antennas. *Downlink antennas can transmit global, hemispherical, zone or spot beams. Ku-band satellite operators often employ zone or spot beams having smaller geographic coverage and higher power levels.*

The term effective isotropic radiated power (EIRP) is derived from the word isotropic which means equal in all directions. Effective isotropic radiated power means the power levels that would be received at any location if an antenna were radiating equally in all directions. Therefore, a 37 dBw isotropic radiated power reading means that a perfect antenna would direct 37 dBw or 5012 watts per square metre in all directions. The reason that a transponder having rather limited power can apparently have such a high EIRP stems from the fact that this power is not radiated equally in all directions but is concentrated in a narrow beam aimed at the earth below. Ku-band transponders having a total power of 50 watts have EIRPs as high as 48 or 49 dBw when this power is directed into a zone beam.

EIRP levels refer to the power of signals measured at the satellite downlink antenna. In the example given above, 5012 watts per square metre would be directed towards a selected location on earth below. The 50 watts total power would be concentrated into an area much less than one square metre when measured at the downlink antenna.

The downlinked signal then spreads out in a cone-like beam as it travels to the earth below. This weakening or dilution in power as it moves away from the satellite is called the "free space path loss." The greater the distance to the target, called the "slant range," the greater are the free space path losses.

Contributing to the losses incurred on the homeward voyage are absorption by molecules of the atmosphere. Water vapor is the main culprit in the attenuation of Ku-band downlinked signals. In fact, during a severe rainstorm, the power received on the surface of the earth at higher Ku-band frequencies can be reduced by more than 10 dB which equals a factor of 10! Rain fading is a serious design concern for Ku-band broadcasts and is explored in more detail in Chapter III.

Free space path loss and atmospheric absorption explain why a 50 watt Ku-band transponder is detected on earth at a strength of less than roughly one billionth of a billionth of a watt per square metre! Detecting a transmission from a 50 watt transponder is like clearly seeing a typical light bulb from a distance of 35,803 kilometres (22,247

SATELLITE DESIGN AND OPERATION

BASICS

miles). Or it is like receiving a relatively powerful CB radio transmission, which is designed to have a 15 to 25 kilometre range, from 35,803 kilometres away. Note that Ku-band satellites operate at a frequency more than 400 times higher than CB radios.

Satellite Operation - Audio and Video Channels

Uplinked microwaves have powers of fractions of a millionth of a watt when received by a geosynchronous satellite. In the heart of the spacecraft, typically about the dimensions of a medium sized truck, these signals are amplified many thousands of times, shifted to a lower frequency range and then broadcast back to earth. If the same frequency were used for both the uplink and downlink circuits, an uplink, which is usually also designed to receive as well as transmit satellite broadcasts, could experience annoying feedback between both types of signals. The frequency downconversion that occurs in a communication satellite also reduces interference at the earth receiving station. If the same frequency were used, some of the uplinked signal could mistakenly be detected by receiving stations near uplink sites along with the desired signal. This would cause feedback between the uplinked and downlinked signals just as a loud ringing is caused by feedback in a sound system when a microphone is too close to a speaker.

Ku-band satellites have followed in the developmental trail of the more evolved C-band systems. In order to fully explore the operation of the newer, higher frequency spacecraft, it is instructive to examine their C-band counterparts, in part because standards have been well established for such circuits. C-band spacecraft typically use an uplink having a 500 megaHertz bandwidth spanning the range from 5.925 to 6.425 gigaHertz and the same bandwidth shifted to the lower downlink range of 3.7 to 4.2 gigaHertz. Ku-band satellites typically uplink signals in the vicinity of 14 to 14.5 GHz and downlink in a portion of the range spanning from 10.95 to 12.75 GHz.

All the power to carry out these electronic functions is provided by the sun's energy which is captured by arrays of solar cells, also known as photovoltaic cells. The illustration of Intelsat V series spacecraft shows these arrays as large wing-like structures. The smaller central features are the receiving and downlink antennas.

Video Transponder Formats

The number of television channels, telephone conversations or the amount of data transmitted is related to satellite electronic design. The early C-band Western Union spacecraft such as Westar I and II relayed 12 television programs simultaneously on signals having one polarity. The Intelsat IV series of satellites, seven of which were launched in the 1971 to 1975 period, also transmitted 12 separate video signals. The RCA series of Satcom spacecraft and subsequent American C-band broadcast satellites were engineered to transmit twice as many video channels in the same bandwidth.

American C-band broadcast satellites thus evolved from 12 to 24 transponder spacecraft. The 500 megaHertz microwave band can be subdivided into twelve 40-megaHertz segments plus a remainder of 20 megaHertz. Since a 36 megaHertz bandwidth is more than sufficient to broadcast a high quality television picture, Western Union designed their pioneer satellites to manage 12 channels each having a 36 MHz bandwidth with 4 MHz guard bands between each channel to eliminate the possibility of crosstalk. Each channel was handled separately on-board the satellite by a device called a transponder.

Engineers who designed the Satcom I spacecraft were somewhat more creative. They doubled the number of channels which could be relayed by this 500 megaHertz total bandwidth by a technique called frequency re-use. All even channels were transmitted earthward with horizontal polarity, all odd channels with vertical polarity. and the frequency centres of these cross-polarised channels were offset from each other for further security against crosstalk. Since each TVRO was equipped to detect only vertical or horizontal polarity at one time, there could be some overlap between the frequencies used for the odd or even channels without interference occurring between the channels of opposite polarity. Of course, when an earth station is required to simultaneously detect all 24 transponders, it must be capable of detecting both horizontally and vertically polarised signals at once.

SATELLITE DESIGN AND OPERATION

BASICS

TABLE 1-8. C-BAND SATELLITE CHANNEL CENTRE FREQUENCIES & RANGES

Transponder Number	Downlink Frequency (MHz)	Frequency Band (MHz)
1	3720	3702-3738
2	3740	3722-3758
3	3760	3742-3778
4	3780	3762-3798
5	3800	3782-3818
6	3820	3802-3838
7	3840	3822-3858
8	3860	3842-3878
9	3880	3862-3898
10	3900	3882-4008
11	3920	3902-3938
12	3940	3922-3958
13	3960	3942-3978
14	3980	3962-3998
15	4000	3982-4018
16	4020	4002-4038
17	4040	4022-4058
18	4060	4042-4078
19	4080	4062-4098
20	4100	4082-4118
21	4120	4102-4138
22	4140	4122-4158
23	4160	4142-4178
24	4180	4162-4198

The standard C-band centre frequencies used in North American and some other domestic satellite systems are listed in Table 1-8

The North American C-band format is by no means the only one in use. Intelsat spacecraft have evolved to the point that the transponder frequency format is a mixture of C and Ku-band transmissions with each transponder spanning a variable bandwidth. For example, the Intelsat V-A series of spacecraft serving the Atlantic ocean region (see Figure 1-25) transmits C-band global, hemispherical and zone beams and Ku-band zone and spot beams over a mixture of 36 MHz wide C-band and 72 and 77 MHz wide Ku-band transponders. The Ku-band spot beams transmit horizontally and vertically polarised signals from two 72 MHz, two 77 MHz and from two 241 MHz wide transponders. The other beams relay a mixture of left-hand circularly polarised (LHCP) and right-hand circularly polarised (RHCP) signals. To further increase the complexity and flexibility of this satellite, a switching matrix on-board allows C-band and Ku-band uplink signals to be translated to either band for downlinking.

Although Intelsat spacecraft which are designed for extremely flexible use represent some of the more complex formats encountered, a whole new breed of domestic dual frequency band satellites is evolving. For example, the two in-orbit Mexican Morelos vehicles have twelve 36 MHz and six 72 MHz C-band horizontally and vertically polarised transponders and four 108 MHz wide circularly polarised Ku-band transponders.

TABLE 1-9. A SAMPLE OF KU-BAND SATELLITE CHARACTERISTICS
(Audio and Video Services)

Satellite Series	Frequency Band	Number of Transponders	Bandwidth (MHz)	Format
Intelsat VA F11	0.950-11.698	2	72	H/V
		2	77	H/V
		2	241	H/V
Telecom F1A/1B	12.500-12.750	6	36	H/V
Aussat I/II	12.250-12.750	15	45	H/V
Eutelsat 1 F1	10.950-11.200	12	72	H/V
	11.450-11.700			
Anik C1/C2/C3	11.700-12.200	16	54	H/V
GStar A1/A2	11.700-12.200	16	54	H/V
Morelos F1/F2	11.700-12.200	4	108	Circular
SBS 1/2/3	11.250-12.200	10	43	H
Spacenet I/II	11.700-12.200	6	72	H

Today, earth station equipment designed to receive C-band satellite television broadcasts has become relatively standardised in the Americas. Transponder centre frequencies are spaced at regular intervals across the 3.7 to 4.2 GHz band and linearly polarised signals with 36 MHz bandwidths are used. Half transponder broadcasts occupying 18 MHz bandwidths are occasionally encountered when receiving Intelsat broadcasts but most

SATELLITE DESIGN AND OPERATION

conventional equipment is capable of detecting and interpreting these signals. The GSTAR series of satellites broadcast on sixteen 54 MHz wide transponders, 8 vertically and 8 horizontally polarised, for a total of 32 half transponder channels.

Ku-band transmissions have a variety of centre frequency spacings, channel bandwidths and either linear or circular polarised signals (see Table 1-9 and 1-10). Any one type of satellite receiver would be hard pressed to have the capability of processing signals from all worldwide Ku-band spacecraft.

Figure 1-25. Intelsat V/VA Satellite. *Intelsat V satellites are three-axis, spin stabilised spacecraft. They are nearly 7 metres (22 feet) high and weigh 1800 kilograms (4000 pounds). They span more than 15 metres (50 feet) when their solar panel "wings" are deployed. These spacecraft can cross-strap between C and Ku-band transponders. This means that signals uplinked in one frequency band can be downlinked on the other band via matrix switching networks on board the spacecraft. (Courtesy of Ford Aerospace)*

BASICS

TABLE 1-10. VARIATIONS BETWEEN KU-BAND SATELLITES
(EXISTING and DECOMMISSIONED)

Satellite	Location (degrees)	EIRP (dBw)	Channels	Bandwidth	Polarisation (MHz)
Aussat 3	164 E	42-51 spot 35-41 conus	15	45	H/V
Aussat 1	160 E	42-51 spot 35-41 conus	15	45	H/V
Aussat 2	156 E	42-51 spot 35-41 conus	15	45	H/V
Morelos F1	113.5 W	44	4	108	Circular
Morelos F2	116.5 W	44	4	108	Circular
Anik C1	107.5 W	46.5	16	54	H/V
Anik C2	112.5 W	46.5	16	54	H/V
Anik C3	117.5 W	46.5	16	54	H/V
Anik B 1	109 W	46.5	6	72	H
Japan-BSA	110 E	55	2	80/50	H/V
Gorizont 9	53 E	28 global	1 31 N.hemi 33 zone 42 spot	34	H/V
Eutelsat 1-F1	13 E	34 Eurobeam	12 40 spot 42 spot	72	H/V
Eutelsat F4	10 E	36 Eurobeam 39.8 multi-user	12	72	H/V
Eutelsat 1-F2	7 E	36 Eurobeam 42 spot	12	72	H/V
Telecom F1a	8 W	47	6	36	H/V
Telecom F1b	5 W	47	6	36	H/V
TDF-1A	19 W	63.8/63.9/64	3	27	RHCP
TV-SAT A	319 W	65.8	3	27	LHCP
Intelsat VA-F11	27.5 W	44 W.spot	1	72	H
			1	77	
			12	41	
		41 E.spot	1	72	
			1	77	
			1	241	
Intelsat VA-F10	24.5 W	44.1 W.spot	1	72	H
			1	77	
			1	241	
		41.1 E.spot	1	72	
			1	77	
			1	241	
Intelsat VA F12	60 E	44.1 W.spot	4	72	H/V
		41.1 E.spot	22	41	H/V
ASC-1	128 W	42	6	72	H
Spacenet 1	120 W	43	6	72	H
GStar A2	105 W	45	16	54	H/V
GStar A1	103 W	45	16	54	H/V
SBS-4	101 W	47-49	10	43	H/V
SBS-1	99 W	47	10	43	H
SBS-2	97 W	47	10	43	H
SBS-3	95 W	47-50	10	43	H
Satcom K1	87 W	44-53	16	54	H/V
Satcom K2	77 W	44-53	16	54	H/V
Spacenet 2	69 W	44.6	6	72	H

SATELLITE DESIGN AND OPERATION

BASICS

The 54 MHz bandwidth satellites can broadcast either one TV channel or two simultaneous channels each having a 22 MHz bandwidth. When this is done the total power must be cut in half and split between the two channels and uplink power then backed off by an additional 3 or 4 dB to avoid interference between these signals. This is known as intermodulation distortion and is caused by a "beating" between the two adjacent signals of nearly identical frequency. A TVRO might have no trouble receiving the signal from such a transponder until it was "loaded" with two signals. The 72 and 108 MHz wide transponders can manage up to 2 and 4 channels, respectively.

Audio Transponder Formats

Audio signals are also relayed within the transponder bandwidth. When a TVRO receives and processes a satellite transmission, the recovered raw signal is contained in a band of frequencies from near zero to about 10 MHz. Video signals occupy only a portion of this region. For example, the video signal on the NTSC format, which is the standard for North American television, occupy the region between zero and 4.6 MHz (see Figure 1-26 and 1-27). All the remaining space can be used for audio channels, some of which carry stereo or monaural sound that matches the television picture. Other audio can have an identity not related to the video signal.

Audio signals are often transmitted on "audio subcarriers." Most American C-band television broadcasts relay the audio information on subcarriers in the 5.0 to 9.0 MHz range. A 6.8 MHz subcarrier or occasionally, 6.2 and 6.8 MHz subcarriers are commonly used. Stereo sound is often modulated onto subcarriers at 5.58 and 5.76 MHz.

Audio information can also be relayed in other formats. In the single channel per carrier mode, known as SCPC, each audio channel is transmitted on a carrier totally separate from the video signal. For example, the American National Public Radio Network dedicates the entire C-band transponder 3 on Westar IV exclusively to their audio broadcasts. Over 300 SCPC audio channels are transmitted over North American satellites. Another type of audio format embeds the information into the video signal. This is known as sound-in-sync and is explored in more detail in the section on scrambling in Chapter IV as well as in both *World Satellite TV and Scrambling Methods - The Technician's*

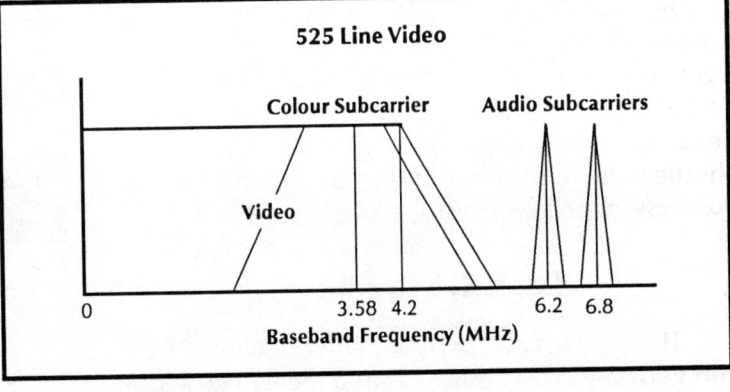

Figure 1-26. Standard 525 Line NTSC Video and Audio Format. *The video bandwidth in a standard NTSC satellite broadcast ranges from 0 to 4.2 MHz with the colour subcarrier falling at 3.58 MHz. The remaining bandwidth, ranging from 5 to 8.5 MHz and typically broadcast with 75 microsecond pre-emphasis to reduce noise, is reserved for audio subcarriers not necessarily associated with the video information being displayed on the television screen. These non-video-related audio subcarriers generally have a more narrow bandwidth and lower power than the standard audio subcarriers.*

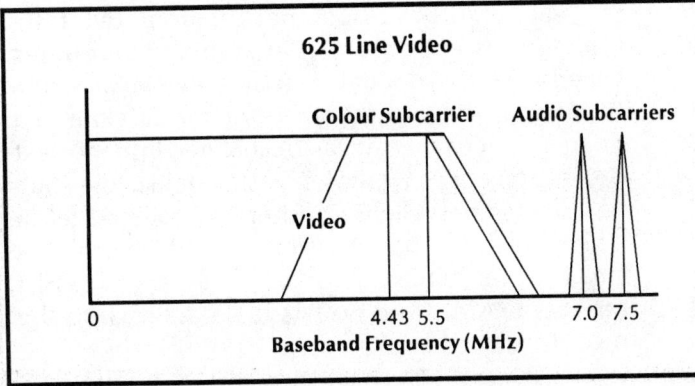

Figure 1-27. Standard 625 Line PAL Video and Audio Format. *The video bandwidth extends from 0 to 5.5 MHz with the colour subcarrier centred on 4.43 MHz. With the exception of certain South American countries, PAL systems employ this type of frequency allocation format in most regions of the world. The audio band extends from 6 to 9 MHz with common subcarrier frequencies of either 6.5 and 6.65 MHz. Three different audio pre-emphasis standards, 50 microseconds, 75 microseconds or J17, are common.*

SATELLITE DESIGN AND OPERATION

Guide and *Satellite* and *Cable TV - Scrambling and Descrambling*. Another book, *The Hidden Signals on Satellite TV*, by Thomas Harrington also studies this and other topics in detail and is an excellent source of information about non-video transmissions. These three books are listed in the reference section at the end of this manual.

I. SATELLITE LAUNCHING AND MAINTENANCE

The excellence of the worldwide satellite communication system depends upon launching long-lived vehicles into stable geostationary orbits. Since Sputnik's successful flight in 1957, man has been able to lift heavier and more sophisticated payloads into orbit. During the 1960's booster rockets were inadequate to position a satellite directly into the required geosynchronous orbit. As a result, extra rockets in the body of a telecommunication vehicle were used for repositioning the spacecraft from its initial, low circular orbit into its final geostationary position. The engine and fuel of the satellite then accounted for nearly half its total weight.

Today, similar techniques for boosting satellites into space are used even though current communication vehicles having greater masses can be lifted into space by much more powerful crafts. Earlier Thor-Delta launch rockets were replaced by more powerful boosters such as the Delta-2914, the Delta-3912 and the Atlas Centaur, and then by America's Space Shuttle. The European Space Agency's Ariane rockets as well as the Space Shuttle have launched the lion's share of worldwide geosynchronous satellites, although the Space Shuttle was phased out of the commercial launch business in the late 1980s (see Figure 1-28).

Launch Sequence

The order of events by which a satellite in lifted into geostationary orbit is termed the launch sequence (see Figure 1-29). Satellites are normally launched in the direction of the earth's orbit to incorporate its rotation velocity of 1,690 kilometres per hour. The closer the launch site is to the equator, the higher this velocity. Therefore the closer the launch site is to the equator, the less power is required to put the satellite in orbit.

A standard launch sequence consists of the following stages: launch, transfer orbit, plane change, drift orbit and geostationary orbit.

Figure 1-28. ArianeSpace Launch Preparation. *(Courtesy of WV Publications, Inc.)*

Launches have been performed in two fashions: a conventional rocket launch, sometimes called an ELV (expendable launch vehicle) or a shuttle launch. The rocket transports the satellite to a point in space where it can be injected into the next orbital stage (see Figures 1-30 and 1-31). The shuttle launch differs slightly as there is a "sub-stage" in which the shuttle has to assume a "parking orbit." A small rocket section, known as the payload assist module (PAM) then boosts the satellite into the next stage.

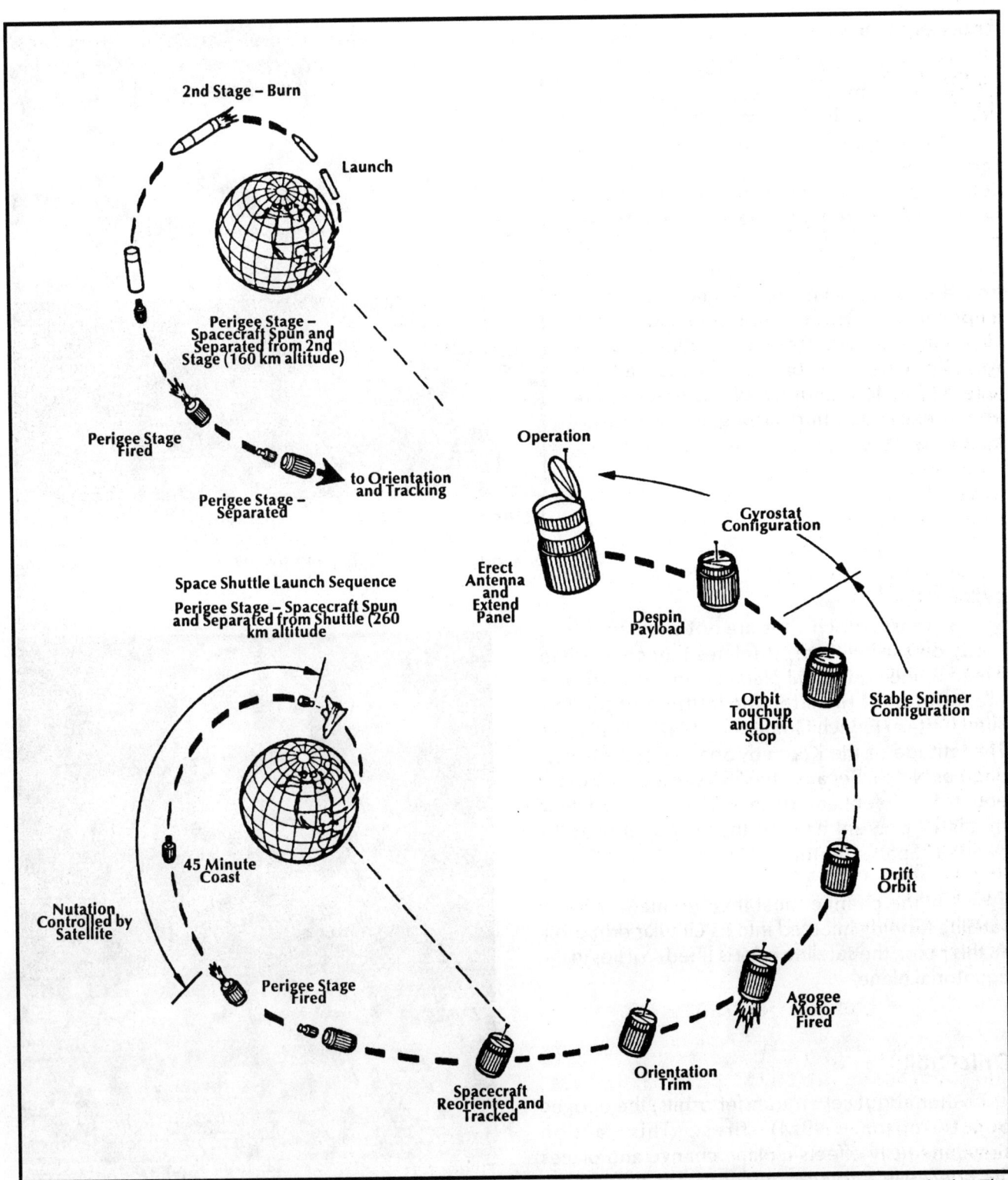

Figure 1-29. Satellite Launch and Deployment. *Satellites are launched via either Space Shuttle, Ariane or Delta vehicles. This figure shows the sequence from launch to operational deployment. (Courtesy of Hughes Aircraft Company)*

BASICS

Transfer Orbit

The transfer orbit is elliptical which is characterised by its apogee and perigee. The perigee is the point closest to the Earth's surface; the apogee is furthest away from the Earth, about the same distance from the Earth as the geostationary orbit. In a rocket launch, a satellite is injected into the transfer orbit approximately twenty-six minutes after lift-off. The transfer orbit serves to move a satellite from a lower into a higher circular orbit via an elliptical path. This method is known as the Hohmann transfer after the German physicist who developed the concept. The main advantage of the Hohmann transfer is that less energy is expended in boosting a satellite into geostationary orbit.

Figure 1-30. ASC Release. *The American Satellite Company communications satellite ASC-1 rises from the shuttle cargo bay at 6:54 A.M. August 27 1985. This vehicle is a hybrid dual-band communications satellite. (Courtesy of NASA)*

Plane Change

As most launch sites are not situated directly on the equator, satellites launches are inclined to the equatorial plane by an angle equal to the latitude of the site. The latitude of the ESA launch site in French Guinea is 5.23 degrees North. The latitude of the Kennedy Space Center is 28.3 degrees North. Because the ESA site is closer to the equator, an identical rocket can launch a payload that is 17 percent heavier than is possible at the Kennedy Space Center.

A plane change must then be made when a satellite is being injected into its circular drift orbit. At this point, the satellite orbit is tilted so it lies in the equatorial plane.

Drift Orbit

After about seven transfer orbits, the apogee boost motor (ABM) fires. This action simultaneously effects a plane change and places the satellite in a circular drift orbit. The drift orbit is an approximation of the geostationary orbit. It is a temporary stage in which all systems are tested before a satellite is placed into service.

Figure 1-31. Release of Satcom K1 Satellite. *The Satcom K-1 communications satellite is shown here spinning away from the protective shield in Columbia's aft cargo bay and beginning its journey into space. It is headed toward several maneuvers which will eventually put the spacecraft in its desired position of 85°W longitude. (Courtesy of NASA)*

LAUNCHING AND MAINTENANCE

Geostationary Orbit

The drift orbit is next slowly adjusted to the final stage in the launch sequence, the geostationary orbit. A satellite is then positioned at its operational longitude, sometimes referred to as its orbital slot or station, by its reaction control system which is an array of small jets used to adjust orbital position. This system compensates for various factors that affect the satellite's orbit. These are principally gravitational effects of the Moon and Jupiter as well as the perturbations caused by the fact that the Earth is not a perfect sphere.

The basic geostationary orbit can be easily derived on a standard scientific calculator. The force of gravitational attraction on a body in orbit around the Earth is given in the following equation

$$F_g = (M_s M_e G)/R^2$$

where M_s is the mass of the satellite in Kg, M_e is the mass of the Earth in kilograms, G is Newton's gravitational constant, ($G = 6.67 \times 10^{-11}$ N M^2/Kg^2). R is the radius of the orbit in metres.

The centrifugal force on the object is given by

$$F_c = R \omega^2 M_s$$

where ω is the angular velocity expressed in radians per second. As the orbit is circular, $F_g = F_c$. Substituting from the above equations:

$$(M_s M_e G)/R^2 = R \omega^2 M_s$$

This equation can be rearranged to produce a value for R

$$R^3 = (M_e G)/\omega^2$$

For a geostationary orbit, the angular velocity, w, must be equal to the angular velocity of the Earth, ω_e. The time taken by the Earth to complete one revolution is approximately 24 hours, equal to 86,400 seconds. One revolution is three hundred and sixty degrees which is 2π radians. The equation for the time taken to complete a revolution is given by

$$T = 2\pi/\omega$$

This equation can be rearranged to give a value for ω:

$$\omega = 2\pi/T$$

Substituting values into this equation gives a value of 727,000 radians per second for ω_e. Substituting this value and the mass of the Earth which is approximately 5.98×10^{24} kilograms into equation 4.4 gives

$$R^3 = 7.54 \times 10^{22}$$

therefore $R = 4.2251 \times 10^7$ metres which equals 42,251 kilometres.

Stationkeeping

Communication spacecraft would remain stationary forever above a chosen location over the equator if there were no extra gravitational forces from the sun and moon, if solar winds did not sweep past our globe, and if the earth were perfectly spherical. This is not the case, so these unbalanced forces cause any satellite to drift slowly away from its assigned location into an a figure-eight excursion with a 24-hour period that follows a path above and below the orbital plane. Ground controllers must periodically adjust satellite positions to counteract these forces. If not, the excursion above and below the orbital plane would build up at a rate of about 0.9 degrees per year and the satellite also drifts in an east or west direction along the orbital arc. This process of maintaining a satellite within a pre-assigned "window" is known as "stationkeeping."

Communication spacecraft are equipped with small hydrazine gas charged thrusters that are remotely fired by ground controllers whenever stationkeeping is necessary. Most geosynchronous satellites are maintained within one tenth of a degree of their assigned position, essentially in a box having dimension of about 90 miles on each side. Thrusters are fired once every 3 to 4 weeks to adjust their east/west location and once every 30 to 90 days to maintain them close to the arc. The effect of the sun is apparent on stationkeeping because the vertical thrusters must be fired most often, every 30 days or so, when the sun is furthest from the equator. Note that ten times as much fuel is required to keep a satellite on the equatorial plane as that needed for east/west stationkeeping.

There are two "zero-pull" locations at approximately 104.5° West and 75.5° East on the exact opposite side of our globe where these forces balance so that a geostationary body will remain stationary along the arc. The Canadian C-band satellite Anik DI, located at 104 degrees West, is closest to the western hemisphere zero-pull location. However, the rotation of the earth around the sun and the subsequent change in its position relative to the equator causes satellites located even at these two points to have some vertical movement above or below the arc.

The orientation of a satellite must also be maintained in a stable orientation with respect to the earth so its downlink antennas do not stray from their targets. Until recently, most spacecraft were aligned by spinning their whole body in order to create a stabilising gyroscopic effect similar to those forces that keep a rapidly spinning top upright. On such "spinners" the antenna platform was spun in the opposite direction so the reception and downlink antennas would be fixed (see Figures 1-6, 1-32). More recent satellite are three-axis stabilised - they use an internal three-axis gyroscope to perform this task (see Figure 1-33). In order to relay a global beam having a width of approximately 2 degrees, as is typically required to cover a land mass such as half the continental United States, downlink antennas must be pointed to within a 0.2 degree accuracy. This is very important because most satellites, even when accurately maintained at their assigned station, still move in a small figure-eight pattern, usually about 15 miles across. Even this very small movement can impair reception of signals by large, narrow beamwidth Ku-band receiving antennas. Note that for body spinning satellites even small wobbles must be corrected. If not, this motion can evolve into a "flat" spin (as opposed to the proper vertical spin) resulting in a total loss of communications with the spacecraft.

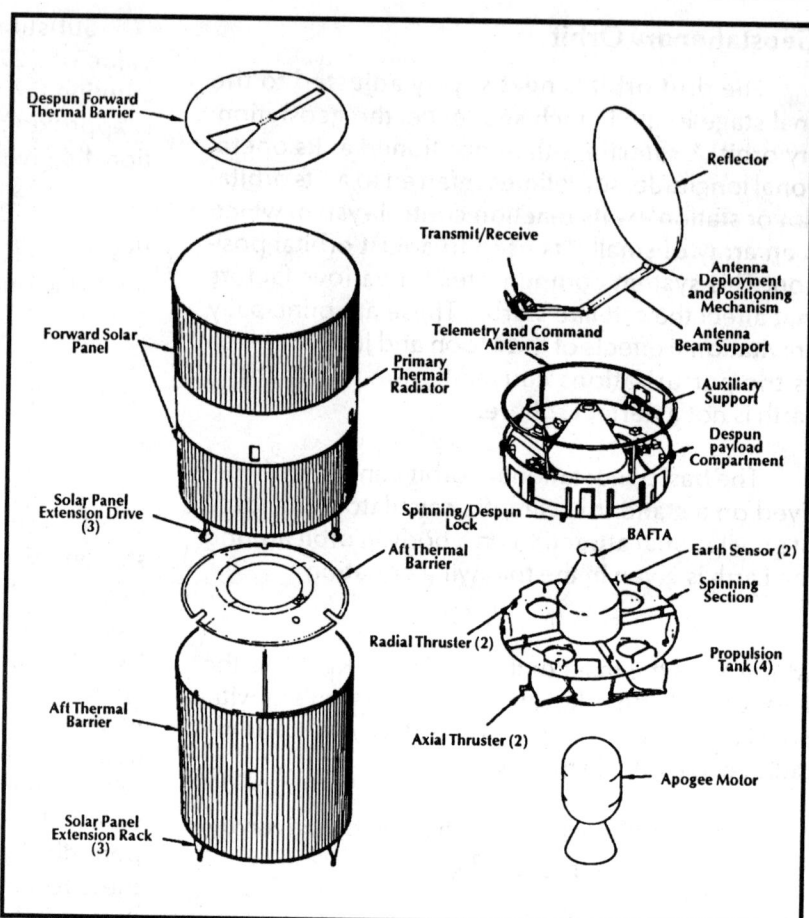

Figure 1-32. The Spin Stabilised HS-376 Satellite. *The HS-376 satellite, used by Aussat and BSB among others, is a spin stabilised spacecraft. The outer section, consists of two cylindrical solar arrays, batteries, propulsion system and most of the attitude controls. The inner section holds the antenna and communications payload platform. The outer section spins to provide stability; the inner section is de-spun, i.e. spins in the opposite direction, so the uplink and downlink antennas can maintain a constant orientation. (Courtesy of Hughes Aircraft Company)*

The solar cell arrays on three-axis stabilised satellites, the large wing-like structures, track the sun to maximise the solar power received. The RCA series of Satcom spacecraft with their large wing-like panels and the newly introduced Hughes HS 601 vehicle are both spin stabilised. Spinners, in contrast, have solar arrays covering their whole external surface so only 40% of the cells are illuminated at once. Spinners typically look like large cans having antennas at one end.

BASICS

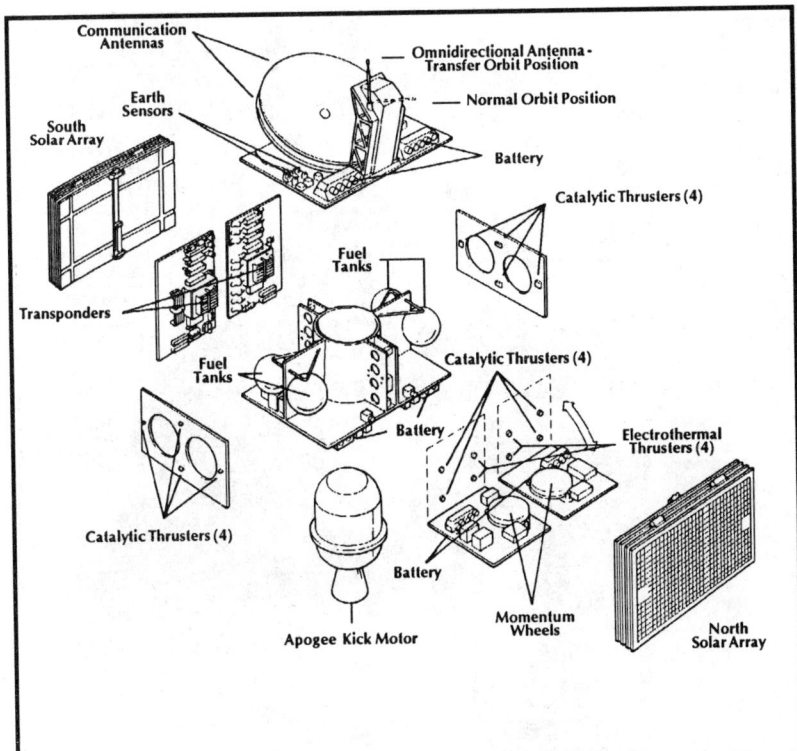

Figure 1-33. Three Axis Spin-Stabilised Satellite. *This diagram shows an exploded view of the RCA model 4000 satellite used by ASTRA. Antenna pointing stability in three axis is provided by momentum wheels, gyroscopes. These allow ground controllers to position the spacecraft in any orientation simply by varying the speed of the appropriate momentum wheel. In this illustration the solar wing are not deployed but still folded into the body of the satellite ready for launch. (Courtesy of Société Européene des Satellites, Luxembourg)*

Although communication satellites are almost always in full view of the sun because they are so far removed from the Earth, during two periods each year, from either February 27th through April 12th or August 31st through October 16th, the satellites pass through the shadow of the Earth. They then experience a series of 45 separate eclipses that gradually increase in length to a peak of about 1.2 hours per day on the equinox dates, March 21st and September 21st. Following the maximum at the equinoxes, the eclipse time decreases. Onboard nickel-hydrogen batteries are then used to maintain design power levels.

Today, both the improved stability and stationkeeping of present-day vehicles permit the use of larger antennas, capable of focusing higher-power microwaves and of housing larger solar cell arrays to increase satellite power.

Satellites do not live forever. Their life expectancy is determined by how long adequate power and stability can be maintained. Repositioning rockets usually run out of hydrazine gas before surviving ten years in space. Also, solar cells, constantly bombarded with micrometeorites and ultraviolet rays from the sun, slowly wear out. Either a 30% reduction in solar cell power or expiration of the hydrazine fuel supply is taken as a signal to retire a satellite from active duty. For example, Satcom III, the time-tested workhorse of American C-band broadcast television, has recently been decommissioned. Transponder powers slowly decayed during the final years of this satellite.

When ground controllers put a satellite to rest it is boosted into a slightly higher, unstable, non-geosynchronous orbit where, in time, it will eventually re-enter the atmosphere and burn up. Some of the decommissioned, pioneer communication vehicles are still resting in "eternal orbits." Recently, a new era in satellite operation began as the Space Shuttle recovered some damaged satellites for subsequent repair. This maneuver has saved 50 to 80 million dollars U.S. in hardware and launch costs and may someday be used to recapture and re-use satellites.

Future Trends in Satellite Design and Operation

Satellite technology is still in its infancy. Developments have proceeded at a rapid pace during the last twenty years. The future will also certainly bring some extraordinary advances in design and operation of the worldwide satellite communication system. Designing, building, launching and operating a satellite communication system is at the forefront

LAUNCHING AND MAINTENANCE

BASICS

Figure 1-34. TV-SAT Communications Satellite. *The TV-SAT satellite is shown here under final construction and testing. (Courtesy of EuroSatellite)*

of the high technology business of the 1990s (see Figure 1-34).

Some interesting developments are outlined below. These include low and inclined orbit vehicles, high power satellites, frequency re-use and multiple frequency satellites, digital compression and other improvements in signal processing, and inter-satellite communications.

Low and Inclined Orbit Satellites

While efforts in the pioneer days of satellite communications were directed towards launching vehicles into stable geosynchronous orbits, today some companies are reversing this trend with both low orbit and inclined orbit satellites. For example, the proposed Motorola Iridium cellular radio land mobile network as well as the Starsys and Orbital Sciences low orbit systems would be capable of using very small, low cost reception terminals be-

cause the distance between the terminal and satellite would be greatly reduced. In addition, the time delay of 0.40 seconds associated with two-way transmissions via a geosynchronous satellite would be virtually eliminated. Low orbit satellites follow a highly elliptical orbital path.

Inclined orbit satellites are geosynchronous spacecraft that are allowed to drift from their assigned locations more than typically permitted. Thus stationkeeping fuel is conserved and, as a result, satellite lifetime is substantially increased. This requires the use of antennas that have a small amount of tracking ability. With the advent of low cost devices to allow receiving antennas to track, the savings in satellite cost because of the extended lifetime is viewed as an economical tradeoff for some communicators. A recent Gorizont satellite that broadcasts the popular CNN has inadvertently been an inclined orbit spacecraft. A small amount of tracking has been necessary.

LAUNCHING AND MAINTENANCE

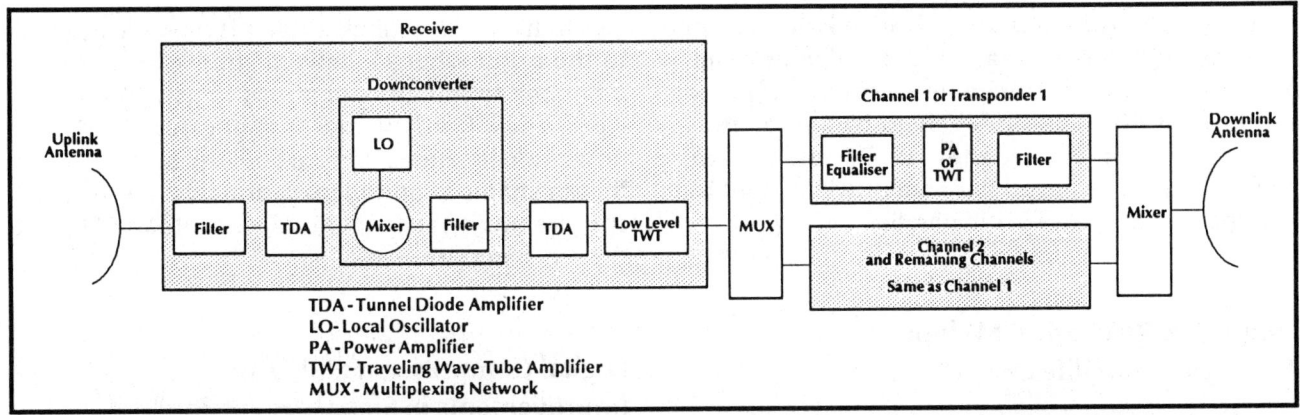

Figure 1-35. Satellite Operation Block Diagram. *This block diagram shows the operation of a typical satellite. An antenna receives the signal, a filter eliminates unwanted frequencies and an amplifier boosts the power before downconversion to the proper downlink frequency range. Following further amplification, a multiplexing network breaks the signal into twelve components for downlinking.*

High Power Satellites

Satellite transponder levels have increased from a few watts to as high as 200 watts or more during the past twenty years. As a result, the diameter of receiving antennas has plummeted from as high as 30 meters to under 40 centimeters for some Ku-band systems. This trend has been quite evident in both C- and Ku-band reception.

Devices known as traveling wave tube amplifiers (TWTs), one associated with each transponder, have provided this power (see Figure 1-34). While earlier TWTs were expensive, difficult to manufacture and subject to failure, recent higher power models have functioned well. In addition, a new option has recently emerged, the solid state power amplifier (SSPA). This device has the advantage of being lightweight, long-lived and reliable. In addition, it does not have the bandwidth limitations characteristic of TWTs. While SSPAs have not yet achieved the power levels of TWTs, especially at the higher frequency Ku-band, this is rapidly changing. The new Intelsat VI series of satellites features a combination of TWTs and SSPAs in both C and Ku-band transponders.

This trend towards higher power transponders is evident from Table 1-11. End-of-life equinox power has increased from 40 watts on Intelsat 1 to 2,200 watts on Intelsat VI.

Satellites have generally been classified as either low, medium or high power. The output power is determined by the TWT capability. Low power spacecraft with powers of 25 watts per transponder have employed hemispherical or global beams to transmit telephone or video programming for cable TV concerns. Medium powered satellites such as the European Astra 1B which relays 55 watts per transponder with a beam centre EIRP of 52 dBw have been used for direct-to-home television and radio broadcasts that can be received by dishes as

TABLE 1-11. TRENDS IN INTELSAT SATELLITE CAPABILITIES

	INTELSAT							
	I	II	III	IV	IV-A	V	V-A	VI
Year of First Launch	1965	1966	1968	1971	1975	1980	1983	1986
Contractor	Hughes	Hughes	TRW	Hughes	Hughes	Ford	Ford	Hughes
Dimensions (width/height, m)	0.7/0.6	1.4/0.7	1.4/1.0	2.4/5.3	2.4/6.8	2.0/6.4	2.0/6.4	3.6/6.4
Spacecraft Orbit Mass (kgs)	68	182	293	2,385	1,489	1,946	2,140	12,100
Communications Payload Mass (kgs)	13	36	56	185	190	235	280	800
End-of-life Equinox Power (watts)	40	75	134	480	800	1,270	1,270	2,200
Design Lifetime (years)	1.5	3	5	7	7	7	7	10
Rated Voice Channel Capacity	480	480	2,400	8,000	12,000	25,000	30,000	80,000
Total Bandwidth (MHz)	50	130	300	500	800	2,300	2,180	3,680

BASICS

small as 60 cms in diameter. Transmissions from high power satellites that generate signals of approximately 110 watts with beam centre EIRPs exceeding 58 dBw can be received by antennas as small as 30 cms. The Hughes HS376 satellite employed by BSB achieved this high power output by running two 55 watt TSTs in parallel.

Frequency Re-use and Multiple Frequency Satellites

There are number of persuasive reasons to use satellites and orbital space as efficiently as possible. First, launching a satellite is an expensive proposition. Clearly, the more communication traffic a spacecraft can manage, the more economical the end result. Second, there is just a limited amount of space in the geosynchronous orbit. If satellites are spaced 2 degrees apart, a total of just 180 satellites can be in orbit at once (see Table 1-12). Third, competition for frequency allocations can be fierce so there is strong motivation to use the available bandwidth as efficiently as possible. A number of solutions have been developed in response to these issues.

Regulators and technical innovators have been progressively using higher and higher frequencies for satellite communications. C- and Ku-band systems are now extensively used; Ka-band frequency satellites that operate in the 20/30 GHz range are being developed and, in some cases, being tested.

Satellites can be co-located in the geosynchronous arc if they operate in different frequency bands and are therefore effectively invisible to each other. Intelsat spacecraft and numerous domestic satellites are dual band, that is they have both C and Ku-band transponders. In time, clusters of non-interfering, co-located satellites will receive information from just one uplink facility. In addition, numerous co-located spacecraft could further conserve resources by sharing a single platform, positioning system and power supply. This could result in increased pointing accuracy for the downlink antennas thus enhancing circuit performance.

The "re-use" of frequency allocations is a common technique in extending the range of a communication system. The use of dual polarization signals is one example of frequency re-use; this has effectively doubled the channels a given satellite can transmit. Downlink antennas have developed to the point where localised spot beams can be downlink to multiple locations re-using the same frequency band in each location. Ku-band downlink antennas are particularly suited to this "beam shaping" and localisation. Using localised spot beams effectively increases satellite EIRP at the receiving site.

Digital Compression and Other Improvements in Signal Processing

Digital compression techniques that allow more information to be carried per transponder are now being commercially introduced. Use of a satellite circuit can be practically increased by a factor or 5 to 10, an impressive gain. One newly introduced digital compression system can transmit up to 12 video channels in the bandwidth previously used to transmit one analogue video broadcast.

Other techniques also exist or are being developed to use satellite capacity more efficiently. Some of these do not directly relate to real-time video broadcasting. For example, a technique known as time division multiple access (TDMA) allows numerous users to share a transponder on a demand basis. When capacity is not needed, even for fractions of a second, the circuit becomes free for use by another. Similarly, digital speech interpolation (DSI) allows reduction in the bandwidth required for a typical voice circuit by using the channel only when speech, not silence, is present.

Inter-Satellite Communication

Satellite communication circuits will be used more effectively by employing communication directly between satellites with microwave or laser beams. This method is very effective in outer space because there is no atmosphere to weaken signals bouncing between satellites. In fact, the ITU has already allocated frequency bands for inter-satellite links. Using such relays could make communication between very distant countries less expensive by eliminating an intermediate link. A message uplinked in San Francisco could be received by a American satellite, relayed directly to one positioned over the Mediterranean and subsequently downlinked to Saudi Arabia. The alternative would be to use an intermediate Intelsat satellite or a submarine fiber optic cable, at a potentially higher cost.

LAUNCHING AND MAINTENANCE

BASICS

TABLE 1-12. WORLDWIDE KU-BAND SATELLITE CHART

WESTERN HEMISPHERE

Location	Code	Satellite	Operator	Band(s)
1	D,E	Intelsat V F12	Intelsat	
8	D,E	Telecom F1/II	France	C,K5,X
5	D,E	Telecom F2-1C	France	C,K5,X
11	D,E	Statsionar 11	USSR	C,K
14	D,E	Statsionar 4	USSR	C,K1
18.5	D,E	Intelsat V-F6	Intelsat	C,K1,L
19	K,E	Olympus	ESA	K3,E
19	K,E	TDF 1A/1B	France	S,K3
19	K,E	TVSAT 1/2	W.Germany/86	K3
24.5	D,E	Intelsat VI F2	Intelsat	C,K1
27.5	D,E	Intelsat VI F4	Intelsat	C,K1,L
31	K,E	BSB I/II	Great Britain	K1
34.5	D,E	Intelsat V F4	Intelsat	C,K1
37.5	K,P	Orion I	US/86	K
40.5	D,P	Intelsat VB	Intelsat	
41	D,E	TDRS	AUS-Spacecom/83	C,L,K2
45	D,E	PAS1	US-Alpha Lyracom/88	C,K
53	D,E	Intelsat VB F13	Intelsat/88	C,K1,L
56	K,P	USA-SAT I	US/87	K1,K2
56	D,P	Intelsat IVA,V,VA	Intelsat	
57	D,P	Panamsat 1	US/87	C,X,K2
58	K,P	USA-SAT II	US/88	K1,K2
61	K,P	Satcom K4	US-RCA	
61	D,P	TDRS B	US-Spacecom	C,L,K2
61.5	K,P	Echostar I	US-Echostar	K2
64	D,P	ASC 3/4	US-ASC	C,K2
65	D,E	Brasilsat A1	Brasil	C,K2
67	K,P	Satcom K3	US-RCA	K2
69	D,E	Spacenet II/IIR	US-GTE	C,K2
70	D,E	Brasilsat A2	Brasil	C, K2
73	K,P	Westar A	US-W.Union	K2
75	K,P	Comstar K1	US-Comsat	K2
76	D,P	Spotnet 2	US-Natl Exchange	K2
79	K,P	Martin Marietta I	US-MM	K2
79	D,P	Satcom Hybrid-1	US-GE Americom	K2
81	K,E	Satcom K2	US-RCA	K2
83	D,P	ASC II	US-ASC	C,K2
85	K,E	Satcom K1	US-RCA	K2
87	D,E	Spacenet II/IIR	US-GTE	C,K2
89	D,P	Telstar 402	US-AT&T	K2,C
91	K,E	SBS IV	US-Hughes	K2
93	K,E	GStar 3	US-GTE	K2
93	D,P	Fordsat F2	US-Ford	C,K2
95	K,P	SBS III	US-SBS	K2
91	K,E	SBS II	US-Hughes	K2
97	D,P	Telstar 401	US-AT&T	K2
99	K,E	SBS VI	US-SBS	K2
99	K,P	Galaxy A	US-Hughes	K2
101	D,P	Fordsat F1	US-Ford	C,K2
101	K,P	Galaxy DBS 1 & 2	US-Hughes	K4
101	K,P	DBSC 1	US-DBSC	K2
101	D,P	Contelsat 1	US-Contel ASC	K2
103	K,E	GStar 1	US-GTE	K2
103	D,P	Spacenet IR	US-GTE	C/K2
105	K,E	GStar 2	US-GTE	K2
107.3	K,E	Anik C1	Canada-Telesat	K2
107.3	D,P	Anik E2	Canada-Telesat	K2,C
109	D,E	Anik B	Canada-Telesat	C,K2
110	K,P	USSB 1	US-USSB	K4
110	K,P	STC 1 & 2	US-Comsat	K4
110	K,E	Anik C2	Canada-Telesat	K2
110	K,P	CSC 1	US-CSC	K2
110.5	D,P	Anik E2	Canada-Telesat	K2,C
113.5	D,E	Morelos F1	Mexico	C,K2
114.9	K,E	Anik C3	Canada-Telesat	K2
116.5	D,E	Morelos F2	Mexico	C,K2
117.5	K,E	Anik C3	Canada-Telesat	K2
119	K,P	DVS-1/2	US-DVS/Scott	K2
119	K,P	CSC II	US-CSC	K2
120	D,E	Spacenet 1	US-GTE	C,K2
121	K,P	GStar 1R	US-GTE	K2
123	K,E	SBS V	US-SBS	K32
124	K,P	Fedex B	US-Federal Express	K2
125	K,E	GStar 4	US-GTE	K2
126	K,P	US-Martin Marietta	US-Martin Marietta	K2
128	D,E	ASC I	US-ASC	C,K2
129	D,P	Contelsat 1	US-Contel ASC	K2,C
130	K,P	Galaxy K2	US-Hughes	K2

BASICS

Location	Code	Satellite	Operator	Band(s)
131	K,P	Galaxy B	US-Hughes	K2
132	K,P	Westar B	US-W.Union	K2
134	K,P	Comstar K2	US-Comsat	K2
135	D,P	Spotnet 1	US-National Exchange	K2
136	D,P	Spacenet 4	US-GTE	
136	K,P	GStar A3	US-GTE	K2
147.5	D,P	Colsat 2	US-ColC	
148	K,P	USSB 2	US-STC	K2
148	K,P	DBS Spacecraft	US-various	K2
148	K,P	DBSC 1	US-DBSC	K2
148	K,P	Echostar II	US-Echostar	K2
157	K,P	SSS	US-Synd.	
160	K,P	ESDRN	USSR	
166	D,P	SCS 2	US-SCS	
170	D,P	Statsionar 10	USSR	C,K1

EASTERN HEMISPHERE

Location	Code	Satellite	Operator	Band(s)
180	D,E	Intelsat V F8	Intelsat	
177	D,E	Intelsat V-F3	Intelsat	
174	D,E	Intelsat VA-F10	Intelsat	
167	K,P	Pacstar	New Guinea Sat.	
164	K,E	Aussat A3	Australia	K5
160	K,E	Aussat A1	Australia	K5
160	K,P	Aussat B1	Australia	K4/K5
158	K,E	Superbird A	Japan	
156	K,P	Aussat B1	Australia/94	K4/K5
156	K,E	Aussat A2	Australia	K5
154	K,E	JCSat 1	Japan-JCSC	K4/K5
150	K,E	JCSat 2	Japan-JCSC	K4/K5
132	D,E	CS-3A	Japan	
110	K,E	Yuri BS-2B	Japan	
108	K,P	Yuri BS-3A	Japan	
90	D,E	Statsionar 6	USSR	C,K1
63	D,E	Intelsat V F11	Intelsat	L,S,C,K1
60	D,E	Intelsat VB F15	Intelsat	L,C,K1
57	D,E	Intelsat V F7	Intelsat	
47	K,P	Zohreh 3	Iran	
41	K,P	Paksat 2	Pakistan	
41	K,P	Zohreh 4	Iran	
40	D,E	Statsionar 12	USSR	C,K1
38	K,P	Paksat 1	Pakistan	
34	K,P	Zohreh 1	Iran	
33	K,P	Eutelsat II F2	ESA	K1,K4
32	K,P	Videosat 1	France	S,K5
26.5	K,P	DFS 2	W.Germany	S,K3,K5,E
26	K,E	Arabsat 1B		
23.5	K,P	DFS 1	W.Germany	S,K3,K5,E
22	K,P	Sicral	Italy	
19	K,P	GDL 6	Luxembourg	X,K5
19.2	K,P	Astra II	Luxembourg-SES	K1
19.2	K,E	Astra IA	Luxembourg-SES	K1
19	K,E	Arabsat 1A		
17	K,P	SABS	Saudi Arabia	
16	K,P	Sircal 1A	Italy	
16	K,E	Eutelsat 1 F1	ESA	K1
15	D,P	AMS 1 & 2	Israel	C1,K2,E
13	K,E	Eutelsat II F1	ESA	K1
10	K,E	Eutelsat 1 F3/F5	ESA	K1
7	K,E	Eutelsat 1 F4	ESA	K1
5	K,E	OTS-2	ESA	K1
4	K,E	Eutelsat I-F2	ESA	K1,K4
1	K,P	GDL 5	Luxembourg	U,X,K5

KEY TO TABLE

SYMBOLS:
- E - Existing
- P - Proposed
- D - Dual Band
- K - Ku-band

BAND DESIGNATIONS (GHz):

U	0.3 to 1.0	L	1.0 to 1.17
S	1.7 to 3.0	C	3.4 to 4.2
X	6.5 to 10.7	K	11.0 to 11.7
K2	11.7 to 12.2	K3	11.7 to 12.5
K4	12.2 to 12.7	K3	11.7 to 12.5
K5	12.5 to 12.75	K6	13.4 to 14.2
E	20 to 60		

II. COMPONENT THEORY AND OPERATION

A. INTRODUCTION

Home satellite systems designed to receive satellite broadcasts have evolved rapidly during the past twenty years. However, the phenomenal growth in the C-band home satellite market in North America in the last six years has been an especially tremendous impetus in creating an industry which can now produce and market increasingly sophisticated components at ever lower costs.

Ku-band equipment is very similar in theory, operation and installation techniques to the more familiar C-band systems which are presently widely employed in communication systems throughout the world, particularly in homes and cable TV headends in Canada, Mexico and the United States. Therefore, it is very useful to examine the theory and operation of Ku-band systems in the light of our broad experience with the more familiar, lower frequency equipment. Recall that this manual draws on and expands upon some materials from our earlier book, The Home Satellite TV Installation and Troubleshooting Manual. The underlying theory and operation of all the components required to assemble a viable Ku-band system are explored in this chapter. Selecting equipment and designing an efficient Ku-band TVRO are examined in Chapter III and IV.

B. ANTENNAS

The purpose of a microwave antenna is to collect and concentrate signals coming from any chosen satellite while ignoring extraneous signals and noise. These seemingly simple design objectives, which have been the subject of a unique scientific and engineering discipline for many years, must be mated with economic considerations. Therefore, a technically excellent antenna must also be moderately priced, aesthetically pleasing, durable, easy to install and virtually maintenance-free.

The majority of microwave antennas in use today for receiving satellite broadcasts have designs based upon combinations of circular or parabolic shapes. Microwave signals reflected from surfaces incorporating such designs are concentrated to a "focal point" or to a series of focal points. Recently, a new variety of antenna, the flat plate antenna has been introduced with some success.

At the microwave frequencies involved in satellite television, reflector antennas are the best suited to the application. Stacked arrays such as Yagis, more common in lower frequency applications, are not used because of excessive feeder losses. Although cables function well at 800 MHz they behave like wet strings at 11 GHz. Individual beam antennas such as helicals can be used for reception and transmission up to 4 GHz but at higher frequencies, losses are simply too great. Reflectors are used in Ku-band satellite broadcasting primarily because so much gain can be attained

COMPONENT OPERATION

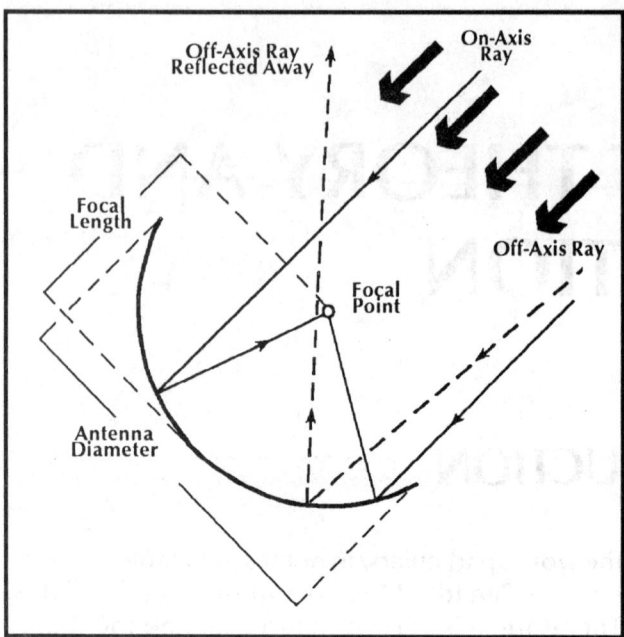

Figure 2-1. Parabolic Antenna Geometry. *In the prime-focus configuration all rays are directed via one reflection to a feedhorn at the focal point. Off-axis rays theoretically miss the focal point.*

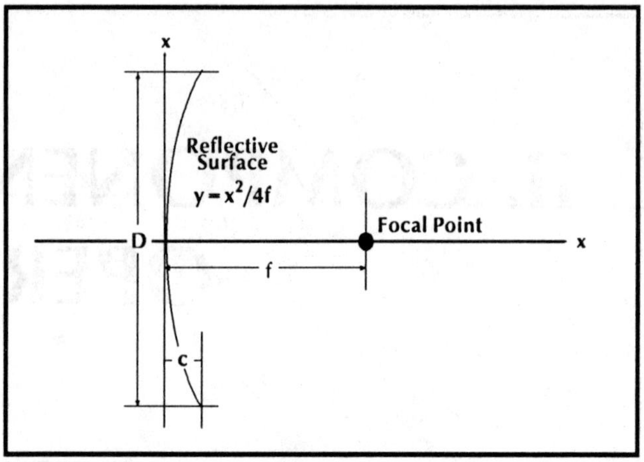

Figure 2-2. Parabolic Reflector Geometry. *The defining parameters of a parabolic reflector include its diameter (D), focal length (f) and depth of curve (c). The gain is calculated from the antenna diameter, reflector efficiency and radiation frequency (see Appendix B for more details).*

from a relatively small area. For example, a parabolic reflector of only 37 centimetres in diameter can achieve a gain of 30 dB at 11 GHz.

Single-Focus Antennas

Prime Focus Reflector

The most familiar antenna, the prime-focus parabola, theoretically focuses all signals arriving parallel to its axis to a single "prime focus." Any incoming microwave radiation from off-axis directions will be reflected away from this focus (see Figures 2-1 and 2-2).

In practice, a prime-focus antenna or "dish" deviates from theoretical performance for three reasons. First, the feedhorn or waveguide, which is mounted at the antenna focus and which is designed to collect and concentrate incoming signals, occupies a finite area in the vicinity of this point. As a result, it will also intercept some microwave radiation arriving from directions slightly off the main antenna axis. The is termed aperture blockage. Second, the reflective surface does not perform exactly as geometrical ray-tracing methods would dictate because the intercepted microwaves behave according to the principles of waves. As a result, a portion of the incoming signal spreads around the antenna and feedhorn edges. This effect is well known in optics as diffraction. Also, because satellite signals have a wave-like nature, the feedhorn cannot be localized at just a single point but must have finite dimensions. Third, irregularities and imperfections in the shape of the antenna surface will cause reflection errors so that some off-axis signals are detected while some targeted signals pass by unobserved.

Cassegrain Reflector

The Cassegrain or "back-fire" antenna similarly has a parabolic reflective surface but also redirects the incoming satellite signals via a second reflector, known as a hyperbolic subreflector, down a waveguide to a low noise block downconverter (LNB) mounted in a rear position (see Figures 2-3 and 2-4). This type of antenna is often more expensive than the ordinary prime-focus variety and is usually installed in commercial TVROs. The Cassegrain geometry can be very useful in hot climates because the LNB is mounted behind the antenna where it is protected from the direct rays of the sun and where it remains relatively cool and functions more efficiently.

COMPONENT OPERATION

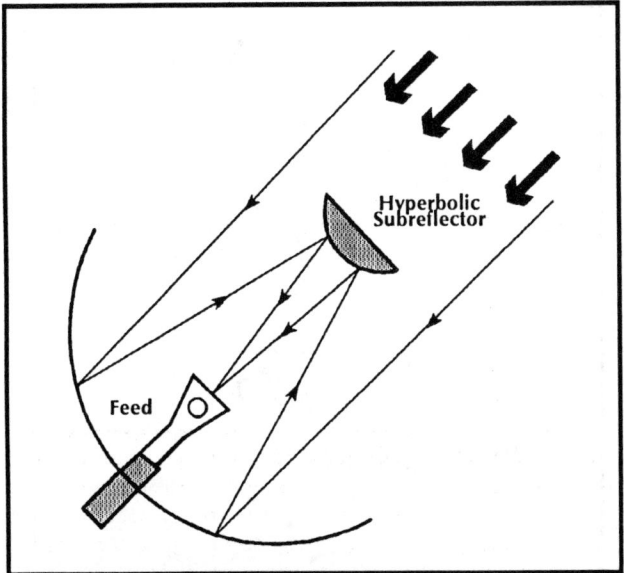

Figure 2-3. Cassegrain Antenna Geometry. *A Cassegrain parabolic antenna employs a second reflector, the hyperbolic subreflector, to direct microwaves to a feed located behind the reflective surface.*

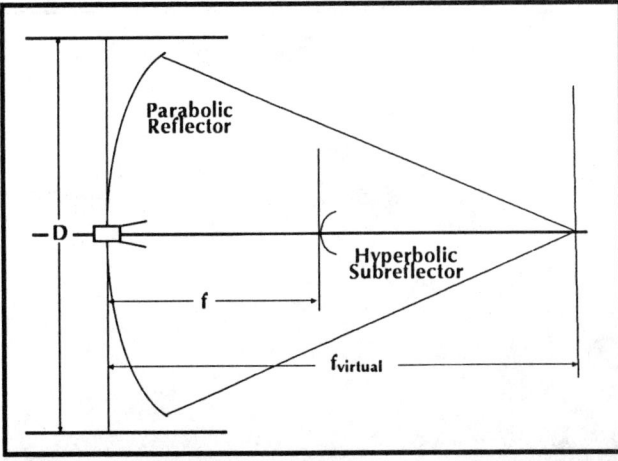

Figure 2-4. Cassegrain Feed Geometry. *The Cassegrain feed uses a hyperbolically shaped subreflector to allow system electronics to be sheltered behind the dish. The subreflector creates what is known as a virtual image so the feed performs as if it sees a longer focal length ($f_{virtual}$ in diagram) than is actually the case.*

Offset Fed Reflector

The surface of an "offset fed" antenna is essentially a section of a larger prime-focus parabola (see Figures 2-5, 2-6 and 2-7). The feed assembly which is still located at the focus of the larger "parent" antenna appears to be offset from that portion of the reflective surface in use. This design eliminates blockage of incoming signals by the feedhorn/LNB structure (see Figure 2-8). Offset fed antennas are more commonly encountered in Ku-band than in C-band TVROs because the shorter focal length and smaller dimensions of both the feedhorn and LNB permit the offset supporting arm to have more stability.

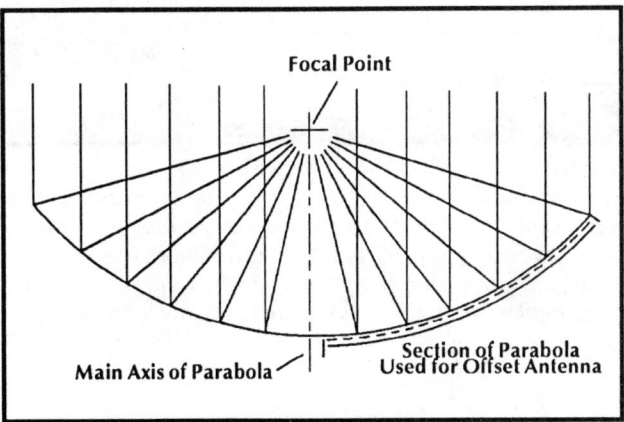

Figure 2-5. Offset Antenna Ray Tracing. *This illustrates how an offset antenna is cut from a section of a larger prime focus antenna. The offset feed is still located at the focus and all incoming microwaves are reflected to this point.*

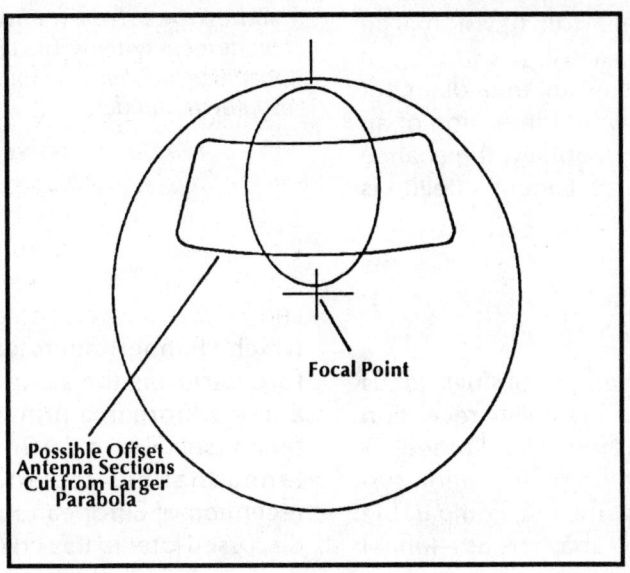

Figure 2-6. Offset Fed Antenna - Front View. *This diagram shows the various shapes that can be cut from the "parent" prime focus antenna. It is clear that an offset fed reflector can easily assume an elliptical or rectangular shape. The critical difference lies in the design of the feed system.*

ANTENNAS

COMPONENT OPERATION

Figure 2-7. The Pico Kid. *The 4 x 7 foot (1.2 by 2.1 metre) rectangular shape of the Pico Kid was designed as if it were cut from a section of a much larger 14 foot (4.3 metre) antenna having a 0.25 f/D ratio. The feedhorn was specially designed to properly detect signals from this smaller rectangular surface. This antenna is not commercially available.*

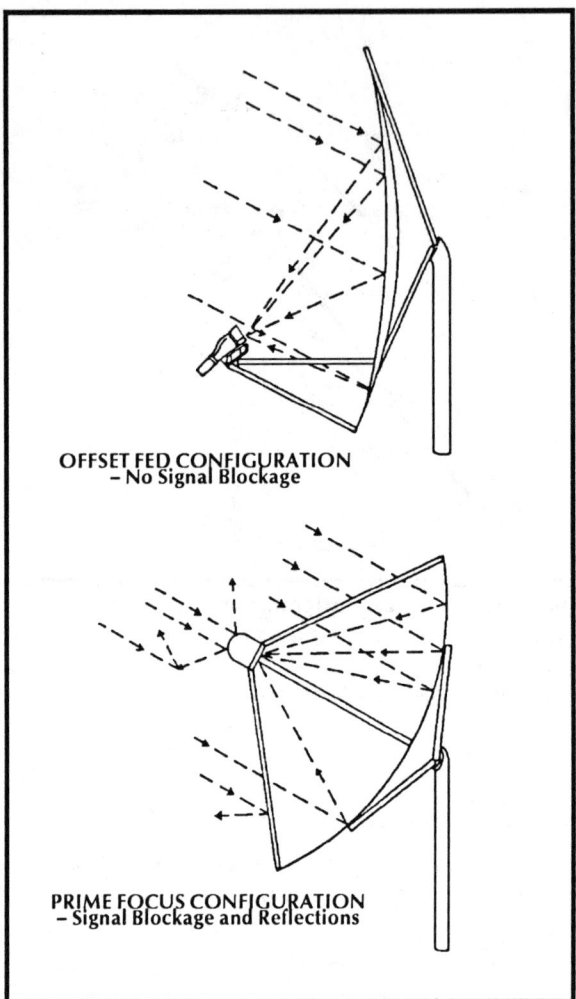

Figure 2-8. Prime Focus versus Offset Fed Antenna - Feed Blockage. *This reflector configuration can have improved efficiency because it avoids "shadows" which are cast by feed structures in prime focus parabolic designs. Antenna noise temperatures are usually lower than those for parabolic reflectors of the same size because the feed is aimed more towards outer space and less towards the warm earth as compared to typical prime focus systems. This tilted design also avoids snow loading problems common in far northern and southern latitudes.*

When correctly oriented, an offset fed reflector offers an advantage in addition to the elimination of aperture blockage. The width of vision of an antenna known as its beamwidth (examined in more detail below) decrease as antenna diameter increases. Therefore, if the wider dimension of an offset antenna is oriented horizontally, the chance for detection signals from an adjacent satellite is minimized (see Figure 2-9).

Horn Reflector

The horn reflector, another type of single focus antenna, has rarely been used in satellite reception systems but is often encountered in land-based, C-band, telephone microwave communication systems. The whole body of this antenna, comparable in shape to the front end of a tuba, acts as a funnel which channels microwaves to its focus. It therefore performs the same function as the reflector and feedhorn in a prime-focus or Cassegrain antenna (see Figure 2-10). One innovative horn antenna that has recently been introduced for reception of European Ku-band DBS broadcasts is discussed later in this chapter (see Figure 2-51).

COMPONENT OPERATION

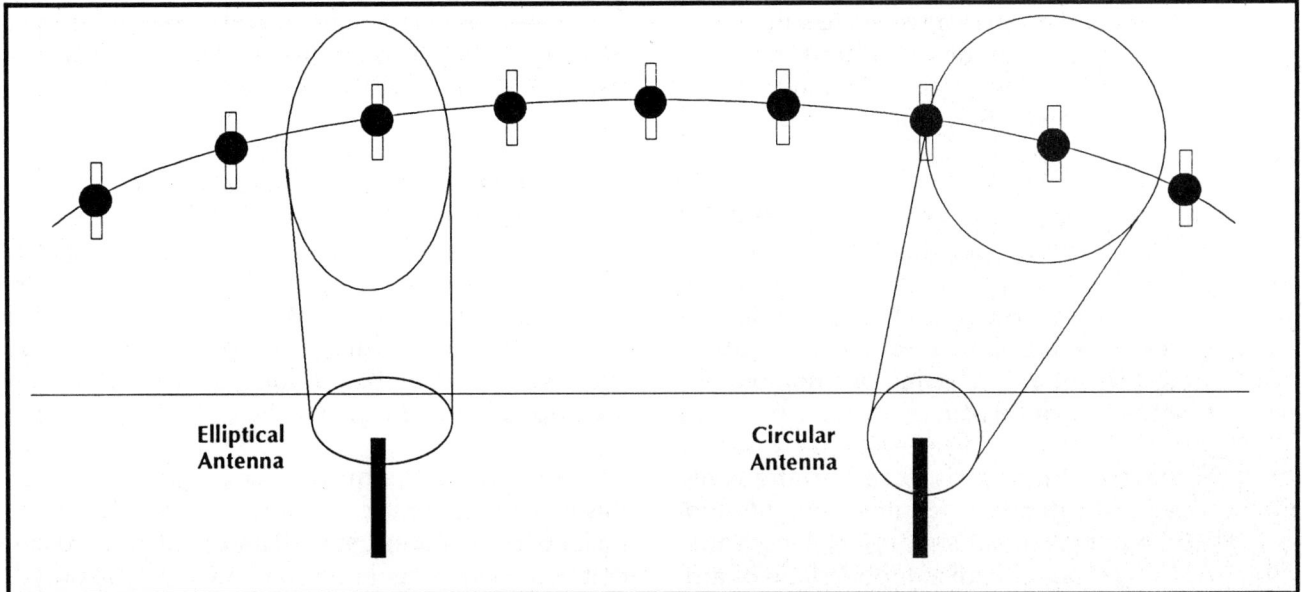

Figure 2-9. Compressed Beamwidth of an Elliptical versus a Circular Profile Antenna. *The beamwidth pattern of an antenna with an elliptical profile is determined in part by its shape. If the dimension parallel to the arc of satellites is longer than that at right angles, the field of view of the dish would be compressed into a vertical elongated ellipse. This occurs because beamwidth decreases as antenna diameter increases. In contrast, a dish having a circular profile would have a symmetrical beam pattern. The chance that interfering signals would be detected from adjacent satellites decreases as the dimension oriented along the arc of satellites increases, as in the elliptical antenna above.*

The Flat Plate Antenna

Flat plate antennas have not yet been widely used in home satellite systems. The operation of passive flat plate antennas is based upon Fresnel reflection while that of active designs upon solid state circuit elements.

Active Flat Plate Antenna

The active flat plate, rarely seen in home satellite systems, is more common in military and civilian radar systems were system cost is not a serious constraint. This is a familiar situation in other technical areas of communications because numerous microwave components have been products of military research.

The military objective in producing a flat plate antenna was originally to create a device suitable for use on high speed aircraft. Clearly, it would have been impossible to install a parabolic dish and its gearing mechanism for use on a jet traveling at speeds in excess of Mach 2, twice the speed of sound. Although radar is often detected via onboard radomes, aircraft can carry only a limited number of such devices.

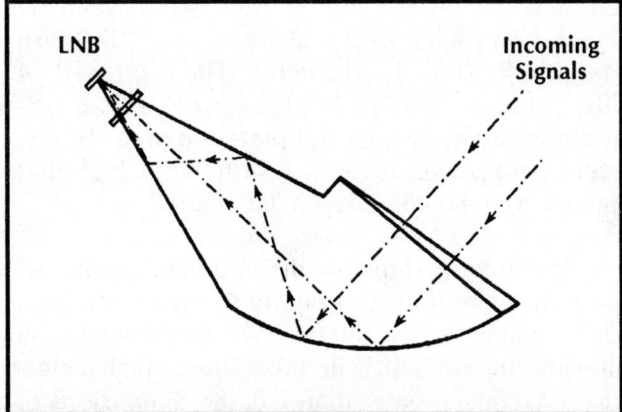

Figure 2-10. Horn Reflector Geometry. *This type of antenna geometry is very effective in capturing microwave signals. Although it is relatively more expensive than prime focus antennas because large amounts of materials must be used, it performs effectively in those situations where high levels of terrestrial interference are encountered.*

ANTENNAS

COMPONENT OPERATION

An advanced flat plate antenna has the obvious advantage that it can be contoured to match the surface on which it is installed. In addition, flat plates can be light-weight because they are constructed primarily of plastic.

Military organizations hold many of the patents on flat plate antenna technology. These usually describe the flat plate antenna as a layer of dielectric material that separates two thin metal layers. In essence, a flat plate antenna is a double-sided printed circuit board having one non-etched side that acts as a ground plane.

Flat plate technology is based on the well-known theory of antennas, radiating elements and tuned lines. A patch conductor, of given length and width, etched on one side of a double sided board can be forced to resonate at a particular frequency. However, early researchers discovered that although a patch could be resonated, its Q factor would be too high to allow it to transfer or radiate the energy into free space. (The higher the Q factor, the more the resonance is concentrated on one frequency). The inverse applied to reception. The signal could not efficiently be transferred from free space to the patch.

In order for a patch to work properly it had to have two characteristics, (1) a low loaded Q factor to enable it to radiate efficiently and (2) provision of a good impedance match to the input/output transmission line. It also had to have a reasonable bandwidth and acceptable gain, both affected by these two parameters.

An American researcher, Lincoln Charlot, investigated the possibility of obtaining efficient radiation from a patch and was subsequently granted U.S. Patent No. 3803623 for his work. He discovered that a direct match to the feeding transmission line could be obtained by connecting it to a specified point on the edge of the patch. An impedance match in the range 20 to 151 ohms could be obtained on an elliptical patch with an eccentricity of 0.65. From his patch design, he measured a gain of 6 dB and a loaded Q of 12. This meant that the patch was 95 percent efficient. A gain of 11 dB, lower than the sum of combined gains, was obtained from a 2 x 2 array of patches. The main reason for the loss in actual versus theoretical gain was the combined loss of the patches and the interconnecting transmission lines. This problem with phasing an array of patches is that the losses increase with each patch added so that beyond a certain number of patches, the increase in gain is more than counteracted by losses.

Arrays are inherently directional in their ability to detect signals. It was found that by shorting various sections of the patches to ground, the beam heading could be changed. Most military flat plate antennas are phased arrays that scan an arc by shorting various sections of their component patches to ground using PIN diodes under microprocessor control. PIN diodes are simply electrically operated two-way switches.

At present, the most discouraging aspect of this type of antenna is the substrate cost. Those familiar with the fibreglass G10 and SRBP printed circuit board laminates know that the G10 type board is usable up to about 4 GHz and the SRBP to 25 MHz. With microstrip technology, the impedance of the track is directly affected by the effective dielectric constant and the dielectric thickness of the board. This means that for any usable results, the board has to have a low dielectric thickness as well as a low dielectric constant. Plastics and teflon are ideal for this application.

The principal disadvantage of flat plates, the difficulty of realising high gains because of dielectric and conductor losses, has effectively limited such antennas to applications where the incoming power level is relatively high. In some uses, such as in radar systems where power levels are substantially higher than in satellite television broadcasting, the use of flat plates is quite feasible. Nevertheless, flat plate antennas can now be used in receiving signals from high power satellites such as the British Marcopolo (BSB 1) spacecraft. The proposed 64 dBw European DBS power level is more than adequate for an advanced flat plate antenna. Today, even Astra power levels are sufficiently high that signals can be received with flat plates.

Production of microstrip antennas is more demanding than that of ordinary printed circuits. A slight variation of conductor width could drastically alter line impedance. With substrates of high dielectric constants, greater than 7.0, the dimensions of the elements are reduced. This would subsequently make a production defect or variation more serious. Plastics with lower dielectric constants are favoured for microstrip antenna substrates. This factor makes the elements larger and also somewhat reduces the loss factor. These losses limit the

ANTENNAS

number of elements that can be used and hence the overall gain of the antenna.

While flat plate antennas are yet in their developmental stages, innovations are certain to come. For example, it should be theoretically possible to integrate a downconverter into a flat plate antenna. This may occur in future models. For those readers who wish to delve further into this subject, the following books are recommended:

The Foundations For Microstrip Circuit Design - T. C Edwards (John Wiley & Sons Ltd.)

Microstrip Antenna Theory & Design - J. R James, P. S Hall and C. Wood Peter (Peregrinus Ltd.)

Microstrip Antennas - Bahl & Bhartia (Artech House)

The Squarial

The Squarial antenna, developed by Fortel Technology Ltd. of the U.K. is an interesting example in the evolution of active flat plates (see Figure 2-11). It is not really an active antenna but a lies at the transition between active and passive designs. A plastic model of this product was unveiled by BSB (British Satellite Broadcasting) at a press conference in late 1988. While the subsequent financial difficulties experienced by BSB somewhat tarnished the reputation of the Squarial, it still holds promise as an innovative product.

The Squarial concept was originally based on a design of a military low profile radar antenna. While the cost of this particular military antenna was subject to strict production criteria and controls, it was not the best suited for mass production. The individual cost of the average military antenna exceeds the retail price of a typical Ku-band TVRO system.

The Squarial consists of three main sections; the top feedhorn structure, the waveguide assembly and the low noise converter. The "front-end" of the dish is a matrix of pyramidal feedhorns which channel the signal to the individual antenna elements (see Figure 2-12). These elements are carefully laid out so that they can be connected by equal transmission lengths to the amplifier, the LNC. Thus when signals from each element reach the LNC they combine in phase and voltages sum. The waveguide is composed of metalised, injection-molded plastic, a factor underlying the potentially competitive cost of the Squarial.

Figure 2-11. The BSB Squariel Flat Plate Antenna. *(Courtesy of WV Publications Ltd., London)*

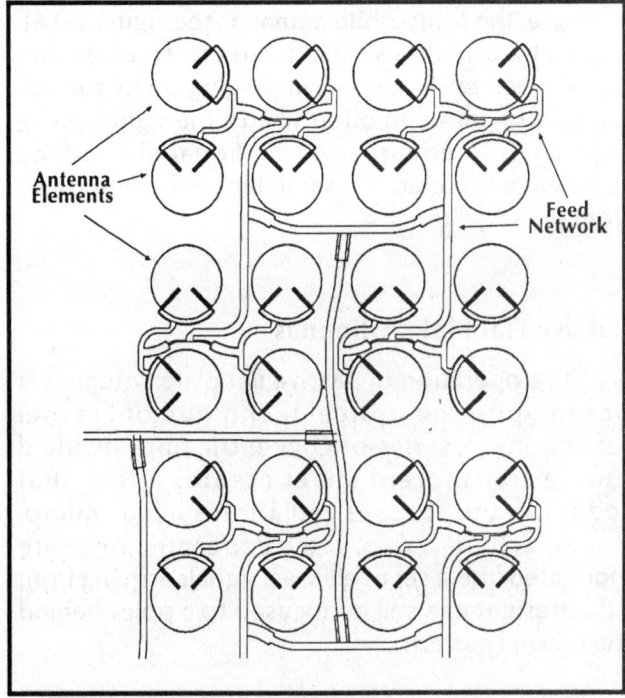

Figure 2-12. Squarial Antenna Elements. *The antenna elements of the Squarial antenna, the round disks, are connected to the LNC by equal lengths of conductor. This path structure is crucial because signals from each small antenna must reach the LNC having equal phases.*

COMPONENT OPERATION

The elements are spaced closely enough together, 0.9 times the wavelength of the incoming signal equal to 22.5 mm for reception of European Ku-band broadcasts, so that there are no "dead spaces" on the antenna surface. This serves to maximise antenna gain. This antenna cannot achieve more than about a 42 dB gain because as its size increase the losses incurred in feeding signals from each element to the LNC also increase (see Figure 2-13). Beyond a dimension of approximately 200 cm, the transmission losses add up faster than the increases in gain.

When the version of the Squarial designed for BSB was for reception of only right hand circularly polarised signals, this antenna is capable of detecting all senses of signal polarity, both circular and vertical.

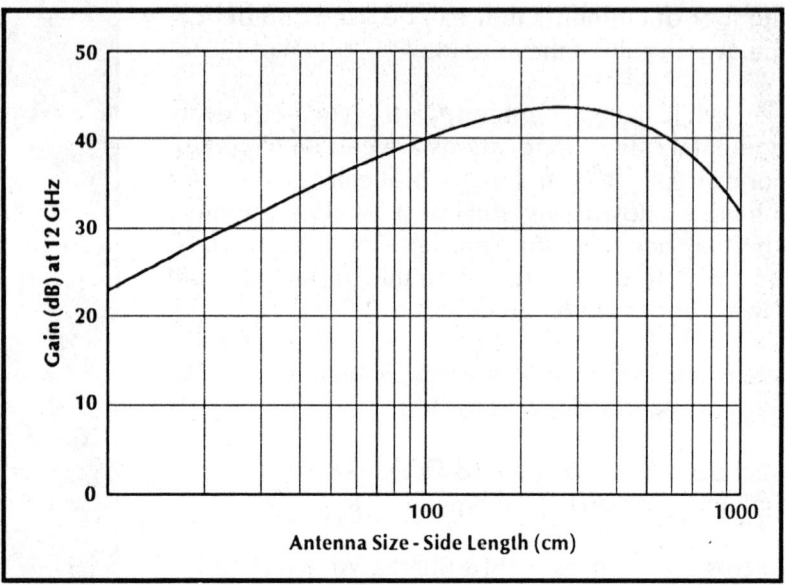

Figure 2-13. Squarial Gain versus Dimensions. *As the size of the Squarial antenna increases so does the gain. However, the losses incurred when the signal travels from each element to the LNC also increase. Beyond a square of about 200 cm in dimension, gain actually decreases as antenna size increases.*

This innovative antenna derives its low cost from a design which avoids using microstrip technology, characteristic of active flat plate antennas. The Squarial is also not a tracking flat plate as, for example, the Matsushita antenna (see Figure 2-14). It must be oriented so that it directly faces the targeted satellite. All signals impinging on its surface must be in phase. In other words, the signal reaching two different areas of the Squarial's surface must have traveled the same length in their paths from the satellite.

Passive Flat Plate Antennas

The operation of passive flat plate antennas is based upon the optical technique of Fresnel reflections. A series of concentric rings overlaid onto a transparent sheet creates a lens that redirects and focuses radiation such as microwaves, light or X-rays. If the concentric rings are elongated into a set of ellipses, signals arriving from off-boresight can still be focused to a point behind the Fresnel pattern.

If a negative pattern is used whereby concentric rings are made transparent and the space between the rings is opaque, then a reflective sheet laid behind this assembly will reflect satellite signals

Figure 2-14. The Matsushita Flat Plate Antenna. *The flat plate antenna, a recent introduction, is lightweight, does not require a feedhorn/LNB and its support structure, and can be easily packaged and installed. While this brand is non-steerable, flat antennas which electronically target a satellite will probably be marketed within ten years. The two square antennas in this figure each measure 34.5 and 72 cm and weight approximately 2.3 and 9 kg. (Courtesy of Matsushita Electric Works, Ltd.)*

to a feedhorn in front of the antenna. Both these configurations are outlined in Figures 2-15 and 2-16.

Both designs have some advantages. The same elliptical pattern can be used within a radius of approximately 300 miles. Slightly different elliptical patterns can be created for regions further removed. Given the appropriate pattern, both the transmission and reflection version can be installed flush to the mounting surface.

COMPONENT OPERATION

Multi-focus Antennas

Multi-focus antennas, developed originally for use with C-band systems, allow more than one satellite to be simultaneously detected. This is done by having a fixed antenna focus incoming signals to a series of feedhorns (see Figure 2-17). In contrast, single-focus reflectors must be repositioned by actuators in order to receive signals from just one satellite at a time.

Multi-focus designs incorporate variations of spherical and parabolic surfaces. For example, the C-band Simulsat antenna is cut from a rectangular section of a sphere and has feedhorns mounted in a box oriented along the focal line. The 5 meter model can simultaneously detect up to 35 satellites within a 70° arc and deliver a 44 dB gain. The Torus antenna, also designed to receive C-band broadcasts, is a dual curvature reflector having a circular cross section along the arc of satellites and a parabolic contour at right angles. This design also causes incoming signals to be reflected to a focal line.

Figure 2-15. Mawzones Flat Plate Antenna. *(Courtesy of WV Publications, Ltd.)*

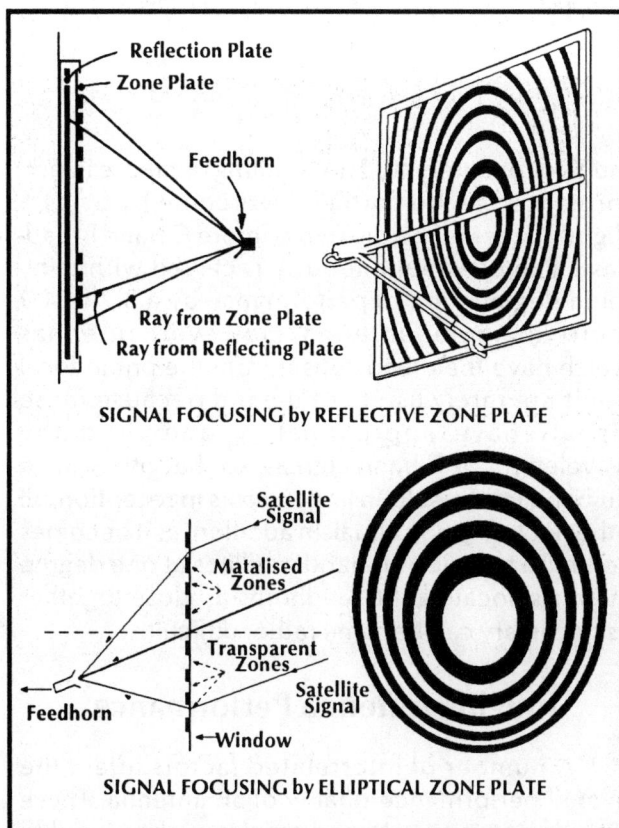

Figure 2-16. Mawzones Flat Plate Antenna Operation. *The Mawzones antenna is design so that impinging signals can either be reflected to or pass through to the feed. (Courtesy of WV Publications, Ltd.)*

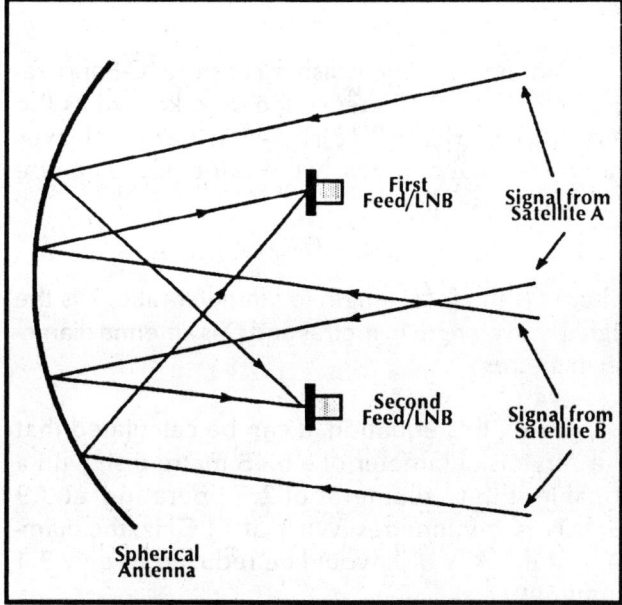

Figure 2-17. Spherical Multi-Focus Reflector Geometry. *A circular multi-focus reflector can be used to simultaneously detect two or more satellites without the need for tracking. This antenna should be at least 4 m or more in diameter to compensate for the lower gain resulting from rays being reflected to an off-axis focal point.*

ANTENNAS

COMPONENT OPERATION

In the past, spherical antennas have been widely used in Europe at C-band frequencies. The noise temperature of a 3 metre spherical antenna is 75°K at C-band frequencies while that of a parabolic dish of the same diameter is just 25°K. (Antenna noise temperature is discussed in the following section). At Ku-band, the disparity is even greater.

In the past American satellites have been better suited for detection with a spherical antenna than European spacecraft because they have been positioned relatively more closely together on the geostationary arc. When a North American installer powers up an LNB and moves it in the vicinity of the focal point, a second or third signal can usually be detected. In Europe, until recently, relatively few satellites had been orbited. They were clustered in groups rather than evenly spaced. However, today more satellites broadcast video entertainment. This includes a cluster of Astra 1A, Arabsat 1A, Eutelsat 1F1, Eutelsat 2F1, Eutelsat 1F5 and Eutelsat 1F4 spanning the arc from 19.2°E to 14.7°E. The usable azimuth range of the average spherical antenna is about 35°. This type of dish must be aimed at a point near the centre of the group of satellites to be received.

Aiming a spherical dish is a easier at C-Band frequencies. The diameter of the disk, known as the Airy disk (see Figure 2-18), in which the microwaves can be received in nearly the same phase can be calculated as follows:

$$d = f\lambda/D$$

where f is the focal length to diameter ratio, λ is the signal wavelength in metres and D is antenna diameter in metres.

Using this equation, it can be calculated that the Airy disk diameter of a 6.48 metre dish with a focal length to diameter of 2.5 operating at 3.9 GHz is 10 centimetres. While at 11 GHz, the diameter of the Airy disk would be reduced to only 3.4 centimetres.

A prime-focus parabolic antenna can also be used in a multi-focus configuration where one or two extra feedhorns and LNBs are offset on either side of the central feed. In this case, the reception of signals is not as "clear" because reflections from different parts of the antenna travel slightly different distances to reach these additional feeds. Multipath distortion also occurs in the operation of other multi-focus antennas. The resulting decrease in performance can be partially overcome by using a slightly larger reflector. Two or more C-band broadcasts can be simultaneously received with minimum degradation of performance by a 3.7 to 4.9 metre diameter antenna. However, antennas which have their feeds offset from the prime focal point are rarely used at Ku-band because these signals have approximately one third the wavelength of C-band signals, so that offsetting a Ku-band feed results in larger errors in reception, all other factors being equal. In addition, as it becomes common to space Ku-band satellites at one degree intervals, locating the feedhorns as close together as necessary can become rather difficult.

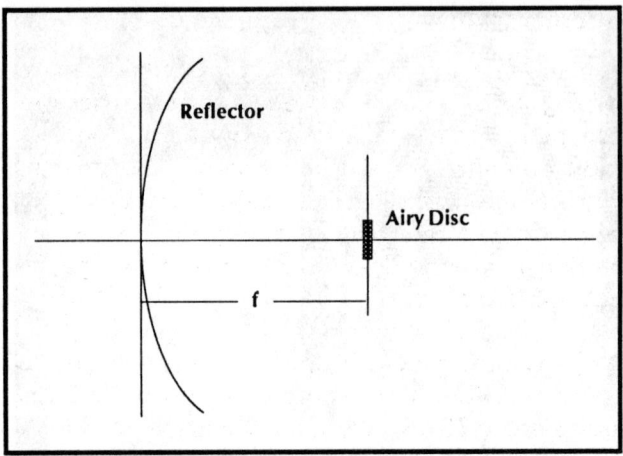

Figure 2-18. Spherical Antenna and Airy Disc. *A spherical reflector has been used in multi-satellite C-band reception in commercial applications such as cable TV headends. This type of reflector is generally too noisy for Ku-band reception. The feedhorn is usually located at a distance from the reflector of about 48% of the sphere radius. The Airy disk diameter defines the region in which the feed can be placed without losses causing excessive reception degradation.*

Judging Antenna Performance

A number of interrelated factors affect the overall performance quality of an antenna. These include antenna gain and efficiency, beamwidth, side lobe patterns and noise temperature. These performance criteria are in turn determined by the overall design of an antenna/feedhorn system. For example, the "depth" of an antenna, expressed by the distance between the feedhorn and the reflective surface, and the type and extent of reflector surface irregularities all affect these parameters.

Antenna Gain

Antenna gain, G, is the factor by which an incoming signal is concentrated when it reaches the focal point. Gain is measured compared to an "isotropic antenna". An isotropic antenna radiates energy is all directions just like a light bulb hanging by a wire does. A microwave reflector concentrates the signal into one direction like a car headlight directs a beam of light.

Gain is measured in decibels relative to an isotropic antenna. For example, a gain of 1,000 is equivalent to 30 dBi (see Table 1-8 and Appendix A for more details on decibels).

Antenna gain depends upon three parameters. First, as antenna surface area increases, more microwave radiation can be intercepted and therefore gain increases. If the reflective surface area is doubled, so is its gain. For example, a 2 metre antenna has 4 times the area and thus 4 times the gain of a 1 metre dish. The surface area used to calculate gain is the total effective area aimed directly towards any particular satellite. Therefore, an antenna with an offset feed which is angled perhaps as much as 30° off the satellite line of sight has less effective "intercept" area.

Second, gain increases with the square of the microwave frequency. This can be conceptually explained because higher frequency signals do not spread out like waves in water but are more easily focused into straight lines like beams of light. To illustrate, Ku-band antennas receiving broadcasts with triple the frequency of C-band signals have gains 9 times higher.

Third, gain is determined by how accurately the surface of an antenna is manufactured to match its designed shape. Surface errors can be either random or systemic imperfections. Random errors, small distortions in the reflective surface, can cause substantial decreases in gain (see Table 2-1). Therefore, an antenna that has a large number of deep welts or ripples in its surface will behave more poorly than one that is smooth and more closely approximates its designed shape. Signals striking a dent will be reflected away from the focal point. This loss of gain with surface imperfections becomes more important at higher frequencies. Therefore, an antenna which may be perfectly adequate at C-band may not have a tight enough surface tolerance for quality performance at Ku-band.

Gain reductions caused by random surface errors are measured by a quantity known as the root mean square (RMS) deviation which characterizes average surface irregularities. The methods used to calculate this and other parameters are presented in Appendix B.

TABLE 2-1. LOSS OF GAIN WITH SURFACE IRREGULARITIES

Average Surface Inaccuracy RMS deviation		Percentage Loss of Gain	
(inches)	(cm)	C-band	K-band
0.03	0.01	1	3
0.13	0.05	6	16
0.25	0.10	11	30
0.38	0.15	16	41
0.51	0.20	21	51
0.64	0.25	25	59
0.76	0.30	30	65
0.89	0.35	34	71
1.02	0.40	37	76

Systemic surface errors are caused by design or assembly problems other than random irregularities. If reflector curvature is smooth but inaccurate or if an antenna is incorrectly designed or assembled so that panels do not fit together properly, systemic errors will occur. Systemic gain reductions can also be caused by "springback" or "metal memory" in metal antennas when ribs or panels "remember" their original flat shape and straighten out a little. In fiberglass antennas springback, as well as shrinkage, can cause displacement of the internal reflective surface. Mesh antennas can have gain reductions since there are often flat areas of reflector between supporting ribs. Systemic errors affect large portions of the reflective surface and can cause a double-focus to occur. Like random errors, the negative effect of such design flaws is much more pronounced at higher frequencies where reflections are more "specular," namely where reflections adhere more closely to straight line paths.

Antenna Efficiency

Antenna efficiency is a measure of the percentage of intercepted signal actually captured by the reflector/feedhorn system. An ideal antenna would have a 100% efficiency. But typical efficiencies

COMPONENT OPERATION

TABLE 2-2. ANTENNA GAIN IN dBi at 12 GHz

Antenna Diameter				Antenna Efficiency			
Metres	Feet	100%	80%	70%	60%	55%	50%
0.3	0.98	31.49	30.52	29.94	29.97	29.59	29.18
0.45	1.48	35.07	34.10	33.52	32.85	32.47	32.06
0.5	1.64	35.96	34.99	34.41	33.75	33.37	32.95
0.6	1.97	37.55	36.58	36.00	35.33	34.95	34.54
0.9	2.95	41.08	40.11	39.53	38.86	38.48	38.07
1.2	3.94	43.58	42.61	42.03	41.36	40.98	40.57
1.6	5.25	46.07	45.10	44.52	43.85	43.47	43.06
2.0	6.56	48.00	47.04	46.46	45.79	45.41	44.99
2.5	8.20	49.94	48.97	48.39	47.72	47.35	46.93
3.0	9.84	51.53	50.56	49.98	49.31	48.93	48.52
3.5	11.48	52.87	51.90	51.32	50.65	50.27	49.86
4.0	13.12	54.03	53.06	52.48	51.81	51.43	51.02

range from lows of 40% for inferior products to 70% or even higher for excellent quality antennas. Offset fed parabolic antennas can have efficiencies in excess of 80% because there are no structures between the target and the reflector surface to impede the capture of incoming radiation. The current efficiency record of 84% is held by a 20 GHz military antenna.

Efficiency is determined by a number of factors including (1) the surface accuracy of an antenna, (2) losses that occur when microwaves are not perfectly reflected but are absorbed by its surface, (3) aperture blockage caused by absorption and reflective losses from the feedhorn, LNB and its supports located in the path of the incoming signal and, (4) "spillover" which is examined below in the section on feedhorns. An offset fed antenna does not have efficiency losses due to aperture blockage.

Antenna gains ranging from the theoretical maximum of 100% to as low as 50% efficiency are listed in Table 2-2 for Ku-band signals. When reading these gain figures remember that a 3 dB difference means a doubling of gain; 1 dB means a 26% difference; and 0.1 means a 2.3% difference. Small decibel changes translate into large variations in performance.

The calculated gains for 50% efficient reflectors are not half of those having 100% efficiencies because the numbers in Table 2-2 are expressed in decibels, not the raw signal concentration factors. For example, a perfectly efficient 1 metre antenna concentrates signals by a factor of 12,566; a 50% efficient dish would have a half this gain, equal to 6,283. However, the higher gain translates into 41.98 dB and the lower into 38.97 dB, a difference of roughly 3 dB as would be expected.

Beamwidth and Side Lobe Patterns

The field of vision of an antenna is characterized by its beamwidth and side lobe pattern. This pattern can be measured by rotating a microwave source across the field of view of a test antenna. The power of detected microwaves plotted against its angle towards the source will then define its beamwidth and side lobes (see Figure 2-19). This pattern is, in essence, a "fingerprint" of the quality and performance of the antenna/feed system.

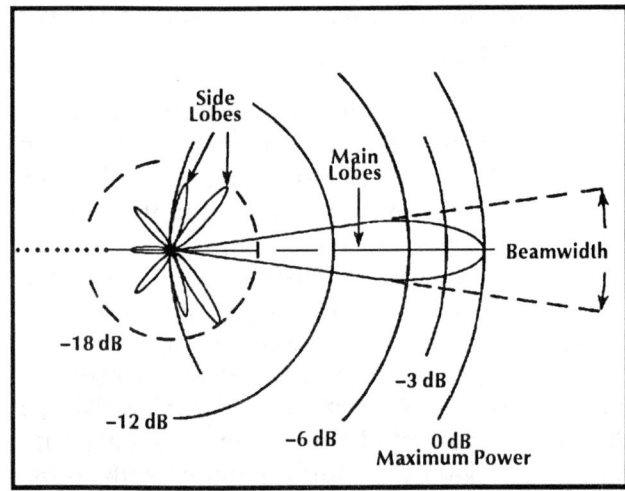

Figure 2-19. Calculated Antenna Lobe Pattern. *Although most of the radiation detected for a parabolic antenna is concentrated in the main lobe, this calculated pattern shows that some noise can be detected via side and back lobes in the full circle around an antenna.*

Antenna beamwidth is a measure of how narrow a region of space can be targeted. This is a very important factor considering that satellites separated by 2° or less and situated at a distance in excess of 35,803 kilometres (22,247 miles) appear to be very closely spaced together. A plot of the "vision" of an antenna shows that most of the received power is contained in the main lobe. The

TABLE 2-3. THEORETICAL 3 dB BEAMWIDTHS DEGREES (at 12 GHz)

Antenna Diameter (metres)	Beamwidth (degrees)
0.5	3.50
1.0	1.75
1.5	1.17
2.0	0.88
2.5	0.70
3.0	0.58
3.5	0.50
4.0	0.44

beamwidth is defined as the width of this main lobe between "half power" points where detected signal power has dropped by 50% or 3 dB. Beamwidth can be calculated from antenna diameter and signal wavelength (see Appendix B and Table 2-3).

The beamwidth determines how well an antenna detects off-axis radiation. For example, if the beamwidth is 1.2° then at 0.6° at either side of the main antenna axis signals will be captured at half the power of those on target. In this case, a second equally powerful microwave source located at 0.6° off-axis would interfere with clear reception of the primary signal.

The exact location and relative magnitude of the side lobes is an important determinant of antenna performance. Side lobe patterns indicate how well an antenna will detect radiation arriving from off-axis directions. In order for the dish to ignore off-axis signals and noise, the side lobes must be substantially reduced relative to the main lobe (see Figure 2-20). For example, if a second Ku-band satellite happened to be 2° removed from the target along the geosynchronous arc and if the first side lobe were located at 2° and reduced in power by only 5 dB, the interference between signals

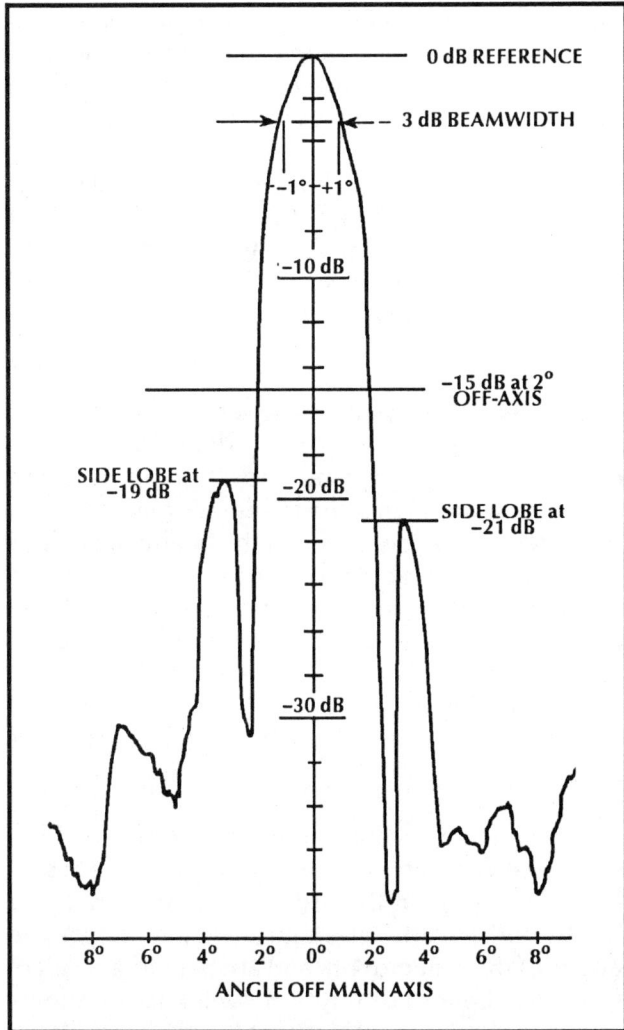

Figure 2-20. Antenna Lobe Patterns. *This graph shows the relative power levels, expressed in relative terms, a 90 cm antenna detects from its boresight or main axis to its periphery. The taller and more narrow the central portion is relative to the side lobes, the better an antenna will target and pinpoint a satellite in the sky. In this case, the 3 dB or half power beamwidth is 2.2° and the side lobes located at 4.5° on either side of the main lobe are reduced in power by 19 and 21 dB. Signals arriving at 2° off axis are reduced to approximately 12 dB relative to the main beam.*

simultaneously detected from both satellites would be very annoying. However, if this side lobe were positioned at 3°, the second satellite would probably have no influence on the primary broadcast.

Thus the difference between an excellent and a poor quality antenna can easily be seen in their lobe patterns. A wide beamwidth or a low ratio of

COMPONENT OPERATION

the main lobe to the first side lobe would characterise a poor quality dish. Such an antenna would be more susceptible to detecting off-axis noise and other interfering signals. Even if two antennas have identical gains their performance might be quite different. For example, if a square and a round prime-focus antennas were cut from identical larger surfaces so they had the same gains, the magnitude and placement of the side lobes would indicate why their performances might be markedly different.

Antenna lobe patterns depend upon a number of factors including antenna diameter, the frequency of the impinging radiation and surface tolerances. The higher the frequency and the larger the antenna, the more narrow the beamwidth. Thus a 2 metre reflector will have one half the beamwidth of a similar 1 metre reflector. Ku-band antennas have one third the beamwidth of C-band systems because the microwave frequencies in use are 3 times higher. Surface imperfection generally tend to widen the main lobe and increase side lobes.

Side lobes are an inherent property of the wave-like character of microwaves and the resulting "spreading" in the vision of an antenna. This "spillover" results when microwaves strike the edges of the reflector rim and are bent as a result of interference and diffraction, well-known effects from the discipline of optics. Microwave spillover actually allows antennas to have the rather astonishing ability to detect radiation coming from directly behind them!

Every antenna has a lobe pattern (see Figures 2-19 and 2-21). The location and magnitude of side lobes as well as the size of the side lobes relative to the main lobe indicate how well power from the surrounding full circle will be detected. These theoretical patterns are altered by real world considerations such as surface tolerances and placement of the feedhorn/LNB assembly and its support. Such considerations underlie the fact that offset fed designs, in addition to having higher efficiencies, may also have "cleaner" side lobe patterns.

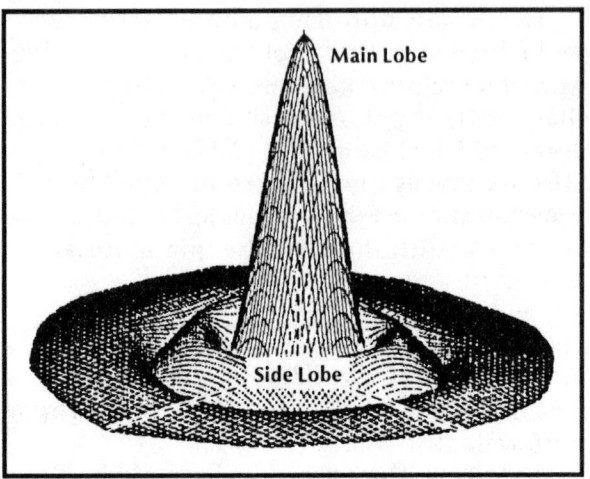

Figure 2-21. 3-D Representation of an Antenna Radiation Pattern. *This 3-D computerised design shows how the main lobe appears through the centre of the donut-like side lobes.*

Antenna Noise

Antenna noise temperature is a measure of how much noise the reflector detects from the surrounding environment. A portion of this noise enters via the antenna side lobes; a smaller amount, predominantly that from outer space, enters via the main lobe. Noise also comes from both man-made sources such as fluorescent lights which emit microwave radiation as well as from natural sources such as the surrounding terrain. In general, a satellite reception system detects mainly the natural, higher frequency sources of noise.

Since the warm ground emits radiation, noise temperature increases as an antenna is pointed at progressively lower elevation angles in regions further from the equator. Larger antennas detect less noise because they have smaller side lobes. Also, offset fed designs which can be inclined steeply towards the horizon benefit both by not collecting excessive snow loads and by having their feeds angled farther away from the "noisy" ground (see Figure 2-22).

Values for antenna noise versus elevation angle and antenna diameter are listed in Table 2-4. The values for a 1.0 metre offset fed and a 1.8 metre prime focus antenna are from Microwave Associates (M/A COM). The others are provided by Andrews Antenna Company. The noise temperatures as well as gains and half power beamwidths have all been experimentally measured on standard models.

COMPONENT OPERATION

This table confirms what would be expected. Antenna noise temperature decreases with reflector diameter because the beamwidth decreases and side lobes are closer to the antenna boresight. It also decreases with elevation angle as the reflector is pointed further away from the "hot" ground. Note that the noise temperature as measured for the 1.0 metre M/A COM offset antenna is lower than that for the 1.8 metre reflector. The feed on an offset antenna is more highly angled towards the sky than one for a prime focus and sees less of the ground. The average noise seen at 90° elevation is only 6°K while that seen at 0° elevation is over 150°K.

TABLE 2-4. KU-BAND ANTENNA NOISE TEMPERATURE
(The 1.0 metre is an offset fed antenna)

Diameter (metres)	Gain (dB)	Beamwidth (degrees)	Antenna Elevation Angle			
			10	15	20	30
1.0	40.0			46.4	34.1	29.8
1.8	44.5			51.7	40.1	36.1
2.4	48.1	0.72		35	31	28
3.0	49.2	0.56		32	28	25
3.7	51.4	0.45		30	26	23
4.5	53.0	0.36		29	25	22

f/D - Antenna Depth

The depth of a prime-focus antenna, quantified by its focal length to diameter ratio, f/D, has a direct impact upon its performance (see Figure 2-23). In general, everything else being equal, deeper antennas having smaller f/D ratios will have smaller side lobes because the feedhorn and LNB are closer to the reflective surface and thus are better screened from the surrounding environment. A similar result can be achieved by attaching a non-reflective shroud around the antenna rim. This is similar to the tactic of using blinders on a horse so that it will see in only one direction, straight ahead. Installing a shroud will result in different antenna performance than lowering the f/D ratio because side lobe patterns of both these configurations will differ.

Figure 2-22. Noise Spillover and Antenna Side Lobes. This drawing shows how an antenna can actually detect microwaves and noise from behind, 180° off its main axis. Noise from the warm ground is diffracted by the reflector edges and is scattered in all directions. Some of this noise manages to enter the feedhorn. As reflector surface inaccuracies increase, higher side lobes are formed. As a consequence, the antenna is more capable of detecting unwanted signals from the ground or from off-axis communication satellites.

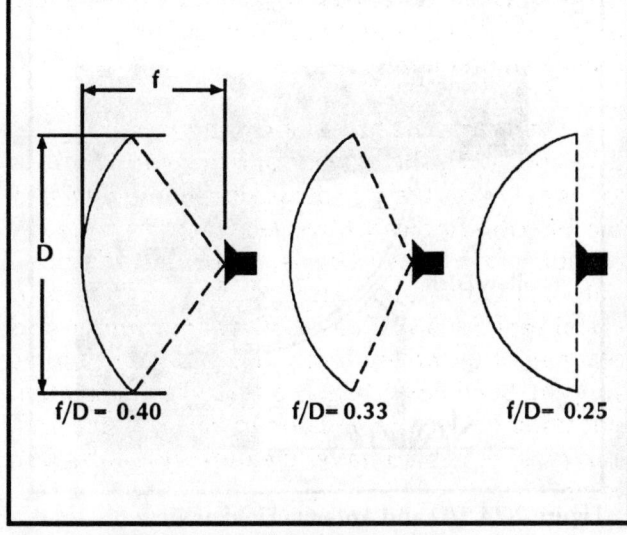

Figure 2-23. Focal Length to Antenna Diameter. A lower f/D means that the feed assembly is located more closely to a given sized reflective surface. At a 0.25 f/D, the feed lines up with the antenna rim so that adequately illuminating the entire reflective surface becomes quite difficult.

COMPONENT OPERATION

The concept underlying the f/D measurement also applies to offset fed antennas. In general, the feedhorn is located at the focal point of the parent parabolic surface even though it may not appear to be because only a portion of the larger parabola is used. Feed design for such antennas can be rather demanding since they must detect microwaves reflected from only a portion of a parabola. Offset parabolic antennas generally have an f/D in the 0.5 to 0.6 range.

Antenna Construction

Antennas can be made from a variety of materials and by numerous manufacturing processes as long as basic requirements are fulfilled. The reflective surface must conform to its designed shape and have good performance over the product lifetime. In addition, antennas intended for residential use must be simple to assemble, aesthetically pleasing, reasonably priced, easy to ship and durable. Some of these latter criteria are not as important for commercial applications.

In locations such as North America where millions of C-band antennas have already been installed for residential and commercial use, the question of compatibility with Ku-band reception arises. Two factors are important in judging whether or not a C-band antenna will function properly in the higher frequency range. First, surface tolerances must be much tighter to maximize antenna efficiency. Both systemic and random surface errors can cause serious performance problems at Ku-band frequencies. Reflectors exhibiting a noticeable warp or having panels that do not line up properly may have a very low efficiency and may have unduly high side lobes at Ku-band. Second, non-solid reflective materials must have openings less than at least one fifth and preferably less than one tenth the wavelength of the K-band microwaves. One tenth of the wavelength of 12 GHz signals is 2.5 mm (0.10 inch). If the openings are too large, antenna efficiency will be extremely low as microwaves "leak" through. The opposite side of this performance coin is that noise of the same frequency originating from the warm ground will pass through the reflective surface when the openings are too large and reach the feedhorn/LNB thus increasing the amount of noise power detected by the TVRO electronics.

Details of design, materials selection and manufacturing will have an impact on antenna durability. A system may function well when new but, in time, after being subjected to environmental stresses, its performance may seriously degrade. Allowances must be made for contraction and expansion of different materials with changes in temperature, for wind loading, for rain, snow, hail and corrosion, for the effects of intense sunlight and for many other specialised environmental factors. Remember, the antenna is the eyes of a satellite reception system and the most critical component needed for quality performance.

Five major classes of antennas are manufactured today: spun aluminum or steel; stamped or drawn steel; fiberglass; expanded or perforated metal or mesh; and formed plastic.

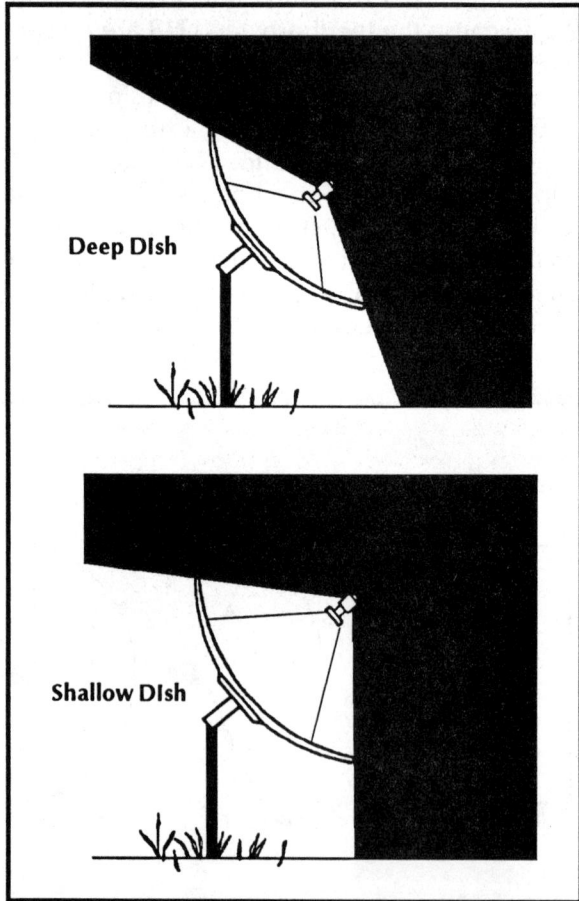

Figure 2-24. f/D and Antenna Field of View.
Antennas with a lower f/D ratio have a more narrow field of view and are therefore potentially less susceptible to noise and interference. When correctly designed, these "deep dishes" have lower side lobes. However, in general, a deep antenna performs more poorly in a dual-band offset fed configuration than a shallow one.

Spinning or Hydroforming

Spun or hydroformed aluminum or steel antennas can have the most accurate surfaces (see Figure 2-25). Spun dishes are manufactured by mounting aluminum or steel, typically of 1.5 mm (0.060 inch) thickness, onto a mold and spinning the whole assembly while pressing it into shape with rollers (see Figure 2-26). When done correctly, this technique results in a highly accurate reflective surface which has been well hardened during manufacturing. However, the spinning process can leave very shallow ridges created by the passage of the forming tool over the antenna surface. It is important that these ridges be substantially less than 1 mm otherwise losses due to surface irregularities can exceed 30 percent.

Spinning has the advantage that several different sized antennas can be manufactured from one mold; each diameter dish so produced has a different f/D ratio. But since these reflectors are manufactured in one piece, the larger ones, over 2 m in diameter, may be difficult to ship using conventional small-package consumer services. Some spun antennas consist of perforated metal. Following spinning the reflector is punched with a regular pattern of small holes. Signal losses through these holes are negligible if their diameters are less than approximately one tenth of the radiation wavelength, i.e. less than 2.5 mm for Ku-band and 7.5 mm for C-band signals.

Hydroformed antennas have potentially the best surface accuracy compared to those which are manufactured or assembled by other processes other than plastic molding (see Figure 2-27). Hydroforming is similar to the spinning method used to shape antenna materials. Water at extreme pressures (typically in excess of 1.5 million pounds per square inch) forces a metal disc to conform to a die

Figure 2-25. Andrew Spun Antenna. *This 1.8 metre, one piece spun reflector is designed to have an accurate surface in order to achieve high gain and efficiency at Ku-band. It has a measured gain of 44.9 dB at 11.95 GHz. (Courtesy Andrew Antenna Company)*

having a precise parabolic shape. The resulting reflector is free of surface deformities, ripples, flat spots or other irregularities often seen when using machine tools that directly contact the metal as in a typical spinning process. As a result, gain is maximised and signal distortion minimised so that performance is optimised.

The hydroforming process begins with a metal disc of diameter equal to that of the final antenna. Although nearly any metal can be used, aluminium is often selected in order to keep the overall weight

Figure 2-26. Spinning a DH Aluminium Antenna. *The fabrication of an aluminium spun antenna begins with a large diameter disk being secured to a mold where it will be spin-formed. The wheel spins at the rate of 250 rpm. It takes only a little more than three minutes to form the completed antenna. (Courtesy of DH Satellite)*

COMPONENT OPERATION

Figure 2-27. Hydroformed Antenna. *In this process a blank aluminium or stainless steel sheet is placed onto a mold and pressed into a parabolic shape using water to aid in creating a smooth and accurate reflective surface during manufacturing. This particular antenna is also available in a perforated model. The reflector is perforated after it has been hydroformed. (Courtesy of Paraclipse Antenna)*

Figure 2-28. Stamped Antenna. *The stamped offset fed antenna is made of galvanised steel with baked-on paint to extend its life expectancy. The polariser is ferrite and LNB is a HEMP. (Courtesy of Maspro)*

as low as possible. The disc is held in place around the edges of the parabolic die so that water forced into a bladder presses it against the metal die. This bends and stretches the metal into its final form.

The metal being hydroformed can never "back-up" during manufacturing. Once the hydroforming begins the product is formed evenly until it reaches the die surface. Because a standard die is used, every antenna can be identical. This repeatability in hydroforming then translates into lower manufacturing costs. This process allows use of materials of any gauge, ranging from 1.5 to 3.0 mm (0.070 to 0.120 inches), without the need to adjust tooling. In contrast, in spinning the forming tool must be adjusted so that a chosen thickness can be attained.

The spring-back of the antenna material can be calculated and this measurement incorporated into the die shape so that once the hydroforming is completed an accurate, parabolic shape is attained. This translate into manufacturing repeatability. An interesting test of this consistency of surface accuracy is to stack hydroformed antennas ontop of each other: one manufacturer claims that 100 of his 2.44 metre (8-foot) hydroformed antennas stack to a height of only 1.2 metres (4 feet). A comparable product such as a spun antenna would form a substantially higher stack.

Stamping

Stamped or drawn steel antennas are manufactured by stamping or pressing flat sheet metal into shape. The required tooling can be quite expensive because massive, thousand ton presses are needed. In addition, each mold is capable of producing reflectors of only one size. Nevertheless, antennas with accurate surfaces can be produced if the die is true and manufacturing techniques are excellent (see Figures 2-28 and 2-29).

Fiberglass

The challenge in manufacturing a fiberglass antenna is to embed wire screen, aluminized mats or other reflective materials in the fiberglass resin so that the metallic substructure retains an accurate parabolic shape. In an improperly manufactured antenna delamination and warping of this embedded reflector may occur months or years into its life-

COMPONENT OPERATION

Figure 2-29. Stamping Reflector Panels. *In this process an aluminium blank is placed over a tooled male dye and is stamped with a matching female dye. This equipment is the same as a thousand ton press used for stamping automobile fenders and doors. (Courtesy of Scientific Atlanta)*

Figure 2-30. Hand Layering of Fiberglass Antenna. *In this process, either screen mesh, flame spray or Romaglass which is a type of fiberglass embedded with small reflective metal particles is placed in a wood or aluminium mold. Fiberglass is then hand layered in 4 or 5 successive coats. Following 24 hours of curing, the antenna is removed from the mold.*

time. While this effect might be minimal at C-band, the tighter surface tolerances required for Ku-band signals may result in serious performance degradation. Some C-band fiberglass antennas also may not function adequately at Ku-band if the embedded metal has excessively large openings. One drawback of fiberglass antennas stems from the fact that the signal must enter the fiberglass, strike the reflective surface and exit a second time through the fiberglass. This path results in additional losses.

Fiberglass antennas are generally manufactured by one of three methods. In the hand layering or laminating process, a fiberglass resin called gelcoat is smeared onto fiberglass cloth laid in a mold (see Figure 2-30). The gelcoat often has pigment to give the surface colour and always has inhibitors to protect it against ultraviolet damage from intense sunlight. Metal is then flame sprayed onto the antenna. This results in a strong, lightweight but relatively expensive unit.

In sheet-molding, polyester resin, calcium carbonate and chopped fiberglass strands are mixed together and are then pressed onto a metal screen. The resulting product is lightweight but is usually brittle and more easily damaged than other varieties of fiberglass antennas.

In the thermal injection molding process, the mat and screen are set in a closed mold and the

Figure 2-31. Compression Molded Reflector. *This antenna is manufactured by thermal compression injection molding whereby liquid fiberglass under pressure is injected over an reflective aluminium screen layered inside of a mold. These machines typically use one thousand tons of pressure. This technique repeatedly produces rugged, durable reflectors. A three segment, 8 foot (2.4 metre) Prodelin antenna is shown here. (Courtesy of M/A COM)*

resin is injected under pressure. Antennas manufactured by this process are strong, impact-resistant and usually have good cosmetic appearance (see Figure 2-31).

ANTENNAS

COMPONENT OPERATION

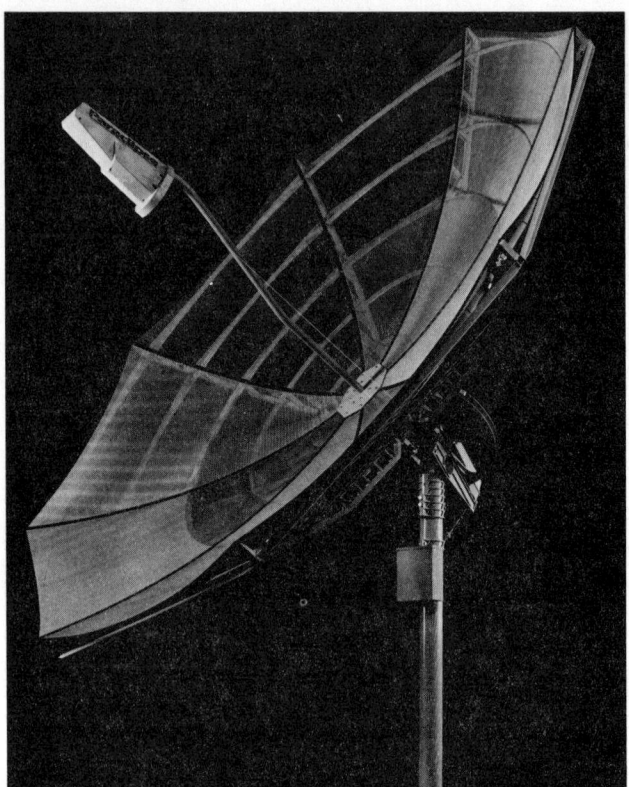

Figure 2-32. Paraclipse Wire Mesh Reflector. *The ribs for wire mesh antennas are manufactured either by welding pieces together or by bending an aluminium extrusion to conform to a parabolic shape. Bending is accomplished by using a hydraulic press and matching male/female dyes. The expanded mesh is attached to the ribs by using clips or screws, or by sliding it into grooves in these shaped extrusions. If the mesh hole size is substantially larger than 2.5 millimetres (1/10 inch) or if it does not adhere to an accurate parabolic shape it will not performed satisfactorily at Ku-band. (Courtesy of Paraclipse Antenna)*

Wire Mesh or Expanded Metal

The fourth type of antenna is constructed from wire mesh or expanded metal. These metal petals are held together and attached to an extruded rib structure by a series of clips or by insertion into channels in the ribs (see Figure 2-32).

Mesh antennas have been very popular in C-band TVRO systems because of their aesthetic appeal, low visibility, light weight and lower overall wind resistance but they can be flimsy if not correctly designed. Some varieties can also be rather time consuming to assemble. Note that perforated metal can be used in place of wire mesh or expanded metal petals in such an antenna design (see Figure 2-33)..

Figure 2-33. Perforated Antenna. *Perforated antennas can be manufactured either by stamping, hydroforming or spinning. The hole diameter should be less than 2.5 mm (1/10 inch) for adequate performance at Ku-band. This Winegard QuadStar QD0750 2.3 metre antenna is constructed with 1 mm thick (0.040 inch) perforated aluminum with a hole diameter of just under 2 mm. (Courtesy of Winegard Company)*

Molded Plastic

A recently patented process is now used to produce a molded plastic antenna (see Figure 2-34). This antenna can be manufactured with either a clear or opaque reflector surface. This process has a number of potential advantages including surface tolerances that can exceed those of dishes produced by hydroforming, light weight and the potential of being produced from recycled plastic stock.

Selecting an Antenna

Judging which type of antenna to purchase and use is based upon many factors including aesthetics, performance, weight, ease of assembly, wind loading, durability, ease of shipping and cost.

ANTENNAS

For example, spun aluminum models can have excellent surface accuracies but generally are produced in one piece so those having larger diameters are not easily shipped. Wire mesh antennas which many people find very attractive can be broken down and shipped in small boxes but conform less accurately to a parabolic shape than other varieties of antennas, especially if their panels are not held securely and accurately in place. Antennas with fewer petals usually conform more accurately to their designed shape. In general, mesh reflectors, especially those designed for operation at C-band, have quite low efficiencies at Ku- band. While, stamped aluminum antennas are very durable if the reflector wall thickness is sufficiently thick, these and other solid reflectors are more easily buffeted by winds than a wire mesh antenna.

It is important to realize that microwave antennas are also capable of concentrating solar energy. So the choice of colours and surfaces can be critical. During periods when the sun is nearly directly along the antenna axis, a brightly painted or smooth metal surface can reflect enough solar energy to melt polariser caps or fry cables. When painting an antenna, an optically rough paint must be used so that the sun's rays are scattered, not focused to generate heat. It is also better to use paints which are non-metallic since rough spots or bumps in the surface coating could cause reflection errors.

Wind Loading

In regions of the world where wind gusts often exceed 120 km per hour, wind loading of satellite receiving antennas is an important design consideration, especially for larger diameter reflectors. In the worst cases, the reflector may be twisted or pulled from its mount causing structural damage or even injury to persons. Wind forces can also cause even the most securely mounted, narrow beamwidth Ku-band antennas to waver off target with the resultant deterioration in reception.

Forces exerted by winds can be powerful. For example, a solid 2.5 metre antenna can experience as much as 360 kilogrammes of force under the impact of a 100 km per hour wind. Even a relatively small 1 metre antenna will experience as much as a 60 kg force at this same wind velocity. Wind loading calculations are presented in more detail in Appendix C.

Figure 2-34. The Molded Plastic Antenna. *This 1 m molded plastic antenna can be produced in a transparent or opaque form. It has potentially a surface accuracy that even surpasses the hydroformed variety. The feed and support will also be made from plastic in the near future.*

In most circumstances, mesh and perforated antennas are subjected to lower wind loading forces than are solid reflectors. This is especially the case when the wind blows directly into the reflective surface. However, when the wind blows parallel to the lip of an antenna, forces are comparable to those on a solid surface. The maximum wind loading on both solid and mesh varieties occur in just this situation, when the wind "spoons" along the reflective surface.

Ku-band antennas often utilise offset feeds in order to increase antenna efficiency. However, these reflectors are tilted at much steeper angles towards the horizon. While this provides an extra benefit in reducing snow loading, it also aims the antenna into the wind which most often follows a path parallel to the ground. Nevertheless, this change in inclination usually results in a decreased wind loading relative to a prime focus antenna pointed at a higher elevation. Gusts perpendicular to the face of an offset fed antenna have less effect because a prime focus model has its surface angled more directly into the wind. Winds approaching parallel to the reflector surface will also have less effect because an offset antenna generally has a more shallow curvature than does a prime focus.

C. ANTENNA MOUNTS

The purpose of a mount is to accurately and securely aim an antenna towards any chosen communication satellite. While many smaller Ku-band dishes permanently aim towards just one communication satellite, more flexible mounts must also be capable of targeting of any other spacecraft in the geostationary arc.

Using a mount that provides stability and pointing accuracy is critical in designing and installing a reception system. Many Ku-band antennas have narrow beamwidths and thus target a very small portion of the sky. A reflector having a diameter even as small as one metre which has a 1.75° half power beamwidth targets just one fifteen thousandth (1 in 15,000) of the visible sky from any point on the surface of the earth. Larger antennas have proportionately more restricted visions. A wind blowing across the sail-like surface of an antenna that results in a 0.01° deviation causes the centre of its vision to scan 75 km (47 miles) at the geosynchronous arc. This, in turn, leads to a decrease in signal strength and broadcast quality.

Pointing accuracy is not necessarily as important an issue for systems installed in European nations. The direct broadcast satellites are spaced 6° apart (see Table 2-5). Antennas as small as 30 cm with relatively wide beamwidths can be used for reception of high power transmissions. A 30 cm antenna has a theoretical half-power beamwidth of 5.8° (see Table 2-3). In contrast, Ku-band satellites in North America are sometimes spaced 2° apart, so a deviation of just over one degree might cause a second satellite to be swept into view.

Figure 2-35. A Rigid Antenna Mount. *The ten mount arms on this Prodelin fibreglass antenna are solidly attached to the reflector in ten places with two 13 mm (1/2 inch) bolts per arm. This rigid design counteracts the tendency of the reflector surface to sag when the antenna is aimed at low look angles. Any flexing would result in a loss of gain and an increase in side lobes.*

Nevertheless, mounts should be rigid, strong and firmly attached to both the antenna as well as its underlying supporting structures. This is particularly true for many Ku-band antennas which are roof or eave mounted, as is often the case in dense urban areas, and which are subjected to substantially higher wind loads than the more familiar, larger, ground-mounted C-band systems.

The design of the antenna-to-mount attachment is very important. Since the mount must be securely bolted onto a reflector, any mismatch can cause stresses and possibly warping. This would result in increased side lobes and impaired reception. Fiberglass or a thin-walled metal will bend when placed under such stress. The points of attachment must be properly located and sufficient in number to evenly distribute the weight both under static and wind or snow loaded conditions, and to prevent antenna warping under these loads (see Figure 3-35). In addition, the mount must be strong

TABLE 2-5. DISTANCE BETWEEN SATELLITES

Angular Spacing (degrees)	Separation Between Satellites (kilometres)	(miles)
0.5	369	230
1.0	739	459
2.0	1,477	918
3.0	2,216	1,377
4.0	2,955	1,836

COMPONENT OPERATION

enough to bear forces which include, in some geographic regions, potentially massive ice and snow build-up. Collars which secure the mount to the support pole must also be very solid in order to prevent twisting away from the correct north/south orientation. Pivot points which permit tracking movements must incorporate ball-bearings, bushings or other well-designed mechanical joints in order to prevent excessive wear over time.

Types of Mounts

There are two principal classes of antenna mounts: those that track the arc with two movements or degrees of freedom like the azimuth- elevation (az-el) mounts and those that require only one axis of motion like the polar mounts. Either of these can be fixed in position, if so desired.

Polar Mounts

Polar mounts were developed by astronomers who realised many years ago that it would be easier to keep a telescope sighted on a celestial body if it could be swiveled along a single axis to exactly counteract the earth's rotational motion. This polar axis, or hour angle axis as it was originally called, is parallel with a line passing through the north and south geographic poles. To achieve this orientation, the polar axis is set equal to the latitude of the installation site. For example, in the American city of Denver which is located at 40° latitude, the polar axis angle is set equal to 40° as measured from the horizon. At the equator where the latitude is 0°, the polar axis angle equals 0° because the arc of satellites is tracked by rotating along a circle directly overhead. At both these and all other locations, this setting aims the polar axis exactly along a north/south line.

Once the polar axis has been correctly aligned, a second adjustment called the declination offset is required. This was not necessary when viewing the stars because they are, to all intents and purposes, at an infinite distance. A polar mount with zero declination points in a line directly away from the earth (see Figures 2-36 and 2-37). The declination offset adjustment then tilts the view of a satellite antenna slightly down towards the geosynchronous arc. The greater the distance to this arc, the smaller the required declination offset. Given the location of the geostationary orbit, this angle ranges from 0° at the equator to over 10° in far northerly or southerly populated regions. The need for a declination adjustment alters the tracking motion of the polar mount from a circle to that of a flattened ellipse.

Even perfectly adjusted polar mounts always have a small tracking error in scanning across the

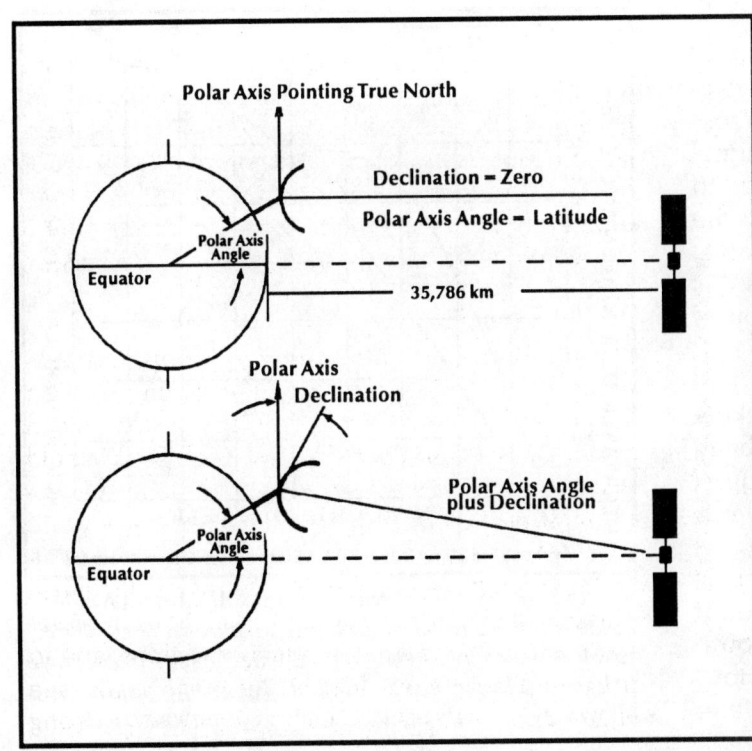

Figure 2-36. Alignment of a Polar Mount. *The polar axis angle which is set equal to the latitude points an antenna along a plane parallel to the one through the equator. Setting the declination offset angle lowers the field of view to the geosynchronous arc of satellites. Declination would not be necessary if the satellites were at an infinite distance, as the stars effectively appear to be.*

COMPONENT OPERATION

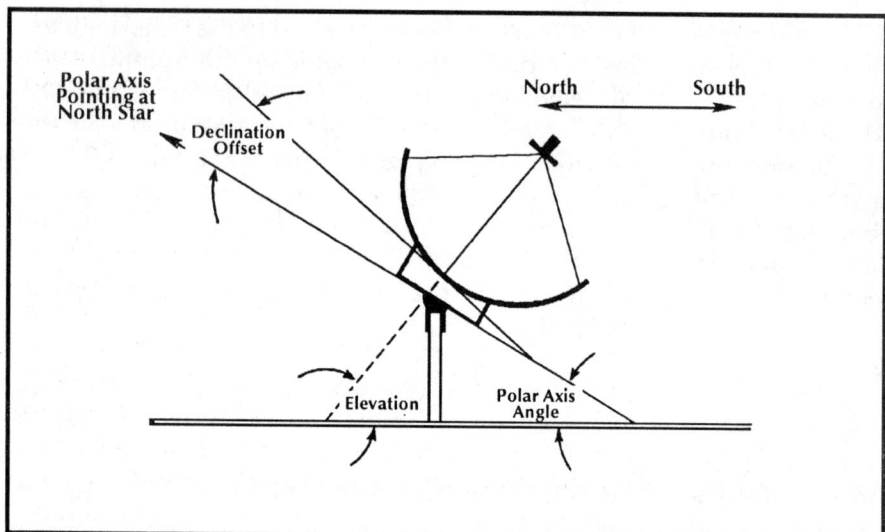

Figure 2-37. Polar Mount Geometry. *This drawing of a polar mount shows all the details of setting true north/south alignment, elevation and declination offset angles..*

entire belt of satellites. If a spacecraft in the centre of its sweep is accurately targeted, then those at the far ends will be slightly above the sight of the receiving antenna. If the end satellites are down the antenna boresight, the centre one will be slightly below its main axis. There is no way to completely avoid this inaccuracy. However, this error can be controlled at less than 0.1° with a well-adjusted, conventional polar mount. This inaccuracy is comparable with the 0.1° "stationkeeping" motion of a satellite and has been more than adequate for aligning C-band receiving antennas.

The Modified Polar Mount

Ku-band antennas have beamwidths one third that of equal diameter units used for C-band reception and require improved tracking accuracy across the entire geostationary arc to ensure optimal reception of broadcasts. Fortunately, a method to improve the alignment of a polar mount across the entire arc from 0.1 to less than 0.05° is available. The modified polar mount geometry reflects this fine tuning of the polar axis and declination offset angles.

The modified polar mount reduces tracking error as follows. Assume that the angles have been correctly adjusted in a conventional polar mount. The polar axis is then tilted slightly forward towards the arc of satellites. Recall that normally this axis points along a true north/south line. The declination offset angle is reduced by an equal amount (see Figure 2-38). A satellite located due south would be unaffected by this net zero change. However, when the antenna is rotated towards either due east or west it is inclined exactly as it would be if the polar axis had been still pointing parallel to the north/south line. Thus the slight inclination in the polar axis is effectively neutralised at due east or west. However, the slight decrease in declination causes the antenna to point higher than normally would be the case except when aimed directly south. In effect, the most southerly satellite is targeted perfectly, while the antenna's main axis is aimed below the easterly or westerly spacecraft less than would normally occur with the conventional polar mount adjustments. Comprehending this geometry usually takes some careful visualisation.

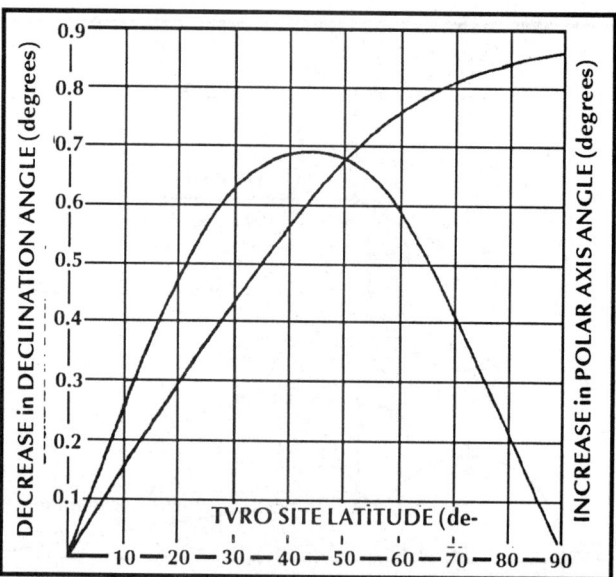

Figure 2-38. Modified Polar Mount Adjustment Angles. *In the modified polar mount, the polar axis is tilted slightly towards the arc of satellites and the declination offset angle is lowered by an equal amount. These fine adjustments allow very accurate tracking of the geostationary belt.*

COMPONENT OPERATION

Professionally designed mounts for multi-satellite Ku-band systems to be installed in the 1990s must certainly incorporate the fine adjustments that are necessary for improved tracking accuracy. With such accurate mechanisms, the modified polar setting is straightforward by following the detailed tracking procedures outlined in Chapter V. A complete description of the instruments required to facilitate this process can be found in Chapter IX.

Az-El Mounts

An az-el antenna mount is conceptually simple to understand and install but generally more difficult to operate. Any communication satellite is located by first moving the mount to the correct azimuth angle, which is along a plane parallel to the surface of the earth, and then by rotating up to the required elevation angle. C-band az-el mounts have found common use in motor homes and recreational vehicles in Canada and the United States. This has occurred because they have the ability to hunt in two directions. They can therefore compensate for the random orientation of antenna bases resting on vehicles which are parked on a variety of terrains. There is no tracking error so each satellite can be perfectly targeted. However, more complex and usually costlier control devices are required for adjusting both axes (see Figure 2-39).

Az-el mounts were historically used with C-band commercial systems in out-of-footprint locations where a much larger antenna was required. In such cases, az-el mounts were easier to install and were usually more strongly secured than polar mounts would have been. The installation of a polar mount by fine tuning the north/south orientation can be quite cumbersome given that massive, large antennas have relatively narrow beamwidths.

Today the mechanical designs have improved to the point that most large C-band antennas can incorporate polar mounts (although see Figure 2-40). Although the problem associated with bulkiness rarely occurs with smaller Ku-band antennas, az-el mounts driven by computer controlled robot arms are available. These have the important advantage of being amenable to fine tuning by use sophisticated feedback circuits. As would be expected, the choice between such robot az-el and modified polar mounts for Ku-band systems will ultimately be resolved in the marketplace on the basis of cost-performance tradeoffs. It is interesting to realise that az-el mounts that are stabilised in both axes with gyroscopes must be used moving ships. Maintaining a supporting pole in a perfectly vertical orientation would be much more difficult.

Figure 2-39. Az-El Robot Arm. *Even the most accurate satellite polar mount has some tracking error. Nearly perfect accuracy can be better obtained by using azimuth-elevation tracking. However, this type of system requires multiple motors and sensors so that off-the-shelf actuator controllers cannot be used. A complete, self-contained system having sensors, controls and motors to drive an antenna in both directions is necessary. An older variety of motorised az-el mount, the NITEC robot positioner, targets a satellite with a 0.2° accuracy. The maximum allowable antenna weight is approximately 68 kilograms (150 pounds) without the use of counter balances. The azimuth travel range is 180° and the elevation can be adjusted from 10° to 50° above the horizon. The programmable controller selects from a menu of up to 32 satellites. (Courtesy of NITEC Advanced Technologies)*

COMPONENT OPERATION

Figure 2-40. The Gibralter Az-El Mount. *This rigid, motorised az-el mount is powered by a 36 Vdc motor and is supported by a 40.5 cm (16 inch) steel pipe. Stamped aluminium dishes of 3, 3.6, 4.2 and 5 metres are available from the same company. (Courtesy of DH Antenna)*

D. ACTUATORS

Actuators provide the mechanical drive and control to allow an antenna to scan the arc of satellites under remote control. In the early 1980s there were only a limited number of satellites in the orbit over North America so that most antennas were either fixed on one target or hand-cranked between communication spacecraft. Most antennas for receiving C-band television broadcasts at cable TV headends in Canada and the United States are still aimed at just one satellite. However, as more satellites began transmitting television programming, antennas had to be moved from one spacecraft to another in order to receive each satellite. Manually moving a three or four metre dish in, for example, cold and wet weather was not an attractive proposition.

Today, most C-band satellite TV installations feature either linear or horizon-to-horizon actuators. While many Ku-band satellites are fixed, those that do feature actuators have similar features. Linear actuators employ a motor-driven, worm-gear assembly which attaches between a support on the antenna and another on the mount. Horizon-to-horizon designs have their entire gear mechanism encased at the fulcrum of the antenna and have the ability to track satellites across the whole visible sky, between both horizons. Robot driven az-el mounts have two encased gear assemblies for tracking in both directions.

Monitoring Antenna Position

Actuators used in the early more primitive days of the C-band home satellite industry in North America had very simple customer interfaces. An east or west button or switch would make or break a DC voltage sent to a motor which would move the gears and therefore the antenna. Position was maintained by the actuator internal clutch assembly. When some customers damaged their actuators and occasionally their antennas by inadvertently driving motors past mechanical limits, programmable east and west electronic limits were introduced. Even so, this problem still occurs at times. Even those "modern" actuators having pro-

COMPONENT OPERATION

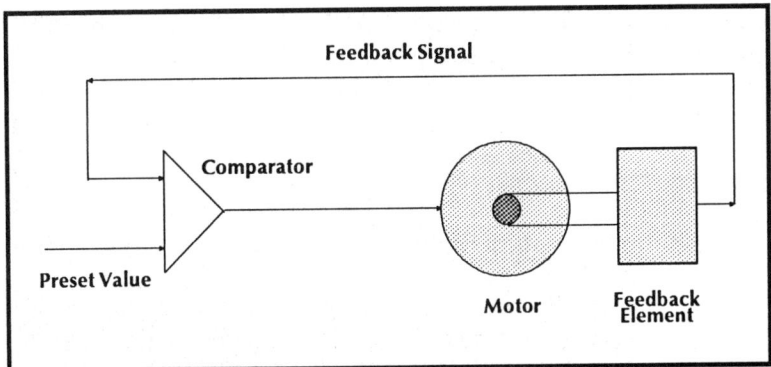

Figure 2-41. Closed Loop Feedback in Antenna Positioning. *This illustrates a closed loop feedback system as created by a circuit including the actuator, controller and feedback sensor. The sensor can be a potentiometer or a reed, optical or Hall effect switch. A preset voltage is compared with the output of the feedback element. The actuator motor is powered until the feedback value equals or exceeds the reference value.*

grammable limits sometimes "forget" their east/west boundaries and can subsequently damage actuator jacks or reflectors, most often during power surges or "brown-outs."

More advanced positioners are part of a closed loop control system (see Figure 2-40). The indoor controller supplies power to the motor in the positioner, while the feedback element in the positioner relays a signal back to the controller. The character of the signal depends on the position of the dish.

A satellite is selected from the front panel of the indoor controller. The position of a particular satellite is ascribed a value, either a number or voltage depending on the feedback system in use. Power is applied to the positioner until the the signal returning from the positioner's feedback element matches the value stored in the indoor controller.

Most positioners on the market have designs that are as simple as the above description. There are no truly servo-motor systems using feedback circuits on the market. In military or space applications where fractional degree accuracy is essential servo systems are used but they can be complex and expensive.

Feedback Element Types

The feedback element is the actuator component that generates what is known in control theory as the error signal. There are a number of types in use including resistive, reed switch, optical and Hall effect feedback (see Figure 2-42).

Resistive Feedback

Resistive feedback was one of the earliest methods used. A ten turn voltage-dividing potentiometer, which is a continuously variable resistor, is attached to the motor by a set of gears. The action of the motor also rotates the potentiometer, which is generally connected between five volts and ground. The voltage at the potentiometer slider is used as a feedback signal to the controller.

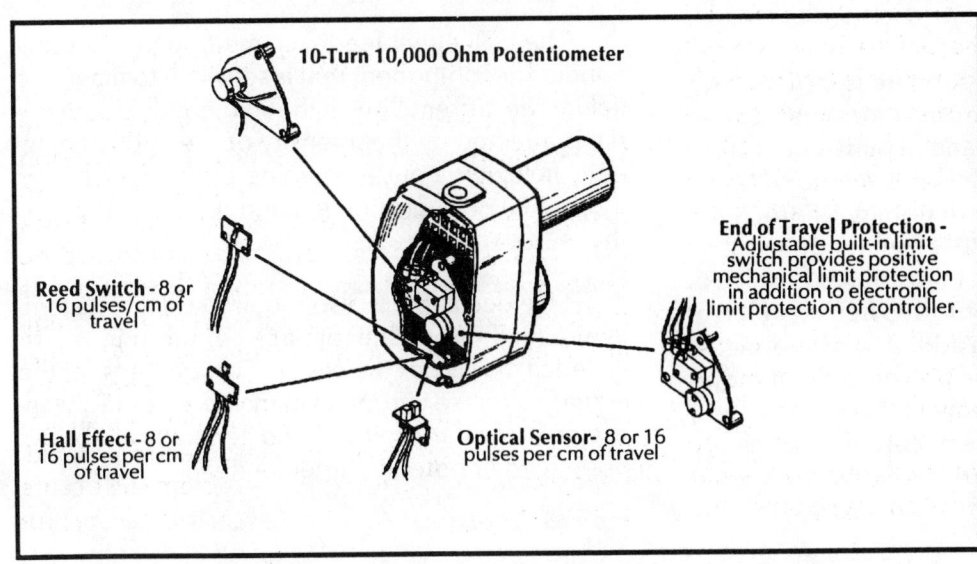

Figure 2-42. Actuator Sensor Positions. *Four types of sensors have been used to monitor antenna position. The mechanical end of travel protection illustrated here is an important design feature. (Courtesy of Pro-Brand International)*

ACTUATORS

COMPONENT OPERATION

In the indoor unit, the reference voltage is set by a preset potentiometer, which is generally accessible through the back panel of a satellite receiver. When this reference voltage is selected via the front panel switch, it is fed into a comparator. The comparator compares the reference voltage to the feedback signal or voltage. It sends power to the motor by means of a relay switch as long as the two voltages are not equal. When the reference voltage is equal to the feedback voltage, i.e. when there is zero error in control theory terminology, the comparator output falls and switches off the relay as well as power to the motor.

This method can be less accurate than one based on a microprocessor which uses digital pulse counting, because potentiometers can have resistances which may drift or change over time as water or dirt affect the contact surfaces.

Reed Switch Feedback

A reed switch is composed of two very thin slivers or reeds of metal encapsulated in a glass tube. When the switch is placed in the vicinity of a magnet, the slivers touch and create a short circuit. When the magnet is removed, the slivers spring open. The feedback signal in such a counter is a pulse rather than an analog voltage. One end of the actuator's limit is defined as count zero and the other limit as the maximum count.

The magnet is attached to the motor by a set of gears so when power is applied to the motor, the magnet rotates. As a result, the reed switch closes once every rotation. The pulses thus created are relayed to the indoor control circuitry via a small diameter wire where a continuous count is maintained.

The position of each satellite is defined as a specific number of pulses from zero. When a satellite is selected at the front panel, a pulse count value is loaded in from the controller's memory. Using digital logic, the controller determines if the pulse count of a different satellite is greater or less than that of the present target. It then applies positive or negative voltage to the motor. When a sufficient amount of pulses has been added to or subtracted from the original count, the power to the motor is switched off, usually by a relay. If the motor is finely geared, this motion can be accurate to within 10 mm or less.

This method is subject to electrical interference so it requires the use of shielded cable on the feedback run to the indoor controller. While ordinary cable can be run, shielded cable decreases the likelihood of false pulses changing the position of the dish.

Optical Feedback

The optical counting element uses a light emitting diode and phototransistor to generate the feedback pulse. It is essentially similar in operation to the reed switch feedback method with the only difference lying in the method used to generate the pulse. These are created when contact between a beam of light and a photodetector repeatedly turns on and off as the antenna moves, an action similar to that of an automobile timing light.

A disk is attached to the motor by a set of gears. The LED and phototransistor are fixed in position facing each other on opposite sides of the disk so that the beam from the LED is interrupted. One or more holes inside the perimeter of the disk allow the beam to pass to the phototransistor once or more times each revolution. As a result, the phototransistor generates pulses which are fed to a buffer circuit before being transmitted down the feedback line to the indoor controller. This buffer circuit is designed to ensure that the pulse has sufficient voltage and the proper polarity.

As this system is also a pulse counting system, the use of shielded cable is necessary. It is the most accurate of the systems as any number of holes can be cut into the disk to increase resolution.

Hall Effect Feedback

The Hall effect feedback method uses a semiconductor component that is sensitive to magnetic fields. The current flow in the semiconductor material is affected by the intensity of the applied magnetic field. It is a pulse counting system which uses the same type of rotating magnet arrangement as the reed switch.

Shielded cable is also required on this system. While Hall effect sensors are common in North America, this is not the case in Europe. Most of the actuators on the European market are from Taiwanese or Far Eastern sources and generally are based upon reed or potentiometer feedback.

COMPONENT OPERATION

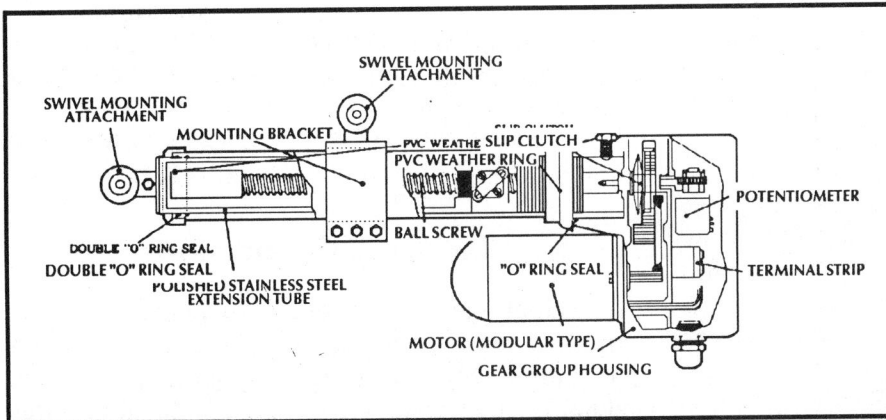

Figure 2-43. Internal Mechanism of a Linear Actuator. *The linear actuator arm bolts onto the mount at one end and onto the antenna at the other. A dc motor extends or retracts the inner arm to track any preprogrammed satellite as selected by the indoors controller. (Courtesy of Prosat)*

Mechanical Construction

Linear Actuators

Linear actuators have a telescoping arm which moves within a fixed external tube (see Figures 2-43, 2-44 and 2-45). Gears which drive the internal arm can employ either acme or ball screw designs.

The acme jack is constructed from a threaded shaft which moves in a threaded collar. It is simple and lower in cost than the ball screw mechanism. The ball screw is similar but has ball bearings which replace the threaded collar that moves in the tube threads. This mechanism has a much lower frictional loading than the former so that a motor of a given power will transfer more force in moving an antenna. This increased force and the smoother ball bearing movement reduces the chance of the shaft seizing up during cold weather or after long periods of non-use.

Figure 2-44. A Linear Actuator. *This linear actuator is mounted to a 2.6 metre spun aluminium antenna. The points of attachment incorporate flexible joints to avoid potential binding.*

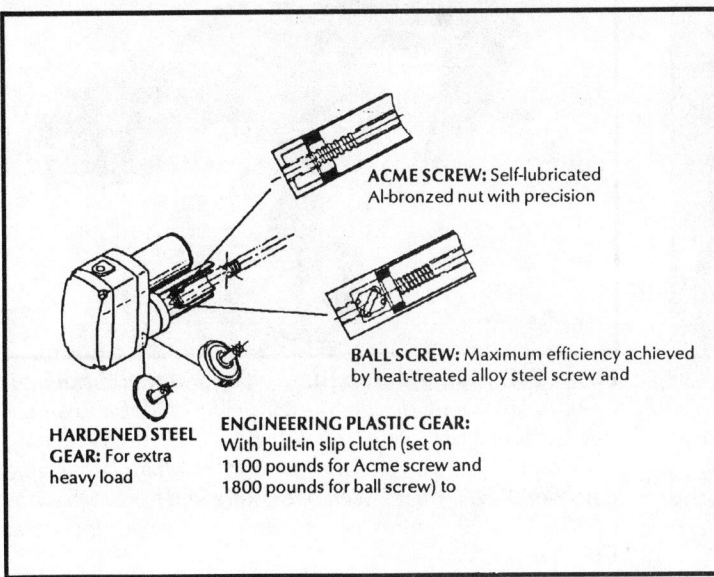

Figure 2-45. Acme and Ball Linear Jack Tubes. *An acme screwjack is constructed from a threaded screw shaft which moves in a threaded collar. This system can move up to 360 kilograms (800 pounds). Ball screwjacks are similar except ball bearings replace the threaded collar which grips the threaded shaft. This system can generate up to 700 kilograms (over 1500 pounds) of force. (Courtesy of Pro Brand)*

ACTUATORS

COMPONENT OPERATION

As a general rule, an actuator arm should never have to push or pull an antenna through an angle of less than 30° as measured between the arm and the back of the reflective surface. The larger this angle, which depends upon the design of the mount and the length of the actuator arm, the less the lateral force on the arm during antenna aiming. Proper installation reduces the chance that the internal rod could be bent and damaged. This type of actuator is limited in its motion to about a 100° sweep across the geosynchronous arc.

Horizon-to-Horizon Actuators

Horizon-to-horizon actuators are attached to the base of an antenna. They must be solidly constructed to generate the torque necessary to rotate an antenna and to allow minimal movement in winds or under other loads such as heavy snows. These are designed to move the reflector between horizons unlike linear actuators which have a more limited range (see Figures 2-46 and 2-47).

Electrical Connections

Electrical connections between the actuator motor and indoor equipment have typically been accomplished with a cable having four or five conductors (see Figure 2-48). Two heavier gauge wires carry the 36 Vdc power required to drive the motor. 14 gauge wire is typically adequate for runs up to a maximum of 50 metres; 12 gauge wire must be used for longer distances. Two 20 gauge conductors and often a ground wire are used to connect the counter to the control circuitry. This extra ground wire is essential in electronically controlled actuators to ensure proper operation of the microprocessor.

User and Receiver Interfaces

Controllers which operate and direct the mechanical equipment at the antenna are either one of two basic types. The simplest manual variety featured two buttons for east and west movements. The position of any satellite is identified by the

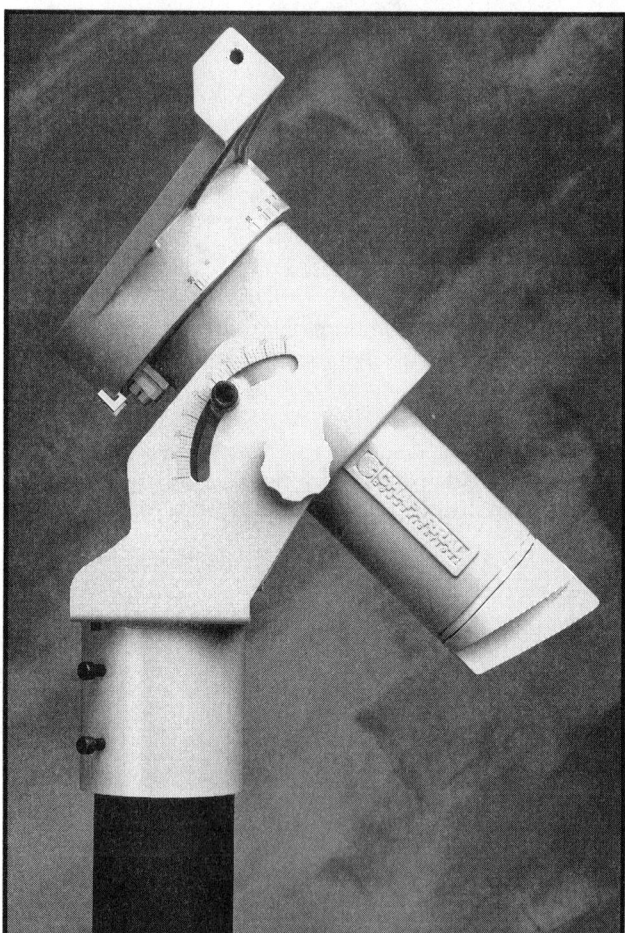

Figure 2-46. Polarmount Horizon-to-Horizon Mount. *The Polarmotor™ is designed for prime focus or offset fed parabolic dishes. It tracks the full 180o arc is 50 seconds, has a pointing accuracy of 0.2° based on 5 counts per degree of rotation. It uses an optical encoder for counting that is compatible with either reed switch or a Hall effect circuitry. (Courtesy of Chaparral Communications)*

Figure 2-47. SuperMount III. *The SuperMount™ horizon-to-horizon mount, designed exclusively for Ku-band antennas, features a completely enclosed gearbox, heavy bearing construction, 36 Vdc motor and built-in adjustable east/west travel limits. It weighs 6.5 kilos and has dimensions of 25 x 22 x 22 centimetres. (Courtesy of Jaeger Industrial Company)*

COMPONENT OPERATION

count on a digital read-out. More sophisticated units have programmable memories that allow an operator to easily locate any satellite at the push of a button. Antenna position is displayed by either a numerical counter or a digital read-out listing satellite code or name. Many units are capable of memorizing the locations of over 70 satellites. Both types of controllers should have east and west limits to prevent possible damage to motors, arms and antennas.

Some brands of actuator positioners are built into satellite receivers; some exist as stand-alone units. For example, the Chaparral Monterrey line of receivers marketed in North America, which are both capable of deciphering either Ku or C-band satellite broadcasts, use one line-of-sight, infrared controlled, hand-held remote to select all actuator and receiver functions. These and other similar integrated units display satellite selection next to transponder channel numbers on the receiver or on the TV screen read-out that can be monitored from other sets elsewhere at the site. Many actuators are stand-alone units which can be interfaced with numerous brands of satellite receivers. For example, the Houston Tracker V has a UHF hand-held, remote control which can be used out-of-sight of both the receiver and actuator. It controls antenna position as well as channel selection.

Improving Actuator Designs

Experiences in the C-band home satellite industry have provided invaluable guidance in designing and manufacturing actuators for Ku-band antennas. Three general conclusions can be drawn from five years of hands-on experience in the North American home satellite market. First, time has proven that the most trouble-free sensor is a reed switch. Second, the use of mechanical limits is an important backup to any other type of end-of-arc limit to prevent damage caused by antenna overtravel. Third, poor weatherproofing and the resultant water ingress has been the major cause of drive failure. (These topics are examined in greater detail in Chapters V and IX).

There are two notable differences between actuators designed for driving antennas for Ku and C-

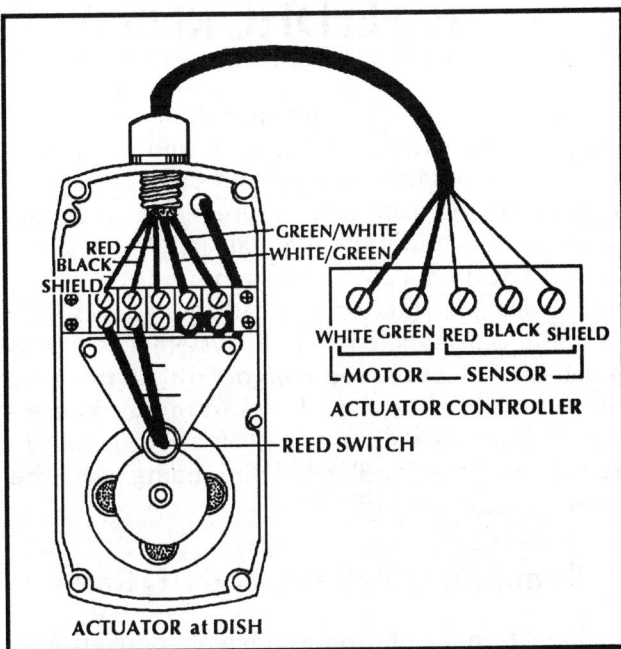

Figure 2-48. Electrical Interface Between Controller and Actuator Motor. *Five wires connect a control box with the actuator motor. Two of these send a voltage of typically 36 Vdc to the motor over a relatively heavy #14 gauge or larger cable. A three-conductor, #20 gauge shielded cable relays the counting pulses or appropriate voltage from the sensor to the indoors controller. In this case, a reed switch sensor is incorporated into the motor.*

band reception. First, Ku-band antennas are generally smaller and lighter than their C-band counterparts and therefore actuators which generate lower forces and, in the case of linear actuators, which have shorter arms can be used. Second, Ku-band antennas that have more narrow beamwidths for a given diameter reflector have less margin for tracking errors that could be caused by inadequate mechanical connections. The need for improved stability and greater aiming accuracy is the reason that some C-band systems having relatively large antennas (perhaps 3 metres or more in diameter) with extremely narrow beamwidths may be difficult to aim and control if converted to operate as dual-band reception systems. There can be no excessive play in jack arms.

E. FEEDHORNS AND POLARITY SELECTION

The feedhorn has the important function of collecting microwaves reflected from the antenna surface while ignoring noise and other unwanted signals coming from off-axis directions. This must be accomplished with minimal signal losses and without adding a significant amount of noise. A poorly designed feedhorn can add as much as 20°K of noise to a satellite receiving system. Feedhorns must also select the required polarity to properly detect a broadcast while rejecting unwanted signals. Recall that Ku-band signals can be either circularly or linearly polarised depending upon the satellite being viewed.

Illumination Patterns and f/D Ratios

The term feedhorn was originally derived from uplink antenna jargon. An uplink feedhorn, which in the primitive days of satellite communications actually looked like an inverted horn, "fed" microwaves onto and "illuminated" an antenna surface below. Before the advent of powerful computers, it was easier to design a receiving system by considering it as an uplink working in reverse. Consequently, the terms feedhorn and illumination have persisted as descriptors of the microwave collection device on a receiving antenna.

Therefore, a feedhorn is said to "illuminate" an antenna surface even though it actually receives reflected microwaves. The detected illumination pattern describes its field of view. A perfectly illuminated feedhorn would collect radiation coming from nowhere but the antenna surface; it would reject microwaves from all other sources.

In practice, prime-focus feedhorns illuminate antenna central regions most strongly and are progressively less able to detect radiation at increasingly greater off-axis angles. Similarly, feedhorns employed with an offset fed reflector are designed to be most capable of collecting microwave energy from the reflector's centre. The precise illumination pattern in all cases is an important design criteria. On one hand, a feedhorn which illuminated just the central portions of an antenna would introduce little environmental noise into the system but would miss some of the signal originating from the reflector edges and hence result in lowered system gain. On the other hand, a feedhorn which over-illuminated an antenna would take advantage of all the

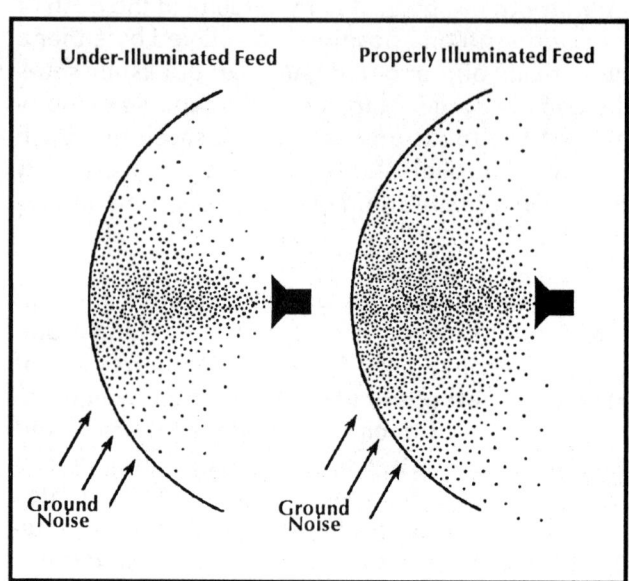

Figure 2-49. Feedhorn Illumination Patterns. *This drawing demonstrates how a reflector can be both under and over-illuminated by not selecting the correct feedhorn for a particular antenna f/D ratio. A poor match results in a lowering of signal-to-noise ratio. Under-illumination causes a loss of gain, while over-illumination causes detection of excessive noise from beyond the antenna rim.*

available gain but would introduce too much ground noise (see Figures 2-49 and 2-50). Over-illuminating an antenna could be a serious problem because the ground on a typical cool summer day emits noise at a "hot" 290°K.

Feedhorn designers have determined that the optimal "edge taper" for maximal signal collection and noise reduction occurs when the received power at the edges of a prime focus antenna is between 10 and 12 dB lower than that seen at its centre. Since the relative distance of a feedhorn from the reflector surface is determined by the f/D ratio, one designed for use with deep antennas will over-illuminate shallow antennas and vice versa. However, most prime-focus feedhorns can be used over a range of f/Ds and still yield near optimal performance. In general, feedhorns for use with offset fed reflectors must be specifically designed for the particular configuration since the illumination pattern is not symmetrical.

COMPONENT OPERATION

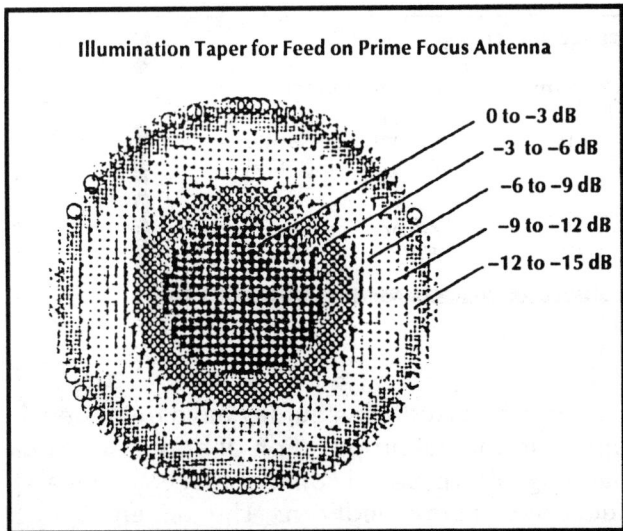

Figure 2-50. Feedhorn Illumination Taper. *This figure shows the illumination taper of a prime focus feedhorn. The design objective is to maximally illuminate the centre of the reflector and to allow detected power to taper off to approximately 15 dB at its outer edge.*

minated. All the incoming signal falling on a flat plate is usable. This contrasts with a conventional antenna which does not maximally use the signal reflected from its outer perimeter. The feedhorn illumination taper strategy decreases both the amount of ground noise as well as the signal detected but still can effectively increase the overall signal-to-noise ratio. Nevertheless some of the incident signal is lost. Furthermore, a flat plate antenna is not oriented towards a portion of the surrounding ground and therefore is not nearly as susceptible to detecting ground noise.

Horn Antennas and Feed Illumination

A recently introduced horn antenna designed for reception of European high power satellite video broadcasts, like the active flat plate antenna, uses its entire aperture at 100% illumination (see Figure 2-51). This antenna looks towards the "cool" noise environment of space rather than at the relatively "hot" earth. A specially designed feed system result in relatively low antenna noise temperature and no signal spill-over. Beyond approximately 30° off the antenna axis side lobes are effectively reduced to zero.

The efficiency of an antenna can be altered by simply changing the illumination pattern of the feed. If no illumination taper were used, i.e. 0 dB edge taper, the entire surface of the dish would be equally illuminated and efficiency would be near 100%. Although the gain would be increased the additional ground noise detected would result in a lower signal-to-noise ratio, the final determinant of system performance. In contrast, using an unduly high illumination taper of, for example, 25 dB, would result in substantially lower antenna efficiency as well as detected noise. The end result would still be the same, a reduced signal-to-noise ratio.

Feedhorns like antennas can also be characterized by their 3 dB beamwidth. The 3 dB beamwidth of a feedhorn is the cone in which detected signals are attenuated by less than 3 dB, a 50 percent reduction in power.

Flat Plate Antennas and Feed Illumination

Flat plat antennas that do not require feedhorns have the advantage of being 100% illu-

Figure 2-51. Horn Antenna with 100% Illumination. *The Innova™ horn antenna operates in the European 10.95 to 11.7 GHz FSS band. The 34 cm model has a gain of 31 dB and half power beamwidth of 5.2°. It is 100% illuminated like an active flat plate antenna. (Courtesy of Innova Corporation)*

COMPONENT OPERATION

TABLE 2-6. STANDARD FEEDHORN SIZES

Nomenclature			Internal Dimensions		Performance	
International Standard 153 IEC-R	American	British	Width (cm)	Height (cm)	Cut-Off Frequency (GHz)	Operating Range (GHz)
40	WR-229	WG-11A	5.817	2.908	2.577	3.3 - 4.9
120	WR-75	WG-17	1.905	0.953	7.869	10 - 15

Waveguides

The front end of all types of feedhorns consist of waveguides, hollow rectangles or tubes of conductive materials, usually metals. Waveguides must be used because at frequencies above approximately 1 GHz microwave energy is sharply attenuation when conducted down simple wires.

Recall that microwaves, a form of electromagnetic radiation, consists magnetic and electrical components. These "fields" travel in planes at right angles to each other. The electromagnetic energy interacts with the conductive walls of a waveguide.

In order for a waveguide to transmit a signal with low loss it must have a dimension at least as great as half the signal wavelength. For example, 12 GHz Ku-band signals which have a wavelength of approximately 2.5 cm will propagate with low losses down a rectangular waveguide of 2.5 cm on a side. However, these signals will be sharply attenuated by a waveguide with dimensions of 1.0 cm on a side. The cut-off frequency of a waveguide, the frequency below which a signal is sharply attenuated is determined from its dimensions. Thus, the cut-off frequency of a 2.5 x 2.5 cm waveguide is 12 GHz. Below this frequency the wavelength exceeds 2.5 cm. Since the wavelength of a signal increases with decreasing frequency, a 5 x 5 cm waveguide would have a cut-off frequency of 6 GHz.

The inherent ability of waveguides to attenuate signals below a cut-off frequency can be used to discriminate between horizontally and vertically polarised signals. For example, a rectangular waveguide of dimensions 1cm in height by 5 cm in width would preferentially allow vertically polarised signals to propagate.

Standardised waveguides have been developed for optimal propagation of both C- and Ku-band signals. These are commonly seen as integral sections of many feedhorns. The dimensions of these components as well as both the American and British nomenclature in use are presented in Table 2-6.

The Feedhorn/LNB Probe

The true antenna in a satellite reception system is the probe located either in the low noise amplifier or at the mouth of the feed. This finely tuned device of precisely the correct dimensions and position serves to convert the microwave energy within the feedhorn/LNB to an electrical signal.

Varieties of Feedhorns

Over the years four types of feedhorns have been developed for use in Ku-band satellite television: the pyramidal, the circular, the conical and the scalar feedhorn. The scalar feedhorn is the most widely used of the four but with the advent of low cost ASTRA systems in Europe, the conical feedhorn is now assuming a more central role. In addition, a rather novel feed, the Polyrod lens feed, has recently been developed and is being used in European satellite reception systems.

Pyramidal Feedhorn

The pyramidal feedhorn was extensively used in the early days of European Ku-band satellite television (see Figure 2-52). It is the easiest feedhorn to fabricate and offers a good match over a wide bandwidth. A major reason for its use was its availability. Most of the early Ku-band downconverters were modified 10.695 GHz radar modules and these usually were mated with a 5 dB pyramidal feedhorn.

A pyramidal feedhorn can be designed to have a different response in each plane relative to its axis. This factor allows a designer to carefully tailor the response of the feedhorn to the dish in use. Thus a pyramidal feedhorn need not illuminate an antenna in a pattern that is symmetrical about its central axis. The f/D operating range for a pyramidal feedhorn ranges from 0.45 to 0.9 and its dimensions are directly related to the design f/D ratio. This means that a pyramidal feedhorn designed for a dish with an f/D of 0.45, for example, will not work as well on a dish of 0.33 f/D.

This type of feedhorn has the disadvantage of exhibiting an uneven response. The E-plane (horizontal electrical plane) 3 dB beamwidth and the H-plane (vertical magnetic plane) 3 dB beamwidths are obtained using the equations given below.

E-plane 3 dB beamwidth in degrees = $29.4 \lambda/d$

H-plane 3 dB beamwidth in degrees = $50 \lambda/d$

where λ is the free space wavelength of the microwave signals and d is the waveguide diameter. All dimensions are expressed in metres.

The minimum diameter value of a circular feedhorn is 0.65 times the signal wavelength. The 3 dB beamwidths for this value correspond to $44°$ and $78°$ for the electric (E) and magnetic (H) fields, respectively. These minimum values correspond to f/D ratios of 0.69 and 0.43 respectively. The minimum f/D ratio for optimum illumination is the average of these two values, namely 0.56. This type of feedhorn is very rare and, to date, only one manufacturer has used it in Europe.

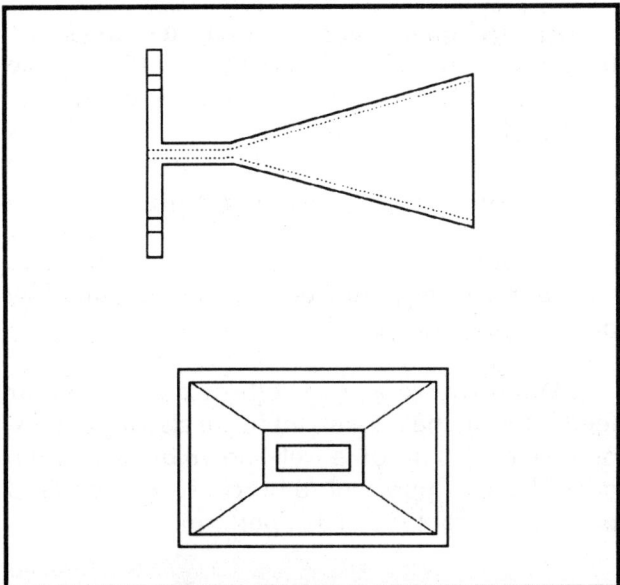

Figure 2-52. Pyramidal Feedhorn. *The pyramidal feedhorn is a section of flared waveguide. This type of feedhorn is specifically designed for each antenna.*

Circular Feedhorn

The circular feedhorn is simply a section of circular waveguide (see Figure 2-53). The diameter of the waveguide is governed by one factor, the cutoff wavelength. The waveguide in a circular feedhorn must pass the lowest frequency in use. Remember that waveguides are metal, hollow pipes of circular, rectangular or other cross sectional shapes that transmit microwaves. They can be compared in operation to fiber-optic cables that relay light. Thus, a horn-type feedhorn used in conjunction with an offset antenna is itself a waveguide. One most useful and interesting property of waveguides is that they rapidly attenuate a signal having a wavelength longer than one half their internal dimensions.

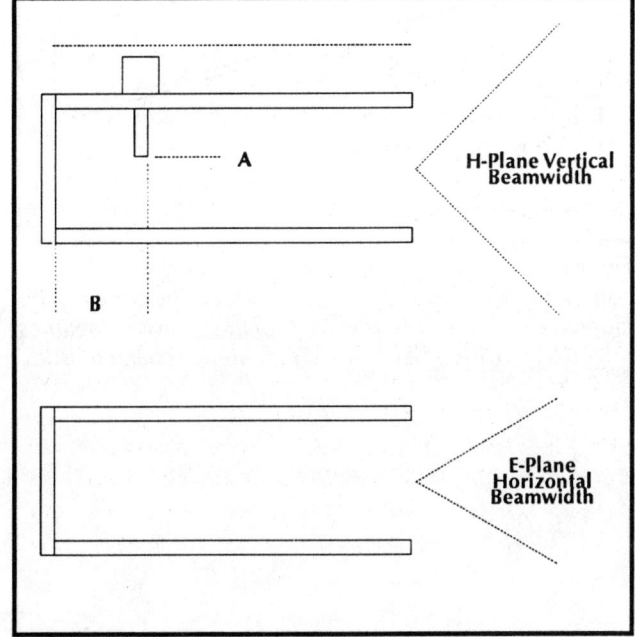

Figure 2-53. Circular Feedhorn. *The circular feedhorn is essentially a piece of circular waveguide with a flange. It does not have an equal response in both planes. In units of wavelengths, the probe length, A, is approximately 0.25, the distance of the probe from the rear of the feedhorn is 0.2 and the diameter of the waveguide is greater than 0.59. The magnetic field (H) and electric field (E) plane 3 dB beamwidths are equal to 50 and 29.4 times the wavelength divided by the waveguide diameter.*

COMPONENT OPERATION

Conical Feedhorn

The conical feedhorn is a design which essentially merges a pyramidal feedhorn with a spherical feedhorn (see Figure 2-54). A circular feedhorn has an uneven E-H response while the E-H response of the pyramidal feedhorn can be tailored to suit the application. This kind of feedhorn is commonly used with offset fed dishes. Readers with knowledge of astronomy will probably have heard of the coma effect associated with offset fed reflectors. This effect is caused by a constructive and destructive pattern of wave fronts resulting from phase delays which occur at the pick up point. A similar, though not as apparent, effect occurs with offset fed dishes. The conical feedhorn design tends to compensate for this effect.

Scalar Feedhorn

The scalar feedhorn is the most common type in use today. It consists of a section of circular waveguide with a set of concentric "scalar" rings (see Figure 2-55). The circular waveguide, the front-end, is dimensioned to detect both senses of linear polarisation. While the central circular portion of this feedhorn has an uneven E-H response, the scalar rings balance this response. These rings also serve to reduce the amount of microwave energy reflected back to the antenna. Typically about 1.5% of the energy entering a scalar feed, equivalent to a loss of 0.08 dB, is lost due to reflections.

The f/D operating range of a fixed ring scalar feed generally falls within 0.33 to 0.45. In some brands of scalar feedhorns, the position of the rings can be adjusted to extend this range from 0.28 to as high as 0.5 to match the f/D ratio of an antenna (see Figure 2-56). In general, the additional expense of an adjustable ring version is justifiable. According to some authorities, an improvement of as much as 0.5 dB in carrier-to-noise ratio can be gained by matching a feed to a dish.

While there are excellent quality scalar feedhorns on the market, there are also many inexpensive copies of some well-known brands. When possible, it is important to purchase a recognised brand to guarantee suitable performance.

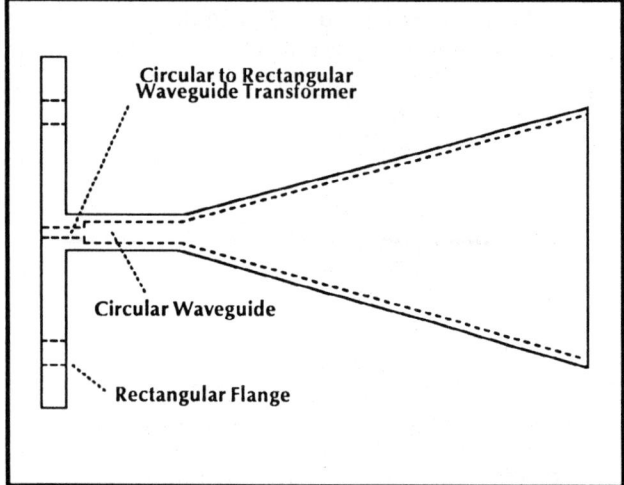

Figure 2-54. Conical Feedhorn. *The conical feedhorn is commonly used with offset fed reflectors. The cone can be tapered to suit the dish. The inside of the cone is sometimes ridged to provide a better match to antenna characteristics.*

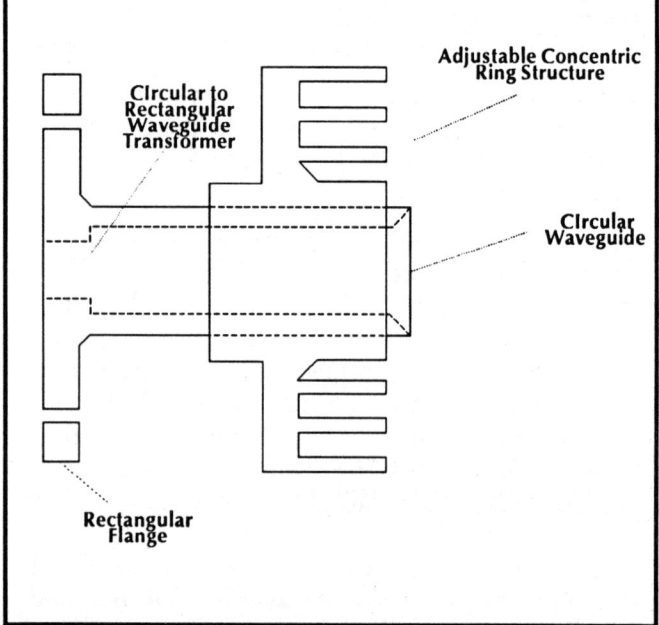

Figure 2-55. Scalar Feedhorn. *This popular design uses a ring structure to balance the response of the circular waveguide it feeds. The ring structure optimises the feedhorn/dish response.*

Figure 2-56. Scalar Feedhorn Assembly. *A scalar feedhorn consists of a circular waveguide surrounded by scalar rings. The adjustable portion of the circular section of waveguide that protrudes in front of the scalar rings effects the illumination pattern of the feed and thus the applicable range of f/D ratios.*

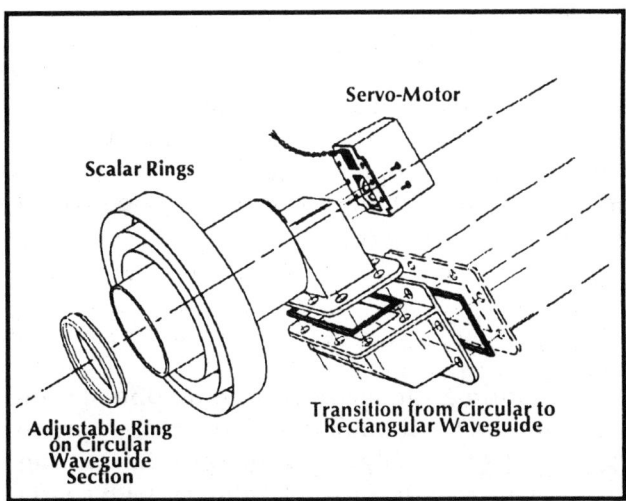

Polyrod Lens Feed

The Polyrod lens feed, developed and patented by Marconi, replaces a conventional scalar feedhorn. It is essentially a microwave lens that has been designed to have a radiation pattern with relatively sharp edges. The lens is capped with a cover known as a radome and fits directly onto the input to an LNB (see Figure 2-57).

The conical lens structure with its tapered wedge profile introduces a 90° phase shift to convert circularly to linearly polarised signals. An LNB probe is all that is necessary to detect this signal. It converts both senses of circularly into linear polarised signals (see Figure 2-58).

The design objective has been to reshape the illumination pattern of a dish so less ground noise is detected and more useful signal is collected. It designers claim an increase in system efficiency from typically 60% to 80% that translates into the need for an antenna with 25% less surface area.

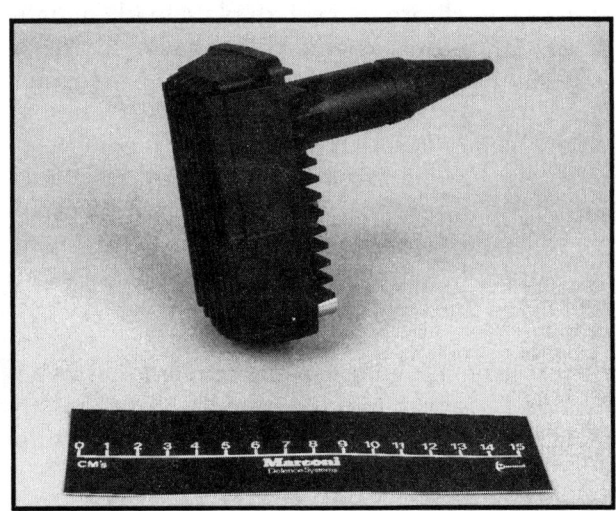

Figure 2-57. The Polyrod Len Feed. *The polyrod lens feed is a small microwave lens that fits directly on to an LNB, as pictured here. The resulting feed/LNB assembly is small and compact. (Courtesy of Marconi)*

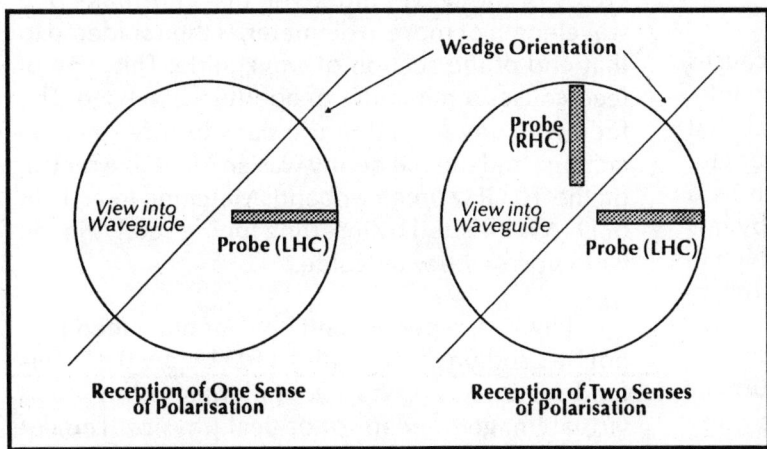

Figure 2-58. Reception of Circularly Polarised Signals with Polyrod Feed. *The polyrod feed can receive either sense of circularly polarised signals via one or two probes as illustrated here. In the configuration at the right, pin diodes are used to switch between LHC and RHC probes.*

COMPONENT OPERATION

Feed Configurations

There are essentially four types of TVRO feed configurations in use today; the prime focus feed, the Cassegrain feed, the splash plate feed and the offset feed. Of the three, the prime focus feed is still the most common.

Prime Focus Feed

In a prime focus system the feedhorn is installed at the focal point of the reflector. A prime focus feed is generally employed on dishes with an f/D ranging from 0.33 to 0.6. Antennas with an f/D exceeding 0.52 which are rarely encountered in satellite reception systems, generally use a pyramidal feedhorn or a splash plate feed configuration. At the other extreme, dishes having an f/D in the 0.25 to 0.28 range require use of a Cassegrain feed. Properly illuminating these dishes with extremely high or low f/Ds with a scalar feedhorn located at the prime focus can vary from difficult to downright impossible. While an adjustable ring type scalar feedhorn can be pushed to operate with an f/D of 0.28, it should be used only as a last resort instead of a Cassegrain feed.

Installing a feedhorn at the prime focus of an antenna is generally quite simple. Note that, in practice, the focal point is located just slightly inside the mouth of a feedhorn.

Cassegrain Feed System

A Cassegrain feed system uses a subreflector assembly in conjunction with a conical feedhorn. A pyramidal feedhorn is sometimes substituted for the conical although its use is generally restricted to terrestrial applications.

The hyperbolically shaped subreflector causes a dish to behave like one with a longer focal length. The conical feedhorn is then designed to match that of a dish with such a "virtual" longer focal length. The subreflector is interfaced with the dish by a technique known as dual distortion whereby it is matched to the outer edge of the antenna which itself is not part of the parabolic curve of the reflective surface.

A Cassegrain feed configuration has a number of distinct advantages and disadvantages compared to a prime focus feed. If the conical feed illuminates an area beyond the boundaries of the subreflector, it only sees sky noise, not ground noise as would be the case had it been installed at the prime focus. Since the level of sky noise is lower than that of ground noise, the negative effect on C/R (carrier-to-noise ratio) is not as severe. If, however, as in the American situation, satellites are closely bunched together with a separation $3°$ to $4°$, interference from adjacent satellites or "overlook" can cause problems. In addition, any subreflector system blocks the aperture of an antenna. This problem is exacerbated with smaller dishes. As a rule of thumb, the area of a subreflector must be approximately 30 percent of the aperture of an antenna for 1 dB loss to occur.

Designing and installing a home-made Cassegrain system can be difficult for an experimenter. To date, only one domestic system manufacturer has used this feed technique in Europe. Commercial versions of this type of feed are preset so an installer need only bolt the LNB onto the back of the antenna.

Splash Plate Feed

There are essentially two types of splash plate systems: the "penny" feed (Figure 2-59) and the flat plate feed (Figure 2-60). Both of these systems have been widely used on C-band systems. Only experimenters have used them at Ku-band.

The penny feed is restricted to use with dishes having an f/D in the range 0.25 to 0.32. The feed consists of a section of rectangular waveguide and a disk or splash plate. A slot of dimensions one twentieth of a wavelength in depth by half of a wavelength in length is cut in each of the wide edges of the waveguide. The disk, normally one wavelength or more in diameter, is then soldered to that end of the section of waveguide. This type of feed causes a minimum of aperture blockage. The feed was developed in the days before decimal coinage and the old penny was an ideal size for use on the 10 GHz amateur band. Matching to an LNB or LNA is effected by inserting tuning screws in the wide side of the waveguide.

Unlike the penny feed, the flat plate feed cannot be used on dishes with an f/D below 0.35. This feed system is a textbook implementation of the virtual image concept of optical physics. The flat

FEEDHORNS AND POLARISERS

COMPONENT OPERATION

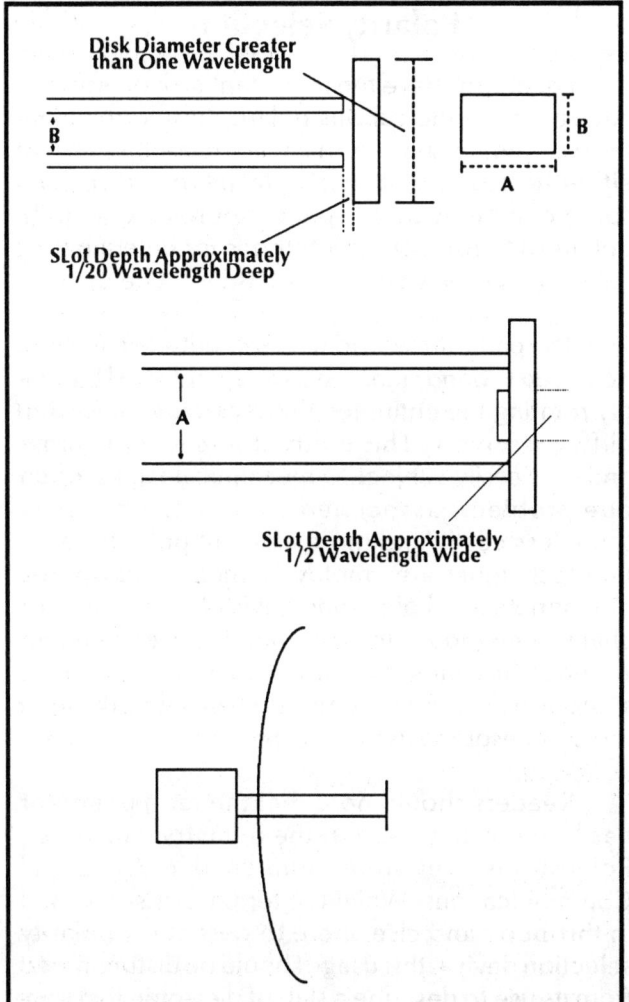

Figure 2-59. The Penny Feed. *The Penny feed was originally assembled using a pre-decimalisation English penny and a section of waveguide. While this technology is not widely used, early Ku-band experimenters tested some systems with Penny feeds. This configuration has the advantage that all electronics can be situated behind the dish. A x B equal 22.86 x 10.16 mm and 19.05 x 8.53 mm for WG16 (WR90) and WG17 (WR75) flanges, respectively. electronics can be placed behind the antenna, as shown in the lower illustration. Dimensions A x B are 22.86 x 10.16 mm and 19.05 x 9.525 mm for WG16 (WR90) and WG17 (WR75) flanges, respectively.*

plate acts like a mirror, creating a virtual focal point at the same distance from the subreflector as the subreflector is from the actual focal point. To date, this feed has been used only in commercial C-band TVROs. None but experimenters have mated it with Ku-band systems because of both the loss due to diffraction around the subreflector. In addition, the flat plate feed is difficult to set up properly.

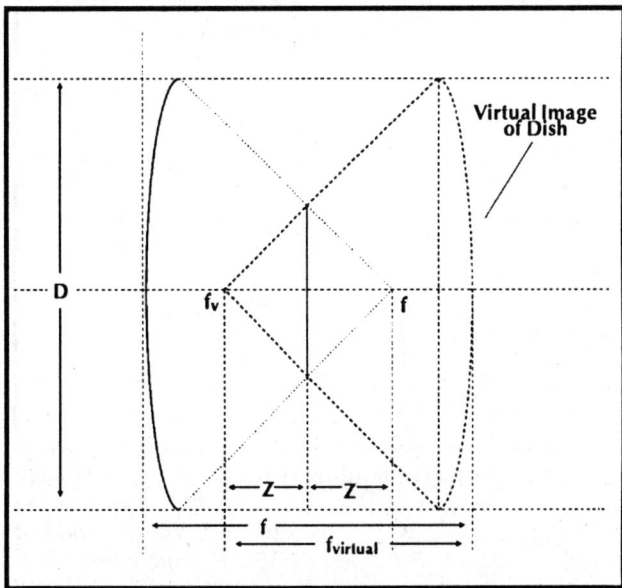

Figure 2-60. Flat Plate Subreflector. *The flat plate creates a mirror image of the dish for the feedhorn. The position of the feedhorn allows the electronics to be mounted behind the dish. The feed also aims at the sky and therefore detects mainly noise from deep space. This structure was popular in the early eighties but has fallen into relative disuse.*

This feed configuration as well as the Cassegrain have a distinct advantage in cases where the electronics are too heavy for a tripod mount as they can be installed on the antenna itself. If weather conditions are harsh, the electronics can be mounted behind the dish with a piece of waveguide being fed through the dish.

Offset Feed

Many smaller antennas reflect signals into a feed that is offset from its boresight. This reduction in aperture blockage can be important, especially in smaller dishes. Offset feeds are usually conical matched to a particular antenna system design.

FEEDHORNS AND POLARISERS

COMPONENT OPERATION

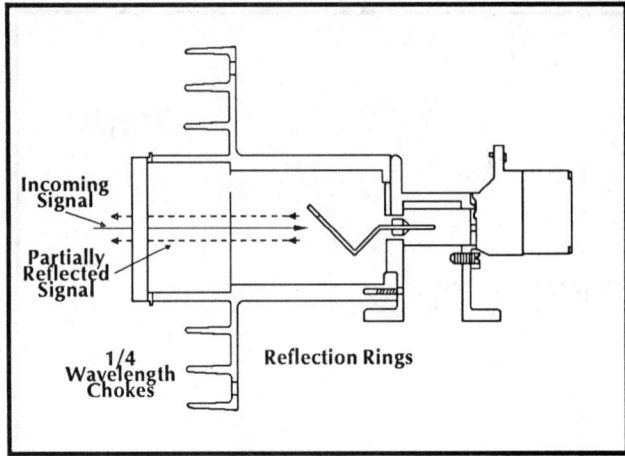

Figure 2-61. Voltage Standing Wave Ratio. *The voltage standing wave ratio, VSWR, represents how much of the incoming signal is reflected back towards the dish and lost from the feedhorn. The closer the VSWR approaches 1 to 1, the more efficient is the feedhorn. VSWR ratios that fall below 1.5 to 1 are acceptable, but a high quality feedhorn should have a VSWR below 1.2.*

Voltage Standing Wave Ratio (VSWR)

Feedhorns collect and channel the electric and magnetic fields of electromagnetic waves into an LNB. Reception is optimised when both these fields are detected with equal efficiency. In order to accomplish this objective, one of the feeds is selected for either prime focus or offset fed operation. Scalar feeds for Ku-band operation are 1/3 the size of those for C-band reception, because the wavelength of Ku-band radiation is 1/3 that of C-band signals.

Once the incoming microwaves have been captured, they are channeled along the feedhorn throat, a carefully designed waveguide. Both the "front-end" and the waveguide portion of a feedhorn must be manufactured to precise dimensions to allow best reception of satellite broadcasts. Engineers have defined a quantity called the voltage standing wave ratio, the VSWR, which measures how much of the incoming radiation is reflected back towards the antenna and lost (see Figure 2-61). The closer the VSWR approaches 1 to 1, the more efficient is the feedhorn. A perfect feedhorn which captured all the incoming signals would have a VSWR ratio equal to 1 to 1. In general, VSWR ratios that fall below 1.5 to 1 are acceptable but a high quality feedhorn should have a ratio less than 1.3 to 1 or better.

Polarity Selection

Feedhorns have the important task of discriminating among the various polarisation formats. Although most Ku-band broadcast satellites relay either horizontally or vertically polarised signals, some communication spacecraft relay circularly polarised transmissions. Methods for detecting circularly polarised signals are considered below.

The earliest method to select between linearly polarised C-band signals was accomplished by simply rotating the entire feedhorn and low noise amplifier assembly. This method was cumbersome and was easily subject to mechanical failure often due problems associated with freezing. Today three less costly and more efficient polarity selection techniques are employed: mechanical, ferrite or magnetic, and pin diode devices. While the mechanical method is most commonly used in North America, the most familiar in Europe is magnetic selection. The PIN technique has recently seen more widespread use in Europe.

Readers should note that the term polarotor has been widely used in the industry. However, Polarotor is a registered trademark of Chaparral Communications. While the term polariser is used in this book and elsewhere to describe a polarity selection device, this usage should be distinguished from its use to describe a slab of dielectric that is inserted in a circular feedhorn to allow reception of circularly polarised signals.

Mechanical Rotors

Polarity selection can be accomplished with a small, lightweight metal probe that is housed in the circular waveguide in the throat of a feedhorn (see Figures 2-62, 2-63, 2-64 and 2-65). The mechanical probe, which is rotated between horizontal and vertical polarity orientations, is shaped like a hook. The stem of this hook is attached to a servo motor that accurately controls its position. The probe section of the hook is housed in a piece of circular waveguide. This design allows it to detect signals of either polarity.

The straight stem of the hook is encased in a plastic dielectric material. This section passes through a piece of rectangular waveguide that is closed at one end. The signal from the probe section of the hook which radiated into this waveguide

FEEDHORNS AND POLARISERS

Figure 2-62. Schematic of a Mechanical Polariser. *A mechanical polariser rotates a probe between horizontal and vertical positions to detect horizontally and vertically polarizer signals. This microwave energy is then re-radiated into the LNB where it is again detected by a resonant probe. Both senses of polarisation can enter the circular portion of the waveguide at the entrance to the polariser.*

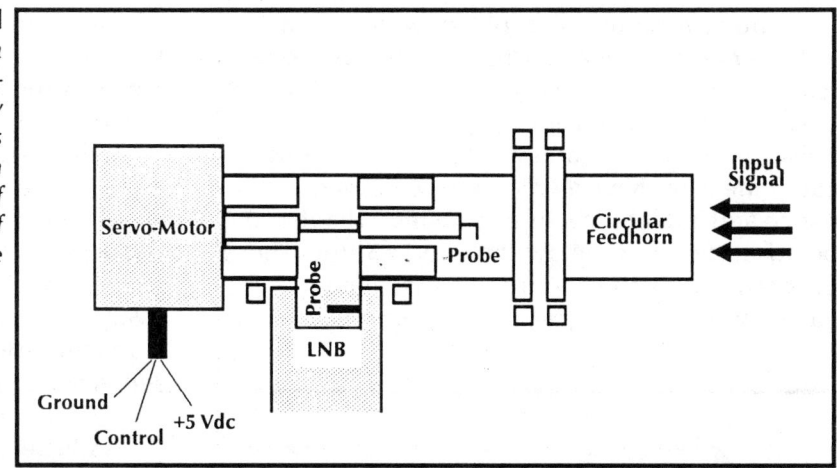

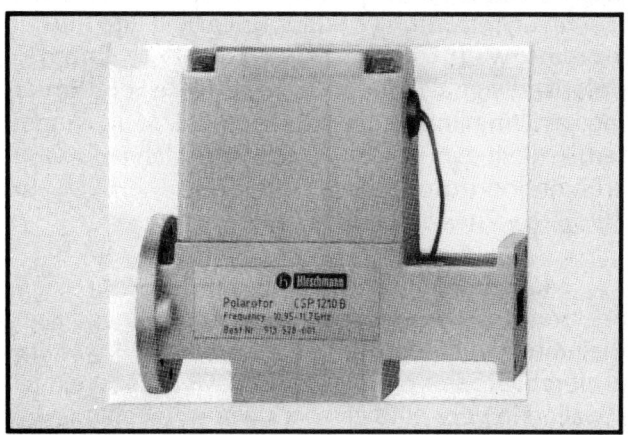

Figure 2-63. Hirschman Polariser. *This polariser operates in the 10.95 to 12.75 GHz band and has a cross-polarisation isolation of at least 25 dB. (Courtesy of Richard Hirshmann GmbH & Co.)*

retains the same polarity in the rectangular waveguide. Therefore moving the position of the probe through 90 degrees allows reception of either polarity.

Either a servo motor or a dc motor is used in mechanical polarisers. The servo motor can position the probe over a 180° range of motion and permits fine tuning at either polarity setting. The dc motor moves the probe through a full 360° albeit somewhat more slowly. Most satellite receivers available today have built-in controls for interfacing with one or both types of mechanical polarisers.

The position of the servo motor is controlled by the technique known as pulse width modulation whereby a satellite receiver continually relays a train of pulses to the polariser. Information which determines the orientation of the probe is determined by the pulse width. The on-time for the vertical pulses is longer than for the horizontal pulses while the off-time is identical for both.

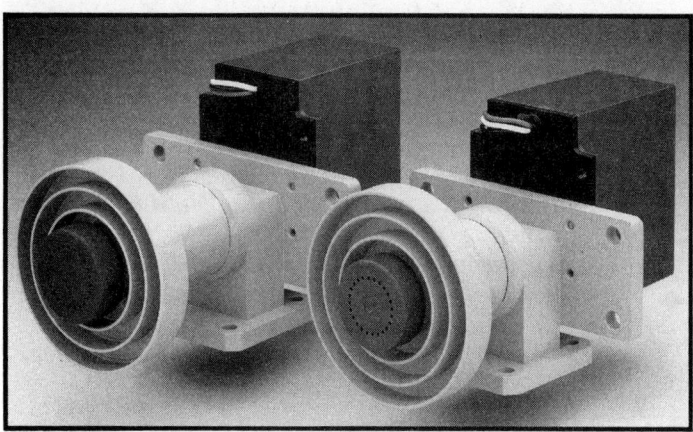

Figure 2-64. Polarotor I-Ku. *These feedhorns features servo motors for switching between horizontal and vertical polarities and infinite skew control. The internal probes are rotated by a controller-generated +5 Vdc TTL logic pulse of 0.8 to 2.2 millisecond duration having a 17 to 21 millisecond repetition rate. The model on the left is optimised for 11 GHz operation; the one on the right for 12 GHz. Both can be used on antennas with f/Ds ranging from 0.33 to 0.45 (Courtesy of Chaparral Communications)*

The most common problem with mechanical polarisers occurs when an insufficient pulse voltage reaches the device. This generally is caused by excessive losses in an overly long run of thin wire. The pulse shape deteriorates over long cable runs because the wires have elements of resistance, capacitance and inductance. As a result the voltage at the polariser can drop below the rated +5 Vdc. These problems are usually cured by installing heavier gauge wire.

Figure 2-65. Echostar Offset Mechanical Polariser. *This feed is designed to be used in an offset configuration. It has an average return loss of 17 to 22 dB and mechanical polarity selection. (Courtesy of Echosphere Corporation)*

Ferrite or Magnetic Devices

Ferrite devices are solid-state and have no moving parts (see Figures 2-66, 2-67 and 68). Polarity is changed via the interaction of the microwaves with a magnetic field. The magnetic selector consists of a ferrite rod held in the centre of a circular waveguide by a plastic dielectric support. The current passing through a large coil wound around the outside of the waveguide produces a magnetic field whose orientation depends upon the direction of the flow of charges. The polarity of the detected signal, in turn, depends upon the orientation of the magnetic field which rotates the plane of polarity virtually instantaneously.

Ferrite polarisers can be somewhat frequency dependent wherein the amount of rotation applied to the input signal varies with its frequency. However, models based Faraday rotation in an axial magnetic field provides polarity rotation that is essentially frequency independent.

A ferrite polariser rotates the plane of linear polarisation, typically over a range of about 45°. However, retrofitting a quadrapole field into the device allows it to convert circularly to linearly polarised signals as well as to provide rotation of the resulting linearly polarised signal. Modifying a ferrite device increases the cross-polarisation discrimination from the normally encountered 20 to 30 dB to values as low as 15 dB.

Magnetic polarisers have no moving parts and are therefore not subject to the faults that affect their mechanical counterparts. As a result, these are preferable to servo or dc motor devices for systems installed in very cold climates where mechanically driven probes often seize up due to condensation and freezing of water. However, failures that do affect the magnetic polarisers are generally catastrophic. The most common breakdown occurs in the coil. In a short circuit failure, the interface can sometimes be damaged. In an open circuit coil failure, the receiver interface generally remains unharmed. Magnetic polarisers are also easily damaged by a sudden impact which tends to splinter or crack the ferrite core. If a magnetic unit does not work after a fall, this is usually the cause. This damage generally necessitates the complete replacement of the unit.

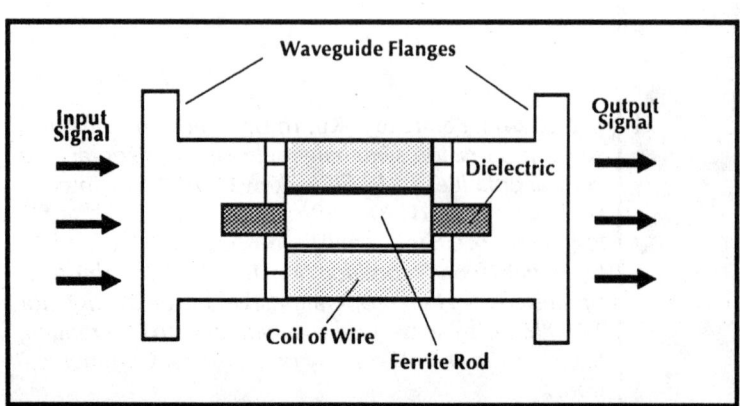

Figure 2-66. Ferrite Polariser Schematic. *A ferrite polariser consists of a short section of waveguide that contains a cylinder of ferrite material. The ferrite is surrounded by a wire coil that generates a magnetic field when a current is present. Stubs of dielectric material at each end of the ferrite cylinder are added to match impedances between the various waveguide transitions. All dimensions of this apparatus are carefully selected to propagate only certain modes of the input signal. The configuration diagrammed here allows 45° rotation of the plane of a linearly polarised signal.*

COMPONENT OPERATION

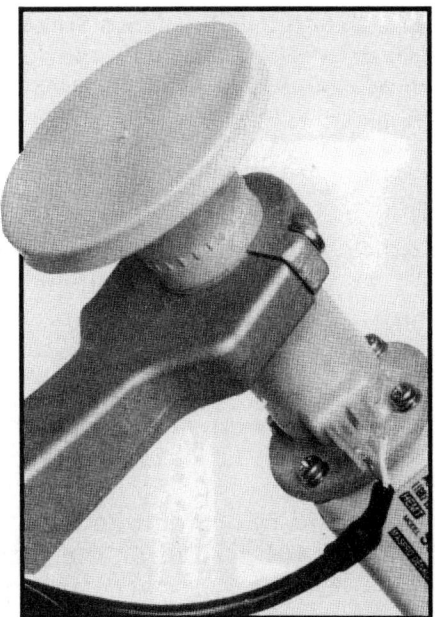

Figure 2-67. Maspro Micro Ferrite Polariser. *This unit, supplied with Maspro's BSK64E antenna, use a ferrite device to select polarity so has no moving parts. It has a maximum insertion loss of 0.2 dB and a minimum cross-polarity isolation of 20 dB. (Courtesy of Maspro)*

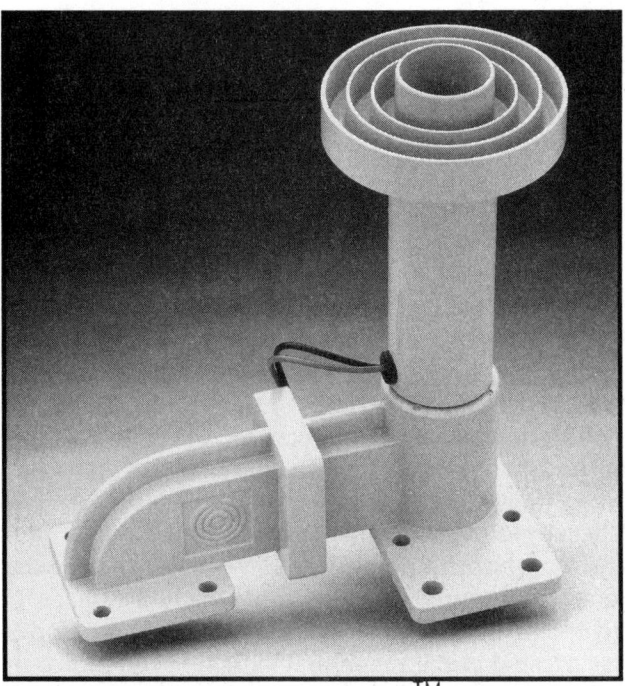

Figure 2-68. The Twister. *The Twister™ is a ferrite, dual band feedhorn with outputs designed for two frequency bands, 10.95 to 11.7 GHz and 11.7 to 12.75 GHz. Cross polarity isolation is 25 dB, insertion loss is less than 0.2 dB and cross-polarity isolation is a minimum of 25 dB. (Courtesy of Chaparral Communications)*

Magnetic devices have lower losses than mechanical or PIN diode polarisers, typically about 0.1 to 0.3 dB. A ferrite polariser modified to convert circularly to linear polarised signals has losses at the upper edge of this range. While most of the more recent European ASTRA satellite receivers have integral interfaces for magnetic and mechanical units, older models have only the mechanical selector input and therefore require a separate magnetic interface. This retrofitted interface is controlled by the original mechanical interface on the receiver.

Pin Diode Polarity Switching

PIN diode switching is perhaps one of the most noise ridden methods of polarity switching. It is essentially electronic switching between two probes which are mounted in circular waveguide. One probe detects horizontally polarised and the other vertically polarised signals (see Figure 2-69). It is not popular since only two fixed directions of polarity, based on the orientation of two probes, can be selected. This design does not permit fine tuning of polarity.

Switching between the two probes is effected by changing the supply voltage on the centre conductor of the cable feeding the LNB. Two versions are employed: voltage polarity switching and voltage level switching. In voltage polarity switching, which is not widely used, the polarity of the supply voltage on the LNB cable is changed. For one polarity, the centre conductor would be $+V_{cc}$ and the shield would be 0 Volts. For the opposite polarity, the shield would be $+V_s$ and the centre conductor would be 0 Volts. This requires that the input socket must be isolated from the LNB casing. On some early commercial Ku-band LNBs, PIN diode switching was used via a separate connector on the LNB. The quoted noise figure on many of these LNBs was in the region of 2.8 dB. The calculated and, in some cases, measured noise figure was closer to 3.5 dB. This disparity has caused many installers to distrust the manufacturer's quoted specifications for PIN diode switches.

FEEDHORNS AND POLARISERS

COMPONENT OPERATION

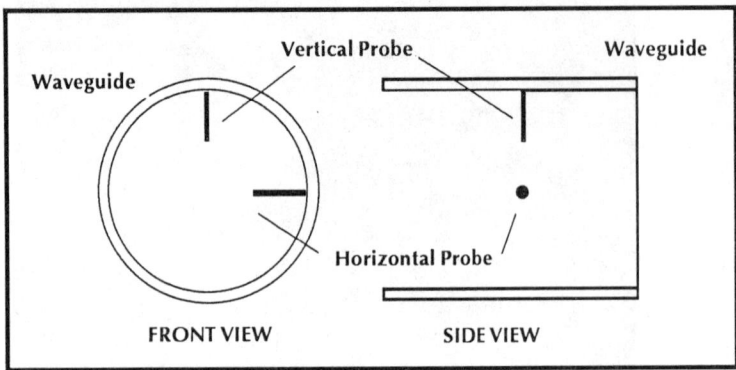

Figure 2-69. Pin Diode Polariser Schematic. *This polariser selects polarisation via two probes, one vertical and one horizontal contained in a waveguide. Signals from the probes are each amplified and fed to two pin diode attenuators. One sense of polarity is then attenuated and one is selected for detection.*

Voltage level switching is one of the most common forms of polarity switching on the economy-type ASTRA receivers. A voltage drop of about 5 Vdc is employed. One polarity is received at 18 Vdc and the opposite polarity at 13 Vdc. Which sense is received is merely a question of which PIN diode is biased "on" and which is biased "off." Each diode has a "blocking" voltage that must be exceeded before the diode can be switched on. For example, the blocking voltage could be 8 Volts. By using resistive voltage dividers, one diode could be set to switch on at 18 Volts and the other at 13 Volts.

PIN diodes are not generally mounted between the probes and the GaAsFET amplifier. (GaAsFET amplifiers, gallium arsenide field effect transistor amplifiers, are discussed in more detail below in the sections F and H later in this chapter). They are used as attenuators so the switch analogy may be slightly misleading. A PIN diode switched LNB will generally have two front-end GaAsFET amplifiers. The outputs of these amplifiers are connected via capacitors to a combining network. Each amplifier output has a PIN diode acting as a voltage controlled attenuator. The required polarity is then selected by attenuating the opposite polarity signal from that amplifier.

Orthomode Transducers (OMTs) and Dual LNB/Feeds

Orthogonal mode transducers (OMTs), also known as orthomode transducers, can simultaneously detect both vertically and horizontally polarised signals. Such devices are commonly found in commercial C-band installations such as satellite master antenna and cable television headends because they allow viewing of all satellite transponders at one time. Until recently OMTs were less commonly used in Ku-band TVROs where fully loaded vertical/horizontal polarity downlink schemes were not encountered as frequently.

Two designs of orthomode feedhorns are available. One is based upon the property of waveguides that prevents the passage of microwaves having wavelengths longer than twice their internal dimension. For example, if a signal of 1.0 cm wavelength entered a waveguide having a 2 by 0.3 cm cross-section, waves polarised in the direction of the 2 cm dimension would pass through while those polarised at right angles to this direction would be rapidly attenuated. A dual feedhorn uses two waveguides of the required dimensions joined at right angles to each other. One arm transmits horizontal polarity; the other vertical polarity. Two LNBs are used to independently amplify each of these signals.

A second type of feedhorn designed to simultaneously detect both polarities, a dual

Figure 2-70. Seavey Orthomode Feedhorn. *This Seavey orthomode feedhorn is designed to simultaneously receive horizontal and vertical polarities without the need for any mechanical movement. (Courtesy of Seavey Engineering)*

FEEDHORNS AND POLARISERS

LNB/feed, has no waveguide section but has two independent probes and two LNBs built into one housing. It is designed so each probe/LNB detects signals of only one polarity.

There are basically two forms of OMTs: the circular waveguide T-type and the circular-to-rectangular T-type. The latter is the more common. The circular variety can be identified by its "Y" shaped appearance. The circular waveguide splits the incoming signal from the dish equally into its two arms, each containing horizontal and vertical polarised components. The required polarity is then selected by terminating the arm with a rectangular flange. It is normal practice to include a matching transformer prior to the flange. This device is essentially a piece of ridged or tapered waveguide that transforms the impedance of the circular waveguide to an impedance matching that of the rectangular waveguide as closely as possible.

The circular-to-rectangular T-type OMT looks like a lower case letter "h." The rectangular OMT waveguide rejects any signal with wavelength greater than half of its internal dimension.

Detection of Circularly Polarised Signals

Circularly polarised signals differ from linear formats because the vibrations of electric and magnetic fields do not lie in one plane but follow a circular path as they travel through space. Although right-hand circular polarisation (RHCP) has been favored in the past, left-hand circularly polarised (LHCP) signals today are being more regularly transmitted from communication satellites. For example, LHCP signals are relayed in C-band zone beams from the Intelsat V and VA series of spacecraft serving areas of Central and South America. North American Ku-band TVROs will soon have the option to receive circularly polarised broadcasts when a new 100 watt per transponder satellite to be positioned at 101° W becomes operational.

Feeds designed specifically for circular polarisation have rarely been encountered in domestic satellite television. However, many of the experimental C-band units used in Europe in the early eighties employed a helical probe with a rear reflector. The polarity sense received depended on the direction in which the helix was wound.

Circularly polarised signals have one principal advantage over linear formats. Reception of linearly polarised signals requires use of a skew adjustment because the polarity can vary continuously between horizontal and linear. The direction of the plane of polarisation can vary substantially between satellites. For example, the American GSTAR A1 spacecraft has its plane skewed nearly 26° relative to most other North American Ku-band satellites. In contrast, circular polarisations have either left or right orientations. There is nothing in between and no need for skew adjustments.

Feedhorns designed for reception of circularly polarised signals can have relatively simple designs. Two internal probes at right angles to each other detect the incoming signal. The output of one is delayed one quarter wavelength relative to the other and the signals are added to detect one sense of polarisation (LHCP or RHCP); reversing the delay detects the other polarity sense. A tiny pin diode switch controlled by a small dc voltage can be used to select polarity. Such a simple device could easily be built into the throat of an LNB.

The most common form of circular polarity feed is a scalar feed that is modified to receive circularly polarised formats by virtue of a dielectric slab inserted into the circular waveguide at an angle of 45° (see Figure 2-71). It acts by "delaying" the circu-

Figure 5-71. Feedhorn Modification for Detection of Circularly Polarised Signals. *A linear feedhorn can detect right and left hand circularly polarised signals if a slight modification is made within the circular waveguide portion of the feed. A teflon slab inserted at a 45° angle in the circular waveguide transforms a circularly polarised signal to one having a linear format.*

COMPONENT OPERATION

larly polarised signal thereby translating it into a linearly polarised signal. Standard scalar feedhorns cannot distinguish between LHCP and RHCP transmissions but this simple retrofit permits detection and discrimination of both circular polarity senses. The dielectric slab, which is simply a properly sized rectangular piece of teflon, inserted into the feed throat accomplishes this feat. A conventional Ku-band feedhorn with skew adjustment is then capable of detecting and discriminating between vertical and horizontal polarities as well as between LHCP and RHCP signals. Some feedhorn manufacturers now sell this product complete with installation instructions. More procedural details are presented in Chapter V.

A scalar feed designed to receive linearly polarised transmissions, when modified, attenuates circularly polarised signals by 2 to 3 dB, cutting the power in half. Since there are RHCP and LHCP channels in the proposed European DBS system, the prospect of adjacent channel interference has been mentioned in some discussions of the ASTRA/BSB compatibility. While some critics have claimed that a dual standard system is impossible, compatibility appears to be feasible. It would require a dual-band LNB and a motor driven antenna. In order to watch BSB, a EuroCypher decoder would have to be obtained. (BSB would probably delay the marketing of a EuroCypher decoder for a year or so in order to properly market their IRD systems). Then it would be feasible to use a linearly polarised feed to receive the BSB satellite. Detected power would be 64 dBw at beam centre and the channels are all on the same polarity. The channel centre frequencies are spaced by approximately 76 MHz.

Feedhorn/Customer Interfaces

Most Ku-band systems can control linear polarity selection in two ways. First, satellite receivers can be programmed for an automatic polarity selection whenever channels are switched. Some more sophisticated units also fine tune channel centre frequency and optimize polarity alignment by employing feedback circuits which maximize signal strength to adjust these parameters. All those brands which are not microprocessor driven can be manually switched between polarity formats. Second, whenever a mechanical feedhorn is used, a skew adjustment allows for fine tuning of the final probe position.

Most satellite receivers can accept the necessary wires, typically three 20 gauge or larger ones, to control polarity selection and fine tuning.

F. LOW NOISE BLOCK DOWNCONVERTERS (LNBs)

The low noise block downconverter, known as an LNB, has the important function of detecting the microwave signal relayed from the feedhorn and converting it to an electrical current, amplifying this extremely weak signal by 40 to 50 dB (a gain of 10,000 to 100,000) and downconverting or lowering its carrier frequency. The downconverted signal is subsequently relayed along cable to an indoor satellite receiver. An LNB is the first "active" or electronic component in the series of steps by which a satellite signal is processed and, in concert with the antenna, is the most important component in determining how well an earth station can function.

LNBs technology for use in Ku-band satellite systems has evolved from the experience gained in the design, manufacture and operation of C-band low noise amplifiers (LNAs). An LNA amplifies the satellite signal, while an LNB downconverts as well as amplifies the incoming signal. Today, C-band systems employ both LNA and LNB technology depending upon the chosen design, but Ku-band systems almost exclusively use LNBs. It is instructive to examine how LNAs evolved in order to more clearly understand the function and operation of LNBs. A more thorough explanation of downconversion is presented in the next section. Following this, LNB design is examined (see Figures 2-72, 2-73 and 2-74).

The power of signals reaching the input of an LNA via a typical 3 metre C-band antenna is less than 10^{-14} watt. Therefore, an LNA must contribute very little noise power or signals will be drowned out in the roar of amplifier internal thermal noise. This feat is made possible by recent advances in

COMPONENT OPERATION

Figure 2-72. Norsat Series 6200 HEMT LNB. *This Ku-band LNB has a 60 dB gain and a minimum gain flatness of 6 dB per 500 MHz. It is available in the following frequency ranges listed below. (Courtesy of Norsat)*

Input Frequency (GHz)	Output Frequency (MHz)	LO Frequency (GHz)
11.70 - 12.20	950 - 1450	10.75
0.95 - 11.75	950 - 1700	10.00
12.50 - 12.75	1025 - 1275	11.475
12.25 - 12.75	950 - 1450	11.30

transistor technology. Without such progress, satellite broadcasting would not exist as we know it today.

Amplifiers used in the early days of radio astronomy were based on standard parametric transistor circuits that were immersed in baths of very low temperature liquid nitrogen or helium. This technique dramatically reduced noise levels by slowing down molecular motion, the source of background noise. The development of the gallium arsenide field effect transistor, known as a GaAsFET, made the modern "low noise" amplifier a reality. These special transistors make an amplifier operate as if it were cooled to near absolute zero where all molecular motion ceases. GaAsFETs, a miracle of modern engineering, are at the core of mass marketed satellite reception systems.

Today there are two types of low noise transistors available to low noise amplifier designers and manufacturers. These are HEMTs and MESFETs, acronyms for high electron mobility transistor and metal semiconductor field effect transistor, respectively. MESFETs have been used since the early 1970s while HEMTs have been commercially available only since 1987. The basic difference between the two devices is that the HEMT performs better at higher frequencies, namely has better noise figure and gain than the MESFET. The comparative performance of HEMT versus MESFET transistors in circuit designs at C and Ku-band frequencies are examined later in this section.

LNAs, LNBs and LNCs

Low noise block downconverters, LNBs, evolved from low noise amplifier designs. In earlier C-band satellite reception systems, the signal was first amplified by an LNA and then lowered in frequency by a separate device called a downconverter. A voltage sent from the satellite receiver then created a mixing frequency that served to select one channel from the C-band range of frequencies, 3.7 to 4.2 GHz, for downconversion and then for relay indoors.

Low noise converters (LNCs) were developed so that both functions of an LNA and downconverter could be combined in one unit. This device both amplified and lowered the frequency of one selected channel to a range that could be processed by a satellite receiver.

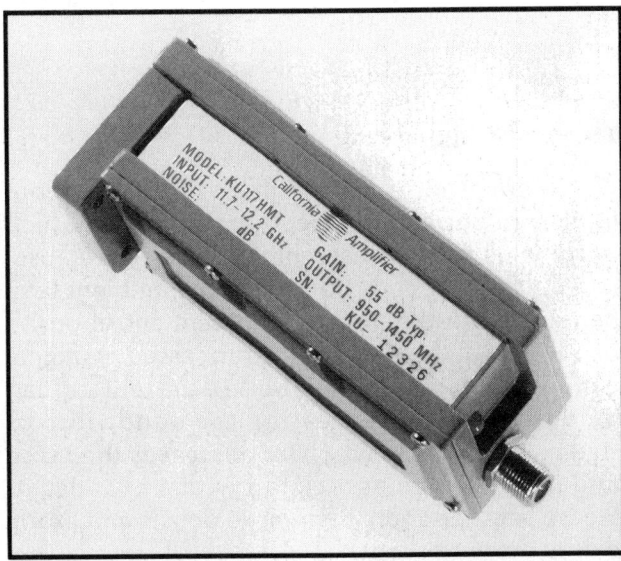

Figure 2-73 CalAmp LNB. *This Ku-band LNB operates in the 11.7 to 12.2 GHz range and has a block output of 950 to 1450 MHz. (Courtesy of California Amplifier)*

COMPONENT OPERATION

Block downconverters were designed to allow the entire 3.7 to 4.2 GHz block of frequencies to be lowered simultaneously to an intermediate range, typically 950 to 1450 MHz having the same 500 MHz bandwidth as the original C-band range of frequencies. Employing the block downconversion scheme eliminated the need to send channel selection voltages to devices mounted outdoors at the antenna and therefore avoided the inherent problems of tuning stability associated with operating electronic components in an environment subjected to temperature and humidity changes or water ingress. Using this electronic design has other advantages that are described later in this section.

LNBs were created to combine the best of both worlds, amplification and block downconversion in one unit.

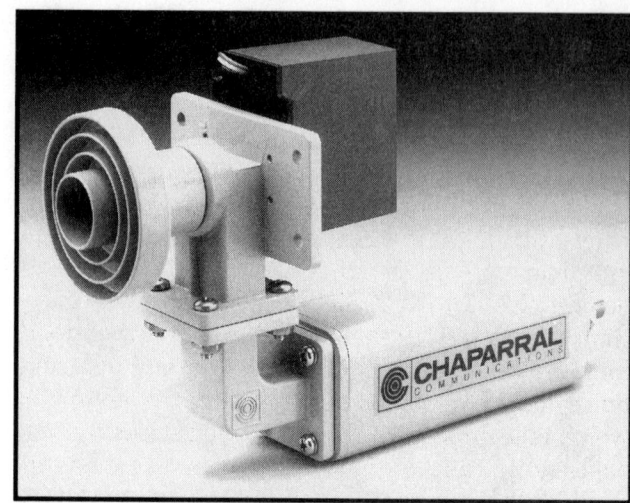

Figure 2-74. Chaparral Feed and LNB. *This HEMT Chaparral LNB is available in 10.95 to 11.7 GHz and 11.7 to 12.7 GHz models having maximum VSWRs of 3.5 and 3.0, respectively. (Courtesy of Chaparral Communications)*

Sources of Noise

An understanding of electronic noise is critical in a study of satellite communications because the very low power signals transmitted are just a step above the noise generated within electronic systems and from the surrounding environment. Noise is caused by molecular and atomic motions which are essentially electrical currents that create electromagnetic waves. Some of this radiation is in the same frequency band as satellite transmissions.

There are two main types of noise in electronic systems: thermal noise and noise associated with the operation of transistors..

Thermal Noise

Thermal noise arises from the physical random motion of conduction electrons and holes which occur in all materials at temperatures above absolute zero. It depends both upon the ambient temperature inside electronic equipment and upon the bandwidth of the signal being processed. As signal bandwidth increases, more noise is present and can be detected. So decreasing the bandwidth of signals entering an amplifier decreases the noise and interference associated with a broadcast. Please see Appendix B for more details and examples of these concepts.

Interference is also considered a form of noise even though it usually originates from some man-made communication device used for a purpose other than satellite broadcasting. However, in those cases where an antenna has an overly wide field of view caused by excessively high side lobes or too wide a beamwidth, interference may originate from a satellite adjacent to the one being targeted. Higher quality LNAs and LNBs are less susceptible to random internal noise but, as a result, are more capable of seeing organised interfering signals from either non- targeted satellites or other man-made sources.

Transistor Noise

Noise within transistors is classified as either shot noise or partition noise. Shot noise is caused by the quantized nature of current flowing across the PN junction in transistors. Like thermal noise, it also depends upon the signal bandwidth. Partition noise is similar to shot noise in being caused by quantized jumps in current. In this case, it is caused by variations in current between the emitter and collector in a transistor.

GaAsFET transistors are inherently less noisy than conventional bipolar transistors. These are therefore used as the "front end" of an LNB.

COMPONENT OPERATION

Noise Temperature and Noise Figure

The noise characteristic of an LNA, LNB, or any other electronic device for that matter, is described either in terms of noise temperature or noise figure. They yardsticks are interchangeable. The noise figure of an amplifier is defined as the ratio of the signal-to-noise ratio at the input of an amplifier to the signal-to-noise ratio at its output (see Appendix B for more details and examples). In other words, the noise figure of an LNA is a measure of the degree by which this amplifier degrades or decreases the signal-to-noise ratio of the satellite signal as it passes through the device. Noise temperature, the other measure of noise is expressed in degrees Kelvin. This scale is based on the fact that at a temperature known as absolute zero, 0°K (equals to minus 273.16 °C or minus 459.72 °F), molecular motion ceases and consequently noise disappears. One of the fundamental principles of thermodynamics states that there is no temperature colder than absolute zero.

It is important to realise that noise figure should be considered separately from gain in system calculations. Once noise has been added to the signal, any subsequent gain will not change the signal-to-noise ratio because both the signal and noise are amplified by the same amount.

TABLE 2-7. NOISE TEMPERATURE AND NOISE FIGURE

Noise Temperature (°K)	Noise Figure (dB)
60	0.819
70	0.942
80	1.061
90	1.177
100	1.291
120	1.508
140	1.711
160	1.908
180	2.097
200	2.278
220	2.452
240	2.619
260	2.780
280	2.935
300	3.085

LNAs for use in C-band systems have typically been rated according to noise temperature while Ku-band LNBs are often described by either noise temperature or noise figure. Some values of both parameters are shown in Table 2-7. Details of the conversion between these two parameters are outlined in Appendix C.

How does substituting an LNB having a lower noise temperature affect system performance? While more detailed system performance calculations are outlined below and in Chapter III, a simple method to calculate this change can be seen in the following example. If a 180°K (2.097 dB) LNB were replaced by one having a noise temperature of 80°K (1.061 dB) in a system which has an antenna that adds 40°K of noise, the ratio of signal to noise power would improve by the relative decrease in noise, namely:

$$(180 + 40)/(80 + 40) = 2.63$$

This factor of 1.42 or 42 percent performance improvement is equivalent to a 2.63 dB change, as calculated by 10 log(1.83). It is interesting to realise that an equivalent 2.63 dB increase in system performance can also be achieved by downsizing from a 1.5 to a 1.0 metre antenna (assuming antenna efficiency is 70% – see Table 2-2). This type of rapid calculation is very useful for comparison purposes. (A computer program that calculates the effects of varying system parameters such as noise temperature has been created to aid the reader in designing satellite reception systems –see the order form for details).

Judging LNB Performance

The central parameter used to judge LNB performance and quality is its noise temperature or figure, assuming that its gain is sufficient and that other important design issues such as waterproofing and lightning protection have been adequately addressed.

Noise Temperature

The LNB is the "front end" of a satellite TV reception system. The noise it adds to the incoming signal sets the noise floor and plays a large part in determining picture quality. LNBs operating in the Ku- band range are now available at reasonable prices and have noise figures in the 1.8 to 0.8 dB range, equal to noise temperatures ranging from

COMPONENT OPERATION

149 to 59°K. The recent introduction of HEMTs (high electron mobility transistors which have extremely narrow gallium arsenide gap dimensions) is ushering in an era of even lower priced, lower noise temperature LNBs. Units having noise figures approaching 0.6 dB, equal to noise temperatures of 43°K, are now available. Prices and noise temperatures for Ku-band LNBs are clearly following the trend set by their C-band counterparts as production volumes increase and technical experience is gained.

It is fascinating to examine how rapidly LNA/LNB technology has progressed in recent years. Commercial GaAsFET low noise amplifiers became available in the late 1970s. At that time, respectable noise temperatures were 220°K while these units had price tags in excess of $3,500. These LNAs were used primarily with antennas having diameters greater than five metres. However, the rapid growth in the home satellite industry changed everything. LNA performance was gradually improved and prices began to decline rapidly. In mid-1985, an 85°K LNA purchased directly from the manufacturer sold for under $100. By 1986, the debate of "isolated versus non-isolated LNAs" was over and the home satellite industry welcomed the improved performance of the non-isolated LNA. During this time, when Ku-band GaAsFETs were introduced, C-band noise temperatures as low as 40°K became common. Now with the recent development of "HEMT" devices, LNB sold at reasonable prices have noise temperatures approaching 20°K.

Improved noise performance of LNAs and LNBs has brought benefits to many end-users living in difficult reception areas. For example, in regions where daily temperatures regularly soar to above 100 °F, products are now available having enough noise figure margin to overcome the associated performance degradation. In weak footprint regions of the world, ultra-low noise LNBs can greatly enhance reception of low power satellite signals. In particular, in those areas that both receive weak downlinked power and are subject to a hot climate, ultra-low noise performance is a must. Note that the amount of noise contributed by an LNB can vary from the rated value which is usually measured at room temperature. An LNB operating at high noon in a desert with a container temperature of 140 °F, generates substantially more noise than at 68 °F. Most manufacturers shield their products from such temperature extremes by using a rubber or enamel paint.

A system designer must understand that LNB noise temperatures vary across the design frequency band. All manufacturers specify the maximum average noise temperature or figure over the design frequency range while some provide these measurements at three or more frequencies on an attached plate or product literature. The overall noise temperature or figure rating should reflect the highest measurement over the range in question. This frequency band varies and might be 10.95 to 11.70 GHz for a TVRO operating in Spain while it can be 11.7 to 12.2 GHz for one in the United States or 12.25 to 12.75 GHz for one in Australia.

Gain

LNB gains today usually range from 60 to 70 dB. As discussed in the following section on LNA design, each GaAsFET and subsequent parametric amplifier stage in an LNB typically contributes between 10 and 12 dB of gain. In addition, downconverter output is usually amplified.

Gain is also very important in characterizing low noise amplifiers. LNAs have been designed with sufficient gain to overcome cable losses as well as the effects of noise contributed within the block downconversion stage. The motivation to produce LNAs with at least 50 dB of gain stemmed from needs of earlier TVROs. A typical 100-foot run of RG213 cable had approximately 13 dB loss at 4 GHz. Connectors contributed another 2 dB of loss.

In modern TVROs, LNB gain must exceed a minimum value. Past a certain point adding extra gain will do little to improve system performance. Tests have shown that this point occurs somewhere between 50 and 60 dB, but 50 dB should be considered a minimum gain, especially when using an under-sized antenna or unduly long cable runs.

Like noise figure, gain also varies with ambient temperature and signal frequency. As temperature increases gain decreases, typically by about 0.6 dB for each 10 °C temperature rise. This variation is of little significance except for operation in extremely hostile environments such as outer space. Small variations in gain also occur across the Ku-band frequency range. A perfectly "linear" LNB would amplify every frequency equally. In general, the variations are small enough to be negligible. Large deviations from linearity which could cause an LNB to become unstable and oscillate are possible at extremely low ambient temperatures below –40 °C.

A perfect LNB should amplify signals only in the designed bandwidth and reject those with higher or lower frequencies. In practice, gain drops off rapidly but measurably at frequencies outside this band. Out-of-band rejection is most important when there is a possibility that other communicators are using similar or adjacent frequency ranges which could interfere with clear reception. While this can be a serious concern for C-band satellite systems, it has not generally been a major factor in Ku-band TVROs. This is slowly changing in some locations as the Ku-band is increasingly being used by other communicators.

VSWR

LNBs, like feedhorns, are rated according to voltage standing wave ratio, VSWR. This ratio, which is a measure of how efficiently the input signal is collected, is found by dividing the power of the input signal by that of the signal actually entering the LNB. When no signal is reflected back towards the feedhorn in a perfect device, the ratio is 1 to 1. VSWRs should be less than 1.3 to 1 in good quality devices.

Electrical Connections

The electrical connections to an LNB are simple. Most LNBs have a F-connector output that is connected to the satellite receiver by a single run of coaxial cable. Note that some LNBs used in European systems still feature N-connectors. Power and signal are relayed via the same cable.

Mechanical Packages

Techniques for packaging LNBs have evolved rapidly. Early designs were large and bulky; some devices weighed nearly 3 kilograms. However, pressures to lower price and improve performance have led to the creation of smaller, more cost-efficient designs. Today, smaller, lighter LNBs are being designed to fit on hybrid C/Ku feeds. This is especially important when combination C/Ku feeds which can be rather bulky or when orthomode feeds supporting two LNBs are used. In these cases, technicians have often installed guy wires to support and centre the feedhorn/LNB assembly.

The electronic components of an LNB must be enclosed in a hermetically sealed box. Water vapor has a very destructive effect on the operation of printed circuit boards and other electronic devices and is carefully avoided by watertight encasing.

The Future of LNBs

Modern C-band LNBs are approaching their theoretical minimum noise figure limits, even with the use of HEMT devices. Noise temperatures as low as 20°K can now be attained with the use of HEMT devices. While thermoelectric or cryogenic cooling can also improve performance beyond 20°K, this technology is cost-prohibitive in the home TVRO market. However, unlike the C-band situation, noise performance at Ku-band continues to improve. HEMT manufacturers are forecasting that a reasonably priced Ku-band LNB with a noise figure below 1.0 dB should soon be available.

LNB Performance Calculations

More detailed LNB performance calculations are outlined in this section. Note that these are based largely upon an article entitled "Advances in Low Noise Amplifier Technology" by Jim Harris, President of Gardiner Communications that appeared in TVRO Technology in 1989. Since this type of calculation is further explored in Chapter III.

Importance of the LNA

Following selection of an antenna with adequate gain, the LNA is the next most important component of the TVRO system. The factor describing the quality of an antenna/LNA receiving system is known a G/T_{sys}. The defining equation for G/T_{sys} is:

$$G/T_{sys} = G_a - 10\log[T_a + (L_f - 1)T_o + L_f T_{lna}]$$

where
L_f = $\log^{-1}[(\text{loss from antenna to LNA in dB})/10]$
T_o = 290°K
T_{lna} = LNA noise temperature
T_a = Antenna noise temperature
G_a = Gain of antenna

This equation clearly shows that antenna gain is most crucial to the TVRO performance, followed by the insertion loss of the feedhorn and the noise figure/gain of the LNA.

COMPONENT OPERATION

Transition from LNAs to LNA/BDCs to LNBs

Earlier TVRO systems typically consisted of a fiberglass antenna, a 50 dB low noise amplifier, 15 to 50 metres of high frequency RG 213 cable and a single-channel receiver operating at a 70 MHz IF. As the home satellite industry grew, price versus performance considerations became very important. This led to a re-design of receivers where the downconverter which lowered frequencies to 70 MHz was separated as an integral component and was located at the antenna. This innovative design permitted very expensive RG 213 cable to be replaced with more economical RG 59 coax. Although some LNAs produced at that time had 40 dB gains, market demand remained strong for 50 dB gain LNAs. When block downconversion was introduced, demand pressured manufacturers to produce an integrated LNB exceeding the price/performance characteristics of a separate LNA and downconverter.

It is interesting to examine the relationship between noise figure and gain as the separate LNA and downconverter evolved into an LNB. The following equations, which are derived in Appendix B, are presented to aid in converting from degrees Kelvin to noise figure (NF) [equation A], NF to degrees Kelvin [equation B], and overall noise figure [equation C]:

$$T = 290 [\log^{-1}(NF/10) - 1] \quad \text{A}$$

$$NF = 10 \log [(T/290) + 1] \quad \text{B}$$

$$F_{12} = F_1 + [(F_2 - 1)/G_1] \quad \text{C}$$

where T = Noise temperature in °K
  NF = Noise figure in dB
  F_{12} = Overall noise figure
  F_1 = Noise figure of first device (LNA)
  F_2 = Noise figure of second device (BDC)
  G_1 = Gain of first device (LNA GAIN)

NOTE: Overall noise figure is expressed as a ratio, not in decibels.

As an example, examine the overall noise figure and gain of an 85°K LNA with a 50 dB gain, connected to a 12 dB gain block downconverter (BDC) with a 15.0 dB noise figure.

	LNA	BDC	LNA/BDC
NF(dB)/T(°K)	1.12/85	15.0/8880	1.12/85
Gain (dB)	50.0	12.0	62.0

Then, using the previous equations, F_{12} can be calculated as follows:

$$F_1 = \log^{-1}(NF/10)$$
$$= \log^{-1}(1.12/10)$$
$$= 1.294$$

$$F_2 = \log^{-1}(NF/10)$$
$$= \log^{-1}(15/10)$$
$$= 31.623$$

$$G_1 = \log^{-1}(\text{gain dB}/10)$$
$$= \log^{-1}(50/10)$$
$$= 100,00$$

$$F_{12} = 1.294 + [(31.623 - 1)/100,000]$$
$$= 1.294 + .0003$$
$$= 1.2943$$

now converting from power ratio to decibels using the definition of 10 log (F_{12}), we calculate that F_{12} equals a noise figure of 1.12 dB.

Clearly, the noise figure of the BDC had little or no effect on overall noise figure performance. This explains why the 50 dB LNA was so popular.

Next, examine an integrated LNB assuming that its noise figure and gain are as follows:

	LNA	BDC	LNB
NF(dB)/T(°K)	1.12/85	15.0/8880	1.18/90
Gain (dB)	32.0	30.0	62.0

The results show that with a lower LNA gain of 32 dB, the second stage contribution to overall noise figure increases. The BDC contributes 5°K to the noise temperature of the integrated LNB. To maintain the same performance of a 50 dB LNA and BDC, one would have to start with an 80°K/32 dB gain amplifier in the LNB. Current LNBs are designed to have 32 to 35 dB gain in the low noise amplifier. This relationship between noise figure and gain is central to the discussion of HEMT and MESFET transistors below.

LNB Design/Performance Considerations

In this section, the use of HEMT and MESFET transistors in circuit designs at C and Ku-band frequencies is analysed and compared.

C-Band LNB

In order to examine the performance differences between HEMT and MESFET transistors, a circuit with each one of the devices is analysed. First, consider a 50°K/65 dB gain LNB utilizing MESFETs. The noise figure of the BDC is again chosen to be 15.0 dB.

	MESFET		
	LNA	BDC	LNB
NF(dB)/T(°K)	0.65/47	15.0/8880	0.69/50
Gain (dB)	35.0	30.0	65.0
	HEMP		
	LNA	BDC	LNB
NF(dB)/T(°K)	0.63/45	15.0/8880	0.65/47
Gain (dB)	37.5	27.5	65.0

Assuming a 0.1 dB insertion loss along the path to the amplifier, the first transistor in the MESFET circuit would require a 0.49 dB NF with 11.5 dB gain to produce a 50°K LNB. Next, if the MESFET is exchanged for a HEMT with the same noise figure, the net effect of the HEMT should be evident. The HEMT has 14.0 dB gain, so the gain in the BDC is reduced to 27.5 dB. Obviously, the HEMT device produces a lower noise figure LNB. The additional gain in the first stage of the amplifier proves to make the difference.

An exercise applied to the above circuits is to reduce the noise figure of the BDC to 12.0 dB. Using equation C on the previous page, the results indicate that the HEMT LNB decreases in noise temperature by one degree Kelvin but the MESFET LNB decreases by two degrees. This shows that there is usually more than one way to improve circuit performance. Improving performance is not just "dropping in" the latest "hot" device and boasting of improved performance. It can be concluded from the prior discussion that to build an ultra-low noise performance system that approaches 30°K, one must use transistors with the lowest possible noise figure and highest gain. Also, attention must be paid to the noise figure/gain aspects of each part of the circuit. However, in order to build C-band LNBs with an average noise performance of 45 to 65°K, HEMTs have no real advantage over the GaAs MESFETs.

Ku-band LNB

Early TVRO Ku-band LNBs were manufactured with GaAs MESFETs. Noise performance was generally in the range of 2.0 to 2.5 dB. "early entry DBS" provided the requirement for these LNBs in the United States. However, this market did not meet its projections and soon there was an excess of Ku-band LNBs. MESFET manufacturers had geared up for a market that did not materialise. As a result, these surplus devices found their way into C-band LNBs.

Good quality MESFETS manufactured today have noise figures of 1.2 to 1.4 dB and Ku-band gains of 8.5 dB compared to 1.1 to 1.4 dB noise figures and 10.5 dB gains for HEMT devices. Comparing designs at Ku- band using HEMTs and MESFETs makes it quite apparent how important it is to have sufficient LNA gain to overcome the noise figure of the BDC. Again, examine the noise figure/gain of a circuit using HEMT and MESFET transistors.

Assume that (1) the Ku-band LNB has a noise figure of 1.8 dB and a 55 dB gain, (2) the noise figure of the BDC is 10.5 dB and (3) the MESFET used in the first stage of the LNA has a 1.3 dB NF and 8.5 dB gain. Assume that the HEMT has the same 1.3 dB noise figure, but a gain of 10.5 dB.

	MESFET		
	LNA	BDC	LNB
NF(dB)/T(°K)	1.72/141	10.5/2964	1.77/146
Gain (dB)	28.0	27.0	55.0
	HEMT		
	LNA	BDC	LNB
NF(dB)/T(°K)	1.62/131	10.5/2964	1.65/134
Gain (dB)	30.0	25.0	55.0

With the noise figure of the two devices equal, the HEMT out-performs the MESFET by 0.12 dB in overall noise figure. This improvement is due to the extra gain in the HEMT device.

In conclusion, HEMT and MESFET transistors both have their places in low noise amplifier circuits. But as is evident from analyses of both C and Ku-band LNBs, HEMT devices have an advantage over MESFET devices in high performance, ultra-low noise circuits.

G. DOWNCONVERSION METHODS

Pioneer satellite earth stations relayed the high frequency microwave signal that had been amplified by an LNA to an indoor receiver where it was was downconverted or lowered in frequency. Relaying such high frequencies required the use of expensive coaxial cable that was characterized by high signal losses. This design constraint mandated that that cabling distances had to be minimised. Therefore, antennas were mounted as closely as possible to the satellite receiver and, in some cases, expensive line amplifiers costing thousands of dollars had to be employed. In addition, installing such high frequency cable was difficult since fitting connectors required expertise and the coax had to have a maximum turning radius and so could not be sharply bent. All in all, early TVRO installations had severe limitations and were quite expensive.

The radical design improvement was to separate the satellite receiver into two components: a downconverter located near the antenna; and the remaining electronics. Then just a short length of cable rated for higher frequencies could be used to relay signals between the LNA and downconverter, which was mounted at the antenna. This less expensive, lower frequency cable could be used to transmit signals more than 100 metres without the use of line amplifiers.

Downconversion and Satellite Receiver Design

The design and operation of Ku-band downconverters and receivers is more clearly understood by examining the somewhat simpler C-band downconversion techniques. Three distinct schemes have been used: single, dual and block downconversion (see Figure 2-75).

The design of C-band downconverters is intimately related to the method used to select individual channels on satellite broadcasts. Each of the 24 downlinked channels typically occupies a 36 MHz bandwidth, although Intelsat spacecraft often use half transponder, 18 MHz bandwidths. Any channel can be selected by translating the 36 MHz band down to one centred at the final intermediate frequency (IF). The majority of C-band receivers use a 70 MHz IF. Thus, for example, if channel 10 having a centre frequency of 3900 MHz had been selected, the downconverter would lower this 3882 to 3918 MHz band to one spanning 52 to 88 MHz and centred on 70 MHz. Both these bands are 36 MHz wide and contain exactly the same audio, video or data information. Similarly, if channel 22 which is centred on 4140 MHz is selected, this 36 MHz band is again lowered to one centred on 70 MHz. Any other channel would be tuned with the same method.

Downconversion is accomplished by a process called heterodyning or mixing. To illustrate, if a 1000 MHz signal is mixed with one of 900 MHz, the resulting sum and difference frequencies of 1900 and 100 MHz are created. If a filter is used to remove the higher frequencies, the effect has been to lower the 1000 MHz signal to one having a frequency of 100 MHz. If "intelligence" had been superimposed or modulated onto this 1000 MHz signal, the final IF of 100 MHz would still contain the original information.

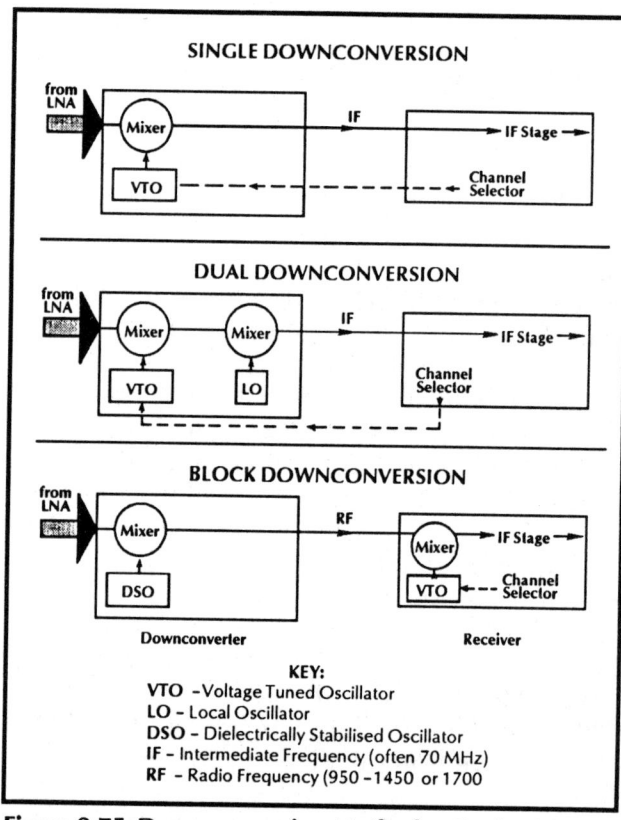

Figure 2-75. Downconversion Methods. *Single, dual and block downconversion methods all accomplish the same end result, converting the extremely high microwave signal to a lower frequency IF and selecting one of the many channels to be viewed.*

COMPONENT OPERATION

The tuner of a satellite receiver selects channels by sending a voltage to a voltage tuned oscillator (VTO) which produces the desired mixing frequency. In single or dual downconversion units the VTO is located in the outdoor downconverter. In block downconversion systems, when either an LNA/block downconverter or an LNB is used, the variable oscillator is a component of the satellite receiver and the block downconverter oscillator is set to a fixed frequency. The details of these processes are described below.

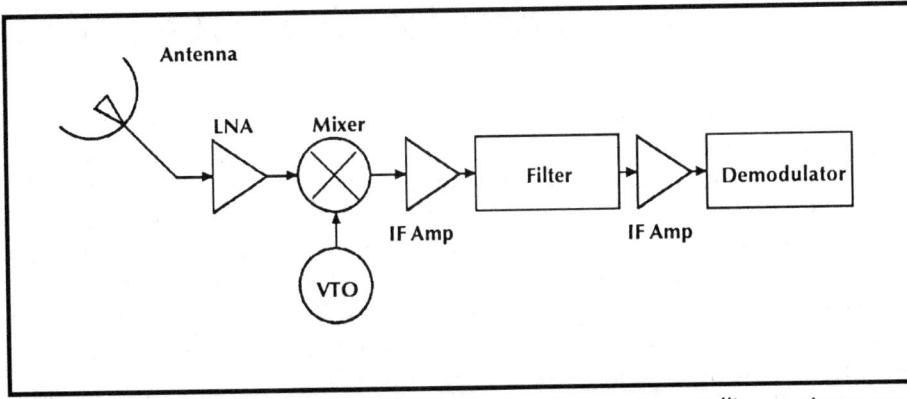

Figure 2-76. Single Downconversion. *Single downconversion satellite receivers were the most widely used in C-band home TVRO systems. These use a voltage tuned oscillator (VTO) to convert the incoming signal directly to an IF such as 70 MHz for transmission indoors. Single downconversion receivers have serious performance limitations when used in a multiple receiver installation because each receiver must have its own downconverter. This requires the use of very expensive microwave splitters and ferrite isolators to avoid interference between the two downconverters.*

Single Downconversion

Single downconverters lower the chosen satellite channel to the final 70 MHz IF in one step (see Figure 2-76). Rotating or pressing the tuner control knob or button selects a voltage to be relayed to the VTO in the downconverter at the antenna. A mixing frequency 70 MHz lower or higher than the centre frequency of the desired satellite channel is produced so that, after mixing, the resultant difference frequency is 70 MHz. For example, if channel 10 having a centre frequency of 3900 MHz is selected, the mixing frequency is set at 3830 MHz and the difference and sum frequencies of 70 and 7730 MHz are produced. The higher frequency is filtered out, leaving the desired 70 MHz IF. If channel 17 having a centre frequency of 4040 MHz is chosen, the VTO generates a mixing frequency of 3970 MHz.

Downconverters can either use high or low-side injection. The previous two examples are of low-side injection. If channel 10 were selected using high-side injection, 3970 MHz would have been chosen to mix with the 3900 MHz centre frequency. When low-side injection is employed, the highest frequency in the signal block appears as the highest frequency in the downconverted block. However, with high-side injection, the highest frequency in the signal block appears as the lowest frequency in the downconverted block and vice versa. This inversion is the reason why video inversion switches have been incorporated on the backs panel of most C-band satellite receivers. (One of the earliest forms of scrambling used was signal inversion.)

Single downconversion systems have a potential problem because the oscillator frequency is 70 MHz above or below the desired channel centre frequency. This signal falls within the C-band range and can leak back into the system to cause interference on nearby channels. (Only channel 24 using high-side injection and channel 1 using low-side injection are immune to this difficulty.) Single downconversion C-band satellite receivers use circuits called image rejection mixers to counteract this effect. Devices called ferrite isolators are also used on the downconverter input to block the exit of unwanted signals into adjacent downconverters.

Dual Downconversion

Dual downconverters attain the final IF in two stages (see Figure 2-77). Both the first and second mixing occurs outdoors in the downconverter. Often 810 MHz is used as an intermediate frequency. For example, if channel 10 having a 3900 MHz centre frequency is chosen, the VTO mixes it with a signal of 3090 MHz, equal to 3900 less 810 MHz. Then the intermediate 810 MHz is mixed with a local oscillator signal of 740 MHz to downconvert the signal to the final 70 MHz IF to be transmitted to the receiver.

COMPONENT OPERATION

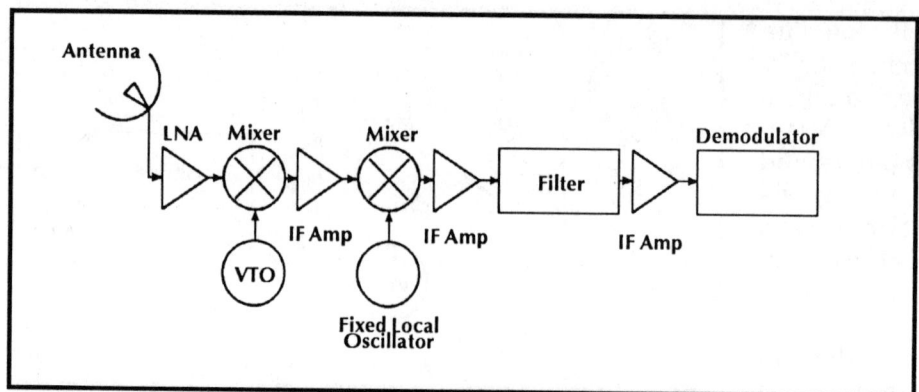

Figure 2-77. Dual Downconversion. *Downconversion to a final IF can be accomplished in two steps. This eliminates potential interference from the internal local osciallator whose signal falls outside the received band of frequencies. This is a useful feature when two adjacent downconverters are installed on the same antenna in multiple receiver configurations. Dual conversion systems have added costs and complexity because an extra stage of local oscillators, mixers, amplifiers and filters is required. When only a single receiver is used, this additional complexity is unnecessary.*

Dual downconversion systems use an oscillator frequency well outside the C-band and there is no concern with interference and the need for image rejection circuitry and ferrite isolators is eliminated.

Block Downconversion

Block downconverters use a fixed frequency local oscillator to lower the whole 500 MHz satellite band to an intermediate range (see Figure 2-78). Until the development of dielectric resonant oscillators (DROs, also known as DSOs, dielectric stabilized oscillators) this was not practical because the frequency produced was not sufficiently stable. By contrast, DROs are stable to within 10 MHz, and can be stable to nearly 1 MHz, over the whole satellite band. Two frequency ranges have been predominantly used, 950 to 1450 MHz, and 440 to 940 MHz, although the former higher range is rapidly becoming the accepted standard. Both have 500 MHz bandwidths and contain all the original intelligence. To attain the higher frequency range, a high-side injection local oscillator running at a fixed 5150 MHz is used in C-band systems. This 500 MHz block is relayed to the satellite receiver where it is mixed with a signal from an internal VTO to select channels by lowering the single channel band to one centred on the final IF. Recall that European broadcast satellites transmit a signal occupying a 750 MHz band so that the block downconversion range becomes 950 to 1700 MHz.

Block downconversion systems have advantages over single and dual downconversion systems. Two or more receivers can independently select channels because each one using its internal mixer and VTO receives the entire downconverted block which contains all the channels on any given satellite. Both single and dual downconversion systems relay only one channel indoors at any time. Block downconversion satellite receivers also are less subject to drifting away from a target channel because channel selection occurs indoors where electronic components are protected from large temperature and humidity swings.

Both of the intermediate frequency ranges have advantages and disadvantages. The lower one, 440 to 940 MHz, was originally selected because it allowed use of off-the-shelf, low cost UHF TV components such as amplifiers and splitters. However, UHF transmitters, two-way, mobile cellular radios and cable TV systems all share this frequency band so that the possibility for interference with the satellite signal exists. Using the higher frequency 950 to 1450 MHz range avoids such potential difficulties but more expensive components having higher cabling losses are required.

Block downconversion can be accomplished by either an LNB or an LNA/block downconverter combination. The only difference is that the LNB combines both functions in one box.

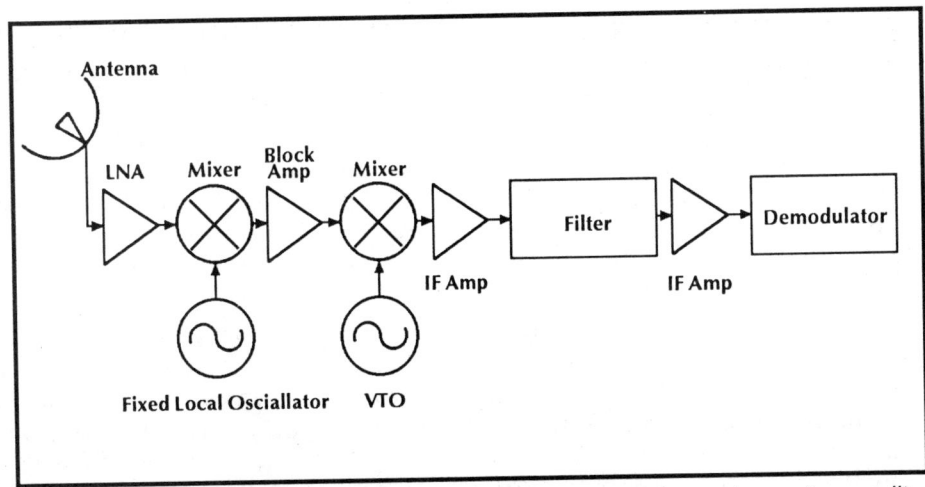

Figure 2-78. Block Downconversion. *This system downconverts the entire satellite bandwidth to an intermediate range, typically from 950 to 1450 MHz or from all or a portion of the 950 to 1700 MHz range. To accomplish this, a fixed local oscillator is located either in an LNB or a separate downconverter at the antenna. Channels are selected by a second downconverter in the satellite receiver. This system provides excellent interference-free performance on all channels and is commonly used for multiple receiver installations. It is the standard in Ku-band TVROs.*

Downconversion in Ku-band Systems

Ku-band LNBs generally use a low-side injection, fixed oscillator to generate typically either a 950 to 1450 MHz or a 950 to 1700 MHz band. However, unlike C-band American satellites which have a conventionally accepted 36 MHz channel bandwidth with 4 MHz guard regions, there is no standard for channel frequencies and bandwidths used in Ku-band broadcasts (see Chapter I for more details).

In Ku-band satellite transmissions, the bandwidth and frequency centres of the entire broadcast as well as individual channels vary. For example, fixed and broadcast service satellites in India, Asia and the Pacific regions have assigned downlink frequencies of 11.70 to 12.75 GHz. Australian DBS satellites use a range of 12.25 to 12.75 GHz. An LNB set for this 500 MHz bandwidth would require an 11.30 GHz low-side injection oscillator to downconvert to a 950 to 1450 MHz band. Direct broadcast services in Europe require an LNB tuned to 10.95 to 11.70 GHz and a satellite receiver with a 700 MHz bandwidth. Fixed satellite services in North America require a low-side oscillator running at 10.75 MHz to downconvert the 11.7 to 12.2 GHz band to 950 to 1450 MHz. In the latter case conventional C-band block downconversion satellite receivers would generally have the capability to process the signal. However, to complicate the issue, some Ku-band LNBs do not conform to this 950 to 1450 industry standard. One popular brand uses a 930 to 1430 MHz range. Also, North American Ku-band satellites may use 43, 54 or even 72 MHz transponder bandwidths with a variety of spacing formats between channels. Even more variable schemes are used by Intelsat and other domestic satellite networks.

Ku-band components must therefore be purchased with the intended use clearly understood. LNBs designed for earth stations in one country are often not interchangeable with those for another application or country.

Downconversion Technical Background

The techniques used for downconversion at C-band are more numerous than those in use at Ku-band because more methods can be employed at lower frequencies. As a result, experimenters have created more devices for C-band than for higher frequency Ku-band systems. Some of the techniques described here were originated as experimental solutions rather than as commercial designs developed in more sophisticated company laboratories.

COMPONENT OPERATION

Many of the original C-band experimental designs mirrored the then standard commercial practice of downconverting a single channel. Standard intermediate frequencies for commercial applications were 70 MHz and 134 MHz. As discussed above, the 70 MHz IF became popular with experimenters primarily because lower cost phase lock loop (PLL) oscillators were available at this frequency. The main design concern, apart from oscillator stability, was image frequency rejection. As the image frequency of the channel selected was only 140 MHz away, just twice the IF, some designs had problems with interference.

Microwave Integrated Circuits (MICs)

In Europe, many of the early downconversion and tuning designs used American microwave integrated circuits (MICs). MICs are self-contained circuits that are designed for higher frequency operation. They are small circuit boards encased in a metal container. Elements such as capacitors, inductors or resistors can be created by conductor tracks having varying patterns, thickness and width. Phase shifting can be accomplished by varying the track length; resistance can be decreased by increasing the width. The most common of these were integrated Schottky barrier diode ring mixers and voltage tuned oscillators. Two of the better-known are the Taylor Howard and the Clyde Washburn receiver designs.

The Taylor Howard design used dual downconversion to improve performance. This technique was described in a manual also included a bipolar silicon LNA design. The amplified C-band signal was fed to a Varil DBM 500 mixer which downconverted it to an 800 MHz first IF. This first IF signal was amplified and fed to a bandpass filter fabricated with microstrip techniques. It was tuned using some UHF trimmer capacitors. The filter output was amplified and fed to an Engelmann Microwave MLP102A mixer where it was downconverted to the second IF of 70 MHz. The voltage tuned oscillator used for the C-band conversion was the Avantek VTO8240 and the VTO used to downconvert the 800 MHz signal was the Avantek VTO8290. The C-band oscillator was used for tuning the channel. The second VTO was preset to convert the 800 MHz signal down to 70 MHz. Automatic frequency control voltage was applied to the C-band oscillator.

While the Clyde Washburn design used a downconversion process similar to the Taylor Howard circuit, it was more commercially oriented than the latter. The C-band mixer was a Varil DBM 500 and the second mixer was a Hewlett Packard HP 5082-9200. The C-band VTO was a Watkins-Johnson V-802. The second and fixed oscillator was an Avantek VTO 8060. The first IF filter was an off-the shelf model K&L Microwave 4B120-820-54 PP. This design change simplified the downconverter setup procedure and made the unit easier to fabricate. The original Washburn design was dated 1980 but yielded a performance that was ahead of its time.

The cost of MICs was initially high and has not yet substantially fallen. Except for some pioneer experimental designs, their use has been limited to high quality home or commercial receivers. In contrast, domestic receivers tended towards more economical solutions.

3 dB Hybrid Mixer

The simplest and the easiest unit to fabricate was the 3 dB hybrid mixer, essentially a balanced mixer and sometimes incorrectly called the RETRACE mixer (see Figure 3-79). This device was widely used by satellite television experimenters due to its ease of construction. The theory of balanced mixers is covered in most textbooks on telecommunications and is not reiterated here. The mixer is basically a 3 dB hybrid with two Schottky diodes. The local oscillator is fed in through one port and the signal through the other. The isolation of this mixer is high and the bandwidth is typically fifty percent of the centre frequency. Its principal disadvantage, like that of a diode mixer, is its relatively high losses. The 3 dB mixer must therefore often be preceded by a low noise amplifier.

The Image Rejection Mixer

The image rejection mixer, developed by David Barker and used commercially in the United States, is an elegant solution to the image frequency problem (see Figure 3-80). It is a single stage conversion mixer that converts directly from C-band to the 70 MHz IF.

The incoming signal from the LNA is split into two phases by a 3 dB hybrid used as a phase splitter. The two signals are then fed to identical 3 dB hybrid

DOWNCONVERSION METHODS

mixers driven by the same oscillator. The outputs from the two mixers are combined either directly or via a transformer. The direct method uses an LC (inductor/capacitor) arrangement on the output of one of the mixers to permit cancellation to be optimised. The required signal should theoretically be in-phase and reinforce whereas the unwanted image frequency signals should be 180 degrees out-of-phase and therefore cancel. The image rejection varies from 20 to 30 dB.

This method was commonly used in Europe as an experimental design. It had high losses and required use of an LNA for proper operation. The amount of microwave substrate (microwave PCB material) required made the design expensive. The most reliable experimental design of this type of mixer was published in Television magazine by Hugh Cox.

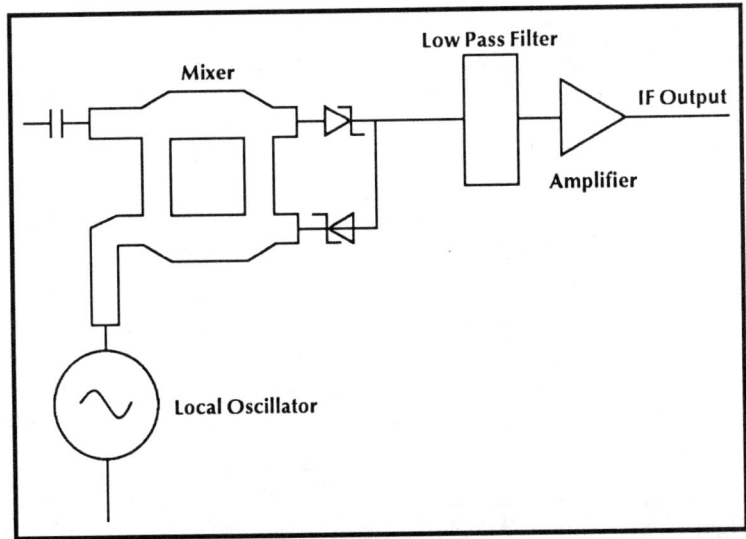

Figure 2-79. 3 dB Hybrid Mixer. *The 3 dB hybrid mixer forms a balanced mixer with the aid of the two Schottky diodes. The local osciallator input is well isolated from the signal input. The local oscillator frequency is set at the normal frequency.*

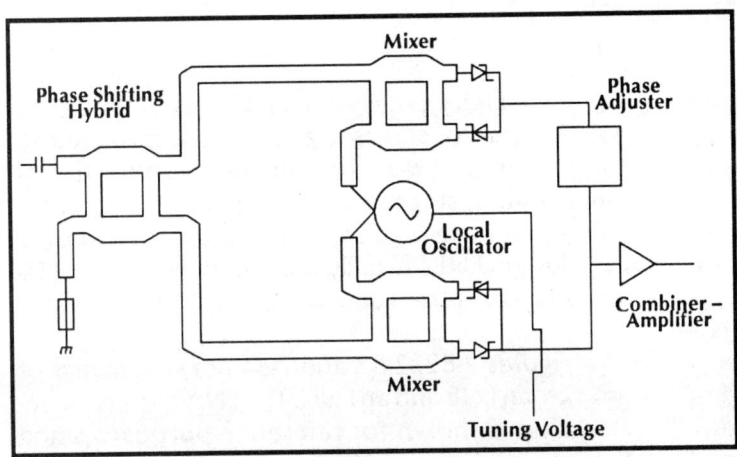

Figure 2-80. Image Rejection Mixer. *The image rejection mixer uses phase shifting to cancel image signals. The circuit can be implemented on microstrip so that there is little setup required. Losses in the mixing process necessitate strong preamplification.*

The Antiparallel Diode Mixer

The antiparallel diode mixer has been widely used in amateur radio applications (see Figure 2-81). Its principal advantage is that it employs an oscillator operating at half the normal frequency. The most popular version of this mixer, which was developed by Hugh Cox in the early eighties, was marketed both as a do-it-yourself kit and as part of a ready-built system. The system which included an LNA and receiver and sold for under 500 pounds Sterling, was in itself a breakthrough because at the time there was virtually no competition that could match its price.

The antiparallel diode mixer consists of two pieces of microstrip. One of these tracks is grounded at one end. The other track is an open circuit. The two pieces of track are joined by two antiparallel Schottky diodes. Antiparallel means that the diodes were mounted in parallel but oppositely polarised, i.e. one was mounted anode-cathode and the other was mounted cathode-anode.

The open circuited track appears as an open-circuited half wavelength resonator at the signal frequency. The IF is taken from the centre point on this track. The other track appears as a short-circuited quarter wavelength resonator at the oscillator fre-

COMPONENT OPERATION

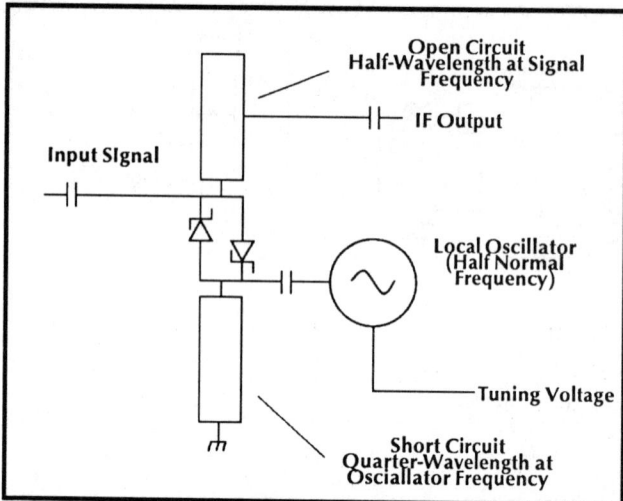

Figure 2-81. Antiparallel Diode Mixer. *The anti-parallel diode mixer uses an oscillator set at half the normal local oscillator frequency. This originally made it a good choice for C-band operation as ordinary UHF tuner electronics could be used to fabricate the oscillator.*

quency. The mixer has good isolation between the RF port and the local oscillator (LO) port. The LO mixer level was typically 0 dBmV.

Single Conversion

Single conversion from C-band to 70 MHz was used by experimenters but rarely in commercial applications due to problems with image frequency interference and oscillator stability. The 3 dB hybrid mixer was the most common converter employed to accomplish this task.

While some early domestic receivers used single conversion, they gradually lost popularity as lower cost, C-band techniques were perfected. These inexpensive methods allowed use of the superior dual conversion methods that significantly reduced the problem of image frequency response.

H. LNB DESIGN

LNBs for Ku-band satellite receiving systems use either a circular or rectangular type of input flange and waveguide. The rectangular variety, known either as a WR75 in America or a WG-17 in the U.K., is evolving into the standard while the circular type is becoming less common. Occasionally, screw-driven adjustments are built into the waveguide portion of an LNB to allow slight changes in its internal dimensions. These permit the precisely manufactured LNB to be even further optimised to maximum performance on sophisticated equipment. Such tuning minimises the amount of signal lost to reflection from the LNB input.

Each LNB typically consists of two or three GaAsFET transistor stages organized in a cascaded arrangement, two or three conventional amplification stages and the downconversion circuitry. A voltage regulator is also included in the circuit design. LNBs usually draw between 80 to 150 milliamps of current and operate at 15 to 24 Vdc.

Figures 2-82, 2-83 and 2-84 are schematics of an LNA, an LNB and an LNC. The LNA and LNC construction is shown for reference purposes since they are rarely encounter anymore.

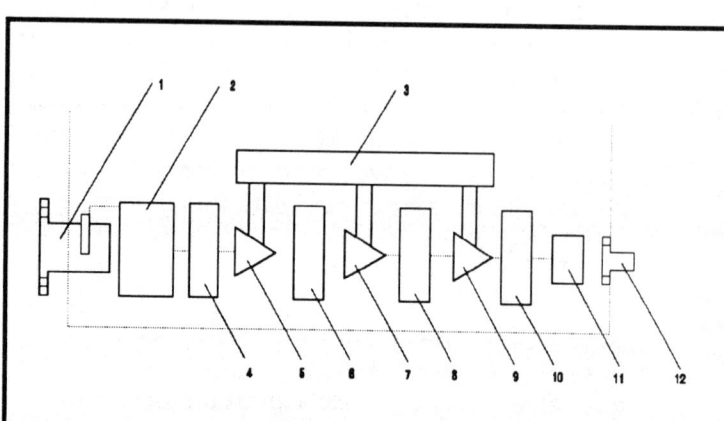

Figure 2-82. LNA Schematic. *The microwave signal entering the WR75 waveguide is detected by the small probe and transferred into the LNA where it enters several stages of GaAsFET transistors. Each stage contributes about 12 dB of gain as well as noise to the signal for a total output of 48 db.*
 1 - Waveguide to microstrip transition
 2 - Isolator
 3 - GaAsFET power supply
 4, 6, 8 & 10 - Impedance matching circuit
 5, 7, & 9 - 1st, 2nd & 3rd GaAsFET amplifier stage
 11 - dc block
 12 - output connector

COMPONENT OPERATION

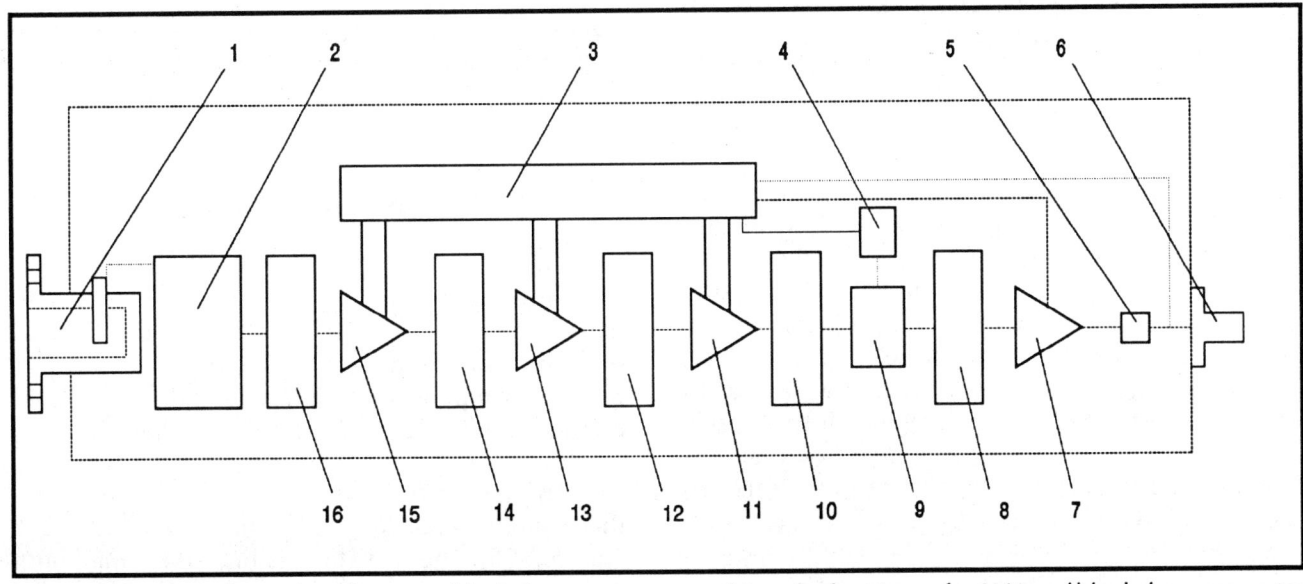

Figure 2-83. LNB Schematic. *A low noise block downconverter combines the functions of an LNA and block downconverter in one case. The LNB has a single female F-connector output which feeds its output over coaxial cable to the satellite receiver.*

    1 - Waveguide to microstrip transition   7 - Block IF amplifier
    2 - Isolator   8 - Block IF filter
    3 - Power supply   9 - Mixer
    4 - Dielectric resonant oscillator   10 - Bandpass filter
    5 - dc block   15, 13 & 11 - 1st, 2nd & 3rd GaAsFET amplifier stages
    6 - Connector   12, 14 & 16 - Impedance matching circuits

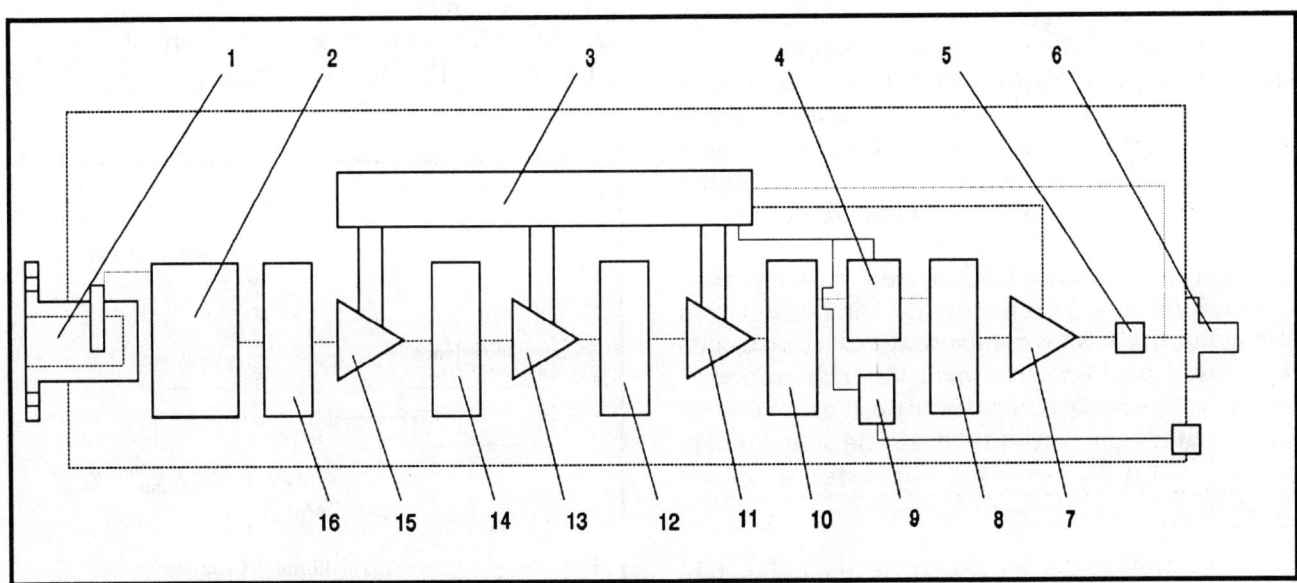

Figure 2-84. LNC Schematic. *A low noise converter, essentially a one channel tunable converter, combines the functions of an LNA and a single conversion downconverter into one casting. This configuration is less expensive to manufacture than a separate LNA and downconverter, but only one receiver can be used per antenna because just one channel centred on 70 MHz is transmitted indoors. This configuration is not capable of driving more than one receiver at a time.*

    1 - Waveguide to microstrip transition   7 - 70 Mhz IF amplifier
    2 - Isolator   8 - 70 MHz IF filter
    3 - Power supply   9 - Voltage tuned oscillator
    4 - Mixer   10 - Bandpass filter
    5 - dc block   15, 13 & 11 - 1st, 2nd & 3rd GaAsFET amplifier stages
    6 - Output connector   12, 14 & 16 - Impedance matching circuits

COMPONENT OPERATION

Isolator

The isolator, found in earlier more expensive LNBs, is now rarely used. It was employed to ensure that signals would travel into the LNB and not be reflected. To accomplish this objective, most isolators have losses of about 0.5 dB in the forward and about 30 dB in the reverse direction. By eliminating reflections that would change the VSWR and subsequently the impedance match designers were attempting to ensure circuit stability. In the worst case, a change of the impedance at the input of an LNB amplifier could turn the amplifier into an oscillator. While this was often seen in home-made C-band amplifiers, it was relatively rare in experimental Ku-band LNBs.

Waveguide To Microstrip Transition

In the waveguide to microstrip transition the microwave signal is transferred from the input waveguide and converted into an electrical form on the printed circuit track in the downconverter. While numerous designs to accomplish this transition are possible, the two most common are the probe and the wedge. The most familiar of the former is known as the monopole or pin probe. This probe is placed approximately a quarter of a wavelength from the closed or short circuited end of the waveguide. With some versions, the probe is stepped to provide a better match between the waveguide and microstrip track impedances.

Some LNBs use a "DC-shorted" probe to prevent high voltages from nearby lightning strikes from frying internal circuit components. Of course, a direct strike would destroy any LNB. The probe has been set at precisely the optimal position for impedance matching in order to maximise signal reception and should never be readjusted even if it appears to be slightly bent.

The wedge type of transition is far easier to fabricate as it can be cut or punched from the same material as the waveguide. The wedge, when fitted into the waveguide divides this element in two. Therefore, the wavelength of the incoming signal is larger than the cutoff wavelength of the two waveguide sections created by the wedge. This forces the signal to travel down the slope of the wedge to the microstrip track. The accompanying diagrams illustrate both types of transition. The wedge transition was used in the earlier FOUP 11KF Mitsubishi radar module. Experimenters have widely used this device in a modified form as a downconverter.

Low Noise Amplifier

The low noise amplifier section is crucial to LNB operation. If not included, the signal would have an extremely low power at the mixer input and the same noise temperature as that of the mixer. In order to compensate, a much larger antenna would be required for clear reception of satellite signals.

The majority of Ku-band LNBs incorporate three stages of GaAsFET low noise amplifiers (see Figure 2-85). The first stage is biased for minimum noise and low gain. The second stage is biased for low noise and slightly higher gain. The third stage is biased for maximum stable gain. Modern GaAsFETs can be used at frequencies up to approximately 30 GHz. The more common silicon transistors cannot be used effectively at the frequencies involved due to their high noise temperatures and low gains.

A GaAsFET amplifier consists of four sections: the input matching network: the GaAsFET transistor, the output matching network and the biasing arrangement. The input matching network matches

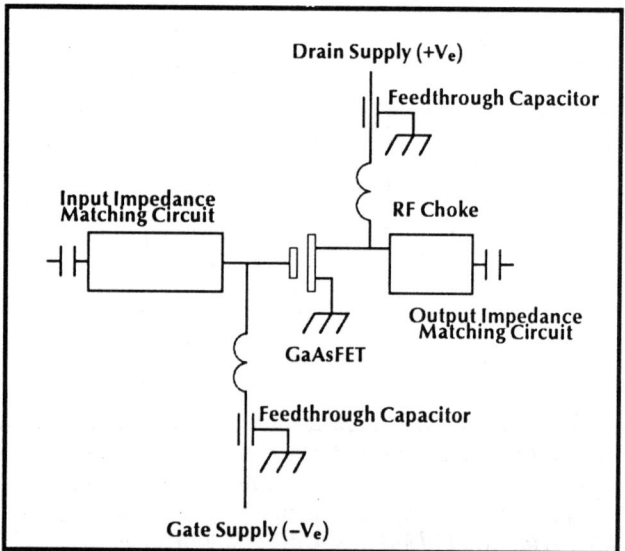

Figure 2-85. GaAsFET Amplifier Schematic. *A GaAsFET amplifier has input and output matching circuits that are fabricated in microstrip. They match the impedance of the signal source to the input of the GaAsFET and the output impedance of the GaAsFET to the output impedance, typically 50 ohms.*

the impedance at the input of the amplifier to the input impedance of the GaAsFET. The next stage, the GaAsFET, is the active component and provides the gain. The output matching network matches the output impedance of the GaAsFET to the output impedance of the amplifier. In circuit design impedances are analysed as consisting of real and complex components. The complex component is generally eliminated by a stub, a piece of short or open-circuited microstrip track that is a precise fraction of a signal wavelength. This causes it to function as a capacitor or inductor that is grounded at one end. The real component is matched to the respective GaAsFET impedance by the use of a quarter wavelength transformer. Depending on the design, the output impedance of the GaAsFET is matched to either 50 ohms or the input impedance of the GaAsFET of the next stage of the amplifier.

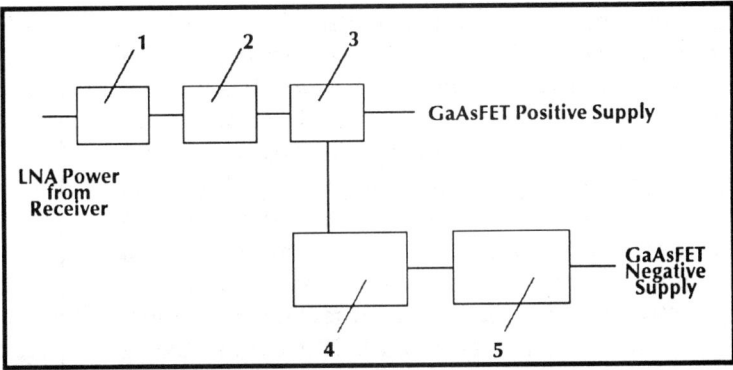

Figure 2-86. Typical LNA Power Supply. *In a typical LNA power supply the main variance between various brands lies in the method of negative rail generation. Some LNAs use a 555 timer chip and a negative voltage doubler while others use a dedicated IC to generate the negative voltage.*
1 - Transient suppresser
2 - 7812 voltage regulator
3 - 7805 voltage regulator
4 - 555 oscillator
5 - Voltage doubler & negative voltage rectifier

The gate and the drain of the GaAsFET are biased via two quarter wavelength inductors of high impedance. These appear as open circuits at signal frequencies. The impedance range for these inductors is between 100 and 200 ohms.

Note that repair of a low noise amplifier should definitely not be attempted without the proper anti-static precautions and equipment. Without correct equipment and knowledge, the end result would probably be additional damage to other circuit GaAsFETs. The symptoms that can be used to identify the faulty stage are detailed later in the chapter on installation and troubleshooting.

GaAsFET Amplifier Power Supply

The GaAsFET amplifier requires a dedicated power supply that must generate approximately +5 and -3 Vdc (see Figure 2-86). While the current drawn by each GaAsFET stage depends on the design, as a rule of thumb, the first stage draws the least current, typically in the region of 10 milliamps or less and the last stage draws the most current. The last GaAsFET stage is generally biased for bulk amplification and the first stage is for the lowest noise. The current drawn by the last stage should not exceed 120 milliamps.

The negative voltage sections of most GaAsFET power supplies are based on oscillator-diode doubler combinations with the NE555 timer often used for the oscillator. The oscillator output feeds the diode doubler. The doubler's output is then fed to a Zener diode regulator network. An alternative approach is to use a dedicated negative voltage generator integrated circuit. This costly approach has been favoured by novice designers.

Bandpass Filter

The signal frequency bandpass filter ensures that only the required block of signal frequencies passes to the mixer. This filter is placed after the amplifier because its relatively large inherent loss precludes it from being located at the input. It is designed so that image frequencies are attenuated by at least 20 dB. While most current LNBs use a microstrip filter, some of the earlier professional models incorporated a waveguide filter. Although the waveguide filter had a cleaner response, it also had a reduced bandwidth, extra losses due to the transitions and was difficult to align. In contrast, the microstrip filter is easier to fabricate and thus is more widely used.

In some more recent LNB designs, the bandpass filter has been incorporated via interstage matching between LNB GaAsFET stages. De-

COMPONENT OPERATION

spite the accuracy of modern printed circuit board production methods, it is still necessary to peak the bandpass/matching sections by hand. The low frequency filter tuning techniques such as varicaps or trimmer capacitors or inductors are not used at microwave frequencies due mainly to their high cost and losses. Bandpass filter and matching networks are designed using microstrip techniques. As a length of PCB track operating at any frequency has inductance and capacitance, capacitors and inductors can be fashioned from PCB track. Peaking these networks of printed inductors and capacitors merely involves removing slivers with a scalpel or soldering on copper foil.

The Mixer

The mixer converts the entire block of frequencies down to a lower range known as the intermediate frequency (IF) block. Of the numerous mixer types in use, the Schottky diode mixer is the simplest. In this device, the signal is coupled to the diode and subsequently a microstrip structure known as a directional coupler mixes some of the local oscillator signal into the diode. A low pass filter then filters the image frequencies from the output. This type of mixer arrangement is used in the FOUP 11KF module from Mitsubishi.

Most of the recent Ku-band LNBs use a single mixing diode. The local oscillator signal is fed through an impedance matching network and then into a bandpass filter. The use of a directional coupler is more common than the direct connection of the DRO (dielectric resonant oscillator). The output from the last stage of the GaAsFET amplifier is fed through a bandpass filter. These two filters are connected to the diode, generally a forward biased Schottky type. The output from the diode travels through a low pass filter and then is fed to the first IF amplifier stage.

To avoid drifting away from the tuned channel, the microwave local oscillator had to be quite stable which implied the use of a microwave integrated circuit (MIC). This MIC was an expensive item and often a thorn in the budget of some of the earlier experimenters. This budgetary constraint resulted in some creative and excellent low cost solutions to achieve frequency stability such as the image rejection mixer and the C-band antiparallel diode mixer. Ultimately, dielectric resonator oscillators were used as a reasonably low cost and stable component.

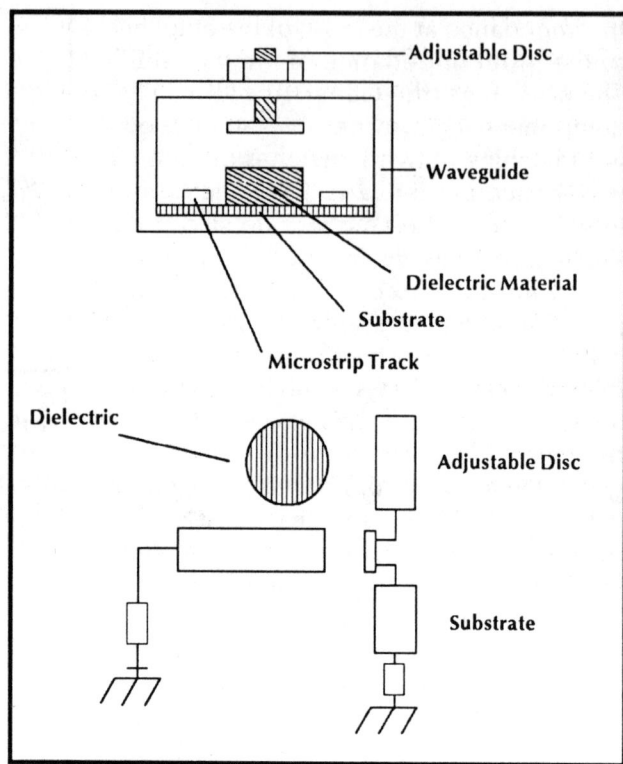

Figure 2-87. Dielectric Resonant Oscillator. *The dielectric resonant oscillator derives its precise frequency from a piece of ceramic material. The higher the frequency required, the smaller the physical dimensions of the ceramic dielectric. These oscillators are extremely stable.*

Dielectric Resonator Oscillator

The local oscillator in virtually every downconverter is a dielectric resonant oscillator (DRO), one of the most stable mass produced microwave oscillators (see Figure 2-87). Some DROs oscillate with less than one megahertz deviation at Ku-band, quite an achievement at such high frequencies. The DRO owes its name to a piece of ceramic or other dielectric material that provides the positive oscillator feedback path. A grounded screw attached to a small metal disc, acting as a capacitor plate, is used to adjust the operating frequency. When the distance between the dielectric and the plate is varied, the capacitance changes.

The DRO operates below Ku-band frequencies. The Ku-band is block downconverted to the 950 to 1750 MHz range, with a DRO frequency of 10 GHz. With the slightly higher 11.7 to 12.5 GHz Ku-band frequencies, a DRO frequency of 10.75 GHz is commonly used. Some manufacturers design their LNBs assuming that the band is split into two 500 MHz blocks; 11.7 to 12.2 GHz and 12.25

to 12.75 GHz. In both cases, the block output is 950 MHz to 1450 MHz. As discussed in the previous section, the technique of using an oscillator frequency lower than the signal frequency is referred to as low-side injection. This results in a translation where the highest frequency in the signal block appears as the highest frequency in the downconverted block.

In C-band LNBs, the DRO operates above the signal frequency resulting in an inversion in the downconverted block so that the highest frequency in the signal block appears as the lowest frequency in the downconverted block. High-side injection is employed because less dielectric material is required for a 5.15 GHz DRO than for a 2.75 GHz DRO. The DRO is therefore less expensive and it occupies less space in the LNB.

IF Amplifier Block

The next stage in an LNB is the intermediate frequency (IF) block amplifier section. This is usually a multistage design preceded by a microstrip bandpass filter used to eliminate image frequencies. There are four types of amplifier designs employed in this section; bipolar, MOSFET, GaAsFET and hybrid amplifier modules. While most LNBs have employed a combination of these techniques, many modern LNBs use GaAsFET stages. The inductors and some of the extremely low value capacitors used in these amplifiers are generally fabricated in microstrip. Therefore, today the power supply is the only LNB section that can be serviced with the bare minimum of specialised knowledge.

I. COAXIAL CABLE AND CONNECTORS

The downconverter output and satellite receiver are connected to each other by a special configuration of conductors called coaxial cable, known as coax. Single wires of copper or aluminium are adequate for conducting electricity at lower frequencies encountered in most familiar electrical devices. However, when higher frequency microwaves are relayed, single strands of metal behave like antennas and can radiate most of the power away before it reaches its destination. With exceptionally high frequency signals that are characteristic of satellite broadcasts, specially designed cable must be used to prevent almost complete loss or attenuation of the transmitted signal.

Coax is composed of two concentric conductors separated by an insulating material called a dielectric (see Figure 2-88). This whole assembly is sheathed in a non-conducting jacket for protection against the elements. The signal travels along the central wire and the external cylindrical conductor is grounded. Using this configuration greatly reduces radiative losses at high frequencies. It is interesting to note that at higher frequencies the signal travels progressively closer to the surface of the centre conductor. If downconversion did not occur in the LNB, extremely expensive coax having a large diameter inner conductor would be required

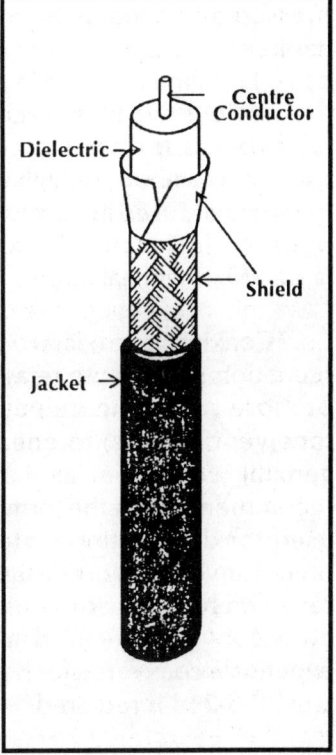

Figure 2-88. Coaxial Cable Construction. *Coax cable is composed of two concentric conductors separated by an insulator called the dielectric. Signals are transmitted on the central conductor while the outer shield is designed to contain the signal and eliminate ingress interference. (Courtesy of M/A COM)*

to transmit the 12 GHz signal since the "skin effect" would be so pronounced.

COMPONENT OPERATION

Types of Coax

A wide range of coaxial cables are available (see Figure 2-89). These fall into three broad categories depending upon the construction of the dielectric sheathing material. These are hardline, conventional coax, and foam or air dielectric coax. Regular coax has one or two pliable metal grounded layers wrapped around a plastic dielectric. Dielectrics in common use include polypropylene, a hard, translucent substance and polyethylene foam, a soft, white material. The outer conductor, which is often a braided copper or aluminium sheath, is occasionally doubled up or replaced with a solid casing to further lower losses and lessen the chance of interference from leakage of unwanted signals into the coax. The degree of cable braiding is rated by what percentage of the dielectric surface is shielded. 67% shielding is a typical value. Completely shielded cables are recommended where high levels of ingress interference from local communicators, whose signals have frequencies similar to those relayed from the LNB to the satellite receiver, may occur.

Foam or air dielectric coax uses foam or compressed air as the dielectric material; it generally has lower loss and is more expensive than ordinary coax. Hardline is similar in construction to either of the previous two types except that it has even lower losses because it has a more rigid, metal sheath and a minimal amount of high quality dielectric. Hardline is usually used in the main trunk lines of cable TV or other similar communication networks such as large SMATV installations.

Coaxial cable has two principal uses in satellite reception systems: to relay the LNB output to one or more satellite receivers and to distribute the receiver output(s) to one or more televisions. In general, a cable similar in characteristics to RG-6 is recommended for the former use. The type of cable used for distribution systems depends upon distances involved and design details. In those very rare cases where an LNA and separate downconverter are used in Ku-band systems, more expensive coax rated for higher frequency use similar to RG-214 is required.

Multi-Run Cables

Coaxial cables are available encased in a common sheath with polariser and actuator control wires. Such a multi-run configuration usually contains two coaxial cables, a 3-wire cable for polariser control and a 5-wire cable for actuator control (see Figures 2-90 and 2-91). Both coaxial cables might be used, for example, in dual-polarity system. Not all the cables are used in every situation.

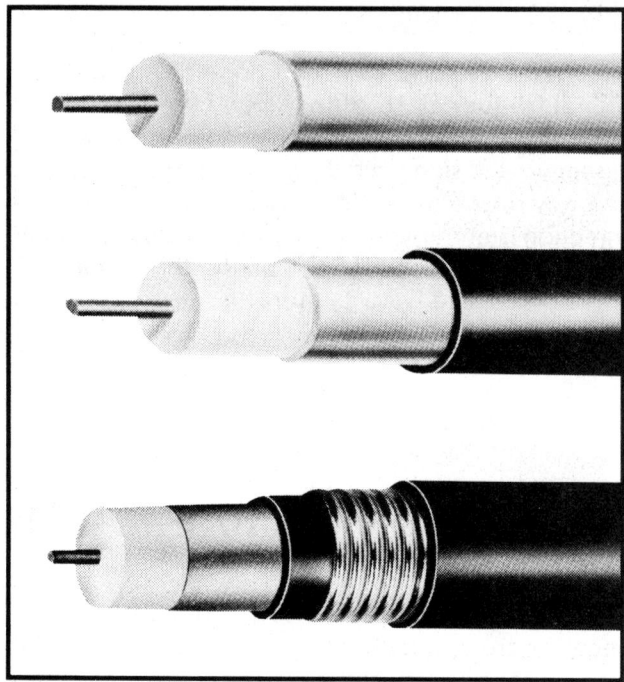

Figure 2-89. Coaxial Cable Types. *These three examples of coaxial cable for high frequency signal conduction are protected from interference to increasing degrees. The middle one has a polyethylene jacket. The lower coax has two metal jackets, one being a corrugated chrome plated steel armor that prevents sharp cable bends. This or any other deformation could cause impedance mismatches and signal attenuation.*

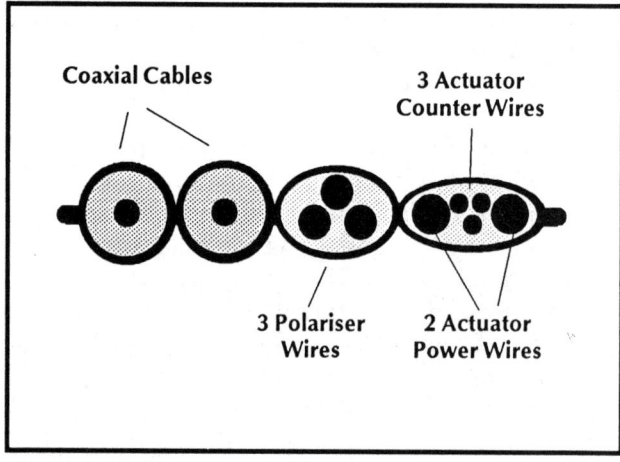

Figure 2-90. Multi-Run Cable. *A standard multi-run cable consists of two coaxial cables, a 3-run cable for polariser control and a 5-wire cable for actuator control. Not all cables are used in all situations.*

CABLE & CONNECTORS

COMPONENT OPERATION

Judging Cable Performance and Use

Coaxial cable is rated according to performance criteria which include characteristic impedance, loss of signal power with distance, and material composition. These factors determine the proper uses of various brands.

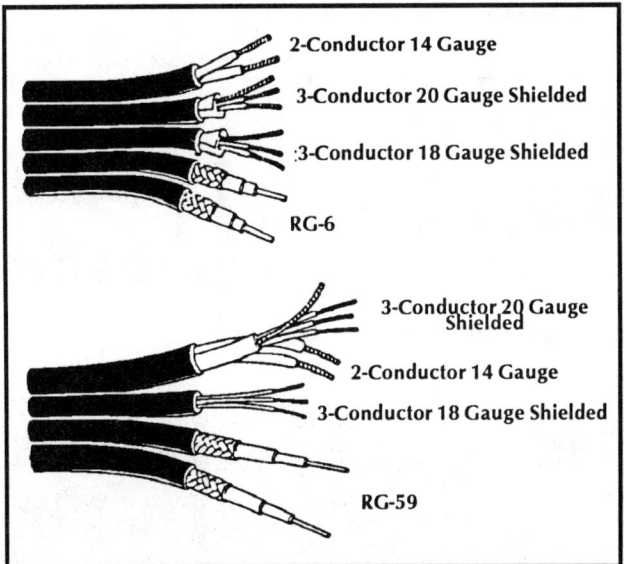

Figure 2-91 Multi-conductor Satellite Antenna Cables. *A ribbon type of flex cable is the commonly used multi-conductor cable in satellite installations. These have all the necessary coaxial cables and voltage carrying conductors to control and receive the signals to and from the satellite antenna. This cable uses dual RG-6 with 67% braid coverage with 20 gauge center conductor with 3-conductor 18-AWG, 3-conductor 20-AWG and 2-conductor 14-AWG wires. All these cables are stranded and shielded with a ground wire. (Courtesy of M/A COM)*

Characteristic Impedance

Every conductor has a specific resistance to current flow which causes power losses. In addition, the interaction between electrical charges on the inner and outer conductors, known technically as capacitance and inductance, causes delays and losses. These factors together determine a value called the characteristic impedance (see Figure 2.92). Cables used in satellite reception and distribution systems are almost always rated at 75 ohms, although a few systems still employ some runs of 50 ohm cable. Higher frequency coax, such as RG-214, has a 50 ohms characteristic impedance.

Knowing the characteristic impedance is crucial in designing electronic systems. Every electronic device also has a characteristic impedance. If the cable impedance does not match that of the device it feeds, substantial reflective power losses, known as return losses, can occur. This mismatch would be similar to sending water from a larger diameter into a smaller diameter pipe and experiencing the resistance to flow that is heard as knocking. Similarly, if the output impedance of an amplifier is not matched to the coaxial line impedance, reflections and the ensuing power losses will occur.

This idea is similar to the concept underlying VSWR, used to characterise internal losses due to signal reflection for both feedhorns and LNBs. If the impedances are matched between the antenna, within the feedhorn components, between the feedhorn and LNB and further downstream to the satellite receiver, losses will be controlled. Any weak point in this chain can seriously degrade performance.

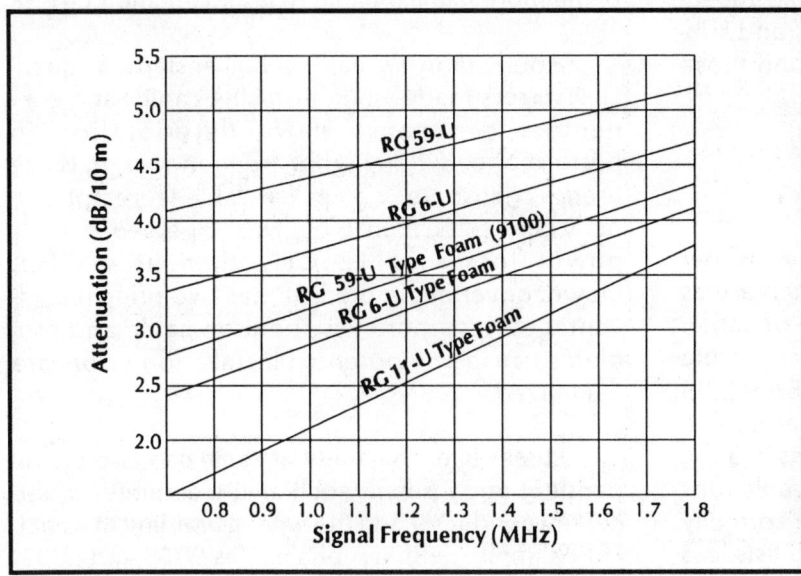

Figure 2-92. Cable Attenuation versus Signal Frequency. *Signal losses per metre of coaxial cable are graphed. Attenuation per metre is measured in dB per 10 m for the range of frequencies typically encountered following the LNB in home satellite systems.*

COMPONENT OPERATION

TABLE 2-8. CHARACTERISTICS OF COMMONLY USED COAX

Cable Type	Signal Loss (dB/10 metres)			Impedance (ohms)
	100 MHz	1450 MHz	4 GHz	
RG-59	1.115	3.609	N/A	75
RG-6A	0.89	2.854	N/A	75
RG-11	0.76	2.297	N/A	75
RG-8	0.62		7.55	50
RG-214	0.76		7.05	50
9913	N/A		3.61	50

TABLE 2-9. LOSSES FOR RG-6 AND RG-11 COAX
(dB per 10 metres)

	Signal Frequency					
	100 MHz	270 MHz	500 MHz	770 MHz	950 MHz	1450 MHz
RG-6	0.89	1.18	0.95	2.13	2.46	2.79
RG-11	0.76	0.79	0.66	1.44	1.71	1.97

TABLE 2-10. COAX LOSSES AND CONSTRUCTION DIFFERENCES
(Belden Coaxial Cables)

Cable Type	Model	Shield Type	Attenuation (dB per 10 m)	
			400 MHz	900 MHz
RG-6	8228	Foil & Wire	1.48	2.26
RG-6	9248	Foil & Copper Braid	1.48	2.26
RG-11	9230	Foil & Wire	1.05	1.71
RG-11	9292	Foil & Copper Braid	1.05	1.71
RG-59	8241	95% Copper	2.33	3.58
RG-59	9275	Foil & Wire	1.77	2.76

When cables are bent too sharply, incorrectly crimped, crushed or otherwise changed in internal dimensions, a local change in characteristic impedance is created. This causes reflections and subsequently signal losses above and beyond those normally encountered.

Signal Losses

Coax is rated according to dB of signal loss per distance the signal travels. Such losses increase as frequency increases for any given type of cable. Therefore, for example, RG-6 which has acceptable losses at 1 GHz is unusable at 4 GHz. RG-6 has 2.85 dB loss per 10 metres at 1450 MHz. Therefore, a cable run of 10.5 metres will result in a loss of 3 dB, a halving of signal power. It is clear that cable runs should be as short as possible and that the correctly rated cable should be installed. Table 2-8 lists loss characteristic of more familiar coax types. Table 2-9 shows a more detailed comparison of losses for two of the more familiar cable types, RG-6 and RG-11.

Attenuation is greater at higher signal frequencies. Errors made when installing cables and connectors can cause more deterioration in performance when higher frequency signals are used. To illustrate, if a coaxial cable were not properly connected there would be 70 percent greater power losses at 12 GHz than at 4 GHz. Downconverting in the LNB has two principal advantages: less money is spent on cable and problems caused by potential installation errors are minimized.

Losses in each variety of cable can also be dependent upon details such as the diameter of the central conductor and the type of braiding material. For example, Belden Alpha cables 9059 and 9803,

CABLE & CONNECTORS

both similar to RG-59, have different diameter central wires. As a result, they have different attenuations at 900 MHz of 35.10 and 33.46 dB per 100 metres, respectively. Table 2-10 shows some examples for Belden manufactured cables.

Installation errors can cause attenuation in excess of rated values. If cables are bent too severely, an impedance mismatch at the sharp bend can occur which will increase reflective losses. To illustrate, RG-8A has a foam dielectric and keeping the centre conductor aligned directly down the middle of the grounded shield is difficult if a wide turning radius is not used, particularly when the coax gets hot. A suggested minimum cable turning radius is 5 cable diameters.

Coax Connectors

The majority of LNBs on the market are attached to coaxial cable by F-connectors, abbreviation for feedthrough connector. The F-connector is generally associated with an LNB output impedance of 75 ohms. This connector which is commonly referred to in Europe as the F584 connector is an American creation that uses the centre of the coaxial cable as the connector pin. Because this wire is easily bent it can potentially cause problems when either the LNB or the receiver is repeatedly disconnected. Nevertheless, there are two distinct advantages to an F-connector: its ease of construction and its ease of removal. It is extremely easy to fit F-connectors to coax, perhaps the reason why this system is preferred by some installers (see Figure 2-93).

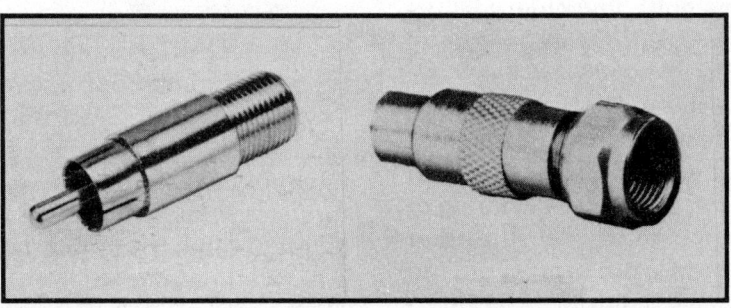

Figure 2-93. Coax Connectors. *F-connectors are commonly used with RG-59, RG-6 or RG-11 coax. RCA male-to-female F-connectors are used to connect an RCA plug to a coax line. RCA female-to-male F- connectors are used to convert coax to an RCA plug. F-81 barrels are occasionally used to splice two F-connector type coaxial cables together. N-connectors, most commonly encountered for attaching to an LNA output and are used on the outputs of some European LNBs.*

COMPONENT OPERATION

The N-connector, used primarily with LNA systems, is generally associated with an output impedance of 50 ohms. This fitting has lower losses and is more mechanically stable though more costly than the F-connector. In addition, it is somewhat more difficult to fit. Some earlier TVROs, installed perhaps by some poorly trained people, suffered from faulty N-connections. N-connector centre pins must be securely soldered to the connector. If this is not properly accomplished, cables usually become disconnected from the connectors.

The F-connector is not ideally suited to the 950 to 1750 MHz European standard IF block. A practice that is becoming more widespread is the use of 75 ohm output impedance LNBs with 50 ohm systems and vice versa. These arrangements do work. The impedance mismatch is apparent only on a spectrum analyser or when the system is operating near threshold. The mismatch is visible as noisy transponders at certain intervals. On a spectrum analyser the mismatch will show up as a ripple throughout the IF block response. While attention to impedance matching is essential in commercial installations, it is not as critical in domestic installations. Therefore it is relatively more common to see 50 ohm LNBs feeding 75 ohm receivers and vice versa.

On many of the more recently introduced European satellite receivers, the tuner module in use is a Mitsumi unit which has an input impedance of 75 ohms. On some of the receivers featuring N-connectors, the input is not 50 but 75 ohms. This fact can be verified by checking the tuner type. The standard is now 75 ohms but a few older, deluxe receivers employ 50 ohm N-connectors.

Power losses can also be substantial where connectors join cables. If these connectors are not installed correctly, impedance mismatches and losses occur. It is important to examine the inside of each connector before mating. This ensures that the centre pin or conductor is not broken off and that it is extended far enough out from the connector to make secure electrical contact but not too far to short out to a chassis and damage circuitry.

Water and Aging

Coax can easily be destroyed by intrusion of water, especially salt water. Also, any leaks at connectors or along the cable body can short out a signal to ground and possibly damage an LNB or receiver.

Underground moisture, which inevitably comes into contact with buried cables, can cause very rapid corrosion of metal components. It is reported that tubular aluminium outer conductors have been almost completely destroyed within 90 days. Even small pinholes in outer jackets can allow this chain of events to occur. Poor installation or cable handling techniques or even the presence of rodents may cause such damage.

Most direct burial cables on the market today are sufficient to do the job. But time will tell whether or not conduit should be used to protect them. Commercial installations employ, when necessary, more expensive coax having a sticky, flooding compound under the jacket which protects against water intrusion. For maximum reliability against rodents, a steel tape armor with over-jacketing or rigid conduit such as gray electrical PVC is suggested.

Summary

To summarize, connectors must be rated to carry the frequencies in use and must be installed to be secure and watertight. In addition, coaxial cable should be carefully selected so that impedances are properly matched and so that the frequency carrying ability is adequate. The distances between the LNB and satellite receiver should be minimised.

The SCART Connector

The SCART connector, also known as the Peritel or Euro-connector, was developed by the French in 1980. (SCART is an abbreviation for Syndicat des Constructeurs d'Appareils Radio Recepteurs et Televiseurs – Syndicated Manufacturers of Radio and Television Equipment). Many different methods to interface audio/video equipment had been used in the past including BNC, phono, 6-pin DIN, 'D' and 'C' connectors. This resulted in substantial compatibility problems between components. The SCART has become the defacto standard and a solution to many of these problems.

The SCART connector consists of 20 spade-shaped contacts fitted into an insulated box and surrounded by a metal skirt that protrudes from a plastic housing. The connecting cable attaches via a threaded collar. The skirt of the connector serves as a ground that is connected the the chassis of the target equipment (see Figure 2-94).

The complementary socket is generally mounted lengthwise in equipment such as televisions or satellite receivers. The SCART has two dimples on one side with corresponding pits in the socket to guide and retain the connector. The receiving socket is usually attached directly onto a printed circuit board.

The SCART connector is now the standard for audio, video and RGB (red, green , blue primary video signals) purposes within European. The pin signal allocation is outlined in Table 2-11.

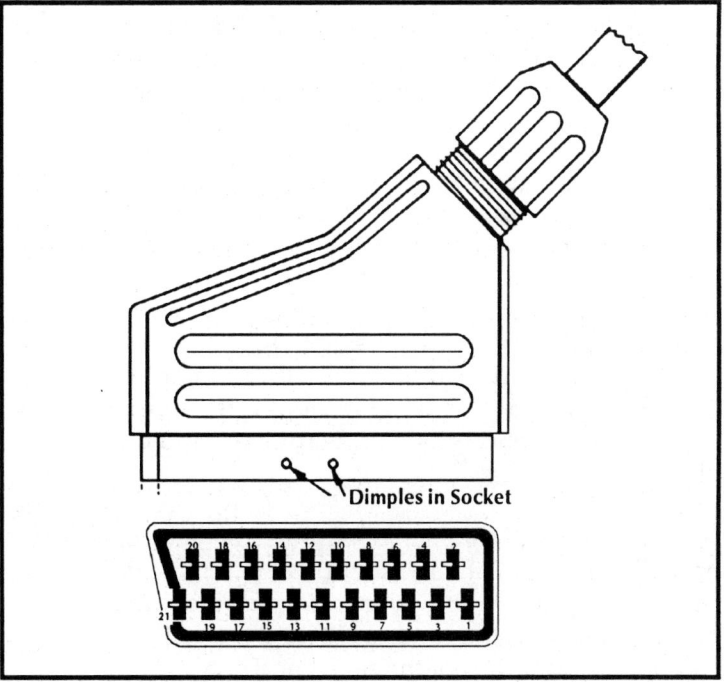

Figure 2-94. The SCART Connector. *This connector features 20 pins in addition to its chassis-connected casing. The dimples serve to align and secure the connector in its socket. Cables are attached via a screw-on clamp.*

Pin Number	Specifications	Signal
\multicolumn{3}{c}{TABLE 2-11. SCART SOCKET CONNECTIONS}		
1	Right Channel Audio Out	0.5 V into 1 K-ohm
2	Right Channel Audio In	0.5 V into 1 K-ohm
3	Left Channel Audio Out	0.5 V into 1 K-ohm
4	Audio Ground	
5	Blue Earth	
6	Left Channel Audio In	0.5 V into 1 K-ohm
7	Blue In	0.7 V into 75 ohms
8	Source Switching	Max 12 V into 10 K-ohms
9	Green Earth	
10	Intercommunication Line	
11	Green In	0.7 V into 75 ohms
12	Intercommunication Line	
13	Red Earth	
14	Intercommunication Line Ground	
15	Red In	0.7 V into 75 ohms
16	Fast RGB Blanking	Variable
17	Composite Video, Blanking and Sync Out	
18	Fast Blanking Earth	
19	Composite Video, Blanking and Sync Out	1 V into 75 ohms
20	Composite Video, Blanking and Sync In	1 V into 75 ohms
21	Socket Ground	

J. SATELLITE RECEIVERS

The function of a satellite receiver is to select a channel for viewing or listening from the available block of information and then to process this signal into a form acceptable by a television, TV monitor or stereo. Gone are the days when satellite receivers were large, clumsy devices whose weight was almost a measure of cost. Modern receivers, which are light and attractively packaged, are the control stations for satellite television reception (see Figure 2-95). Even equipment destined for commercial "headends" is streamlined and efficiently packaged.

As discussed earlier, pioneer satellite receivers had all downconversion functions built-in. When downconverters evolved into separate components installed at the antenna, they were still considered an extension of the receiver. Today, LNBs, even as totally separate devices, must be matched to each receiver's bandwidth and input frequency range.

Figure 2-95. Maspro SRE-90R Receiver. *This unit, designed specifically for the European frequency range of 950 to 1700 MHz, features PLL frequency synthesised tuning, an infrared remote control, 26 user programmable and 24 fixed video channels, 50 user programmable audio channels, parental lockout, 10 favourite channel memory and V/H switch control. It outputs either a SCART decoder interface or a composite baseband for interface with other decoders.*

Receiver Design

A home satellite TV receiver consists of a power supply, downconverter/tuner, final IF stage, discriminator, video and audio processors and a built-in modulator (see Figure 2-96). Most commercial receivers are similar except that they feed their output signal into a stand-alone, commercial modulator in order to improve signal quality.

Modern satellite television receivers owe more to the experimentation carried out on kitchen tables than that conducted in well-equipped commercial design laboratories. In particular, the development of the block downconversion process arose more from necessity than design. While the majority of the early North American experimental work was carried out with systems operating at C-band frequencies, pioneer European development was conducted on 860 MHz signals from the Indian ATS satellite communication experiment.

Power Supply

Power supplies used in satellite receivers are standard components, similar to those found in FM radios and other sound equipment. This device consist of a mains switch, a mains fuse, a transformer, a number of rectifiers, a number of smoothing capacitors and integrated circuit regulators. Occasionally, dc fuses in the 300 milliamp range are included on the LNB supply line.

The electrical operation of a typical power supply is outlined below because the most common and easily repaired breakdown in a satellite receiver occurs in this section. This subject is returned to in more detail in Chapter IX where troubleshooting is explored.

A power supply produces the necessary constant voltage in the following way: AC power from a wall outlet is fed into a step-down transformer. The reduced voltage is then fed through a rectifier circuit to convert the alternating to direct current. The rectifier is nothing more than a series of diodes which allow current to pass in only one direction. The end result is that either the negative or positive voltages are eliminated and the 50 or 60 cycle per second sine wave is converted to a series of either all positive or all negative waves. These ripples are smoothed out by using a series of low pass filters resulting in a relatively constant DC voltage. The filtered output is finally fed into voltage regulators

COMPONENT OPERATION

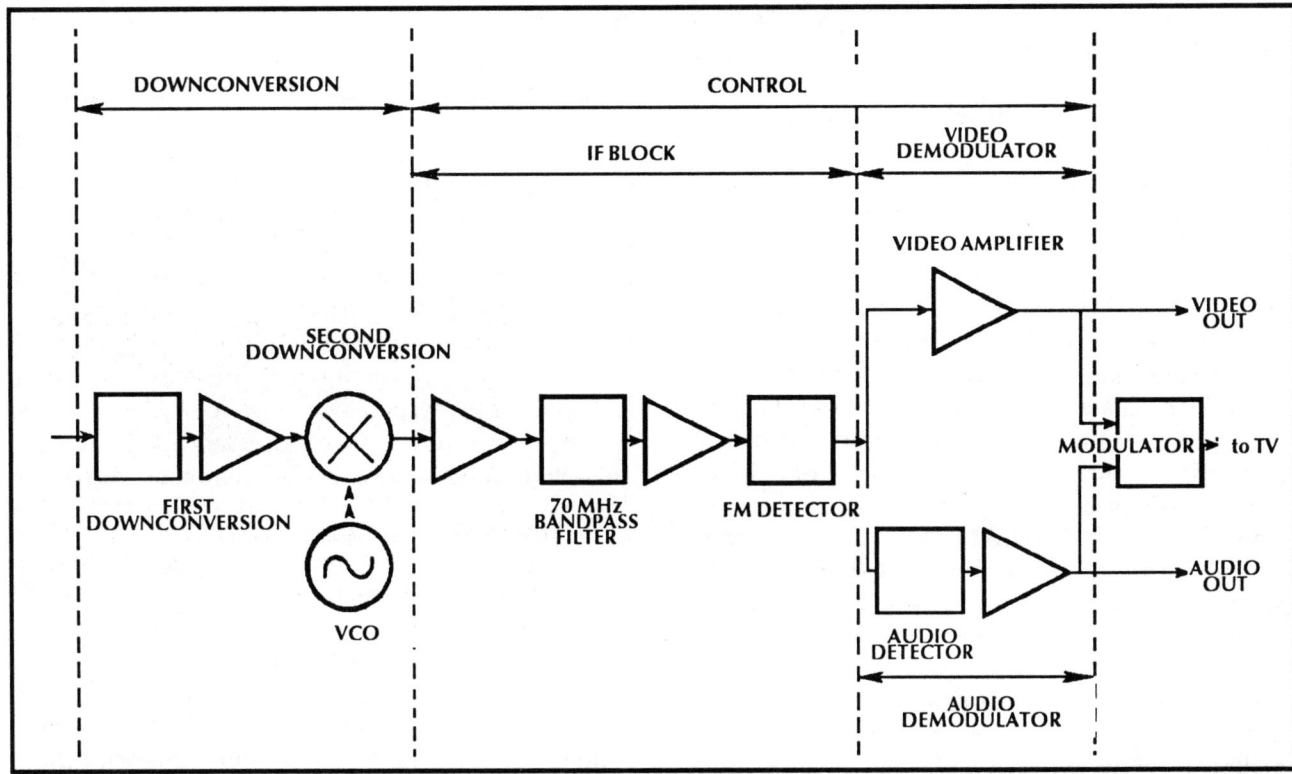

Figure 2-96. Satellite Receiver Block Diagram. *A receiver accepts the IF frequency from the block downconverter and selects one channel by downconverting one portion of the satellite bandwidth to a single IF often of 70, 130, 134 or 612 MHz. This band centered on the IF is then amplified and fed to the detector/demodulator where the FM carrier is removed and the signal is converted to baseband. These signals are than remodulated so that AM video and FM audio can be transmitted to the TV set.*

which provide "clean" power necessary for proper receiver operation.

Voltage regulators are rated at 24, 18, 12 and 5 Vdc. Almost every receiver uses a 12 Vdc regulator because most transistor and integrated circuits are powered by this voltage. Regulators with other voltage ratings are used for a variety of circuit components. LNBs draw about 100 to 200 milliamps at 12 to 28 Vdc while the polariser draws 500 milliamps at 5 or 6 Vdc. These components often have their own built-in regulators. While satellite receivers draw about 1 amp of current, antenna actuators draw up to 6 amps regulated at plus or minus 18 to 36 Vdc.

Downconverter/Tuner

The purpose of the downconverter/tuner stage is to select one channel from the block of frequencies relayed from the LNB and to lower its frequency to the final IF. The core of the tuner is a voltage tuned oscillator (VTO) that mixes its output with the LNB-to-receiver input signal. A signal centred on the IF is then passed on to the receiver's final IF stage.

Choice of IF

The industry standard intermediate frequency of 70 MHz is an arbitrary choice. Other IF frequencies are occasionally used. For example, DX-600 and 700 receivers have a 134 MHz IF, Mitsumi components downconvert to 479 MHz, while Hitachi and Maspro components are designed for 612 and 400 MHz, respectively. Some of the newer receivers being prepared for the Japanese DBS market have an IF of 402.78 MHz.

The 70 MHz standard arose because a circuit called a phase lock loop (PLL) detector, that was required for conventional television receivers, was available when satellite receivers were first developed. These electronic components had originally been used in the telephone microwave industry and operated at a maximum frequency of 35 MHz. The 70 MHz satellite IF was therefore divided by two so that readily available conventional, lower

COMPONENT OPERATION

cost devices could be used. Today, off-the-shelf PLL detectors operate up to as high as 612 MHz. In general, although amplifiers using a lower frequency IF have the advantage of lower noise and improved gain, improvements in video quality have been observed at design IFs above 400 MHz.

Tuner - Selecting Channels

Selecting Ku-band channels is a more complicated matter than is the case with North American C-band broadcasts where centre frequencies and bandwidths are standardised. The design of the tuning section in Ku-band receivers reflects its intended use. For example, in those cases where a given receiver is to be used exclusively to decipher transmissions from just one transponder on one spacecraft, a fixed VTO frequency can be used. Commercial C-band systems are occasionally fixed-channel receivers. Other brands are termed semi-agile whereby channels can be selected by interchanging highly stable, fixed-frequency crystal oscillators. However, today most C-band home receivers and many commercial brands of satellite receivers have evolved to the point that they are stable as well as completely agile. Fortunately the design of Ku-band receivers can be based on experience gained from C-band technology.

Satellite receivers use either rotary tuning, preset or detent tuning, and synthesised tuning to select channels.

Rotary Tuning

Rotary tuned receivers use a front panel-mounted potentiometer configured as a voltage divider to transmit the required voltage to a VTO. One end of its track is connected to the tuning voltage supply, typically 18 Vdc, while the other end is connected to ground. The tuning voltage is taken to the tuning pin on the block tuner via a resistor. In addition, the automatic frequency control (AFC) signal is sometimes fed to the tuning pin on the block tuner. In other receivers, the AFC signal is connected to the AFC input on the IF downconverter module. The Avcom COM-2 C-band satellite is an example of a popular receiver that uses rotary tuning.

The quality of the rotary tuning method depends upon the stability of the potentiometer and the stability of the tuner's VTO. Although most receivers have incorporated AFC circuits to counteract the tendency to drift, this method is a only a partial solution to the underlying problem that results either from poor circuit design or from changes in potentiometer resistance due to temperature variations. Note that such receivers tend to drift most in the first half hour after power up.

While rotary tuning is unsuited to use in commercial receivers, some brands that employ this method incorporate potentiometers of higher quality than those found in the domestic receivers. Such potentiometers typically can be locked in various positions and have more elaborate AFC circuitry than that found in domestic satellite receivers.

Preset and Detent Tuning

Preset tuners select channels via one or more preset resistors or potentiometers. A user then chooses the required channel by means of a switch that selects the tuning voltage from the relevant preset. Since such receivers must be set up before being taken into the field, this method is not a popular option.

Another form of preset tuning, also known as detent tuning, is used in the United States where C-band channels are evenly spaced and follow an alternate pattern with the odd channels being on one polarity and the even channels on the other. This regularity in frequency spacing allows a voltage divider chain to be used to set voltages in selecting these channels. The channel voltages are then chosen via a front panel switch. Two presets are incorporated in the voltage divider chain for optimising the tuning voltages.

Synthesised Tuning

Synthesised tuners, employed in some satellite receivers, create a more stable mixing frequency than that resulting from using resistive dividers. There are two accepted methods to accomplish this result, phase-locked loops (PLL) and voltage-synthesis. The PLL is the most accurate approximation to the 0.005 percent accuracy of a single crystal oscillator. This tuner incorporates a divide-by-256 circuit to divide the VTO oscillator frequency by 256 and lower it to a frequency where it is compared to a very stable crystal oscillator. Any deviation from the correct value causes a DC correction voltage to be sent to readjust the tuner oscillator.

Quartz-locked or voltage-synthesis circuits use a microprocessor to generate a digital signal for each specific channel. This digital signal is converted to an analog voltage to drive the VTO. A more stable mixing signal is generated than with detent tuning methods. Some examples of voltage synthesised C-band receivers are the Houston Tracker V, Luxor 9900 and Chaparral Sierra. The more sophisticated of these brands are really computers that remember which satellites have which centre frequencies and bandwidths. Such high quality synthesised tuners are much less subject to drifting away from the satellite channel centre frequency than are receivers which use detent tuning. It is also important to realise that just because a button is pushed instead of a knob being turned on the front receiver panel to select channels does not necessarily mean that synthesised tuning is being employed.

The key to designing satellite receivers which are capable of automatically processing satellite signals of varying transponder centre frequencies and bandwidths is microprocessor controlled synthesised tuning. This is also the road to producing lower cost receivers since more versatile units will have a much broader market and can be mass produced for use in many countries. Many C-band satellite receivers simply cannot be used with Ku-band broadcasts because either their channel frequencies are fixed as in some synthesised brands or they are not able to off-tune the full 10 MHz from pre-assigned C-band frequencies. This topic is examined in more detail in Chapter VI where retrofitting existing C-band systems to receive Ku-band broadcasts is examined.

The Block Tuner

The first stage of a satellite television receiver is the block tuner. Such a device, operating in a band centred on 860 MHz, was developed during the Indian SITE experiment conducted via the ATS-6 satellite during the 1970s.

Pioneer home satellite television systems used converted UHF television tuners as inexpensive block tuners, a concept originally proposed by Birkill in the American CATV (community antenna television, cable TV) journal, CATJ in 1978. He calculated that an adequate signal could be received at his location with just a 1.5 metre dish, a 2.0 dB LNA and a receiver having a 5 MHz bandwidth. The off-the-shelf UHF television tuner was capable of processing input signals in the 470 to 860 MHz range and generated an output at 35 MHz. This relatively low output frequency allowed use of a standard NE561 phase lock loop circuit. From the published pictures, one can conclude that this technical solution functioned well. Birkill later used the same tuner/PLL concept for reception of weak C-band Intelsat, C-band Gorizont and Ku-band OTS signals.

Satellite television evolved in parallel in North America and Europe. In America, Keith Anderson comprehended the possibilities of Birkill's block downconversion design based upon television tuners. He developed the idea of broadcasting the raw block of 450 to 950 MHz over the air. The receiver at the viewer's location would essentially be a UHF tuner with an FM demodulator. The fact that the transmission format was FM would serve as a low grade form of signal security. The original over-the-air concept specified the broadcasting of signals of both horizontal and vertical polarities. However, due to problems stemming from adjacent channel interference, it was not possible to combine both polarities in the same block of frequencies.

In 1982, Anderson began marketing his 450-950 MHz block downconversion receiver. It was a commercial success and other major manufacturers followed suit. This 450 to 950 MHz block evolved into one of the standards for C-band satellite receiver operation. Subsequently, the American company, Avantek, integrated a block downconverter with an LNA to produce an LNB with an IF block of 940 to 1440 MHz. Another U.S. company, M/A COM, developed an LNB with an IF block of 950 to 1450 MHz. Manufacturers of commercial satellite equipment adopted this block in preference to the lower 450 to 950 MHz block. Within a few years this choice filtered through to the manufacturers of domestic receivers and gradually became the accepted standard.

While the American C-band is 500 MHz wide, the European Ku-band occupies a full 800 MHz and is essentially composed of three separate 250 MHz blocks. While the middle block is unused on most communications satellites such as ECS and Intelsat, ASTRA now transmits signals in this frequency segment.

COMPONENT OPERATION

Early European receivers were converted American models and therefore were limited to processing 950 to 1450 MHz block inputs. Problems related to this limitation in certain brands of receivers were initially solved with a switchable tuner arrangement. This partial fix was later superseded by tuners that were designed specifically for the European market and that covered the entire band from 950 to 1750 MHz.

The Mitsumi Tuner

The most common tuner in use today is that manufactured by Mitsumi. It operates over the full 800 MHz band and features a divide-by-256 prescaler output for frequency synthesis tuning. This tuner has an input impedance of 75 ohms and an output IF of 479 MHz accessible via an F-connector. In some 50-ohm receivers, the Mitsumi tuner is connected to an N-connector socket by an adapter.

The Mitsumi IF Downconverter

The 479 MHZ IF is used directly for demodulation by only a small number of satellite receivers. The majority of brands using the Mitsumi tuner also use the Mitsumi IF converter. The main function of the AFC (automatic frequency control) when applied to this unit, as is the case in some receivers, is to shift the conversion bandwidth. The downconversion bandwidth is wide because it is not properly filtered until the signal is downconverted to the 70 MHz IF. The input to the module is a PCB pin whereas the output is a phono socket. In those cases where terrestrial interference is a problem, a TI filter can be inserted at this point. Note that the receiver circuit diagram should be consulted first because there may be other modifications necessary before insertion of a filter at this point.

Those receivers which use the Mitsumi downconverter and tuner often also feed the 70 MHz signal to a Mitsumi demodulator module. This solution is discussed in more detail below.

Final IF Stage

Once downconversion and amplification has been accomplished, the signal is transmitted to the final IF stage which is composed of a bandpass filter, an amplifier and a limiter. The bandpass filter sets the channel bandwidth at typically 28 MHz or less by selectively eliminating any out-of-band signals. A bandpass filter as narrow as 18 MHz for those satellites relaying "half-transponder" Ku-band broadcasts may be required. An amplifier then restores losses incurred during the downconversion process and strengthens the signal passed to the detector/demodulator. Finally, a limiter converts the signal into a square wave form that can be more easily processed by subsequent satellite receiver stages.

Bandpass Filter

The 70 MHz IF amplifier stage is generally followed by one of three types of bandpass filters; discrete LC, surface acoustic wave (SAWF) or ceramic. Most of the discrete LC filters are variations of a filter originally designed in the United States by Taylor Howard in 1979.

The SAWF is now the most widely used bandpass filter. Although, the ceramic filter was often incorporated into American receivers in the early eighties, its use diminished with the increasing availability of inexpensive SAWFs of reasonable quality. The SAWF has a better response than either of the two other varieties of bandpass filters and has the distinct advantage that no tuning or peaking is necessary.

The bandpass characteristic of the filter, as expressed by its noise bandwidth, has a major effect on signal carrier-to-noise ratio. The noise bandwidth, which is not the same as the 3 dB response of a filter, can be defined as the width of an ideal bandpass characteristic which has the same area and peak value as that under a graph of the actual gain versus frequency response of the filter. However, it is simpler to regard the noise bandwidth as being the same as the 3 dB bandwidth for the purpose of calculating carrier-to-noise ratio.

Although many of the filters manufactured in the Far East use fixed bandwidths of 24 MHz, 27 MHz or 36 MHz, some receivers designed for the international market in the early eighties incorporated an 18 MHz bandwidth for use on the Intelsat half-transponder transmissions. Current European receivers generally allow a user to switch between bandwidths of 27 MHz and 36 MHz.

COMPONENT OPERATION

The ASTRA channel specification is more similar to the American C-band than formats found on any other European television satellites. This factor alone has caused more service call outs than virtually any other problem. Most of the existing multi-satellite receivers in Europe had a nominal minimum bandwidth of 27 MHz. However, when viewing ASTRA, a problem similar to ghosting on terrestrial broadcasts occurs. The video on the selected channel would have a weak background image of the adjacent, cross-polarity channel. The only way to cure the problem has been to replace the ceramic filter with one having the correct bandwidth or to insert a filter between the output of the Mitsumi IF converter and the discrete demodulator. The insertion of the filter seems to be the better option as it can be difficult to obtain a ceramic filter with the correct specification. Note that satellite receivers with ceramic filters have been the worst offenders.

Some receivers on the European market can be switched between narrow and wide bandwidth. The wide bandwidth filter is generally a SAWF and the narrow filter is usually a discrete LC type. In order to modify a satellite receiver to detect ASTRA broadcasts, the LC filter can be adjusted for the more narrow 26 MHz ASTRA bandwidth. A signal generator as well as a sweep generator should ideally be used for this task.

Limiter Circuits

The IF bandpass filter is usually followed by one further stage of amplification before the signal is fed into a limiter circuit. The function of the limiter is to produce a square wave output from a sine wave input because most, if not all, demodulators operate more efficiently with such an input. If the signal entering the demodulator is not properly limited, the video will have sparklies.

There are a number of limiter designs in use. The simplest limiter is a pair of antiparallel diodes: germanium diodes are commonly used though Schottky diodes exhibit a better response. However, this simple diode limiter has a disadvantage in that phase modulation of the signal occurs if its amplitude is too low; thus near threshold performance suffers. Some of the earlier American designs used an overdriven MWA type amplifier to provide limiting. A more common limiter in the receivers of Far Eastern origin is a TTL or ECL logic gate. Usually the gate used is a NAND gate or a NOT (inverter) gate. Another common technique is to divide the 70 MHz signal by two or four via TTL or ECL circuitry. The division circuitry usually acts as an effective limiter because the TTL circuitry is operating above its recommended limit. This prompted some designers to use ECL circuitry for the operation. The ECL logic family is more suited to the high frequency involved though there is little visible improvement. Most of the designs used by experimenters did not use limiters for reasons that are explained below.

Detector/Demodulator

The detector/demodulator circuits process the FM modulated satellite TV signal and strips off the carrier to convert the wideband FM television signal back into the more familiar composite video signal and FM audio subcarrier known as the baseband signal. This has all the original audio and video information contained in a bandwidth of approximately 10 MHz. The baseband signal is used as an input to stereo processors and decoders (but not external modulators or VCRs as explained below).

There are basically three demodulator designs currently in use: the phase locked loop (PLL), the quadrature demodulator and the balanced demodulator. The delay line discriminator was used as a modulator in earlier satellite receivers.

PLL Demodulators

The phase-locked loop (PLL) demodulator, which like the PLL tuner uses a very stable crystal reference signal, is capable of detecting weak signals and of discriminating between the desired signal and interference (see Figure 2-97). However, less expensive PLL circuits can exhibit a tendency to produce slightly fuzzy video. In more extreme cases, bright colours or high contrast scenes will tear or streak and the picture may flicker.

The PLL was the most common demodulator in use up to about 1984. The integrated circuit used for this function was almost invariably the NE564 which had a specified upper operating limit of 60 MHz. One of the earliest designs, published by Taylor Howard, used this IC running at 70 MHz. At that time, most satellite television receivers were also based on this design. However, there was a slight problem with this IC as some of these circuits

COMPONENT OPERATION

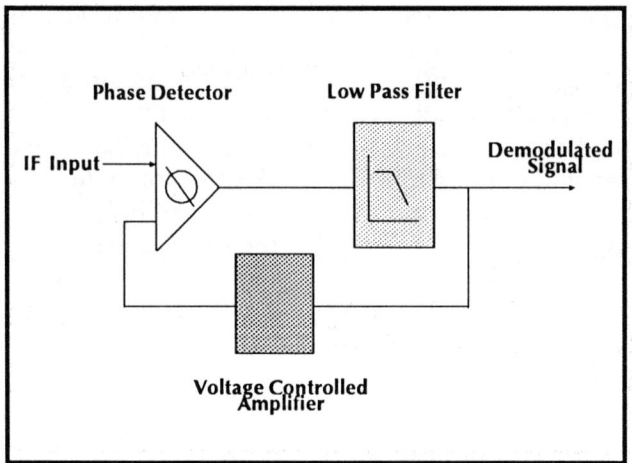

Figure 2-97. PLL Demodulator Schematic. *The PLL demodulator uses a centre frequency generated by a VCO as a reference. The frequency of the VCO is varied or controlled by an input volage. The phase comparator compares the phase and frequency of its input with that of the VCO and generates an error voltage proportional to the difference between the two frequencies. The low pass filter amplifies and filters the error voltage which is fed back to the VCO. This voltage causes the VCO to alter its frequency so that the error signal is reduced. Once locked the VCO frequency nearly exactly matches the input frequency. The error voltage, the demodulator output, follows the rate of change of the input frequency. The amplitude of the output voltage is determined by the frequency deviation of the input signal.*

would not operate at frequencies as high as 60 MHz without overheating. Manufacturers compensated by adhering the IC to heat sinks but the glue used was usually non-conductive. Experience gradually taught designers that not all NE564s could be used at 70 MHz, or even at 35 MHz. When a divider circuit was employed, the ICs ran cooler at 35 MHz or 17.5 MHz. Tracking of the IC also improved at the lower frequencies.

The NE564 IC offered designers two simple methods to reduce bandwidth. If the amplitude of the input signal to the PLL was below a certain level, the PLL narrowed its lock range to cope. This technique was used by experimenters to receive Intelsat C-band global beams using antennas with diameters of just two metres. The lock range could also be varied by adjusting the current on IC pin number two. This technique was used by most of the American designs using the NE564 PLL. The recommended input level to the circuit was in the range of 200 to 900 millivolts. (Note when checking the voltage between pins three and nine, use an oscilloscope or a properly decoupled multimeter. The voltage should be in the region of 1.3 Vdc if the circuit is operating correctly).

A European trade magazine article a few years ago mentioned that the voltage reading should be in the region mentioned above. However, some experimenters were perturbed to discover that when they tried to measure the voltage with an analogue multimeter, the needle shot off scale. As a result, a few even replaced the IC thinking it was faulty. In addition, no data on the PLL was given in the article. Even though the "this is it build it" article was outstanding, it did not mention that this link was the phase detector to the voltage controlled oscillator line.

The other PLLs used for demodulation are the uPC1477, the SL1451 and the discrete type. The discrete approach has been used at 70 MHz and higher frequencies. The receiver designed by Clyde Washburn is probably the best example of a 70 MHz discrete PLL. It was developed in the early eighties, was based on ECL logic circuitry and had a threshold of approximately 7.3 dB. This was an unusually low figure as the average receiver thresholds at that time were 12 dB. The discrete approach has also been used by some of the best European designers, most notably Steve Birkill.

Quadrature Demodulators

Those familiar with the theory of quadrature demodulation and with ICs used for demodulation at low frequencies know that some of the common 10.7 MHz FM demodulator ICs can actually be used at 70 MHz. The CA3089/TDA1200 has been employed by experimenters with good results. This particular IC also offers a delayed automatic gain control facility, a signal strength meter output and an automatic frequency control output though the AFC output was not used in any earlier designs. The quadrature coil was excessively damped to cover a 30 MHz bandwidth.

SATELLITE RECEIVERS

The primary disadvantage when using the quadrature demodulator is its fixed bandwidth response. When the signal drops below threshold the bandwidth remains constant resulting in poor near-threshold performance. The quadrature demodulator can, in some cases, yield better video than a PLL demodulator when the signal is well above threshold. In one test, an NE561 PLL demodulator was compared with a CA3089 demodulator under both strong and weak signal conditions. The results obtained on the weak signal test were as expected. On the strong signal test, the quadrature demodulator out-performed the PLL. This occurred because the PLL could not track the strong signal. However, the NE561 has been superseded by the NE564 which has no problems with tracking when proper design rules are followed. The CA3089 has not been used in any commercial receivers because near threshold response is poor.

Some of the early American designers used the MC1357 quadrature demodulator IC as a video demodulator. This component has better near threshold response than the CA3089 and gives approximately 1 Vdc peak- to-peak video output so none of the designs required a video amplification circuit, as was the case with the NE564. Most of these designs also dispensed with the de-emphasis network. A trimmer capacitor/resistor combination was used between pin fourteen of the IC and ground in order to tailor the high frequency response of the demodulator output. The near threshold performance is very good with a quadrature demodulator. This IC was used in one early European design and experimenters have used the circuit with good results.

Another quadrature demodulator IC used in satellite television applications is the SL1452 from Plessey. The input IF for this IC ranges from 300 MHz to 1 GHz. The input frequency is divided by four and fed to the quadrature demodulator. The operational frequency of the quadrature demodulator is set by an external inductor/capacitor/resistor network. The circuit requires use of a good quality filter prior to the IC in order to function properly. At UHF frequencies this filter is often fabricated in microstrip; alternatively, a SAWF or a cavity tuned filter can be used. The SAWF is the most common. Thresholds of 8 dB are common with good designs utilising this device.

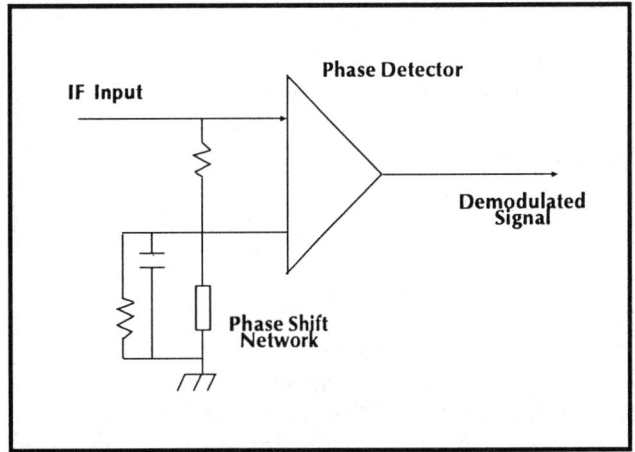

Figure 2-98. Balanced Mixer Demodulator Schematic.
This phase shift network has a response that varies with frequency. The amount of phase shift thus depends upon the frequency. The phase comparator compares the input signal and the phase shifted input and produces an output voltage corresponding to the phase difference.

Balanced Demodulators

The balanced demodulator, employed in many of the earlier American low cost receivers, had its output proportional to the product of the input signal and a carrier (see Figure 2-98). The most common IC for this application is the MC1496. The IF input is coupled capacitively to pin one and pin ten via a delay network. The negative carrier input is pin ten. Pin eight is the positive carrier (return) input. The input signals on these pins are mixed with the out-of-phase input signals on pins one and four. These are positive signal input and negative signal (return) input, respectively. The IC provides a balanced video output with the standard polarity (positive) video available from pin six. Negative polarity video is available on pin twelve. With the early American designs, the video output was taken from pin twelve and the output from pin six was usually taken to an AFC circuit. A typical balanced demodulator is shown in this figure.

Delay Line Discriminator

Another demodulator used in earlier designs was the coax delay line, although this device has not appeared in any quantity in European receivers. It is not as effective in discriminating between the

COMPONENT OPERATION

satellite signal and interference and in detecting signals having marginal strength as is the PLL. However, when signal strength is adequate, this type of demodulator delivers sharp, crisp pictures with well defined colours.

It is essentially a delay line and two diodes. The incoming signal is phase shifted by the delay line and equally divided between the two diodes. The magnitude of the combined output from the diode is proportional to the deviation from the carrier's centre frequency. As the signal frequency goes below the centre frequency the voltage becomes more negative and as the signal frequency goes above the centre frequency the voltage becomes more positive. Some experimental designs for this type of demodulator used a sealed, double balanced mixer (DBM), the type commonly used for frequency conversion in amateur radio designs.

In this circuit, the signal is split using a resistive divider. One signal path leads through the three quarter wavelength delay line, which is usually fabricated from thin coaxial cable. The other signal input is fed to the primary winding of one of the transformers in the DBM. The centre tap of the secondary of this transformer is grounded. This causes the signals at each end of the secondary to be antiphased or 180° out-of-phase with each other. The delayed signal is fed to the primary of the other transformer. The signals from the ends of the secondary windings of each transformer are then fed to a diode bridge. The difference signal, i.e. the video is then taken from the centre tap of the transformer secondary fed by the delay line. The recovered video requires amplification because it is only approximately 250 millivolts peak-to-peak.

The Mitsumi and Other Demodulator Modules

A number of demodulator modules are available for circuit designers. The Mitsumi demodulator is a balanced mixer type. The Mitsumi tuner, converter and demodulator are available to kit builders as a set. The price of a set of these modules is in the region of 70 £ (approx. $110 U.S.).

A less common tuner/demodulator combination is available from ASTEC Ltd. The tuner in this case has an output IF of 612 MHz. The demodulator is a quadrature IC type. While it was used in the mid- eighties by a few UK firms who decided to build their own receivers, its was limited because it had a U.S. bandwidth and IF block input. However, more recent designs use a demodulator operating above 400 MHz.

The UHF intermediate frequency is now preferred by most manufacturers because it is easier to design and fabricate a bandpass filter at UHF using microstrip than at 70 MHz. As discussed earlier, the 70 MHz had been previously selected as the final IF more for commercial than technical reasons. Therefore, the majority of earlier agile American C-band downconverters had a 70 MHz IF. Receivers using this IF were already in production when the block IF concept began to take hold. So engineers designed a set of modules to upgrade their receivers to manage this block IF. This reasonable solution enabled a manufacturer to serve the tunable front end market and the block IF market with only slight product alterations.

The majority of receivers built by experimenters used a standard UHF tuner as the block tuner. Many of these components were upgraded by setting the operating frequency of the UHF tuner equal to the IF output frequency of either the ASTEC or the MITSUMI block tuner. This procedure enabled such receivers to be mated with commercial LNBs.

Most receivers of Far Eastern origin use only the Mitsumi block tuner and the 479 MHz to 70 MHz downconverter and incorporate their own unique demodulator. The IF downconverter is generally followed by a number of amplifier stages. The majority of the American designs employed MWA type modules to provide this amplification. A few of these circuits use discrete transistor amplifiers. Others published in European magazines have used integrated circuit (IC) amplifiers such as the SL560. These IC amplifiers can be unstable and have caused some problems for experimenters. However, most manufacturers have opted for less expensive, discrete amplifiers.

Baseband Processing

The output of the demodulator contains the composite video signal and the FM audio subcarriers. Baseband is the term used to describe this combined output. At this stage in the receiver, the baseband is split into its component signals. The video signal must be passed to the video processing circuitry and the FM audio subcarriers must be fed to the audio demodulation circuitry.

SATELLITE RECEIVERS

COMPONENT OPERATION

Video Filtering

A low pass filter is used to extract the composite video signal from the baseband. Satellite broadcasts relay this signal in one of three video formats, system M, G and I. System M is the American NTSC standard which uses a video bandwidth of 4.2 MHz. This system is rare in Europe and to date is only available on the C-band American Armed Forces Radio and Television Service (AFRTS) and newslink feeds. System G is the European PAL standard and uses a 5.0 MHz video bandwidth. This system is in use by some of the continental European Ku-band channels. System I is the Irish and English PAL standard having a video bandwidth of 5.5 MHz. Most more expensive receivers on the European market can select between PAL I or G. System M is rarely included as it is not generally used by any of the European broadcasters.

De-Emphasis Circuit

FM video signals are pre-emphasised at the uplink and de-emphasised within the satellite receiver. At the uplink, the video signal is passed through a network that attenuates only the lower frequencies effectively boosting the high frequency portion of the signal. At the receiver, a de-emphasis network attenuates the upper frequencies to restore the original video waveform. The objective of this technique is to improve the signal-to-noise ratio by about 2 dB. This method is effective because the noise in an FM satellite transmission increases from the 0 MHz to the upper limits of the baseband signal, about 6 MHz (a triangular noise distribution). Pre- and de-emphasis reduces noise by lowering the higher power, higher frequency noise.

Pre- and de-emphasis circuits are simple RC networks (see Figure 2-99). Time constants of 50 or 75 milliseconds are typical.

Pre-emphasis and de-emphasis are also applied to audio signals. These methods are further discussed at this end of this section.

Clamp Circuit

At this stage, the video signal has been filtered and de-emphasised but still rides ontop of a dispersal waveform, also known as the "dithering" waveform. The clamp circuit removes the dispersal waveform. In accomplishing this objective, the signal-to-noise ratio had to be lowered to accommodate this waveform; an annoying flicker or

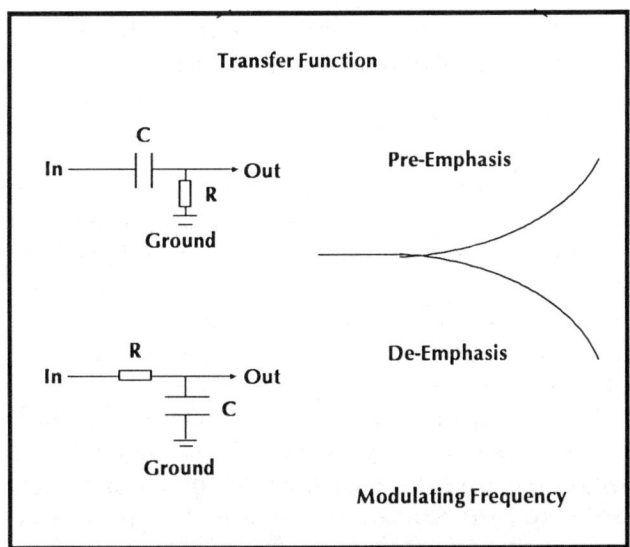

Figure 2-99. Pre-Emphasis and De-Emphasis. *Pre-emphasis and de-emphasis are used on satellite transmissions both to provide a means to improve the signal-to-noise ratio. Higher frequencies are boosted at the uplink and the reverse is accomplished in the satellite receiver to restore the original signal balance. This effectively reduces the high frequency noise characteristic of FM transmissions.*

pulsation would be evident if this waveform were not removed. Note that unnecessary clamping would cause a tearing of the whites and distortion of captions and text in the television picture.

The motivation for including an energy dispersal waveform stems from the earlier days of satellite television when satellite frequencies were shared by terrestrial microwave links. Indeed, this is still the case with American C-Band transmissions. The traveling wave tube amplifiers (TWTA) used on-board pioneer satellites were not very robust. If a transponder was left vacant, the only signal transmitted would be a black or a white level, effectively a high concentration of energy at one particular frequency. This sometimes resulted in failure of the amplifier. The dispersal waveform caused the energy to be dispersed over a wide range of frequencies. This tactic also reduced the potential for interference with terrestrial microwave links which operated in mainly narrowband ranges. Without the dispersal waveform, the satellite signal could have severely affected terrestrial transmissions because its energy would have been concentrated near one frequency.

SATELLITE RECEIVERS

COMPONENT OPERATION

The dispersal waveform is removed from the video signal by the use of a simple diode clamp. There are a few circuits used for this function. Some designs incorporate a Schottky/Zener diode combination though in a few lower cost devices the Schottky diode is replaced by a cheaper 1N916 silicon switching diode. In one unusual design, a field effect transistor/Schottky diode combination was employed. Most receivers on the European market have an option select switch for clamped or unclamped video.

With the launch of ASTRA, the level of technical knowledge of dealers and distributors has been growing. However, there have been some cases of notable ignorance. For example, one dealer reported that a receiver was faulty because the picture was flashing on the screen. The solution was simple. The video clamp/unclamp switch on the back of the receiver had been left in the unclamped instead of the clamped position.

When a satellite receiver is used in conjunction with a descrambler, the video clamp/unclamp switch should be in the unclamped position. On some receivers an internal modification to the clamping circuit is required to enable it to properly drive the descrambler. This modification normally involves replacing the clamping diode with a resistor. The unclamped video is essential for descramblers because some descramblers use the dispersal waveform, which is at frame frequency, for synchronisation. In descramblers that rely on level detection, a clamped signal would produce errors.

In North America, Ku or K-band satellite transmissions do not have a need for a dispersal waveform since in-band terrestrial interference is not a concern. Therefore, the energy dispersal waveform is not present in Ku-band broadcast signals. However, it is added to uplinked signals in satellite communication systems in, for example, Europe and Australia. The standard European dispersal waveform is a 25 Hertz triangular waveform. American C-band systems use a 30 Hz waveform while Russian systems use a 2.5 Hz triangular waveform on some C-band transmissions. When unmodified receivers are used to detect Soviet broadcasts, this difference in dispersal frequency results in a very wide deviation causing some sections of the video to be lost.

In some of more recently introduced satellite receivers, the clamp function can be selected when storing the channel parameters with the remote control or via a back panel switch.

Filtering

Following clamping, this signal is filtered to remove audio subcarriers above the upper edge of the video band and finally amplified.

Audio Processor

The audio processor demodulates the audio information carried by any selected subcarrier. The output from the audio processor is the original audio information contained in a band ranging from 30 to 15,000 Hz.

The audio subcarrier on a satellite TV broadcast differs from that of the standard terrestrial broadcast in that its centre frequency generally ranges from 5.0 to 8.5 MHz. On European Ku-band transmissions there are three main subcarrier frequencies, 6.5, 6.6, and 6.65 MHz but 7.02 and 7.20 MHz are not uncommon. Russian C-band broadcasts use 7.0 and 7.5 MHz; American C-band transmissions use various centre frequencies in the range 5.0 MHz to 8.5 MHz, although the two most common are 6.2 and 6.8 MHz. Most American Ku-band satellites relay audio at 5.51 or 6.20 MHz but the Canadian Anik C3 spacecraft has subcarriers at 5.41 MHz.

Audio Demodulators

Two methods are employed to cover the rather wide audio subcarrier frequency range: the fixed frequency demodulator and the variable frequency demodulator. The most popular method at present is the fixed frequency demodulator/tunable upconverter combination. The variable frequency demodulator was preferred by the early experimenters because the bandwidth could be adjusted by varying the magnitude of the input signal. This variety is also quite easy to construct. The typical design, commercially and experimentally, was the varicap tuned NE564 PLL demodulator.

The fixed frequency demodulator was more common in the early experimental designs of the late seventies and early eighties. Most development occurred in the U.S. when there were basically only two subcarrier frequencies used for satellite TV transmissions. The IC based quadrature

SATELLITE RECEIVERS

demodulator was then extensively used. Its basic configuration was a tuned filter, quadrature demodulator and audio amplifier. The accompanying circuit diagram shows a typical fixed frequency demodulator using a CA3089 IC. The fixed frequency demodulator was expensive and was found in only top-of-the-line receivers.

As the number of audio subcarriers has increased, a fixed frequency demodulator per subcarrier has generally become too expensive. A simple solution has been to insert a tunable front-end before a fixed frequency demodulator. Most FM radios use 10.7 MHz as their IF. The majority of the demodulators at this frequency used ceramic resonators to tune the centre frequency of the demodulator and ceramic filters to adjust its bandwidth. This resulted in a demodulator that required very little setting up and that used inexpensive, readily available components.

The use of ceramic filters has caused some problems with a few brands of receivers which were not fully converted to European standards. These have been and, in some cases, are still sold in Europe. With original 180 kHz bandwidth ceramic filters, distortion occurred in reception of European audio signals. Manufacturers of these receivers now recommend installing a 400 kHz bandwidth filter, but a 280 kHz filter may suffice until the higher frequency filters are widely available. Note that the ASTRA specification now calls for a 180 kHz bandwidth.

The resultant output from the audio and video processors is known as the composite baseband audio signal.

Modulator

A modulator is required to "rebroadcast" the composite baseband audio and video signals onto an AM form which can be interpreted by a conventional television. Over-the-air channels, unlike FM satellite signals, are amplitude modulated. The selection of modulation frequency determines which television channel will receive the satellite programming. Channels 3 and 4 have been typically selected in North American C and Ku-band reception systems while European TVROs modulate in the region of channel E36. More details are provided in section K below.

Judging Receiver Performance

The quality of a satellite receiver is ultimately judged by clarity and fidelity of the television picture and crispness of the sound it produces. These subjective criteria can be partially estimated from the receiver video bandwidth and threshold. These, in turn, are all determined by the care taken in designing and manufacturing as well as by the choice of circuit components.

Signal-to-Noise Ratio and Picture Quality

Signal-to-Noise Ratio

The central factor in determining whether a communication signal is usable is the ratio of the signal to noise power, the S/N ratio. Even if the noise power is quite high, if the S/N ratio is acceptable, a clear picture can be received. Thus, for example, a S/R of 40/1 will be achieved both by signal of 400 watts and a noise power of 10 watts as well as a signal power of 4 watts and a noise power of 0.1 watt.

The concept underlying the use of a ratio of signal to noise power can be clearly understood by imagining a listener in a room full of people. If everyone was speaking and the noise level were high, then someone would have to shout to be heard. Signal power would have to be higher. However, if it were so quiet that everyone could easily hear a pin drop, if the noise level were low, then a whisper at much lower audio power would suffice.

The S/N ratio is expressed either as a ratio or in decibels. The S/N expressed in decibels is given by:

$$S/N \text{ (decibels)} = 10 \log [S/N \text{ ratio}]$$

Picture Quality

Subject experiments have determined that television picture quality is directly related to the signal-to-noise ratio it is fed. For example, in 1959, the FCC (Federal Communications Commission) in the United States, authorised a task group to study the problems associated with television broadcasting and reception. This group was known as the Television Allocations Study Organization, or TASO. In one experiment, a standard television set was viewed by numerous people who were then asked to rate the picture quality as S/N was varied.

COMPONENT OPERATION

TABLE 2-12. SUBJECTIVE RATING OF PICTURE QUALITY versus S/R (TASO Study)

S/N Ratio (dB)	Subjective Rating of Picture Quality
above 45	Excellent quality – near perfect picture
40	Good quality – some interference perceptible
33	Passable quality – interference not objectionable
28	Marginal – poor quality with annoying interference
22	Poor – just-watchable picture

The results shown in Table 2-12 show that for an acceptable picture a S/N of at least 35 dB was necessary. Other studies have confirmed these results.

Video Bandwidth

Satellite video broadcasts are transmitted with a bandwidth ranging from as high as 72 MHz to as low as 18 MHz half transponder formats that have been familiar on some Intelsat spacecraft. Although a watchable television picture can be reproduced with a bandwidth as narrow as 15 MHz, reducing bandwidth lowers picture quality. Using a limited bandwidth allows additional channels to be transmitted over a satellite circuit but also increases its susceptibility to noise.

Picture fidelity worsens as receiver bandwidth is narrowed. Details are slowly lost and colour shades begin to vary. This appears as a "softening" of the video, as streaking in rapidly changing scenes and as the presence of "sparklies" in the saturated red and green colours. However, receivers which are designed to detect a more narrow bandwidth also see less noise power so that "sparklies" are avoided in all but the most saturated colours. Expressed in another way, picture clarity is traded off for fidelity. Subjective studies have shown that the picture quality using a 24 to 28 MHz bandwidth is virtually indistinguishable from that with a 36 MHz bandwidth. Therefore, a bandwidth reduced from 36 MHz to approximately 28 MHz has been adapted as a de facto industry standard for satellite receivers. Home style brands often use a narrower bandwidth to minimise detection of noise.

The judgment of what constitutes reception quality is of course based on subjective criteria. Consumer studies in which picture quality was rated according to the signal-to-noise ratio (S/R) demonstrate that, although judgments vary, satellite receiver threshold can be an excellent indicator.

Receiver Threshold

The threshold of a satellite receiver determines how weak an input signal, as measured by the carrier-to-noise power ratio (C/N), can be before a picture is judged unacceptable. For a given C/N, a receiver will generate an output signal-to-noise ratio (S/N). The video S/N is expressed as a ratio of the luminance signal (in peak-to-peak volts) to the noise signal (in rms volts). The luminance portion of a 1 V p-p video signal is 714 mV p-p.

If the input S/N is plotted against its output C/N, a straight line results over most of this range. This linear relationship means that for a given change in input, there is a proportional change in output. To illustrate, if an input of 1 watt results in an output of 5 watts, then an input of 2 watts creates a 10 watt output. A simple example shows why this linear relation between input and output is important for picture clarity. A photograph of a scene having one area twice as bright as another would not look proper if the reproduced picture had that area only 50 percent brighter, namely, if the reproduction were not linear.

Threshold is defined as that point where the deviation from a linear or straight line plot equals one decibel (see Figure 2-100). Near or just above threshold "sparklies" begin to appear. More exactly, threshold is the point at which the peak white of the demodulated video signal has decreased to the point that it equals the amplitude of the impulses noise spikes. This is observed as white comet-like streaks in the television picture.

COMPONENT OPERATION

TABLE 2-13. PICTURE QUALITY and RECEIVER THRESHOLD
(Threshold chosen as 8 dB)

Carrier-to-Noise Ratio (dB)	Picture Quality
5	Extremely noisy; tearing, noise in audio
6	A little better; sparklies
7	Watchable but sparklies
Threshold	A few sparklies
9	Good picture; sparklies on saturated colours
10	Video tape quality
11	Cable TV quality

Sparklies appear at low C/N ratios because the FM discriminator is no longer able to distinguish between the signal and noise but instead creates it own white noise that is added to the video signal. This is observed on an oscilloscope as sharp peaks of noise exceeding the video signal. If the tuner centre frequency is tuned up or down at low C/N ratios, this can make either white or black sparklies the predominant form of noise, respectively. As would be expected, even if the input C/N is well above threshold sparklies can be forced to appear when the centre frequency causes one polarity of the signal to be dragged into the noise.

Threshold Extension Techniques

Subjective tests of picture quality reflect the degradation seen near threshold (see Table 2-13). Therefore receiver manufacturers have been motivated to develop methods to "extend" or lower threshold and thus make a weaker signal more watchable. Most good quality satellite receivers today have thresholds at C/Rs in the range of approximately 8 dB, the value chosen as a reference in the Table 2-13. Referring to this table, it is clear that a signal of 5 dB would generate an unacceptable picture.

Before discussing threshold extension, readers should realise that noise can be either impulse noise that is responsible for sparklies and thermal noise a more random and uniform noise. The effect of thermal noise differs from impulse noise, the source of sparklies. The result of a high level of thermal noise is a picture that appears uniformly grainy, most often seen in terrestrial AM modulated transmissions. Threshold extension techniques have been developed to counteract the effects of impulse noise in low C/N signals.

There are a number of approaches to extending receiver threshold. Using the best possible circuit design and components that contribute the lowest amounts of noise to a signal being processed is the starting point. A simple method built into some brands of lower cost receivers and featured as an option on others is to reduce video bandwidth. As discussed above, this results in some loss of picture fidelity that is noticeable on a scene with saturated colours.

Three types of active threshold extension techniques are also available, phase-locked look (PLL), dynamic tracking filter (DTF) and frequency feedback loop (FFL). The PLL concept has been discussed earlier in this section (see Figure 2-97). These circuits essentially track only the FM video signal but tend to ignore rapid variations in the carrier and effectively reduce the noise bandwidth of the demodulator circuit.

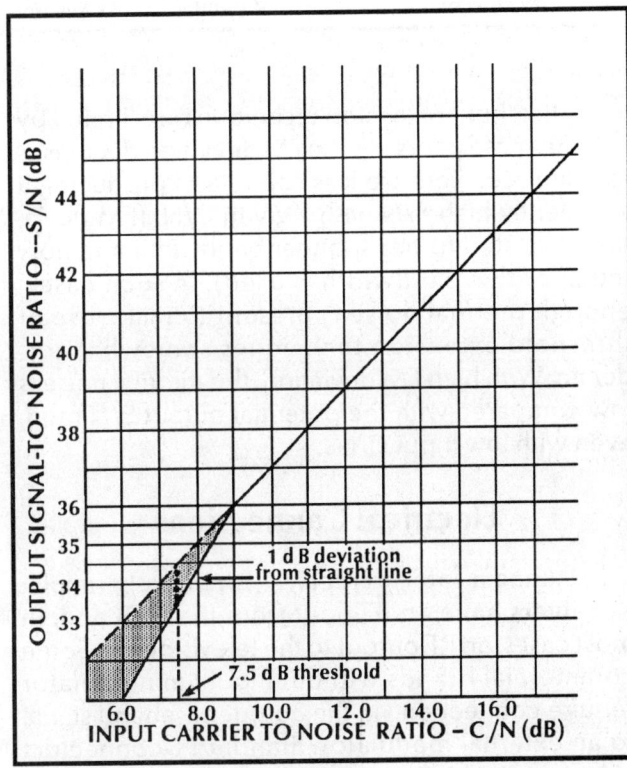

Figure 2-100. Receiver Threshold. *A satellite receiver should provide a sufficient carrier-to-noise ratio to exceed the receiver threshold. Threshold is measured at the point where the departure from linearity is 1 dB.*

SATELLITE RECEIVERS

COMPONENT OPERATION

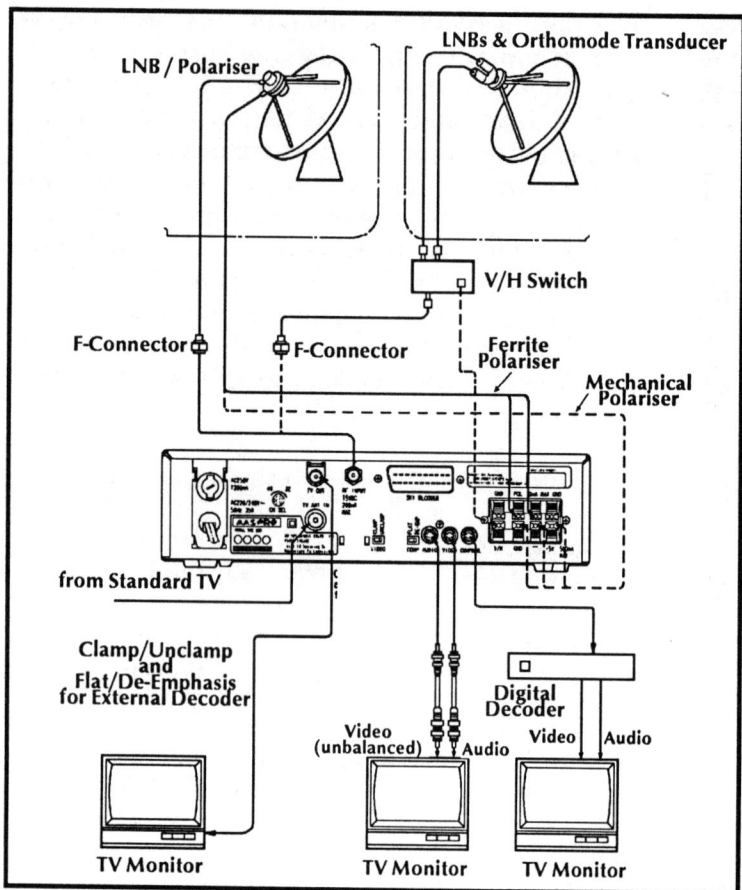

Figure 2-101. Maspro Wiring Diagram. *In wiring a Maspro receiver, an orthomode feed can be connected through a V/H selection switch to which either 0 and 15 Vdc is applied. The necessary output is provided to drive either a mechanical or a ferrite PSD. Note the SCART connector in the centre of the rear panel for connection to a Sky decoder. Other decoders operate via the composite baseband output. (Courtesy of Maspro)*

Threshold extension techniques are limited by a number of factors. As the FM deviation decreases satellite receivers are less capable of producing a sufficiently high S/N with a given C/N. (FM deviation is related to transponder bandwidth and how much of this bandwidth is used). In such cases, enough thermal noise is present to make use of threshold extension techniques somewhat academic. With high FM deviation, the thermal noise is low compared with the potential output S/N ratio, even with low input C/Ns.

Electrical Connections

Wiring a satellite receiver is relatively simple. Receivers have an IF input from the LNB and, in most cases, an RF output to the television set. Some commercial brands without a built-in modulator require connection via the output baseband signal to an external modulator. Standard F-connectors are almost always used in North America and SCART connectors are extensively used in Europe. Interfacing decoders in examined in some detail in Chapter V. Hooking up a non-SCART connected European satellite receiver is illustrated in Figure 2-102.

Receivers that drive a mechanical polariser have three wires, the ground, white/pulse and 5.7 Vdc red/power wires, which are usually attached via screw or lug connections onto a barrier strip. A separate set of appropriate screw connectors for a dc motor polariser are sometimes also provided. Receivers that drive ferrite polarisers require connection of just two wires.

When the receiver has a built-in actuator, two narrow gauge motor wires and three fine counter wires must be hooked up.

Other output terminals are available. The unclamped and unfiltered baseband signal is usually accessible for relay to a stereo processor or descrambler. The clamped and filtered baseband signal, known as the composite baseband audio and video, is also usually available with two audio ports and one video port for use in external modulators, television monitors and VCRs. Some receivers with built-in stereo processors have two

COMPONENT OPERATION

RCA-jacks, also called phono jacks, for connecting the left and right audio channels for relay to a stereo receiver. Often an RCA-jack output is provided to allow connection of an external signal strength meter.

User Interfaces and Adjustments

All satellite receivers feature some standard controls including power on-off switch, channel selection, audio subcarrier selection, polarity and skew adjustment, and usually, video fine tuning. Many other controls are found on other brands. These include switches and buttons to select from among stereo formats, channel scan mode, narrow/wideband audio filters and video bandwidth control. Some receivers come equipped with signal strength meters and, occasionally, frequency centring meters.

The trend has been towards simpler customer interfaces in non-commercial units. Many consumer brands are now available which automatically fine tune both the video and skew and have no customer interfaces for these functions.

K. MODULATORS

A modulator is a "rebroadcast" device which processes the satellite composite baseband signal, the raw video and audio information, into an amplitude modulated form that is acceptable by television sets. This device is essentially an interpreter that imposes the clamped and filtered baseband signal onto an AM carrier frequency that a television has been designed to receive.

Television sets accept AM modulated signals. Amplitude modulation was chosen for over-the-air broadcasts and later for transmission over cable TV networks because relatively high signal power levels are available, as required for clear AM reception, and also because the transmission bandwidth must be maintained in a relatively narrow range to permit a maximum number of channels to be carried in the allocated frequency space. Satellite broadcasts are frequency modulated for precisely the opposite reasons. Power is limited but wide frequency bands are available.

The composite baseband signal, which contains all the video and audio information required to recreate a television picture, can be fed directly into a TV monitor without being remodulated. However, a conventional television receiver has built-in AM demodulation circuits designed for detecting over-the-air AM broadcasts. Unlike a TV monitor, it can accept only remodulated signals. In general, higher quality reception will be attained by using a monitor since these additional modulation/demodulation steps are eliminated. Note that a home video cassette (VCR) also accepts the composite baseband signal and has its own built-in modulator for output to a television. Thus, VCRs are capable of recording higher quality videotapes from the satellite composite baseband signals than from demodulated and remodulated over-the-air broadcasts.

The composite baseband signal can be modulated onto any chosen television channel by simply selecting the correct carrier frequency. For example, North American televisions are capable of handling 57 channels, 2 through 13 on the VHF band and 14 to 64 on the UHF band. Most North American satellite consumer receivers have built-in modulators that feed signals into either channels 3 or 4. European modulator output frequencies are generally in the region of channel E36. If any of these channels are occupied by local over-the-air broadcasts, satellite signals can be modulated onto any other unused frequencies. Then all satellite programs can be selected by leaving the television set to this channel and tuning via the satellite receiver. Home satellite receivers modulate video and audio information onto channels commonly used for microcomputers, video recorders and video games. Whenever necessary, a modulator may be retuned to another channel so that it does not interfere with these other sources.

Numerous varieties of modulators having performances ranging from poor to adequate are available (see Figure 2-102). While some receiver manufacturers incorporate their own internal modulator design, the ASTEC series of modulators is

COMPONENT OPERATION

Figure 2-102. Inexpensive Modulator. *Modulators that are factory installed into home satellite receivers have evolved from the same type that are built into video tape recorders. While they have adequate performance and are small enough to easily fit into home satellite receivers, they are not suitable for SMATV installations which have multiple adjacent channels.*

probably the most widely used in Europe and North America. Note that since modulators in home satellite receivers are fairly reliable, distributors rarely carry spare components.

The UM1286 is the most common ASTEC modulator used in satellite receivers. This device has also been widely employed in computer applications. The UM2301 is a newer version that, to date, has been rarely incorporated into home satellite receivers. Each of these ASTEC modulators differs in construction. The UM1286 has standard lead type components while the UM2301 uses newer surface mount technology. The latter is also more stable than its predecessor. When any type of modulator fails, a UM1286 modulator can be reliably used as either a temporary replacement or a permanent repair.

The UM1286 must have a 0.5 volt peak-to-peak input video signal with a 2.2 Vdc bias on the video input. The required audio level for full deviation is 1.77 volts. The UM2301 requires a standard 1.0 volt peak-to-peak video signal with no DC bias at the video input and a 1.0 volt RMS audio level for full deviation. A common problem with UM1286 modulators, a buzzing sound on the audio, occurs when the DC bias at the video input is incorrect.

Commercial Grade Modulators

Modulators which are nearly always integrated into home satellite TV receivers are adequate for residential purposes. However, a more complex commercial installation, such as a satellite master antenna system for a large apartment development, requires that a higher quality signal be fed into its distribution network. Commercial modulators are of higher quality and are generally more expensive than those built into satellite receivers (see Figures 2-103 and 2-104). In such applications, each receiver is permanently fixed to a given satel-

Figure 2-103. Nexus Commercial Modulator. *Modulators used for commercial applications such as private cable or SMATV systems must have vestigial sideband filtering to eliminate the lower sideband that will cause interference in the lower adjacent channels. These modulators should be crystal controlled to eliminate frequency drift and have an output level of 30 dBmv or more. The Nexus VM5 series modulators are available in NTSC and PAL formats. (Courtesy of Nexus Engineering)*

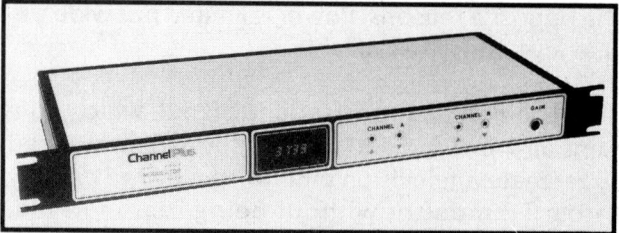

Figure 2-104. Multiplex Dual Agile Modulator. *This commercial modulator features two crystal controlled modulators in a standard 19-inch rack case. This unit accepts video signals from VCRs, satellite receivers, CCTV cameras and other sources and remodulates the output onto UHF or Hyperband channels. Advanced strip line manufacturing techniques with surface mounted circuit devices is used. Various models are compatible with either NTSC or PAL formats. (Courtesy of Multiplex Technology)*

MODULATORS

lite transponder and has a dedicated external modulator tuned to any chosen television channel. Signals from the external modulators are then combined at this "headend" for transmission into a cable leading to the distribution network.

Channel selection in commercial systems can be accomplished in a variety of ways. The modulators can be set to a series of channels tunable directly from each customer's television set. Or all the modulator outputs can be fed, often at mid or super-band frequencies, into a set-top converter. Channels are selected via the converter tuned to channels 3 or 4 in North America or to channel E36 in Europe. This design is similar to those used by many cable TV companies.

Note that some poorly designed home modulators can wreak havoc in the UHF band when connected into a UHF distribution system. Distortion usually appears as intermodulation, caused when the video and audio carriers interfere or "beat" with each other. In some receivers the problem results when the UHF signal is fed to the a back panel connector by a thin piece of coaxial cable. Replacing or rerouting this cable can sometimes cure the problem. Some home TVRO receivers currently on the market have inexpensive, built-in diplexers, devices that allow the user to mix various channels together, that also can cause intermodulation. In these cases, it is better to use an external diplexer to combine the terrestrial and satellite television signals.

L. TELEVISION RECEIVERS AND MONITORS

The purpose of a television receiver is to recreate the original picture and sound creating in the studio as accurately. A picture is "painted" line by line onto a phosphor-coated screen by an electron beam. The scanning caused by an organized television signal produces organized changes in illumination which are perceived as a picture. This is possible because the beam can be controlled by electrical fields at its source in the heart of the television tube.

A black and white (b/w) picture is composed by a single beam while a colour picture is created by the scanning of three beams over three independent grids of blue, green and red phosphor dots embedded into the screen surface. All other colours can be derived from these three basic ones.

The purpose of this section is to provide sufficient background information to allow an understanding of the fundamentals of worldwide television standards. The description of television operation in the next section is based partially upon examples that use the NTSC format. The principles underlying the operation of PAL or SECAM television receivers are basically the same and an outline of the differences between the various formats follow below.

Understanding Television Operation

Scanning

Scanning was the solution to the problem of recreating a complex scene that occurred simultaneously at many points in space, and transmitting it as a sequential stream of information. A

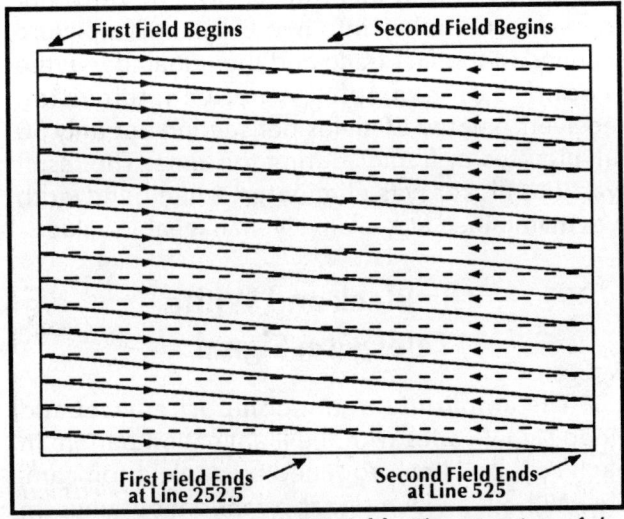

Figure 2-105. Scanning NTSC Fields. *The scanning of the first field begins at the upper left hand corner of the television tube and ends at the bottom centre at line 252.5. The second field begins at the top centre and ends at the bottom right after line 525 is completed. Two of these fields equal one frame or complete picture.*

picture is essentially traced line-by-line onto the phosphor-coated television screen (see Figure 2-105). This scanning begins at the top, left-hand corner of the screen as viewed from the front. The first line is swept across the screen. This "trace" is completed when the right side is reached. No picture information is transmitted during the retrace, known as the horizontal blanking interval, while the extinguished beam moves from the right to left. The second line is then traced and so on. After the bottom trace line is complete, the beam illumination is again suppressed during the vertical blanking interval while it is repositioned at the top of the screen. During both the horizontal and vertical blanking intervals, other information such as captions for the hearing impaired, teletext, digital audio or addressing information for scrambling systems can be embedded into the television signal.

Vertical Resolution

The number of lines that are used to scan a picture determine its vertical resolution. Clearly, as the number of lines are increased, the system becomes capable of displaying a more detailed scene. As few as 405 and as many as 819 lines per frame had been used in the earlier days of commercial television. The lower line number resulted in poor vertical resolution, namely a grainy appearance to the picture, while the higher number required the use of an unacceptably wide bandwidth to transmit the massive amount of information. Today either 525 or 625 scanning lines have become the accepted standard in television systems around the world.

Trace Rate and Horizontal Resolution

The choice of trace rate involved a tradeoff. Ideally, individual pictures or frames should be painted onto the screen as rapidly as possible in order to simulate continuous motion as closely as possible. However, at higher trace rates, the amount of brightness produced on the screen surface decreases because of inherent limitations in the response rate of the phosphor coating. The beam would stay in one location for shorter periods of time as trace speed increased and would therefore have less effect. In addition, a higher frame rate also necessitates use of a wider transmission bandwidth because more variations in intensity are relayed.

The wider the channel bandwidth, the finer can be the changes in signal voltage and, therefore, the greater the number of brightness changes that may be transmitted in each line. The quality of the phosphor screen also determines how well a television can respond to the changes in beam intensity. Therefore, horizontal picture resolution is determined by transmission bandwidth as well as by the design and construction quality of the television set.

Frame Rate and Picture Stability

Pictures or frames must be painted onto a television screen rapidly enough to simulate continuous motion. Originally, black/white transmissions had their frame frequency set at the line frequency of the country in question. Therefore, European and American televisions would have flashed 50 and 60, respectively, pictures per second on the screen. These frame rates would have required an unacceptably wide signal bandwidth. Therefore, engineering committees finally chose a frame frequency of half the line frequency; 30 pictures per second in North America. This difference in frame rate is one of the central problems impeding the introduction of a worldwide HDTV (high definition television) standard.

An ingenious method to eliminate the flicker that would have resulted from using this low frame frequency called interlacing was developed. Frames were broken into two fields. Half a picture scanned in the first trace and the remainder in the second. For example, American television receivers painted 60 fields per second but only 30 full pictures or frames during this time. The result was that flicker was eliminated while bandwidth was maintained at a relatively manageable value.

The Black and White Television Signal

The amplitude of the composite baseband video signal varies with the illumination pattern in each scanned line produced by a television camera. Without a video signal, a random but uniform pattern of dots ranging from black to white called the "raster" is produced. When a signal is imposed, it increases and decreases the intensity of the electron beam. The more intense the electron beam, the brighter the illumination and vice versa.

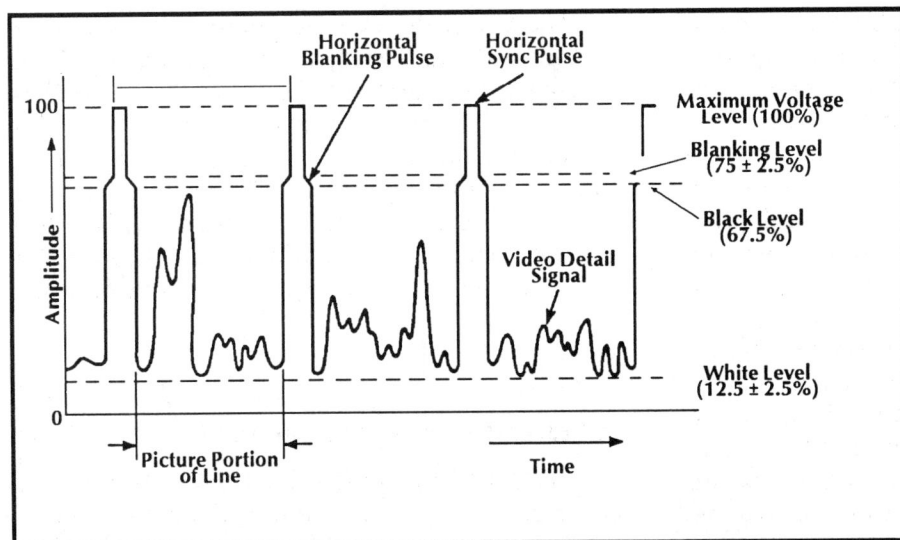

Figure 2-106. The Television Signal. This illustration shows the structure of three lines of a television video signal. Above the 67.5% level, the black level, the beam is shut off. The "blacker than black" level falls between 67.5% and 100% of the maximum voltage level. The picture information is relayed between the black and white levels. Therefore, during the horizontal blanking interval, which contains the blanking pulse and sync pulse, no illumination is produced.

Two reference levels are built into this raw video signal, the reference white and the reference black level. At voltages below the reference white level, the electron beam intensity is at a maximum level and the phosphor screen is as bright as possible. At levels above the reference black level, no illumination is produced (see Figure 2-106).

In order to accurately recreate a television picture, the scanning inserted at the camera must be perfectly synchronised with that at the television receiver. Pulses to precisely synchronise the timing of the scanning are inserted into the non-picture portions of the video signal. Horizontal and vertical synchronisation pulses set the start of the horizontal and vertical traces, respectively. If the vertical sync is not accurate, the television picture will "roll." If the horizontal sync pulse is removed or faulty, the picture will "tear." Blanking pulses are also inserted during the vertical and horizontal intervals to turn the beam off so that retrace lines will not be visible. The electron beam is turned off by these blanking pulses which are simply voltages exceeding the black reference level.

The over-the-air broadcast television signal is composed of two separate carriers, one modulated with the video and the other with the audio information (see Figure 2-107). The audio portion is frequency modulated to reduce noise; the video portion is amplitude modulated to minimize disturbances called "ghosts." Such ghosts result when a signal is reflected off a building or another object and follows a different path from the direct signal to the television receiver. This causes a very weak second image to be displayed on the screen. (A similar though unrelated phenomena occurs on satellite transmissions where the filter bandwidth is too wide so that some of the adjacent channel signal enters the television receiver.) The AM video carrier is modulated by a voltage created in proportion to the variations in illumination of each scanned line produced by a television camera. A television receiver simultaneously detects both audio and video signals, amplifies these waveforms to improve reception and then demodulates them to produce the composite video and audio baseband information. It is the composite baseband audio and video signals which contain all the intelligence, namely the picture and sound information, plus the necessary sync and blanking pulses required to reconstruct the original picture and sound (see Figure 2-108).

The Colour Television Signal

The colour signal is also composed of an FM sound carrier and an AM video carrier, both contained in a 8 MHz PAL or 6 MHz NTSC bandwidth. The video portion has the same sequencing of blanking interval and synchronisation pulses as a black/white (b/w) video signal. This similarity is essential since the colour signal must be able to recreate a picture on a b/w television. The colour systems was designed to be "backwards" compatible with the b/w system so that b/w television receivers detect colour broadcasts as b/w broadcasts.

However, the colour signal must necessarily be more complex than the b/w signal. The amplitude variations of the b/w video signal represent

COMPONENT OPERATION

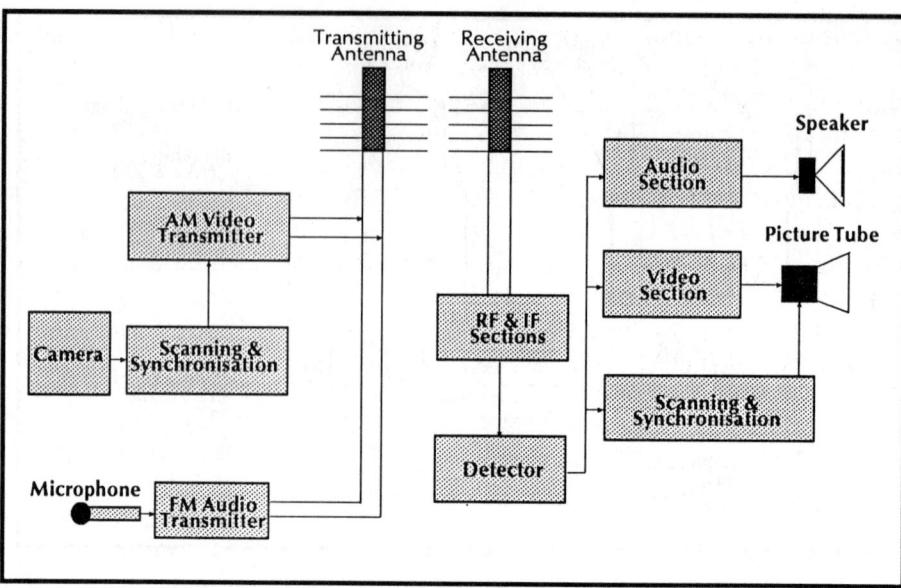

Figure 2-107. The Basic Television Circuit. *The same scanning and synchronisation pulses that drive the television camera are also added onto the video signal for modulation. AM is used for over-the-air and cable broadcasts while FM is used for satellite transmissions. This signal is relayed with the FM audio to a receiving antenna. The television receiver demodulates both signals and feeds them to audio and video sections. Scanning and synchronisation information is extracted and is used to drive the electron beam. The beam intensity is controlled by the video section.*

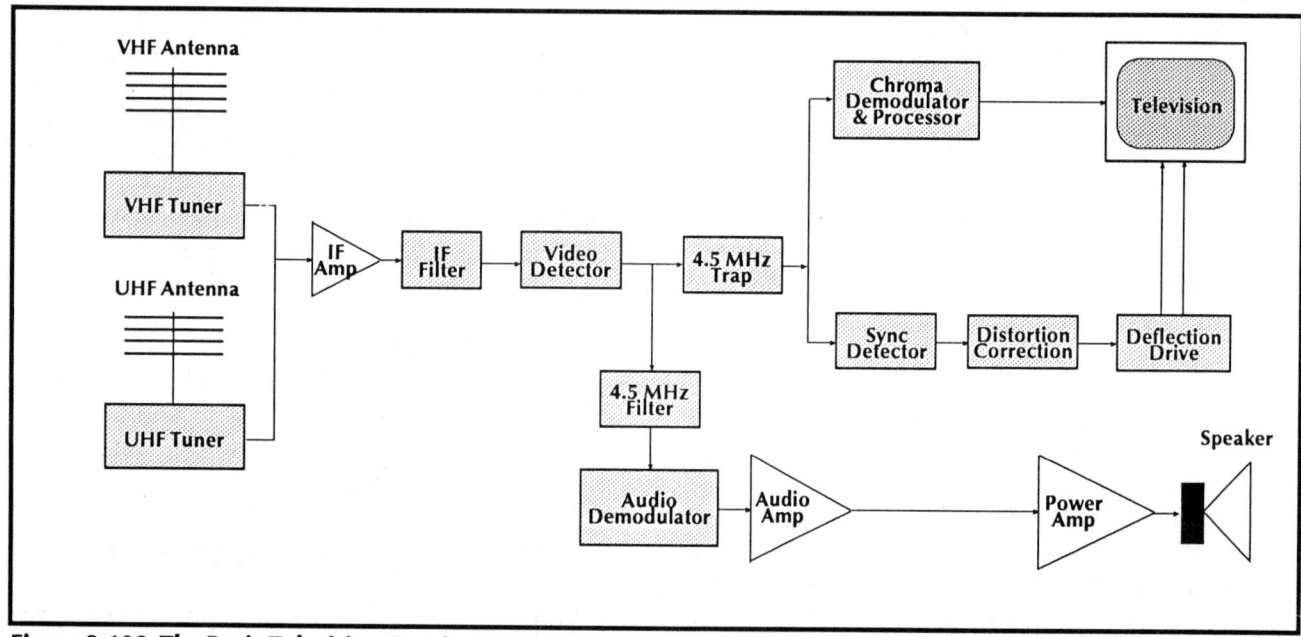

Figure 2-108. The Basic Television Receiver. *A signal enters via either the VHF or UHF terminals and is connected to the appropriate tuner. This circuit component, whether controlled by a rotary switch, push button selector or an electronic remote, is responsible for selecting one channel or frequency band. An NTSC VHF tuner must be capable of spanning the range from 54 to 216 MHz or from 54 to 456 MHz for cable-ready receivers. A UHF tuner accepts frequencies ranging from 470 to 806 MHz. The output of either the VHF or UHF tuner is a common intermediate frequency (IF) of 45.75 MHz centred on any selected 6 MHz channel. This IF is bandpass filtered to remove any unwanted signals from adjacent channels. The signal is fed to a detector or demodulator which extracts the composite baseband signal. The detector also produces an automatic gain control (AGC) feedback signal that maintains the gain of the tuner section at the appropriate level so that the detector is fed the proper voltage. The signal is then filtered to remove the 4.5 MHz audio subcarrier. Prior to filtering, a portion of this video signal is tapped off and sent to the audio detection circuitry. Once the 4.5 MHz subcarrier is isolated and demodulated, it is filtered and amplified. Finally, a variable attenuator is used for controlling the voltage which drives a speaker. The "clean" composite video baseband signal then enters a chroma or colour demodulator which separates the difference signals into the three basic colour signal components. These red, green and blue signals drive the television tube electron beams and display the picture. A sample of this video signal is also fed to the synchronising circuits. The timing pulses, which are an integral part of the signal, are stripped away in the sync separator. These send the correct voltages to the deflection circuitry to produce the organised picture scanning.*

variations from complete darkness to bright white images. But the amplitude variations of the colour signal must be a complex representation of both the illumination and colours of the image originally captured by the television camera. To this end, the colour signal includes a "colour sync burst" which is 8 to 11 cycles of a 4.43 MHz sine wave in both the PAL and SECAM formats (a NTSC broadcasts uses a 3.58 MHz sine wave). This waveform is inserted immediately after the horizontal sync pulse on the horizontal blanking pulse and serves to ensure that the recreated colours on the TV set match the initial scene at the studio. Any variance will cause the demodulated colour to change so, for example, a flesh tone could be portrayed as green or yellow.

A colour camera decomposes a scene into a signal for each of the three primary colours, red, green and blue, from which all other colours can be reconstructed. The luminance (Y) signal, which corresponds to the intensity of the original picture and from which a b/w scene can be composed, is assembled from all three colour signals. The three primary colours produce white light when combined in the proper proportions: 30% red, 59% green and 11% blue. This mixture can be expressed as follows:

$$Y \text{ (white light)} = 0.3R + 0.59G + 0.11B$$

From this equation, it can be seen that any variation in the intensity of the basic colours, will cause a change in the luminance level. This is displayed on a black and white picture as a change in brightness or gray level.

The raw red, green and blue video signals are processed in a matrix circuit that produces three different signals: the luminance or Y signal and two colour difference signals. The Y signal by itself is sufficient to recreate a b/w picture. The colour information is contained in the colour difference signals, R-Y and B-Y, each composed of two of the three basic signals. The R-Y signal, also known as Q or U, has an orange or cyan tone; the B-Y signal, also known as I or V, has a green or purple colour.

Each difference signal is modulated onto a 4.43 MHz PAL/SECAM subcarrier (or a 3.58 MHz NTSC subcarrier). The two subcarriers, which are 90° out-of-phase with each other, are then combined to form the chrominance or colour signal (C). The method used is a form of sideband modulation wherein the central carrier frequency is suppressed following modulation. The television receiver must then recreate the carrier frequency and phase in order to properly demodulate the colour information. This colour burst in the video waveform serves as the reference signal to accomplish this task by allowing the television receiver's colour oscillator to be locked in step with the phase and frequency of the input video signal. Note that the colour burst is purposely chosen to have a frequency identical to that of the colour subcarrier.

The luminance and chrominance signals are then combined or frequency multiplexed to form the composite baseband video signal. This waveform is modulated onto the carrier for transmission to a television receiver (see Figure 2-109). This process is reversed in a TV set so the the red, green and blue signals can be recreated.

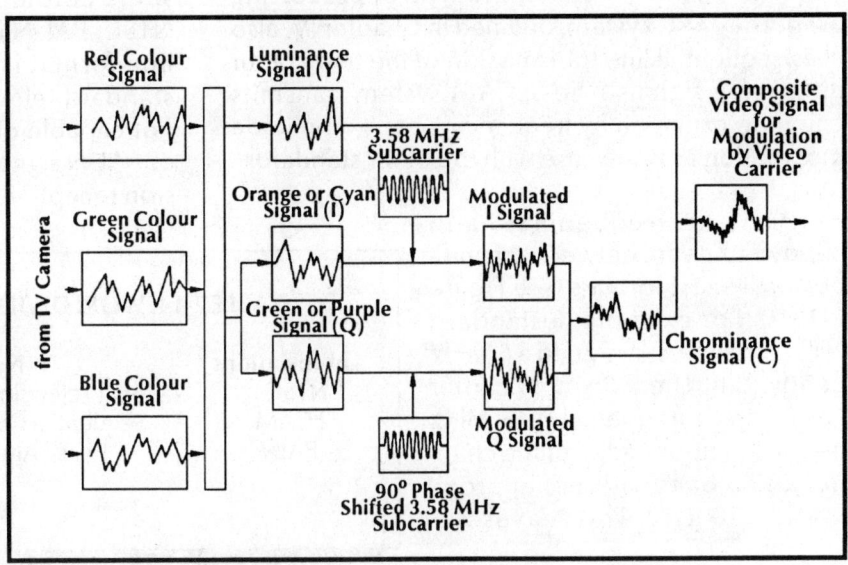

Figure 2-109. NTSC Colour Television Signal. *The three raw colour signals, red, blue and green, from a television camera are combined to create the luminance signal (Y) and two difference signals (I and Q). The luminance information is sufficient to recreate a black/white picture. The difference signals are modulated onto two 3.58 MHz carriers having phases differing by 90° and then are combined to form the colour or chrominance signal. The chrominance and luminance signals are mixed together into the baseband composite video signal for modulation onto a subcarrier. The 3.58 MHz subcarrier must be relayed along with the composite signal so that the raw colour signals can be properly extracted from the difference signals*

M. WORLDWIDE BROADCAST FORMATS

PAL, SECAM and NSTC

Three standard television broadcast systems, the NTSC, SECAM and PAL, were created when colour television was introduced (see Table 2-14). The PAL and SECAM standards evolved from and were improvements upon the original NTSC format. Although these standards all use a similar method of scanning, they differ in the number of lines in each frame and in the way that the colour information is encoded. Their development was based on the line frequency and the channel allocation scheme adopted in each country. The field frequency is matched to the line frequency which is typically either 50 or 60 Hertz.

The PAL system slightly altered the way colour was transmitted in the original NTSC standard to minimise the effects of phase distortion on colour fidelity. When phase errors occur, some difficulty is experienced with recreating the original raw colour signals from the difference signals. In the PAL format, the phase of the colour signal is inverted from line to line. The name, "phase alternating line" stems from this modification of the NTSC signal. As a result, any phase errors experienced will have an opposite effect on sequential lines and the viewer's eye will average out and eliminate moderate amounts of colour distortion.

The SECAM system transmits only one colour difference signal per line. Two lines are therefore required to recover the colour information so that some averaging occurs. The name "sequential colour with memory" reflects the type of processing used. The MAC system, outlined in Chapter IV, also uses sequential line transmission of the two colour difference signals. The SECAM system transmits colour information by frequency, not by amplitude modulation as is the case with the other standards.

Carrier frequencies and bandwidths vary between the various broadcast formats (see Figure 2-110). For example, a standard NTSC channel is assigned a 6 MHz bandwidth. The FM sound carrier has a centre frequency 0.25 MHz below the upper edge of the channel and a bandwidth of approximately 50 kHz. This leaves 5.7 MHz available for the video carrier. The centre of the video carrier is then placed 1.25 MHz above the lower edge of the channel. Therefore, the audio carrier is located at 4.5 MHz above the video carrier. Since most of the video information is contained in the region above the 1.25 MHz video carrier centre frequency, in effect, the composite baseband audio and video carriers occupy a band of frequencies ranging from 1.25 MHz to 5.75 MHz plus 50 kHz or just under 4.6 MHz. This composite signal contains the video signal, the blanking information and the horizontal and vertical synchronisation pulses. The reason for concentrating this signal in the middle of the 6 MHz bandwidth is to minimise adjacent channel interference.

Channel bandwidths of both 7 and 8 MHz are also used in television broadcast systems around the world. Table 2-16 lists the formats and standards for both VHF and UHF broadcasts and the line voltage and frequency in use in various nations. Those cases where a standard for satellite service has been established are indicated. Table 2-17 outlines terminology used in describing the North American over-the-air channel assignments.

Most nations use one of five variations on the three basic video formats in use: PAL-B, SECAM, NTSC, and PAL-N or PAL-M. PAL-N is the standard that will be used in Argentina and may be adapted by other Latin American countries in the future, while PAL-M is used in Brazil. PAL-M incorporates all the advantages of the normal PAL system scaled down to a 4.2 MHz video bandwidth and uses the same colour burst pattern and line number as NTSC. PAL-N is similar to PAL-M except that a 625 line format is used. Unfortunately, many multi-standard televisions and VCRs on the market are not capable of adequately interpreting the PAL-M or PAL-N signal. The result is black and white television reception.

TABLE 2-14. WORLDWIDE TELEVISION STANDARDS

Abbreviations	Name	Year Developed
NTSC	National Television System Committee	1948
SECAM	Sequential Color à Mémoire	1957
PAL	Phase Alternating Line	1961

COMPONENT OPERATION

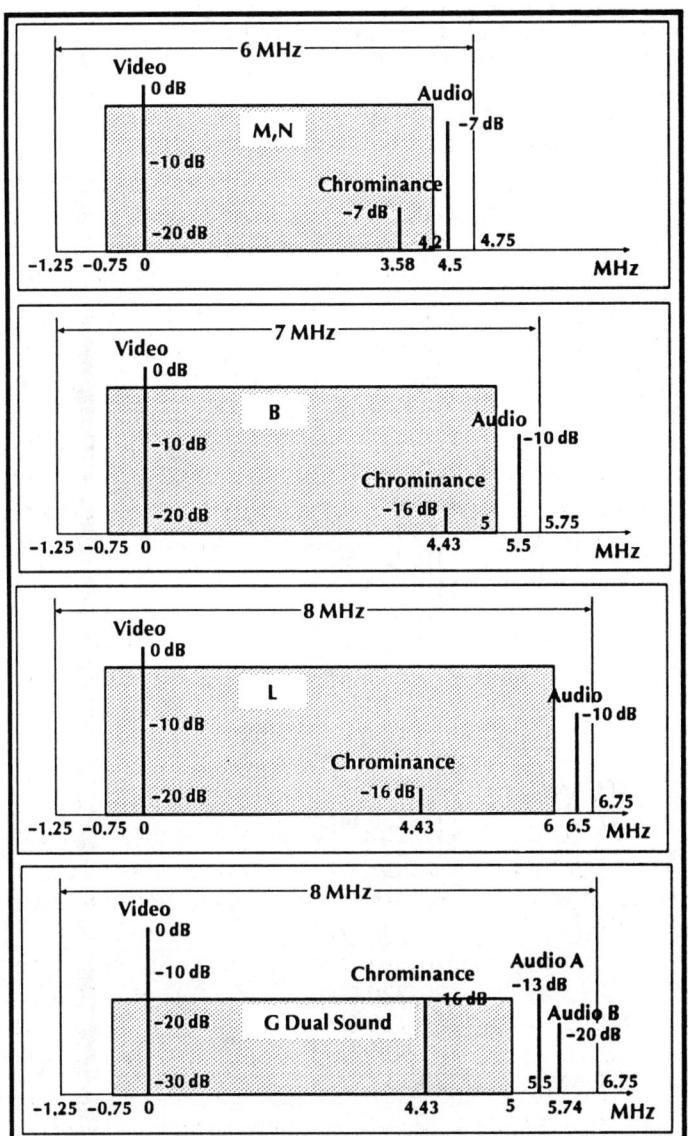

Figure 2-110. Comparison of Worldwide Television Standards. *The basic 6, 7 and 8 MHz wide channels used in the various transmission formats are compared here. The levels of the various carriers and subcarriers are adjusted to minimise interference between each other and between adjacent channels.*

Most eastern European countries and the USSR presently transmit SECAM broadcasts, but production and processing costs are forcing them to consider the possibility of a changeover to PAL. Nearly all televisions now sold in Europe are multistandard and can decipher either PAL or SECAM broadcasts. This design upgrade is not difficult to implement because the colour decoder chip employed is dual standard. Thus only a few extra components are required to give a television receiver a multistandard capability.

Many of the televisions sold in Europe in the last few years are based upon a digital sub-assembly known as DIGIT2000, introduced by ITT in the early eighties. This chip series allows the video to be digitally processed so that effects like picture-in-picture and frame stores can be incorporated in the television receivers.

BROADCAST FORMATS

COMPONENT OPERATION

TABLE 2-15. WORLDWIDE TELEVISION BROADCAST STANDARDS

Country	Broadcast Standard VHF	UHF	Colour System	Satellite Service	Line Voltage - Frequency Standard
Afghanistan	B		PAL		220-50
Albania	B	G			220-50
Algeria	B		PAL	PAL	127/220/380-50
Angola					I220-50
Antigua	M		NTSC		110-50
Argentina	N	N	PAL	PAL	220-50
Australia	B		PAL	PAL	240/415-50
Austria	B	G	PAL		220/380-50
Azores	B		PAL		220-50
Bahamas	M		NTSC		110/220-60
Bahrain	B		PAL		230-50
Bangladesh	B		PAL		220/440-50
Barbados	M		NTSC		110/220-50
Barbuda	M		NTSC		110/220-50
Belgium	B	H	PAL		220-50
Benin	K				220/380-50
Bermuda	M		NTSC		115/230-60
Bolivia	M	N	NTSC		110/220-50
Brazil	M	M	PAL	PAL	110/127/220-50/60
Brunei	B		PAL		110/127/220-50/60
Bulgaria	D	K	SECAM	SECAM	220-50
Canada	M	M	NTSC	NTSC	115/230-60
Canary Islands	B		PAL		110/220-50
Chile	M		NTSC		220/380-50
China	D	K	PAL		220-50
Colombia	M		NTSC	NTSC	110/220-60
Congo	D				220-50
Costa Rica	M		NTSC		120/220-60
Cuba	M	M	NTSC	NTSC	115/220-60
Cyprus	B	G	PAL		240-50
Czechoslovakia	D	K	SECAM	SECAM	220-50
Denmark	B	G	PAL		220/380-50
Diego Garcia	M		NTSC		110/220-60
Djibouri	K		SECAM		220-50
Dominican Rep.	M	M	NTSC		110-60
Dubai (UAE)	B	G	PAL	PAL	220-50
East Germany	B	G	SECAM	SECAM	220/380-50
Ecuador	M		NTSC		110/220-60
Egypt	B		SECAM		220-50
El Salvador	M		NTSC		110/220-60
England	A	I	PAL	PAL	240/415-50
Ethiopia	B		PAL		127/220-50127/220-50
Faroe Islands	B		PAL		220-50
Finland	B	G	PAL		220/380-50
France	E	L	SECAM	SECAM	various-50
Gabon	K		SECAM		220-50
Ghana	G		PAL		230-50
Gibraltar	B		PAL		240-50
Greece	B	H	SECAM	SECAM	220-50
Greenland	B		PAL		220/380-50
Guadeloupe	K		SECAM		220-50
Guam	M		NTSC		110/220-60
Guatemala	M		NTSC		110/220/127-60

BROADCAST FORMATS

COMPONENT OPERATION

Country	Broadcast Standard VHF	UHF	Colour System	Satellite Service	Line Voltage - Frequency Standard
Guyana	K		SECAM		127/220-50
Guinea Republic	K		SECAM		220/380-50
Guinea-Equatorial	B				220/380-50
Haiti	M		NTSC		110/220-50/60
Holland	B	G	PAL	PAL	220-50
Honduras	M		NTSC		110-60
Hong Kong	B	I	PAL		200-50
Hungary	D	K	SECAM	SECAM	220-50
Iceland	B	G	PAL		220-50
India	B		PAL	PAL	230/400-50
Indonesia	B		PAL	PAL	127/220-50
Iran	B	G	SECAM		220-50
Iraq	B		SECAM		220/380-50
Ireland	I/A		PAL		220/380-50
Israel	B	G	PAL		230/400-50
Italy	B	G	PAL	PAL	220/380-50
Ivory Coast	K		SECAM		220/380-50
Jamaica	M		NTSC		110/220-50
Japan	M		M	NTSC	110/220-50/60
Johnson Island	M		NTSC		110/220-60
Jordan	BG		PAL		220-50
Kenya	B		PAL		240-50
Korea (South)	M	M	NTSC		110/60
Korea (North)	D		PAL		220-50
Kuwait	B		PAL		240-50
Laos	M				127/220-50
Lebanon	B		SECAM		110/190-50
Liberia	B		PAL		120-60
Libya	B		SECAM		127/130-50
Luxembourg	C	G/L	PAL/SECAM		110/220-50
Madeira	B		PAL		220/380-50
Madagascar	K		SECAM		110/220-50
Malawi	B				230-50
Malaysia	B		PAL	PAL	230/400-50
Maldives	B		PAL		220-50
Malta	B	H	PAL		240-50
Martinique	K		SECAM		220-50
Mauritania	K		SECAM		220/380-50
Mauritius	B		SECAM		240-50
Mexico	M		NTSC	NTSC	110/220-60
Midway Islands	M		NTSC		110/220-60
Micronesia	M		NTSC		
Monaco	E	G/L	PAL/SECAM		127/220-50
Mongolia	D		SECAM		220-50
Morocco					
Mozambique	B		PAL		220-50
Namibial	PAL				220-50
Netherland Ant.	M		NTSC		127/220-50-60
Netherlands	B	G	PAL	PAL	220-50
New Caledonia	K		SECAM		220-50
New Zealand	B		PAL		230/400-50
Nicaragua	M		NTSC		120-60
Niger	K		SECAM	SECAM	220/380-50
Nigeria	B	G	PAL	PAL	230/400-50
N.Mariana Island	M		NTSC		110/220-60

BROADCAST FORMATS

COMPONENT OPERATION

Country	Broadcast Standard		Colour System	Satellite Service	Line Voltage - Frequency Standard
	VHF	UHF			
Norway	B	G	PAL	PAL	230-50
Oman	B	G	PAL	PAL	220-50
Pakistan	B		PAL		230/400-50
Panama	M		NTSC		110/115/120/126-60
Paraguay	N		PAL		220-50
Peru	M		NTSC	NTSC	220-60
Philippines	M	M	NTSC	NTSC	110/220-60
Poland	D	K	SECAM	SECAM	220/380-50
Portugal	B	G	PAL	PAL	220/380-50
Puerto Rico	M	M	NTSC		120-60
Qatar	B		PAL		220-50
Reunion	K		SECAM		220-50
Romania	D		SECAM		220-50
Rwanda	K				220-50
Samoa-American	M		NTSC		230-50
Sarawak	B		PAL		220-50
Saudi Arabia	B	G	PAL/SECAM	SECAM	127/220/380-50/60
Senegal	K		SECAM		127/220-50
Seychelles	B		PAL		230-50
Sierra Leone	B		PAL		230-50
Singapore	B	G	PAL		230/400-50
Society Islands	K				240-50
South Africa	I	I	PAL	PAL	220/380-50
Spain	B	G	PAL	PAL	127/220/380-50
Sri Lanka	B		PAL		220-50
St. Kitts	M		NTSC		220/60
St. Pierre	K		SECAM		220-50
Sudan	B		PAL		240-50
Surinam	M		NTSC		110/115/127/220-60
Swaziland	B	G	PAL		230-50
Sweden	B	G	PAL		220/380-50
Switzerland	B	G	PAL		220/380-50
Syria	B		SECAM		115/200-50
Tahiti	K		SECAM		127-60
Taiwan	M		NTSC		110/200-60
Tanzania	B		PAL		230/440-50
Thailand	B	G	PAL	PAL	230/380-50
Trinidad/Tobago	M		NTSC		115/230-60
Togo	K		SECAM		127/220-50
Trust Territory	M		NTSC		110/220-60
Tunisia	B		SECAM		115/220/380-50
Turks-CaicosIs.	M		NTSC		110/220/440-60
Turkey	B	G	PAL		220/380-50
Uganda	B		PAL		240/415-50
United Arab Em.	B	G	PAL		220-50
United Kingdom	A	I	PAL	PAL	240/415-50
U.S.A.	M	M	NTSC	NTSC	110/220-60
Upper Volta	K				220/380-50
Uruguay	N		PAL		220-50
U.S.S.R.	D	K	SECAM	SECAM	127/220-50
Venezuela	M		NTSC	NTSC	120/240-60
Vietnam	D		SECAM		120/127/230-50
Virgin Islands	M		NTSC		110-60
Wake Island	M		NTSC		110/220-60
West Germany	B	G	PAL	PAL	220/380-50

BROADCAST FORMATS

Country	Broadcast Standard		Colour System	Satellite Service	Line Voltage - Frequency Standard
	VHF	UHF			
Yemen(South)	B		PAL		230-50
Yemen (North)	B		NTSC		230-50
Yugoslavia	B	G	PAL		220/380-50
Zaire	K		SECAM	SECAM	220-50
Zambia	B		PAL		220-50
Zimbabwe	B		PAL		230/240-50

KEY TO TABLE

Broadcasting System	Number of Lines	Channel Bandwidth (MHz)	Audio/Video Separation (MHz)	Audio Modulation	Line/Field Frequency
A	405	5	3.5	FM	50/10125
B	625	7	5.5	FM	50/15625
C	625	7	5.5	AM	50/15625
D	625	8	6.5	FM	50/15625
E	819	14	11.15	AM	50/20475
G	625	8	5.5	FM	50/15625
I	625	8	6.0	FM	50/15625
K	625	8	6.5	FM	50/15625
L	625	8	6.5	AM	50/15625
M	525	6	4.5	FM	60/15750
N	625	6	4.5	FM	50/1562

TABLE 2-16. NORTH AMERICAN CHANNEL ASSIGNMENTS

Band Designation	Channel Numbers	Frequency Range (MHz)
Over-the-air Channels		
Lo-VHF	2 - 6	54 - 88
M Band		88 - 108
I-VHF	7 - 13	174 - 216
UHF	14 - 69	470 - 806
Cable TV Channels		
Sub-VHF	T7 - T13	5.75 - 47.75
Lo-VHF	2 - 6	54 - 88
Mid-Band	98(A2) - 99(A1)	108 - 120
	14(A) - 22(I)	120 - 174
Hi-VHF	7 - 13	174 - 216
Superband	23(J) - 36(W)	216 - 222
Hyperband	37(AA) - 62(ZZ)	300 - 456
	63 - 64	456 - 648

COMPONENT OPERATION

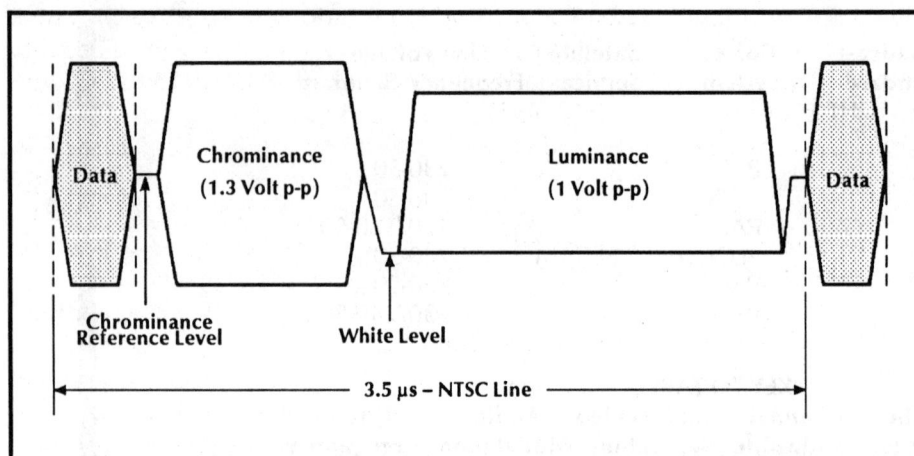

Figure 2-111. MAC Transmission Format. *Data, chrominance and luminance information are transmitted in a compressed form at different times within one horizontal scan line. This technique avoids some of the problems associated with picture distortion experienced with the conventional broadcast formats.*

The MAC System

The MAC (multiplexed analog components) is an entirely new and different broadcast system. In conventional PAL, NTSC or SECAM frequency multiplexed broadcasts, the audio/data, colour (chrominance) and black and white (luminance) signals are multiplexed together into one signal for transmission (see Figure 2-111). The television receiver circuitry must then break this signal down into the original components. While this is an adequate design, various types of picture distortions can and often do occur. In the MAC system the audio/data, chrominance and luminance signals are compressed in time and then relayed sequentially on each scan line. Each of the two colour difference signals are transmitted on alternate lines. This time division multiplexing transmission method avoids any interaction between the various components and can result in better quality reproduction. The MAC system has some other noticeable advantages. The synchronization information requires only 0.2 percent of the total transmission time compared with over 20 percent for conventional systems! The space and bandwidth which has been freed up can be filled with other digital information. The sync is extremely "rugged" meaning that the line trigger point is unlikely to be mistaken by a television receiver so that horizontal picture "tearing" rarely occurs. Colour distortion is minimized and the available colour bandwidth is increased. Since subcarriers are not used with the MAC system, the transmitted video signal-to-noise ratio can be higher so that satellite receivers can be pushed to their low threshold limit and performance can be improved. Antennas up to 20 percent smaller in surface area may be used in comparison to those required for reception of PAL, SECAM or NTSC satellite broadcasts. MAC receiver electronics has been designed so that the raw red, green and blue video signals are available for use with some of the more sophisticated video monitors now available.

Audio channels in the MAC format are designed to be compatible with the highest fidelity sound. Transmission uses variable pre-emphasis or spectral companding (see below for more details) to minimise the amount of channel noise.

A variety of different MAC systems have been created (see Table 2-17). The differences lying principally in the way in which the data and audio information is transmitted. For example, in C-MAC, a possible choice for future European Ku-band DBS (direct broadcast systems), the RF carrier is time multiplexed during the horizontal blanking interval. Up to eight high fidelity audio channels can be relayed. D2-MAC is also being considered for use in Europe. In B-MAC, now being implemented in Australia and used in the Holiday Inn's North American hotel system, data is time multiplexed at baseband allowing accommodation of six audio channels.

In Europe, the dominant standard PAL will likely be replaced by the new MAC format within the next ten years or so. (The structure of the MAC system is examined in more detail in Chapter IV.) The use of DIGIT2000 chip design, mentioned in the previous section, makes the introduction of MAC easier. The colour standard of the composite video signal can be decoded before digitisation so the MAC signal can be input at this point. So it is feasible to design and manufacture an economical PAL/SECAM/MAC multistandard television receiver.

The MAC system has an advantage in being upgradable to high definition television (HDTV). Unlikely other proposed HDTV systems, the customer may not have to buy a new, possibly expensive television receiver. Since the majority of European television receivers are being designed to be capable of multistandard operation, upgrading a television to MAC may consist of simply adding an additional printed circuit board. Since the MAC system uses primarily digital processing it might be possible to introduce a digital HDTV system that would be compatible with existing MAC decoders.

The MAC system is a serious contender to become a world television format. It has already been introduced and is now readily available. Since it is European developed with most system patents being held in Europe, European television manufacturers have a head start compared to American and Far East companies. There is thus the motivation to make the MAC system a European standard. As East Europe and USSR adopts new governments, economic systems and policies, they will be heavily influenced by the technology of the European Community and may therefore choose MAC by default. The wide use of MAC in Europe may possibly force US and Japanese markets into adopting this system. This series of event could effectively result in MAC becoming the dominant world standard. Of course, the familiar NTSC, PAL and SECAM standards will certainly continue to be used as a low cost alternative to the newer systems.

TABLE 2-17. MAC AUDIO FORMATS

	Frequency Multiplexed	Time Multiplexed
Baseband	A-MAC	B-MAC
RF	D-MAC	C-MAC

N. MONOPHONIC AND STEREO AUDIO RECEPTION

High quality audio programs can be transmitted via satellite, either in conjunction with the video or as separate signals. A number of different formats are used to accomplish this task. However, they all share some common methods used to improve broadcast fidelity.

Transmitting High Fidelity Sound

One primary objective in broadcasting sound is to minimise the noise added to the signal en route to its final destination. Due to physical limitations of the FM link, this is usually supported by a second objective, that of masking the noise which cannot be avoided. Most of the annoying noise is high frequency hiss and static.

The type and amount of masking used is related to the type of audio being transmitted. Therefore, for example, background noise is much more noticeable if the program content is a soft bass note played by a cello rather than a raucous electric guitar run burned onto a rock album. Over the years, sound engineers have developed some elegant masking techniques so that an audio signal can be processed within the confines of a given bandwidth and power level to sound as clean and accurate as possible.

Three related techniques can be identified: pre-emphasis and de- emphasis: amplitude companding and spectral companding, a more general form of the first method. There are presently three patented companding methods: the Dolby, which is quite familiar to audiophiles, the dbx and the Telefunken. Surprisingly, none of these improve the overall signal-to-noise ratio of the audio information. They do, however, very effectively reallocate the energy in the signal to a frequency pattern which matches that perceived as clean, high fidelity sound by the human ear.

Pre-Emphasis and De-Emphasis

Pre-emphasis was developed in conjunction with FM radio to counteract the increasing influence of noise at higher frequencies. A signal-to-noise ratio of 60 dB at low audio frequencies may

COMPONENT OPERATION

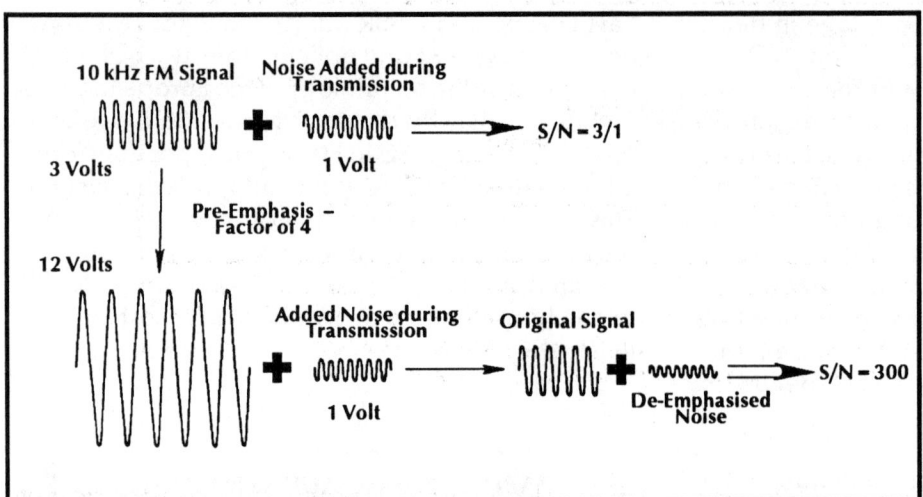

Figure 2-112. Pre-emphasis and De-emphasis. *This illustrates how the signal-to-noise ratio can be markedly improved by pre-emphasising before transmission and de-emphasising after transmission. The amount of pre-emphasis applied to the higher frequency components of an FM signal is progressively increased to counteract the tendency towards an increase in detected noise and the associated reduction in the signal-to-noise ratio.*

be degraded by more than 10 dB in the higher frequency portion of the audible spectrum. In order to remedy the situation, higher frequency components of the audio signal are boosted in level before transmission. The noise added during transmission is then lower relative to the signal. In de-emphasis, the opposite of pre-emphasis processing, higher frequency components are reduced in level by the same amount (see Figure 2-112). This action effectively restores the original signal while eliminating a good portion of the undesirable noise.

All FM radio broadcasts use the standard "75 microsecond" pre-emphasis. This time factor simply defines the relative amount of boosting that occurs over frequencies within the audio spectrum. While most North and South American satellites also use this standard audio processing, both 50 and 75 microsecond pre-emphasis as well as a standard known as J17 are used on European audio transmissions. Note that pre-emphasis is used with both monophonic and stereo broadcasts. Table 2-18 presents some examples of pre-emphasis formats used on European satellite transmissions.

Amplitude Companding

Pre-emphasis has one limitation since it is activated when high frequency information is present but has no masking effect on noise at low frequency, low amplitude signals which are not far above the noise floor. A second method developed to improve perceived audio fidelity is to compress signals at the transmitting end (see Figure 2-113) This means that the higher voltage portions of the signal are attenuated while those nearer zero amplitude are increased in level. This has the effect of beefing up the weaker parts of the signal so that the dynamic range is compressed or "squeezed." A compressed signal has a higher average level and therefore may have more apparent loudness than an uncompressed one, even though the peaks are no higher in value. To illustrate, the patented dbx method typically reduces signals at maximum am-

TABLE 2-18. A BRIEF SAMPLE OF EUROPEAN SATELLITE AUDIO SUBCARRIERS

Channel	Satellite	Subcarrier (MHz)	Type	Bandwidth (kHz)	Pre-emphasis (μsec)
Sky Channel	Astra 1A	6.60	Mono	280	50
		7.02	L Stereo	130	50
		7.70	R Stereo	130	50
Teleclub	Eutelsat 1 F-4	6.50	Mono	280	75
TV-5	Eutelsat 1 F-4	6.65	Mono	900	J-17

COMPONENT OPERATION

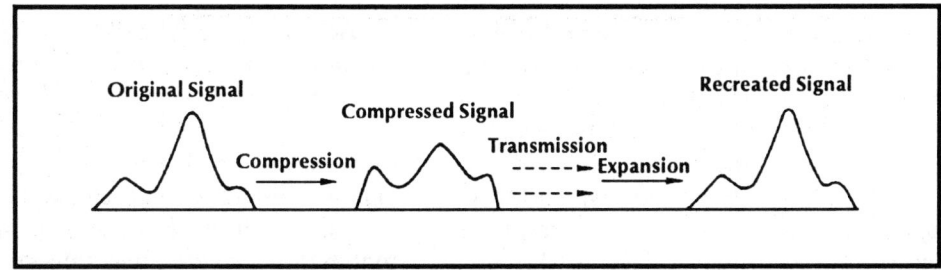

Figure 2-113. Amplitude Companding. *A signal can be compressed so that lower voltage portions are increased and higher voltage portions are decreased in order to increase the signal-to-noise ratio of those portions nearer the noise floor. This is a commonly used noise reduction technique in audio broadcasts.*

plitude by 30 percent, leaves signals at 8.99 percent of peak unchanged, and boosts signals of 1 and 0.1 percent of peak by 3 and 95 percent, respectively.

The mirror amount of amplitude expansion applied at the receiver restores the original signal. This chain of compression and expansion is known as amplitude companding.

Spectral Companding

When using fixed pre-emphasis alone, two potential difficulties still remain. First, a fixed amount of pre-emphasis applied to those audio signals containing predominantly high frequencies could boost them too much. This is known to audio engineers as lack of headroom, namely insufficient capacity for the peaks of the program to be cleanly transmitted. Second, the fixed pre-emphasis could be insufficient to adequately boost the very low amplitude, high frequency signals and therefore may not sufficiently mask the channel noise.

The solution to this communication problem is spectral companding. A spectral compressor continuously monitors the frequency composition of the audio signal and varies the amount of pre-emphasis accordingly. When small amounts of high frequency signal are present, the spectral compressor provides high amounts of pre-emphasis. When high frequency components are strong, it actually contributes de-emphasis in order to prevent high-frequency overload. The end result is that the system frequency response is dynamically adjusted and the transmitted signal consistently contains the optimal masking of channel noise and results in a signal perceived as being of high fidelity.

Satellite Audio Reception

Audio signals can arrive at a satellite receiver in various forms. Typically, audio messages are transmitted as subcarriers modulated onto the main video carrier. For example, NTSC subcarriers are inserted in the 5.0 to 8.5 MHz range with 5.51 MHz being a familiar frequency used on Ku-band broadcasts. PAL and SECAM subcarriers often have 6.60 and 7.20 MHz centre frequencies but, as is the case with NTSC transmissions, other frequencies are frequently chosen (see Figure 2-114).

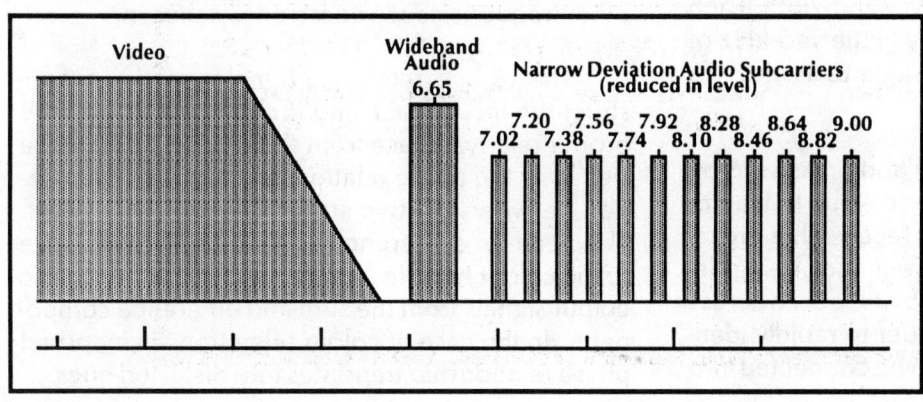

Figure 2-114. Audio Subcarriers. *A relatively large portion of the available bandwidth, in addition to that for the video information, is reserved for audio subcarriers*

AUDIO RECEPTION

COMPONENT OPERATION

It is also not unusual, especially on Intelsat spacecraft, for the audio portion of a television broadcast to be in a single channel per carrier format (SCPC) which is transmitted independently of the video carrier. In some instances, this SCPC signal may be relayed on a second transponder. In these cases, a separate audio receiver is required for reception. The SCPC format is also used for hundreds of other audio transmissions including telephone relays and radio broadcasts.

The audio signal may also be relayed by a technique known as sound-in-sync. This term refers to inserting a digital audio signal into either the horizontal or vertical blanking intervals of the standard television signal. This is perfectly feasible because these intervals, which occupy nearly 20 percent of the transmission time, have little information content. Using sound-in-sync eliminates the need for separate audio subcarriers and allows the transponder power to be concentrated on the video information so that its signal-to-noise power can be increased.

Digital audio signals are transmitted within the blanking intervals in two principal satellite scrambling systems, the M/A COM VideoCypher II and the Oak Orion. Digital audio is also used as an integral component of the television signal in the new Scientific Atlanta MAC broadcast format. These topics are explored in more detail in Chapter IV.

Single Channel per Carrier Audio

Audio messages having their own carriers and frequency assignments can also be broadcast independently of video carriers. Hundreds of single channel per carrier (SCPC) signals can be transmitted over each satellite transponder. By contrast, a single 36 MHz wide transponder can manage at most 20 to 25 audio subcarriers because the video signal requires a relatively wide bandwidth. Each SCPC carrier may occupy from as little as 5 kHz of bandwidth for voice-grade messages to as much as 250 kHz for high fidelity audio.

SCPC signals are detected and processed by an audio receiver, in essence a beefed up FM radio, known as a scanner. This type of receiver has digital read-outs indicating the exact frequency being detected. It generally has the capability to scan across the entire frequency band in order to rapidly identify active carriers. A scanner can be connected into a TVRO at either the block downconversion output or the final IF loopthrough via a DC-blocking, two-way power splitter. If the block output is used, the scanner can tune to any transponder while only one channel can be observed when it is connected via the satellite receiver 70 MHz IF loop. It is essential that the input frequency range of any scanner match that provided by a satellite system. For example, one which was capable of detecting carriers in the 950 to 1450 MHz range would miss those European SCPC broadcasts falling within the 1450 to 1700 MHz upper portion of the available band. However, some brands are compatible with almost any TVRO output. For example, the Icom America IC-R7000 scans from as low as 20 MHz to as high as 2000 MHz.

TVRO Stereo Reception

Presently, there are four different systems used to transmit stereo sound over satellite circuits: multiplex, Warner Amex, discrete and processed narrow deviation stereo. Once this stereo is decoded by either a stereo processor or a satellite receiver with a built-in processor, the baseband composite audio signal, namely the raw audio signal, can be fed into a home stereo system.

Multiplex Stereo

The multiplex stereo system is similar to that used in regular FM radio broadcasts. The left (L) and right (R) audio channels are combined in a circuit called a matrix that produces the sum and difference signals, L + R and L - R. The difference signal is modulated onto a 38 kHz subcarrier. Then the L + R baseband signal, the modulated L - R difference signal and a 19 kHz reference signal are combined or multiplexed together onto the audio subcarrier (see Figure 2-115). Satellite circuits often use audio subcarriers at 5.51, 6.2, 6.6, 6.8 and 7.0 MHz, although other frequencies are occasionally chosen.

The 19 kHz reference signal is used to reconstruct the individual L and R components with the proper relative phase from the sum and difference signals. If the phase relationship is garbled, separation between the two stereo channels will suffer. This need for a reference signal is similar to the use of the colour burst in reconstructing the three basic colour signals from the sum and difference components. In the case of colour television, an incorrect phase relationship translates into distorted hues.

AUDIO RECEPTION

COMPONENT OPERATION

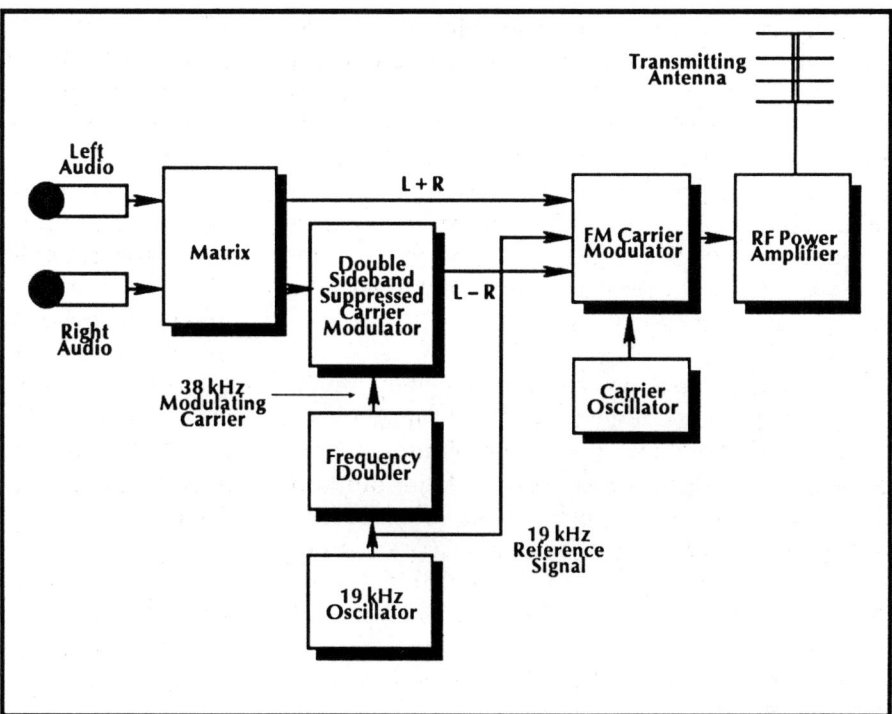

Figure 2-115. Multiplex Stereo. *A multiplex stereo circuit processes the left and right audio channels so that both can be relayed on a single audio subcarrier and then reconstructed at the receiver. The signals are processed into sum and difference components. A 19 kHz reference signal is used at the receiver to adequately separate the two stereo channels. Different forms of multiplex stereo use companding and emphasis to reduce perceived background noise.*

Warner Amex Stereo

The Warner Amex stereo system transmits the sum and difference signals over two separate subcarriers. In North American C-band broadcasts the L + R signal is frequency modulated usually onto a 5.8 MHz subcarrier; the L - R signal onto either a 6.62 or 6.8 MHz subcarrier. Any combination of these subcarriers or others which are located in the audio subcarrier range within the television channel bandwidth can also be used. When signals are demodulated at the satellite receiver, they are combined or matrixed to recover the original left and right channel information.

Discrete Stereo

This is the simplest and most widely used satellite stereo system. The left and right channels are not matrixed but each is simply relayed on a separate audio subcarrier.

Processed Narrow Deviation Stereo

Processed narrow deviation stereo is another form of discrete transmission except that very narrow deviation modulation is used. Deviation is a measure of how much the carrier wave is varied from its centre frequency. For example, a frequency modulated 10 kHz carrier which varied between 9 and 11 kHz deviates less than the same carrier varied between 7 and 13 kHz. There is a trade-off involved here. Transmissions having more narrow deviations and which occupy less frequency spectrum unavoidably contain greater amounts of noise. Therefore, companding techniques are used in conjunction with processed narrow deviation stereo.

There are two principal forms of processed narrow deviation stereo. The Wegener system uses spectral companding while the Lemming also employs a second stage of amplitude companding. This results in an apparent 15 dB improvement in signal-to-noise ratio in the Lemming system. This improvement is in addition to the approximately 18 dB apparent contribution of the first spectral companding stage. Note that this improvement is called apparent because the average signal-to-noise ratio is the same with or without companding. The improvement is in perceived fidelity. The signal is processed or creatively readjusted to match the sensitivity of the human ear.

Processed stereo transmission has one principal advantage over conventional discrete methods. Many more audio subcarriers can be relayed since the deviation is reduced and less bandwidth is used. A satellite receiver must have a narrowband audio filter to receive such audio broadcasts.

COMPONENT OPERATION

III. DESIGNING AND TESTING KU-BAND SYSTEMS

A. OVERALL DESIGN CONSIDERATIONS

The performance quality of a Ku-band system depends upon the overall characteristics of the uplink-to-satellite-to-reception antenna link as well as upon the type of equipment chosen for receiving the satellite signal. It is crucial to design a system that not only performs well when first installed on a sunny day but that will deliver high quality audio and video on a rainy day six years in the future.

The Link Equation

The prime criterion in designing an satellite reception system is to generate a signal-to-noise ratio (S/N) that properly drives a television receiver. Although standards for acceptable S/Ns vary, the signal entering a satellite receiver should have a C/N at least at or above threshold. If so, the receiver would operate in a linear region where a 1 dB increase in C/N results in a 1 dB increase in S/N.

The S/N available at the receiver output depends upon a number of factors beyond the control of a downlink system designer. It varies with the input C/N, the amount of pre-emphasis used at the uplink as well as a factor known as FM improvement. FM improvement increases as the frequency deviation of the FM signal varies around the centre frequency of the carrier signal. If the FM deviation is low as in, for example, a half transponder format with limited bandwidth, even if the C/N exceeds threshold so that sparklies are absent, thermal noise may degrade video quality.

To reiterate, the link equations are used to calculate C/N at the input to a satellite receiver, not the S/N at its output. The value of C/N depends upon both satellite link parameters as well as upon the choice of antenna, LNB, cable, connectors and satellite receiver.

The two most important choices in designing a satellite reception system are antenna gain and LNB noise temperature. The antenna and LNB chosen depend upon a number of factors including variations in satellite output power, downlink antenna pointing accuracy, satellite stationkeeping accuracy, atmospheric attenuation of microwaves, aiming and deformation error of the receiving antenna caused by wind loading, signal scattering due to obstructions caused by wind-blown branches or airplane crossings and changes in local humidity, temperature and precipitation.

The effective isotropic radiated power (EIRP) at any particular location is a good measure of parameters that reflect to satellite operation. But EIRP calculated from satellite design and location may not correspond exactly to what is measured at a receiving site, especially because the output power of a satellite transponder slowly decreases as it ages. Calculated EIRP is based upon an assumption that the downlink antenna is perfect. In addition, published EIRPs do not account for the fact that the downlink antenna has side lobes and nulls between side lobes. Therefore, areas far off axis may receive more power than expected, especially if they happen to be on the target of a side lobe peak (see Figure 3-1). This phenomena is called antenna spillover. For example, a Ku-band satellite downlink antenna transmitting a 43 dB spot beam to Europe may have its first side lobe relaying power far off boresight to the Middle East or to the eastern coast of North America. These effects are usually de-

SYSTEM DESIGN & TESTING

tected only by on-the-spot tests. This information can be quite useful to other local TVRO enthusiasts.

The factors discussed here are described in the "link" equation. Footprint maps are first used to determine EIRP, the signal power aimed towards any geographic location. The analysis then accounts for the weakening of earthbound microwaves as they spread out in space and are partially absorbed by various components of the atmosphere. The receiving antenna captures and concentrates these signals according to its gain. The job of deciphering satellite signals is more difficult the higher the system noise temperature, which is in large part determined by antenna and LNB characteristics.

All these factors can be included in the link equation which is expressed as:

Carrier-to-noise power ratio entering receiver = Signal power leaving the satellite, EIRP

less

Free space path losses and atmospheric absorption

plus

Antenna gain, G

less

Noise introduced by the antenna, LNB and other system components

or, in a more condensed form, in which these symbols simply express the above words in algebra:

$CNR = EIRP - \text{Path Loss} + G - 10 \log kT_{sys} B$

$= EIRP - PL + G - 10 \log T_{sys} - 10 \log B - 10 \log k$

$= EIRP - PL + G - 10 \log T_{sys} - 10 \log B + 228.6$

All the terms in this equation are expressed in decibels. T_{sys} equals the noise temperature of the antenna, low noise amplifier and all other system components. $kT_{sys} B$ expresses how noise is related to noise temperature and system bandwidth. As either noise temperature or bandwidth is increased, more noise enters the sys-

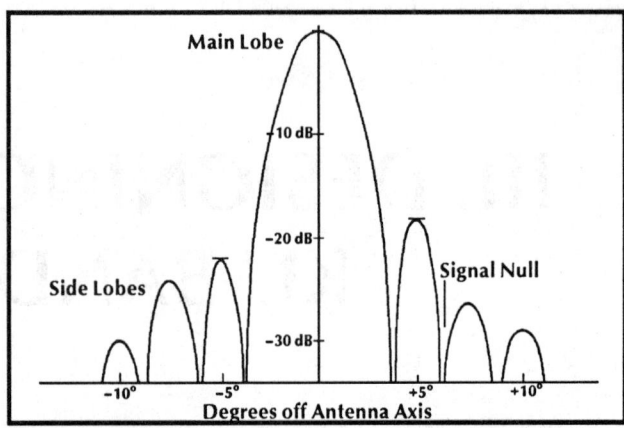

Figure 3-1. Side Lobes and Out-of-Footprint Reception.
A typical downlink antenna radiation pattern is shown in this diagram. The main lobe in the centre has the highest gain while the side lobes powers are reduced by 18 to 22 dB relative to those of the main lobe. Under favourable conditions it is possible to receive a broadcast from a satellite's side lobes in a geographical location far removed from the desired footprint centre. This phenomena is known as downlink antenna spillover.

tem, so this term is subtracted to show its effect on C/N. The term k is a universal constant called Boltzman's constant which is equal to 1.38×10^{-23} joules/°K.

Therefore, if EIRP, path loss and bandwidth are known, the G/T_{sys} required to obtain a desired CNR can be calculated by addition and subtraction. The term G/T_{sys} is shorthand for

$G - 10 \log T_{sys}$

where both gain and noise temperature are expressed in decibels. The G/T_{sys} is the "figure of merit" of an antenna/feedhorn/LNB combination. The "bottom line" G/T_{sys} defines the combination of minimum antenna diameter and LNB noise temperature required so that CNR input is just at the receiver threshold. Most good quality satellite receivers have thresholds in the range of 8 dB. Clearly, each link in the satellite circuit determines the characteristics that all the other components must exhibit.

EIRP

The EIRP of a particular satellite can be easily determined for any geographic location from a published footprint map. Although calculated footprint maps are available for most communication satellite (see Figure 3-2), on-site measurements of EIRPs are available for only a limited number of spacecraft. As discussed above, these can differ

SYSTEM DESIGN & TESTING

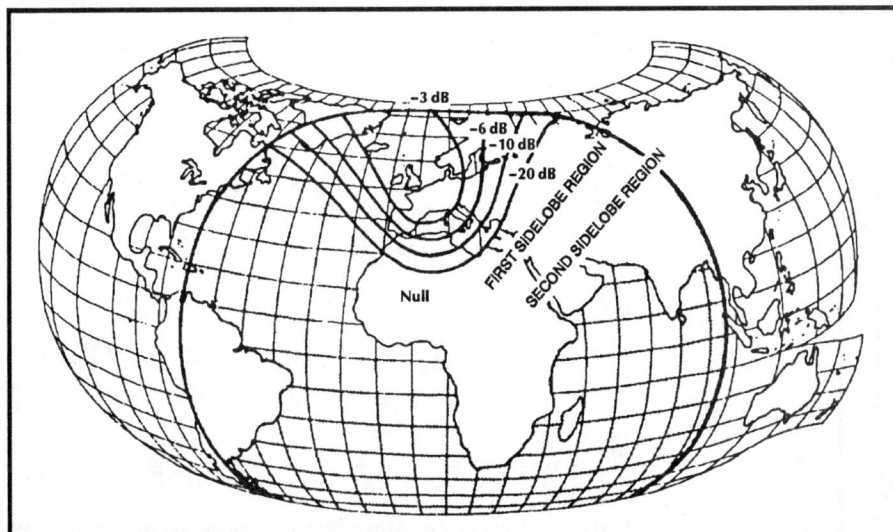

Figure 3-2. Calculated Radiation Patterns from ECS-F1 Western Spot Beam. *The downlink antenna on ECS-F1 resulted in a pattern of sides lobes in out-of-target regions as indicated in this diagram. As the boresight EIRP increases, side lobe levels also increase in power. This phenomena, known as antenna spillover, allows sufficiently strong signals to be detected at an off-boresight location with use of relatively large antennas.*

somewhat from theoretical expectations. Also, the maps show only the power levels radiated from the downlink antenna's main lobe. Power levels of signals from downlink antenna side lobes, the antenna spillover, are rarely published because even small movements in a satellite's orbital position can radically alter their effect. Realise that if the aiming of a narrow side lobe is changed even slightly it will not direct any power to a previously illuminated area on the earth below.

Path Loss

Path loss, the second term in the link equation above, expresses how the power of the downlinked signal is attenuated in traveling from a satellite to a receiving antenna. There are two effects at work. First, microwaves spread out and are weakened on their earthbound journey. It is evident that this must occur since the energy radiated from the small surface area of a downlink antenna ends up blanketing a portion or all of a continental land mass. This free space path loss is simply a function of the distance between a satellite and the receiving antenna (see Figure 3-3). Formulas for calculating path loss and slant range, the distance from any site to any satellite, are given in Appendix B.

Effects of Rain and Atmospheric Attenuation

The second, much less predictable, cause of path loss is absorption of signal by the earth's atmosphere. By far the greatest reduction in power is caused by water vapor and rain; this attenuation is approximately 9 times greater at Ku-band than at C-

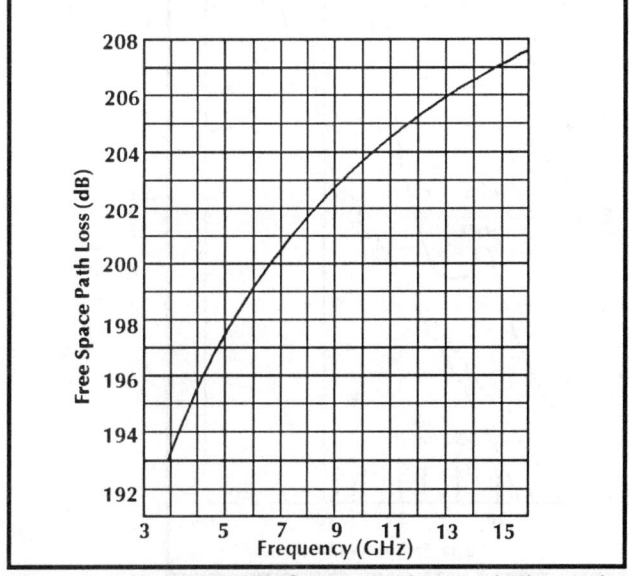

Figure 3-3. Free Space Path Loss. *This graph shows the free space path loss for a site at the equator. As frequency increases so does pass loss. On a clear day it is 196.3 dB at 4 GHz, while it increases to 205.8 dB at 12 GHz. Signal losses increase as the path traversed through the atmosphere increases. Consequently, losses are greatest for nearly horizontal paths at low elevation angles. Additional free space path losses incurred at 45° and 0° elevations are 0.44 dB and 1.33 dB, respectively.*

band for a given amount of water in the downlink path (see Figures 3-4 and 3-5). The higher the humidity or rainfall, the larger is this reduction in power level. A torrential downpour can cause a power loss of 10 dB or more and complete picture wipe-out in a system not designed with an adequate margin of safety. Note that since uplinks operate at even higher frequencies, in the 14 to 16 GHz range, rainfall attenuation can have an even more dramatic effect on this portion of the link.

SYSTEM DESIGN & TESTING

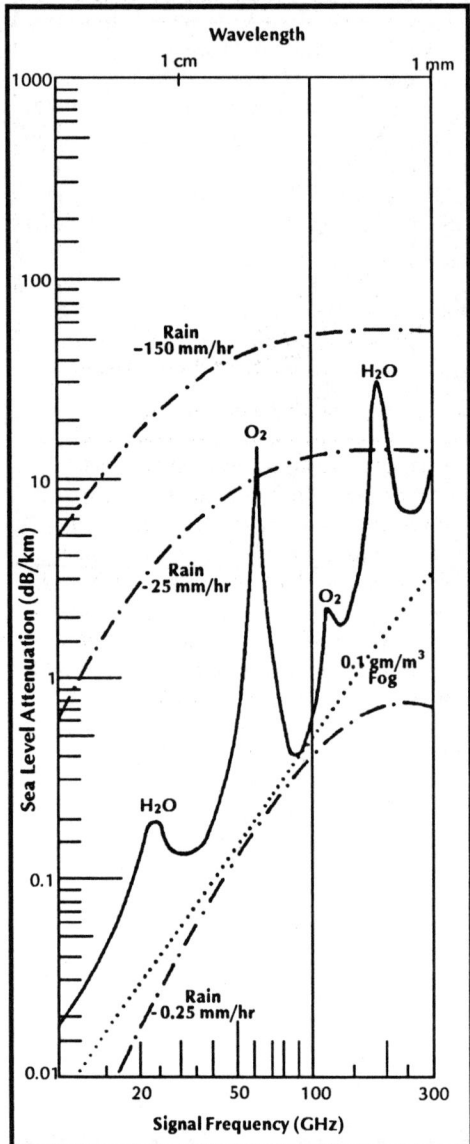

Figure 3-4. Specific Attenuation of Signals by Water Vapour. *Water absorption is a major cause of signal losses at frequencies up to about 20 GHz where resonant molecular absorption of microwaves begins to dominate. At even higher frequencies other resonant peaks for both water and oxygen in the atmosphere have a major effect. Rain droplets can scatter as well as absorb satellite signals as they travel to and from the satellite. This graph also shows signal attenuation in decibels caused by various rainfall rates as well as a characteristic fog density of 0.1 g/m³.*

Engineers have gathered a great deal of data to describe the effects of rainfall on a satellite link. The descriptive measure used is termed the "availability," which is defined as the percentage of time that reception is not disrupted, evaluated during the

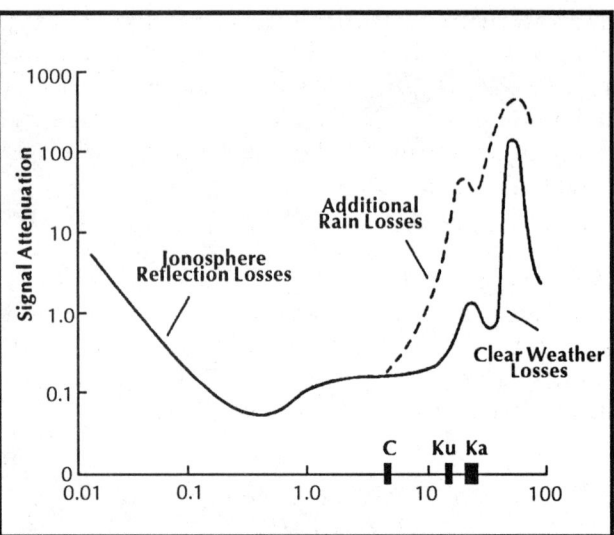

Figure 3-5. Atmospheric Attenuation of Microwaves. *Rain absorbs and scatters microwaves causing signal losses and depolarisation as well as an increase in detected noise temperature. This graph shows the average results of the more detailed absorption patterns outlined in Figure 3-4.*

month when rainfall is heaviest. For example, a 1 metre antenna might produce an acceptable picture 99% of the time. Therefore during the remaining 1% of the time, which equals about 7 hours in the worst month, reception will be poor. By comparison, a C-band system under the same conditions would experience just half an hour of marginal reception. The availability of the Ku-band link could be improved by increasing antenna size or by installing an LNB with a lower noise figure. If reception is poor during the early hours of the morning or at other non-critical times for television reception, it may have little effect on customers. However, when Ku-band links serve telephone networks, CATV headends or computer data lines down-times would be unacceptable even for minutes if reliable backups did not exist.

Regions of the earth's surface can be classified according to the average yearly rainfall in order to estimate availability of Ku-band broadcasts (see Figures 3-6, 3-7 and 3-8. Also see Figure 3-9). This allows a correlation to be made between availability and the extra margin of power required in a satellite receiving system. It is clear, for example, from Table 3-1 that a margin of 1.0 dB above threshold is required at C-band to achieve 99.9% availability but this requirement for extra gain increases to 5.2 dB at 12.0 GHz. These figures are computed at an antenna elevation of 30°. As the reflector is inclined more towards the horizon, losses increase since the

SYSTEM DESIGN & TESTING

view towards any satellite is made through a thicker layer of atmosphere. Table 3-2 shows that in extremely rainy regions, a system requiring high availability, as is usually the case in commercial systems, requires an especially large antenna and a low noise LNB.

Rain also has an additional effect on system performance. On an average summer day, raindrops have a noise temperature of 290°K which is substantially higher than the background temperature of clear sky or space. Therefore rainfall also raises antenna noise temperature.

TABLE 3-1. RAIN ATTENUATION FOR TYPICAL SITE

Availability (% of time)	Attenuation 3.95 GHz	12.0 GHz
98	0.3	0.56
99	0.5	2.6
99.9	1.0	5.2
99.99	1.6	7.5

TABLE 3-2. WORST CASE DOWNLINK RAIN ATTENUATION (dB)

Percent Availability	B	D1	D2	D3	E
99.00	1	1	1	1	1
99.50	1	1	1	2	2
99.80	1	2	3	4	6
99.90	2	3	4	6	9
99.95	3	5	6	8	13
99.98	5	8	10	13	20
99.99	7	11	14	17	25

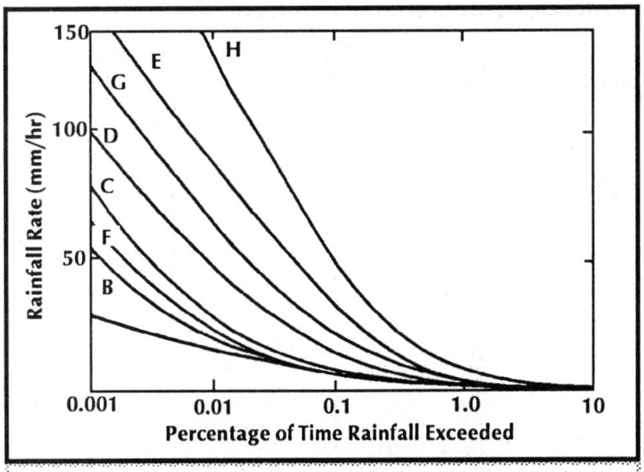

Figure 3-6. Percentage of Time Rainfall Exceeded.
This graph which corresponds to the accompanying map in Figure 3-6 shows the percent of time that rainfall exceeds any given rate as expressed in millimetres per hour. For example, in region B a rate of 5.6 mm/hour is exceeded 0.1 percent of the time. The rainfall rates at 99.9% availability can be used to compute signal attenuation and arrive at the following recommended TVRO performance margins assuming a 49 dBw EIRP:

ZONE	FADE MARGIN (dB)
E	10.0
D	6.3
B	3.7
C	3.3
F	2.0

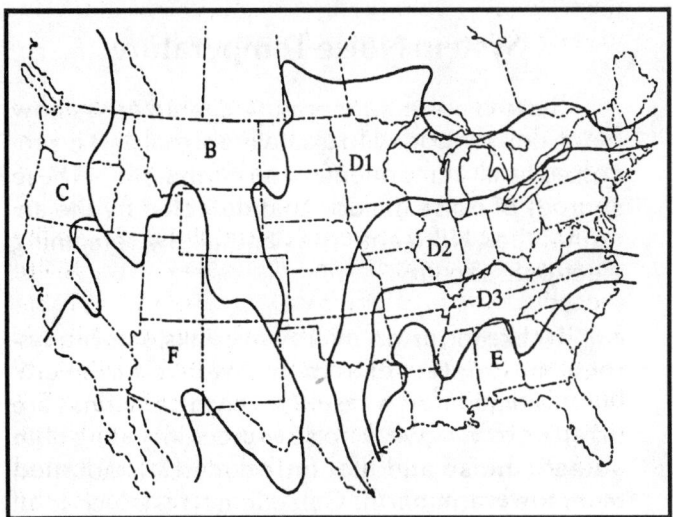

Figure 3-7. U.S. Rainfall Zones. *This map divides the continental United States into rain zone areas. The heaviest signal attenuation at Ku-band can be found in zone E with a rain rate of 35.4 millimetres per hour being exceeded 0.1% of the time.*

SYSTEM DESIGN & TESTING

Some satellites slated for launches in the future are being designed to have footprints that partially compensate for rainfall absorption. Therefore, for example, one spacecraft destined for broadcasting to the eastern seaboard of the United States may direct a stronger beam towards the state of Florida where rainfall is often heavy.

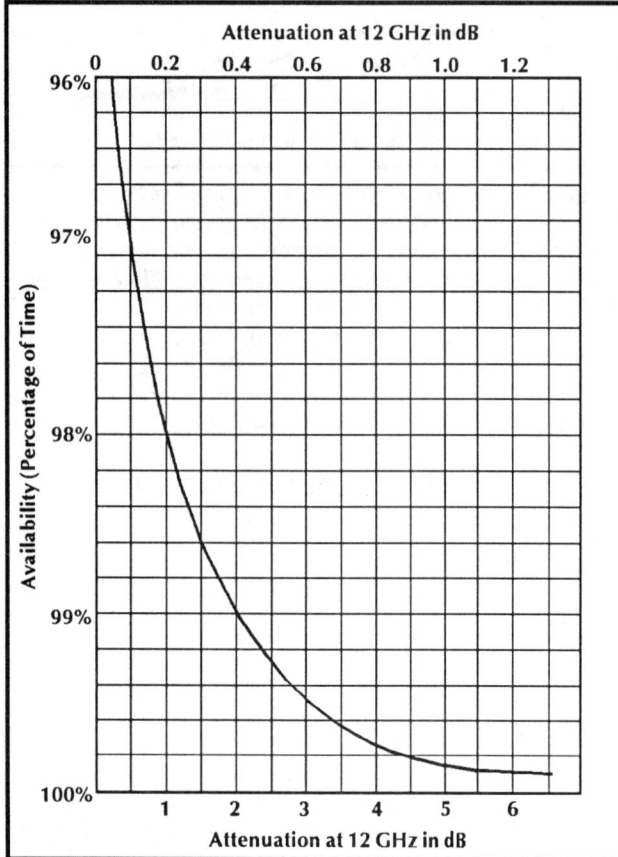

Figure 3-8. Probability of Attenuation. *This graph shows the attenuation at C and Ku-band as the expected availability increases. For example, if a 99.99% availability is desired, the performance margins should equal or exceed the expected attenuations of 1.6 dB at C-band and 7.5 dB at Ku-band. However, for 98.00% availability, much lower C and Ku-band margins of 0.20 and 1.0 dB, respectively, are required. It is assumed here that the antenna is aimed at a 30° elevation angle. If satellites near the horizon at very low elevation angles are targeted, signal attenuation increases.*

Typical Ku-band Path Loss

A typical value for path loss from a Ku-band satellite on a clear day is −205.8 dB (see Appendix B). This is equivalent to a reduction in signal power by a factor of over 3.8×10^{-20} (380 billion billion)! This number reminds us that communication satellites are over 36,000 kilometres away in outer space. Detecting Ku-band satellite signals is similar to attempting to observe a 50 watt light bulb from such an enormous distance.

It is interesting to compare Ku-band path loss with that of a C-band link which is typically about −196.3 dB. The difference between 205.8 and 193.3 dB is 9.5 dB. This means that Ku-band transmissions are attenuated by an additional factor of 8.9 compared to C-band broadcasts.

Antenna Gain

The receiving antenna intercepts and concentrates signals from transponders with power outputs ranging from 20 to 100 watts that have been attenuated by approximately 205.8 dB on their journey earthward. The higher the antenna gain, the better a TVRO will perform. Below a minimum gain, even the best LNB cannot compensate enough to adequately capture the signal essential to reconstruct a satellite broadcast.

Although antenna gain has been examined in detail in Chapter II, one interesting point can be noted here. Since gain increases with the square of the signal frequency, and since Ku-band signals are triple the frequency of C-band transmissions, gain is increased by a factor of 9, equal to 9.5 dB. This increase in gain is comparable to the 8.9 dB increase in path loss at Ku-band frequencies. The difference, 0.6 dB, amounts to just a 15% improvement in performance at Ku-band on a clear day with no extra rain attenuation. It is therefore clear that reflectors with smaller surface areas can be used at Ku-band rather than at C-band only because higher transponder power is available.

System Noise Temperature

System noise temperature characterises how much noise is added to a satellite signal as it is processed by all components of an earth station. These sources of noise include that detected by the antenna, the LNB, cable runs and all the remaining electronic circuitry.

Unlike terrestrial microwave links, satellite systems must detect extremely low levels of microwave radiation. As a result, receiving antennas are sensitive to very weak noise sources including both galactic noise and that introduced by radiation from the warm earth. Galactic noise, which is all

SYSTEM DESIGN & TESTING

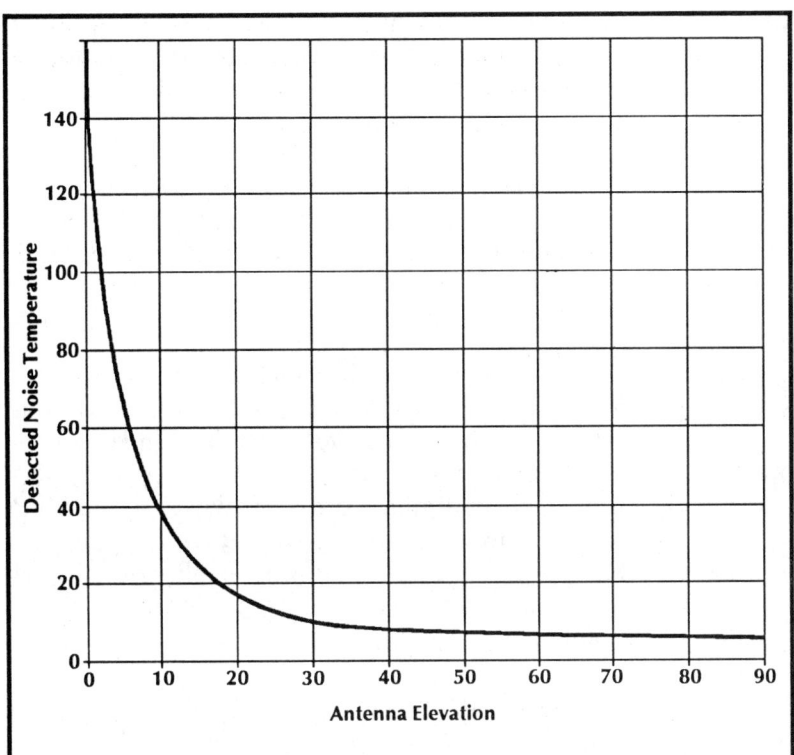

Figure 3-9. Ground and Cosmic Noise Contributions. *This graph shows how noise temperature increases at Ku-band on a clear night as antenna elevation is varied from pointing straight into space to along a line parallel with the horizon. Cosmic noise at Ku-band is approximately 6°K while the average ground temperature assumed here is 290°K. Below 30° elevations ground noise begins to have an effect and detected noise temperature increases. Note that this graph will vary for satellite systems having different side lobe patterns.*

that an orbiting antenna pointed into deep space would detect, typically contributes about 6°K to a Ku-band TVRO.

The noise added to a TVRO by the warm earth at an average temperature of 290°K is detected via antenna side lobes. It depends upon both the antenna's side lobe patterns and its angle of elevation. As a dish is inclined towards the horizon it faces closer to the ground and subsequently detects more ground noise. Since the elevation to satellites at the center of the geosynchronous arc is higher than that of those at the easterly and westerly portions, contributed antenna noise will be lower when the central satellites are targeted (see Figure 3-9).

The noise performance of an antenna can be improved in a number of ways. A higher quality antenna of a given size has lower level side lobes and therefore is susceptible to less noise. A larger diameter antenna also has a "tighter" side lobe pattern and a narrower main lobe so it generally contributes lower amounts of noise. Similarly, a deeper antenna having a lower f/D, all else being constant, will detect lower amounts of ground noise.

The noise introduced by the LNB is simply expressed by its noise temperature or figure. Therefore, for example, a 190°K LNB adds 190°K of noise to a TVRO.

The noise contributed by the cables and satellite receiver is low by comparison. This is because the LNB amplifies both the noise and signal detected by the antenna/feedhorn system. Its output signal is substantially higher in power than the noise added by cables, connectors and the satellite receiver. Therefore, the balance of system components has only a small effect, typically about 5°K, on system noise temperature. G/T_{sys}, the figure of merit for an earth station, can therefore be closely approximated by replacing T_{sys} with the noise contribution of just the antenna and low noise amplifier. The LNB output signal is thus the key determinant of overall performance quality of a TVRO.

Nevertheless, noise generated in cable runs can be a major factor in some situations. For example, when signal attenuation in long cables increases to the point that the signal power is not substantially above the system background noise, performance can suffer. This may occur in the connection between the LNB and receiver or in an extensive SMATV distribution system. If, for example, line amplifiers are not used in appropriate locations, an antenna/LNB system having a high G/T_{sys} may not function properly.

DESIGN CONSIDERATIONS

SYSTEM DESIGN & TESTING

How Small Can an Antenna Be?

Although the link equation is a powerful tool in determining the correct choice of antenna diameter and LNB noise temperature given the EIRP of any communication satellite, one factor not considered in this type of analysis is antenna beamwidth. Even if EIRP were hundreds of watts and the link equation suggested that an extremely small parabolic dish would perform well, antenna beamwidth limits minimum dish size. Therefore, a 0.25 metre antenna would have a 7° beamwidth, much too wide for reception of Ku-band satellites spaced one or even two degrees apart. Such a small antenna would detect relatively strong signals from seven satellites that were spaced one degree apart!

In order to clarify how the link equation can be used to determine required antenna diameter and LNB noise temperature, some sample calculations are presented here. The first analysis shows how to calculate C/N for an antenna with given characteristics. The second works backwards to show how the figure of merit, G/T_{sys} and subsequently antenna diameter can be obtained from an assumed C/N.

Determining C/N

Assume that the free space path loss on a clear day is −205.8 dB. Substituting this value into the link equation yields

$$CNR = EIRP + G - 10 \log T_{sys} - 10 \log B + 22.8$$

Assume also that the satellite receiver bandwidth is 27 MHz. This choice should allow reception of a television picture with good fidelity. (If the bandwidth were further reduced, the C/N would be increased at the expense of picture fidelity as discussed in Chapter II). The link equation becomes

$$CNR = EIRP + G - 10 \log T_{sys} - 51.51$$

The next step is to include the antenna and LNB noise temperatures in this equation. Assume that a 100°K LNB is used with a 2.4 metre (7.9 foot) antenna having a noise temperature of 35°K and a gain of 48.1 dB. These values, which are taken from Table 2-4, are those measured for an Andrews antenna aimed at a 15° elevation.

The noise temperature at this low elevation angle is used because it presents a near worst case value for antenna noise. A reflector would point at such a low elevation in high northern or southern latitudes and near the ends of the satellite arc even as far south as the equator.

Plugging the above values for noise temperatures and gain into the link equation yields

$$C/N = EIRP + 48.1 - 10 \log(35 + 100) - 51.5$$
$$= EIRP - 24.70$$

Therefore, if the EIRP is 45 dBw, the system delivers a C/N of 20.3 dB to the receiver. This is about 12 dB above most receiver thresholds and provides more than enough fade margin during all but the most torrential rain storms. However, if the EIRP were 33 dBw on the edge of a downlink antenna footprint, the C/N would be 8.3 dB which is just above threshold with no margin for rain fade or pointing inaccuracies.

If the 1.0 metre antenna in Table 2-4 had been used at an elevation of 30°, following the same logic yields:

$$C/N = EIRP - 32.63$$

Therefore, an EIRP of 45 dB would result in a C/N of 12.4 dB which is just a few decibels above threshold. This particular antenna would function only under near ideal circumstances near the downlink antenna boresight where EIRP is highest. An EIRP of 33 dBw would result in a C/N of 1.63 dB, well below threshold of any receiver. However, such an antenna is more than adequate when receiving a satellite with a more powerful spot beam with an EIRP of 50 dBw.

Determining G/T_{sys}

Begin with the same assumptions made above for path loss and receiver bandwidth. The link equation can be expressed in two equivalent ways

$$C/N = EIRP + G - 10 \log T_{sys} - 51.51$$
$$= EIRP + G/T_{sys} - 51.51$$

Rearranging the algebraic terms yields

$$G/T_{sys} = C/N - EIRP + 51.51$$

G/T_{sys} can be determined from this equation if both the EIRP and required C/N are known. For example, assume that a particular location has an

EIRP of 41 dBw and that a C/N of 14 dB, 6 dB above a threshold of 8 dB, is acceptable. Then

$$G/T_{sys} = 14 - 41 + 51.51 = 24.51 \text{ dB}$$

Assuming a worst case antenna noise temperature of 50°K and the availability of a 90°K LNB leads to

$$G - 10 \log (90 + 50) = 24.51$$

$$\text{or } G = 45.97 \text{ dB}$$

Referring to Table 2-2, such a gain can be obtained using a 70% efficient 2.0 metre antenna or a 50% efficient 2.5 metre antenna. If the margin requirement were to be dropped by another 2 dB to 12 dB above threshold, then a 70% efficient, 1.5 metre reflector would suffice. However, if the smaller antenna were used, the probability of rain outage would increase.

Recommended Antenna Sizes

The link equation has been used to construct Table 3-3 which lists recommended minimum antenna diameters necessary to attain a given C/N. The assumptions here are that the free space path loss on a clear day is −205.8 dB, antenna efficiency is 65%, satellite receiver bandwidth is 27 MHz, the system noise temperature is 140°K and signal frequency is 11.5 GHz. If noise temperature is lower, bandwidth is reduced or efficiency is increased, the minimum acceptable antenna diameter decreases. These recommendations can therefore be regarded as the worst case scenario.

Four values for C/N were chosen. The lowest, C/N = 8 dB, is a typical minimum value for receiver threshold. This allows no margin for signal fading due to rain, wind or other factors. The highest allows for a 9 dB margin, sufficient for all but the most severe rainstorms.

The effect of increasing the necessary performance margin or the EIRP is quite clear from this table. A very powerful Ku-band satellite radiating 60 dBw would allow the use of small, easy-to-install antennas. However, in weak footprint areas, a relatively large, high performance antenna is necessary to obtain adequate reception.

TABLE 3-3. RECOMMENDED ANTENNA DIAMETER VERSUS EIRP

EIRP (dBw)	Antenna Diameter (metres)			
	C/R=8	C/R=11	C/R=14	C/R=17
30	3.63	5.13	7.24	10.23
31	3.24	4.57	6.46	9.12
32	2.88	4.07	5.75	8.12
33	2.57	3.63	5.13	7.24
34	2.29	3.24	4.57	6.46
35	2.04	2.88	4.07	5.75
36	1.82	2.57	3.63	5.13
37	1.62	2.29	3.24	4.58
38	1.45	2.04	2.88	4.07
39	1.29	1.82	2.57	3.63
40	1.15	1.62	2.29	3.24
41	1.02	1.45	2.04	2.88
42	0.91	1.29	1.82	2.57
43	0.81	1.15	1.62	2.29
44	0.72	1.02	1.45	2.04
45	0.85	0.91	1.29	1.82
46	0.58	0.83	1.15	1.62
47	0.51	0.72	1.02	1.45
48	0.46	0.65	0.91	1.29
49	0.41	0.58	0.81	1.15
50	0.36	0.51	0.72	1.02
51	0.32	0.46	0.65	0.91
52	0.29	0.41	0.58	0.81
53	0.26	0.36	0.51	0.72
54	0.23	0.32	0.46	0.65
55	0.20	0.29	0.41	0.58
56	0.18	0.26	0.36	0.51
57	0.16	0.23	0.32	0.46
58	0.15	0.20	0.29	0.41
59	0.13	0.18	0.26	0.36
60	0.11	0.16	0.23	0.32

(1 metre = 3.28 feet)

Assumptions:
Path Loss = −205.8 dB
Receiver Bandwidth = 27 Mhz
Antenna Efficiency = 65%
System Noise Temperature = 140°K
Carrier Frequency = 11.5 GHz

B. BEAMWIDTH AND SATELLITE SPACING

There is no mistaking the fact that smaller antennas are more appealing to all except the rare individual who thinks bigger always means better. As Ku-band satellite power levels have risen and as satellite reception technology has improved, ever smaller antennas have become capable of producing excellent quality video. However, the spacing between satellites puts a lower limit on antenna size because beamwidth increases as reflector diameter decreases. Beyond a certain beamwidth, more than one satellite can be detected by an antenna.

Satellite spacing was not an issue for C-band systems in the United States until 1980. RCA Americom had been operating two satellites, Satcom I and II, Western Union had three vehicles, Westar I, II and III, and AT&T/GTE operated Comstar I, II and III. There was more than ample orbital space and spacing between satellites was wide. In 1981, the American Federal Communications Commission (FCC) surprised the industry by requesting comments on the possibility of authorizing spacing of satellites along the North and South American portion of the geosynchronous arc at tight 2° intervals. Until then, the accepted policy was 4° and 2° degree spacing for C and Ku-band satellites, respectively.

In 1982 the American FCC approved 2° orbital spacing for satellites operating in both frequency bands. This decision, implemented gradually by reducing spacing to 3° at first, would eventually allow a doubling of the orbital arc capacity. At this time. the FCC also ruled that higher power DBS satellites would be spaced at 9° intervals.

The discussion of satellite spacing has come full circle in North America. As satellite transponder powers have increased and as smaller antennas have become more capable of good quality reception, many are now arguing that all broadcast satellites should be spaced a minimum of 3° apart. This strategy would be central in the attempt to avoid interference from adjacent communication spacecraft.

The Importance of Beamwidth

Antenna quality and beamwidth are critical factors in satellite spacing (see Table 3-4). For example, if the half power (3 dB) beamwidth of a reflector is 2°, then at half of this value or 1° on either side of the main axis, signals are received at 3 dB lower than those detected along the boresight. If a satellite happens to be located just 1° off the target and is transmitting on the same channel and polarity, two garbled pictures and the resulting interference would appear on the television screen.

TABLE 3-4. ANTENNA SIZE VERSUS 3 DB BEAMWIDTH

Antenna Diameter (metres)	Theoretical Half Power Beamwidth
0.5	3.43
1.0	1.72
1.5	1.14
2.0	0.86
2.5	0.69
3.0	0.57
3.5	0.49

However, the situation would be quite different if the antenna detected signals from one degree off boresight at powers reduced by 10 dB, or by 15 dB. At what level is a picture judged unwatchable when interference from an adjacent satellite is just barely detectable? Subjective tests have shown that an interfering signal was first noticeable when its level was 11 dB below the wanted signal. It was annoying when received at –5 dB and quite unacceptable at –4 dB.

This means that a one metre antenna will see interfering signals reduced by only 3 dB at 1.72° off its main axis. Levels will even be higher at 1.0°. However, a good quality 3 metre antenna would have unwanted signals reduced by substantially more than 3 dB at 1.0° off-axis.

Most prime focus parabolic antennas have adequately low side lobes so that off-axis signals are not appreciably detected. However, smaller diameter Cassegrain fed antennas can have side lobes only 10 dB lower in power than the main lobe. Such reflectors are more susceptible to interference from adjacent satellites as well as from terrestrial sources.

SYSTEM DESIGN & TESTING

The Role of Antenna Quality

An antenna with an accurate surface and a well-matched feed assembly is closer to optimal performance and has a smaller beamwidth than one of lesser quality. This is because measured beamwidths are also determined by antenna efficiency. Also, because smaller antennas require substantially less material for construction than comparable larger reflectors, leftover manufacturing dollars can be spent in achieving better surface accuracy. The other side of the coin is that better accuracy can be achieved with more manageable, smaller antennas. The net result is that a well made 1 m antenna may have less than triple the beamwidth of a poorly constructed 3 m unit as is theoretically expected.

Is One Degree Spacing Really One Degree?

The published separation between satellites is measured by their differences in longitude. From the perspective of an observer at the centre of the earth, the angular separation between two satellites directly overhead would be the same as their longitude difference. But an antenna located on the earth's surface pointing straight upwards would see these satellites separated by more than one or two degrees. This spacing decreases as satellites closer to the horizon are targeted but still remains in excess of the nominal value (see Figure 3-10)

When a TVRO antenna is located at other points on the surface of the earth, the perceived separation between satellites will also have values in excess of or equal to the satellite spacing. For example, from Honolulu, Hawaii satellites at 131° and 132° longitude will appear to be not 1.0° but 1.14° apart. It is important to realise that these angles are simply related to the geometry of the situation. The maximum spacings occur when an antenna is aimed due south, while the minimum ones occur when it points towards either horizon (see Table 3-5).

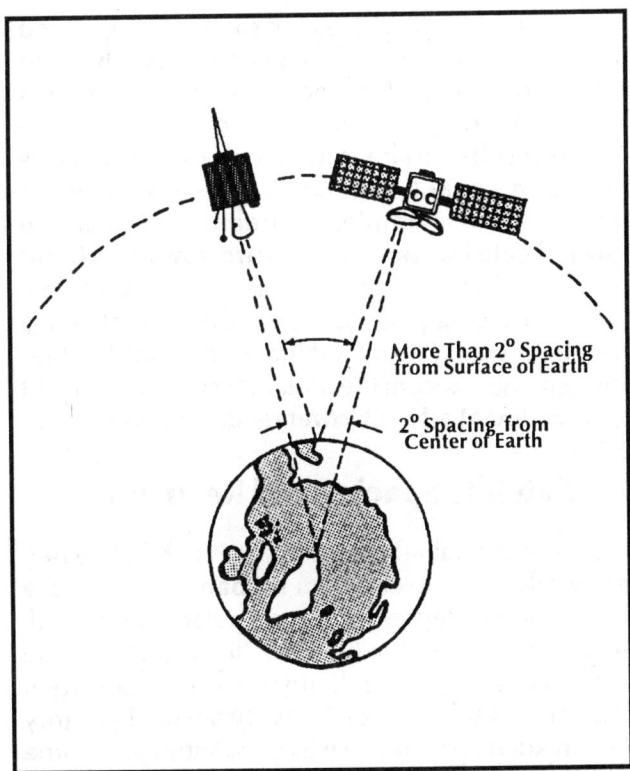

Table 3-10. Variation of Spacing With Site Latitude. Satellites spaced 2° apart are really separated by a slightly larger angle. While not drawn to scale, this polar view of the earth illustrates the point. The published separation between satellites is measured from the center of the earth but an antenna actually lies closer to the arc and therefore subtends a smaller angle. As the site becomes closer to either pole, the angle decreases towards the published angle.

TABLE 3-5. EFFECTIVE SATELLITE SPACING VERSUS LATITUDE
(At one degree satellite spacing)

Latitude	Maximum Angular Separation	Minimum Angular Separation
81.00	1.000	1.000
64.83	1.053	1.000
47.60	1.101	1.000
40.67	1.120	1.000
39.73	1.121	1.000
33.75	1.138	1.000
25.75	1.154	1.000
00.00	1.178	1.000

Maximum and minimum angular separation occur when oriented at the horizon and due south, respectively.

SYSTEM DESIGN & TESTING

The bottom line is clear. If satellites are spaced at one degree intervals it does not necessarily mean that the power levels at the one degree point on a beamwidth diagram tell the whole story. Performance results can be determined only from comparing the exact angular separation between satellites. For example, the beamwidth diagram may indicate that power levels are down by 18 and 20 dB at 1° and 1.2°. So if a dish sited at a 40° latitude is targeting satellites due south, the 20 dB figure might be used, but if it is aimed towards either horizon, signals from the off-target spacecraft might be detected at a 18 dB lower relative power level.

Satellite Spacing and Regulation

In all probability, smaller antennas will be perfectly adequate even when Ku-band satellites are spaced at one degree intervals if polarity formats alternate between satellites. To illustrate, for those satellites using linear polarity formats, if a system were tuned to channel 5 having vertical polarity and an adjacent satellite was transmitting the same channel on horizontal polarity, the feedhorn would hardly detect these interfering signals of opposite polarity. This "cross-polarisation discrimination" adds approximately 20 dB of additional protection from interference. For example, even if a two metre reflector detected adjacent satellite signals at only 6 dB below those from the desired broadcast, the added protection incurred by having the unwanted signals polarised in the opposite sense would bring these interfering levels down to a perfectly manageable 26 dB.

European broadcast satellites employ three distinct 250 MHz wide bands to transmit television signals. Therefore, if adjacent Ku-band satellites were operating in different bands, interference would not be present. This design factor allows more flexibility in allocating spacing of satellites. Today European high powered DBS satellites are being orbited with no less than 6° spacing.

Nevertheless, using an undersized antenna can still present problems even if adjacent satellites are cross polarised. Its gain may not be sufficient to detect the weaker transponders without unwanted sparklies. In addition, a reflector having a wide main lobe and high side lobes is much more susceptible to interference from terrestrial sources as well as other communication satellites. Of course, a judgment of what constitutes adequate antenna diameter ultimately rests upon subjective opinions about quality of reception.

C. INTERPRETING TELEVISION TEST SIGNALS

Television test signals can serve two functions. They enable a technician to perform fine adjustments to the receiving equipment and they are useful in an equipment evaluation by providing a quality measure of both the transmission link and the receiving system. The correct use of test signals allows a knowledgeable technician to step beyond subjective judgments of picture quality gained by simply observing the received picture.

Test signals are transmitted along with the television signal or can be generated at the test bench by devices known as signal generators. Responses to these electronic fingerprints which are viewed on a special type of oscilloscope known as a waveform monitor, are often published in reviews of satellite receivers and indicate the quality of electronic design and operation.

The objective in this section is to briefly explain the standard test signals that are commonly encountered and to outline the types of picture distortion that are indicated by such measurements. This material is based upon an understanding of television operation which has been presented in Chapter II. Interested readers can find additional technical detail in textbooks devoted to this subject or in articles published by manufacturers of test equipment. Ultimately, the most extensive experience must be gained by hands-on experience at the test bench.

Test Parameters and Picture Distortion

Four central parameters used to determine how well a television signal is reproduced at the receiver are:

1. Video frequency response as measured by how much the signal amplitude changes as input frequency is varied.

2. Phase stability versus frequency as measured by how much the phase changes as input frequency is varied.

3. Differential phase as measured by the phase error that occurs when the luminance signal is varied from its black to white reference levels.

4. Differential gain as measured by the change in the chrominance subcarrier amplitude as the luminance component of the video signal is varied from its minimum to maximum value.

Picture crispness depends upon video frequency response of the television receiver. If the high frequency components are attenuated or rolled off relatively more than the low frequency portions of the video signal, then the picture becomes "softer" or even blurred. Colour depth which is perceived as a lack of bold variation between colours also suffers as a result of high frequency rolloff.

Accurate phase reproduction is critical for true colour fidelity because constructing the raw red, green and blue signals from the two colour difference signals depends on maintaining the exact phase relationship established at the studio. Errors in differential phase or variations in phase with frequency can cause annoying changes in hue. Low frequency phase distortion produces streaking in a picture particularly on transitions from black to white and vice versa.

Excessive amounts of differential gain can cause some colours to look washed out, while others might appear to be excessively bright. These distortions can result from changes in the luminance signal with variations in brightness between different scenes.

The Basic Test Signals

The principal test signals include the vertical interval test signal (VITS), the vertical interval (colour) reference signal (VIRS) and the colour bar test signal (see Figures 3-11, 3-12 and 3-13) Although other types of reference signals may be encountered, the correct interpretation of these basic ones provides extensive information about receiver function (also see Figure 3-14). All the test signals are transmitted in the picture portion on lines in the vertical blanking interval.

NTSC test signals are used here as examples. However, the PAL and SECAM test signals are quite similar, so explanations presented below should suffice for all interested readers. Manufacturers of test equipment generally provide excellent information that can be an excellent additional source.

The first portion of the NTSC VITS, usually transmitted on line 17, field 1, is composed of a series of equal amplitude bursts containing frequencies of typically 0.5, 1.5, 2.0, 3.0, 3.58 and 4.2 MHz. The variations in amplitude observed by changes in the height of these pulses establishes the video frequency response of the receiver under test. A burst of peak white, known as the "white flag," is also transmitted and serves as a white reference level. Note that this multiburst waveform test signal as well as all the others are framed at their beginning and end by the sync pulse.

The second portion of the VITS is transmitted in both fields of line 18 in the vertical blanking interval. It is more complex than the first portion found on line 17, field 1 and consists of a number of components. The first section, a staircase, can be used to measure the amounts of differential gain, the variations in gain across the frequency spectrum. Differential phase can be measured by comparing the phase of the reference colour burst with the phase of the 3.58 MHz signal impressed upon the staircase. An instrument called a vector scope is used for this and other similar measurements where phase must be visualised (see Figure 3-15, 3-16, 3-17, 3-18 and 3-19). The spike following the staircase, known as the sine squared pulse, is a good indicator of phase distortion. It can be used to effectively judge the amount of differential phase and gain inherent in any satellite receiver. The next wider pulse is referred to as the chrominance pulse test signal and provides an accurate method to determine gain and delay differences between the chroma and luminance signals. The last component, the square wave, is called the line bar or window test signal. Any tilting of the top portion indicates poor low frequency response which is visible as picture streaking.

The VIRS occupies both fields of line 19. This signal consists of a chrominance reference waveform at the colour burst frequency as well as black

SYSTEM DESIGN & TESTING

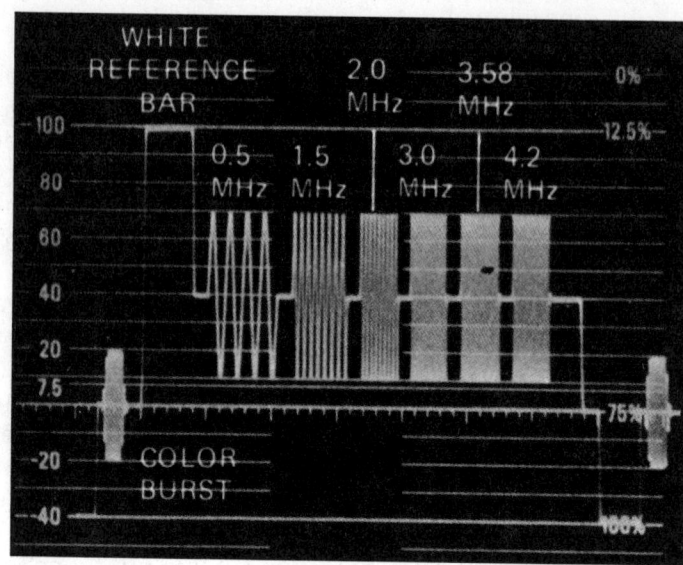

Figure 3-11. NSTC Vertical Interval Test Signal (VITS). *This portion of the VITS, usually transmitted on line 17, field 1, is composed of a series of equal amplitude bursts covering a range of frequencies at typically 0.5, 1.5, 2.0, 3.0, 3.58 and 4.2 MHz. If signal amplitude varies across the frequency spectrum, this will be seen in the height of the bursts. A burst of peak white, known as the "white flag," is also transmitted as a white reference level. Note that the sync pulse followed by the colour burst is seen in the lower left-hand corner of this trace.*

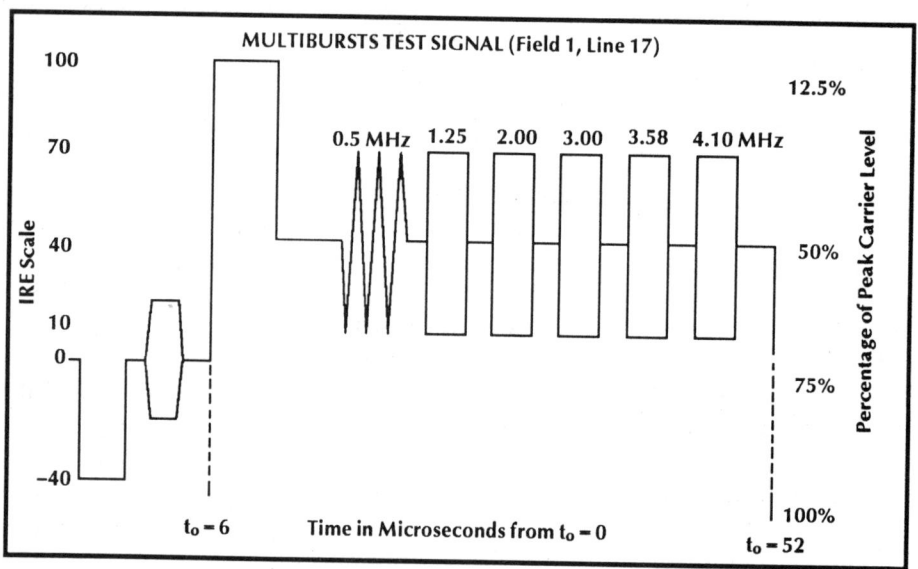

and white reference level signals. This indicates when the chroma reference and colour burst signals are either out of phase or reduced significantly in amplitude. When this occurs colour distortion, which is seen as purple or green skin tones on a television set, is visible evidence.

The colour bar test signal, usually relayed on the second field of line 17, is composed of the six standard colour bars – yellow, cyan, green, magenta, and blue, plus the black and white reference levels. This signal, like the VIRS, is useful for adjusting transmitter output level at the television studio.

Test signals are also quite useful in fine tuning satellite receivers. There are three standard adjustments which can be performed. The video level should be set to one volt peak-to-peak amplitude. The high frequency response is adjusted by equalizing the pulses in the multiburst waveform as well as possible. And the low frequency response is set by minimising distortions as viewed in the leading edge of the white flag portion of the VITS.

TEST SIGNALS

SYSTEM DESIGN & TESTING

Figure 3-12. NTSC Vertical Interval Colour Reference Signal (VIRS). *This test signal usually occupies both fields of line 19 of the vertical blanking interval. It consists of a chrominance reference waveform at the colour burst frequency as well as both a black and white reference level signal. This indicates when the chroma reference and colour burst signals are either out of phase or reduced significantly in amplitude. Picture distortion is visible evidence of these changes in phase and colour levels.*

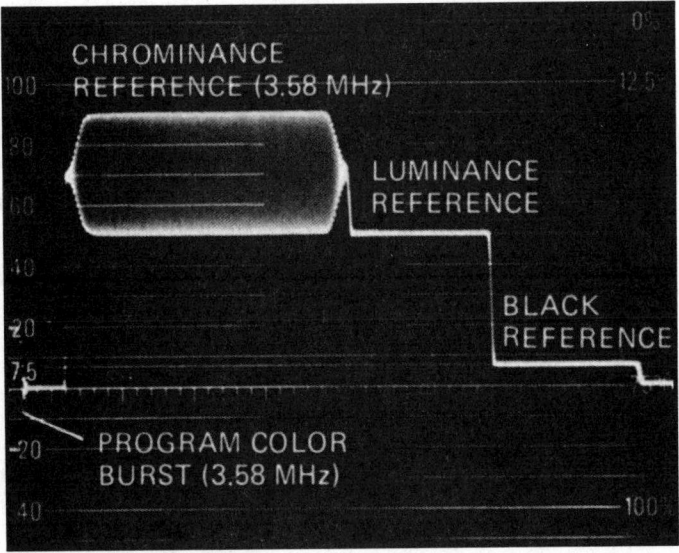

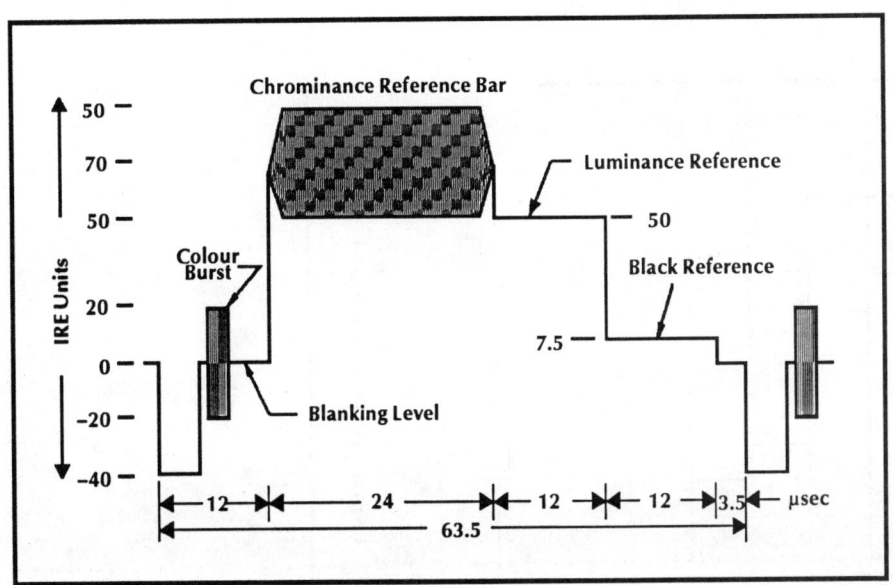

SYSTEM DESIGN & TESTING

Figure 3-13. NTSC Colour Bar Test Signal. *This test signal is usually relayed in the second field of line 17 in the vertical blanking interval. It carries the six standard colour bars, yellow, cyan, green, magenta and blue, as well as black and white reference levels. This signal is useful at the studio for adjusting transmitter outputs.*

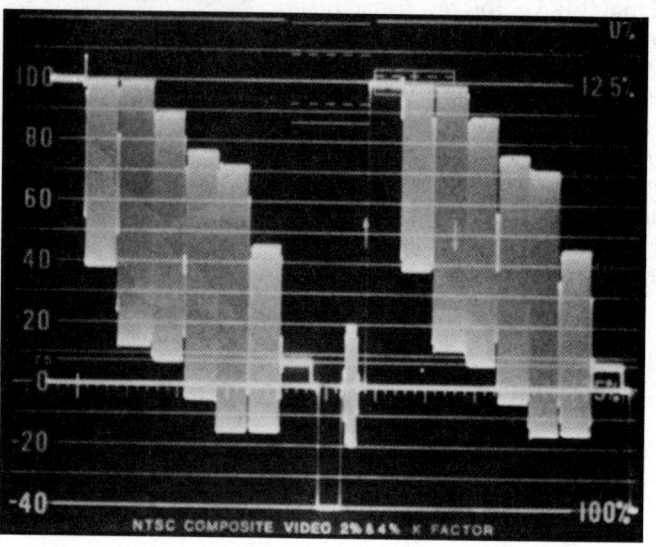

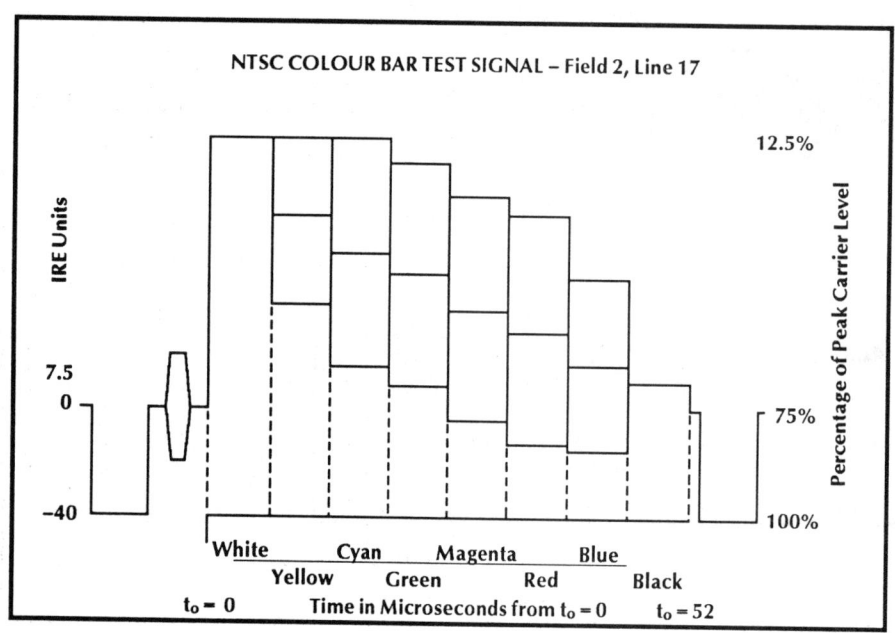

SYSTEM DESIGN & TESTING

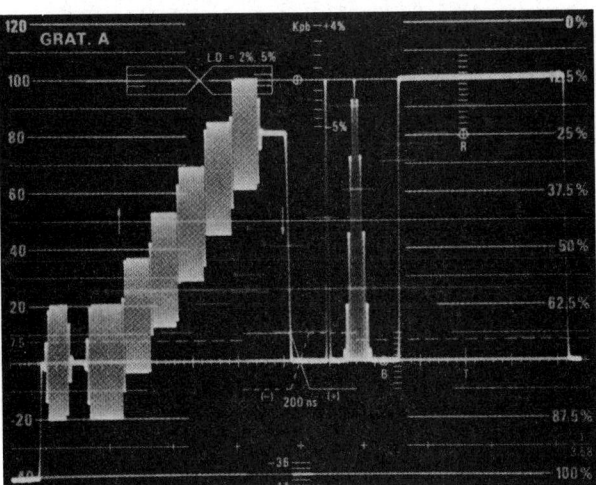

Figure 3-14. NTSC Staircase Composite Test Signal. *This test signal is usually transmitted on both fields of line 18 in the vertical blanking interval. It has a number of components. The first square wave is called the line bar or window test signal. Any tilting of the top portion indicates poor low frequency response visible as picture streaking. The spike which follows, the sine squared pulse, is a good indicator of phase distortion. The next wider pulse is referred to the chrominance pulse test signal and provides an accurate method to determine gain and delay difference between the chroma and luminance signals. The final staircase can be used to measure the amounts of differential gain, or variations in gain across the frequency spectrum.*

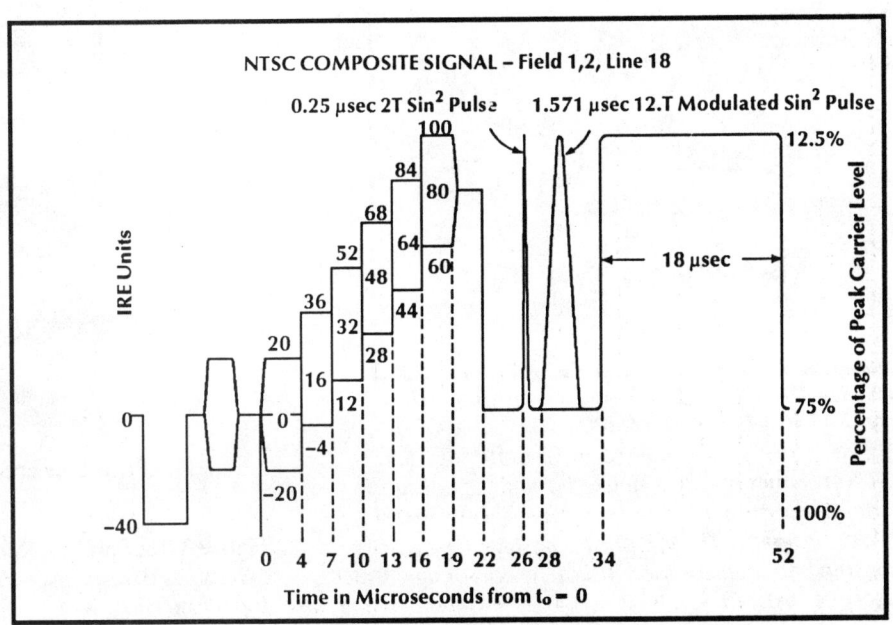

SYSTEM DESIGN & TESTING

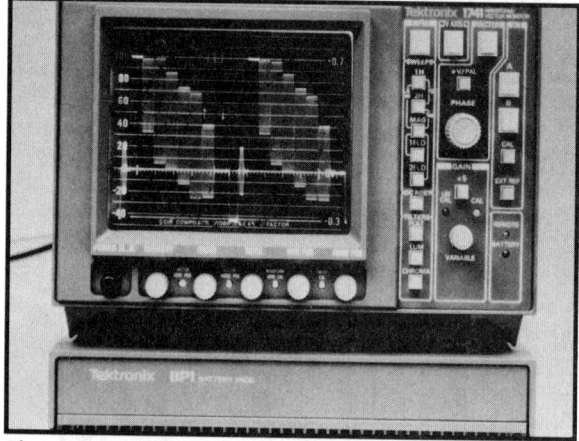

Figure 3-15. Waveform and Vector Monitors. The Tektronix 1740 Waveform Monitor is capable of displaying all the basic waveform monitoring and vectorscope functions in a single, compact package. Two lines of the colour bar test signal are shown in the bottom photograph. The top photo displays the phase relationship between components of the colour signal. (Courtesy of Tektronix)

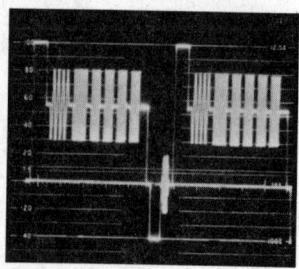

Ideal VITS Multiburst Test Signal

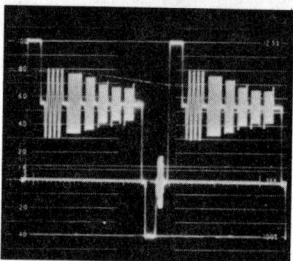

Poor High Frequen cy Response

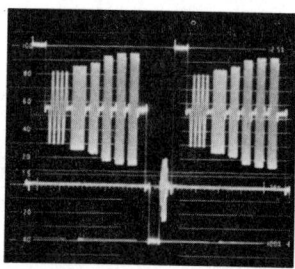

Poor Low Frequency Response

Figure 3-16. Satellite Receiver Frequency Responses. Two types of responses to the VITs are shown here. The top illustration is a perfect frequency response. The second multiburst signal exhibits a poor high frequency response known as "roll off." This condition causes a loss of fine detail in b/w pictures and makes brightly coloured objects appear pale in colour systems. The bottom illustration demonstrates a peaking of higher frequency signal components. While a small amount of mid and high frequency peaking may increase the apparent detail in a picture, excessive amounts will probably allow noise degradation. (Courtesy of Tektronix)

SYSTEM DESIGN & TESTING

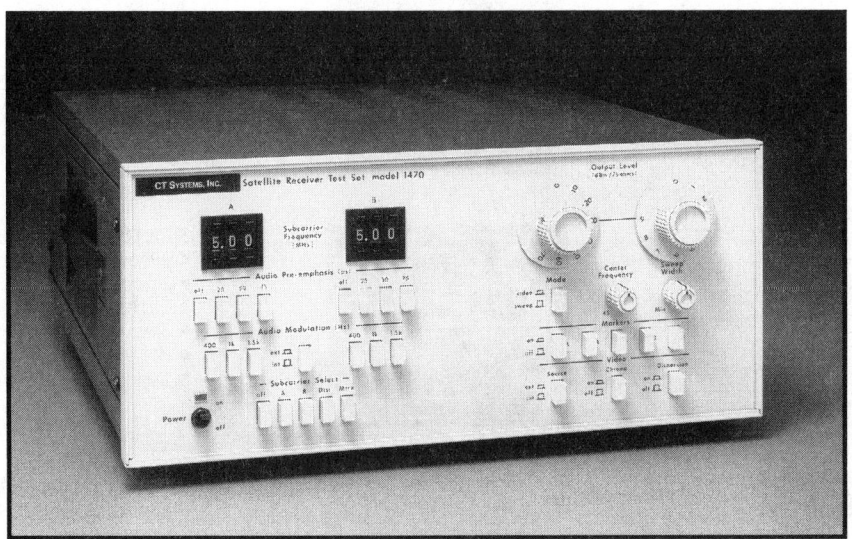

Figure 3-17. CT Systems Model 1470 Test Set. *This instrument is designed to simulate satellite television transmissions for testing satellite receivers. It operates as an IF sweep generator for troubleshooting and diagnosing receiver problems. The IF sweep ranges from 45 to 90 MHz and 0 to -70 dBm with five crystal controlled frequency markets. The 1470 test set also generates an FM video source to verify receiver operation. The 70 MHz FM carrier can be modulated externally or can generate internal colour bar video, dispersion and audio subcarrier signals. (Courtesy of CT Systems, Inc.)*

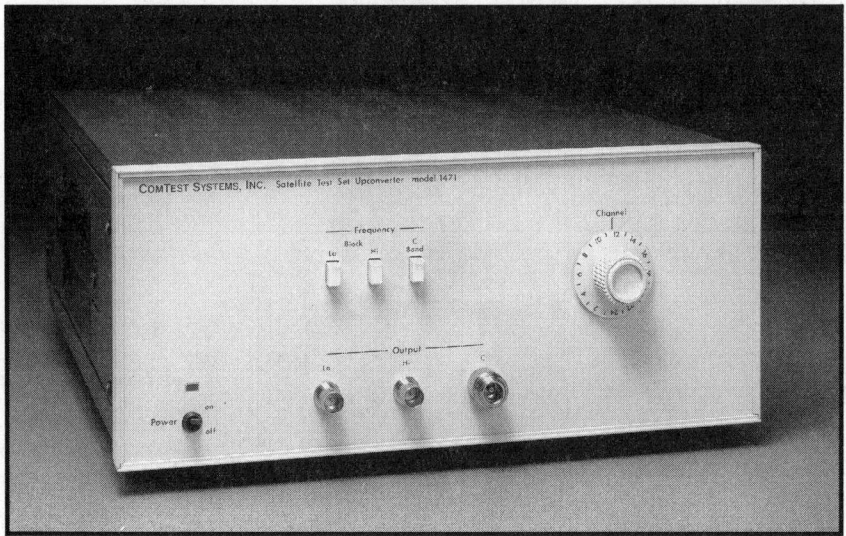

Figure 3-18. CT Systems Model 1471 Test Set. *This unit is operates as an upconverter for the CT Systems Model 1470. It converts to the C-band range for testing low noise amplifiers as well as block IF outputs for testing C and Ku-band receivers. Each of these bands are separate plug-in modules. (Courtesy of CT Systems, Inc.)*

SYSTEM DESIGN & TESTING

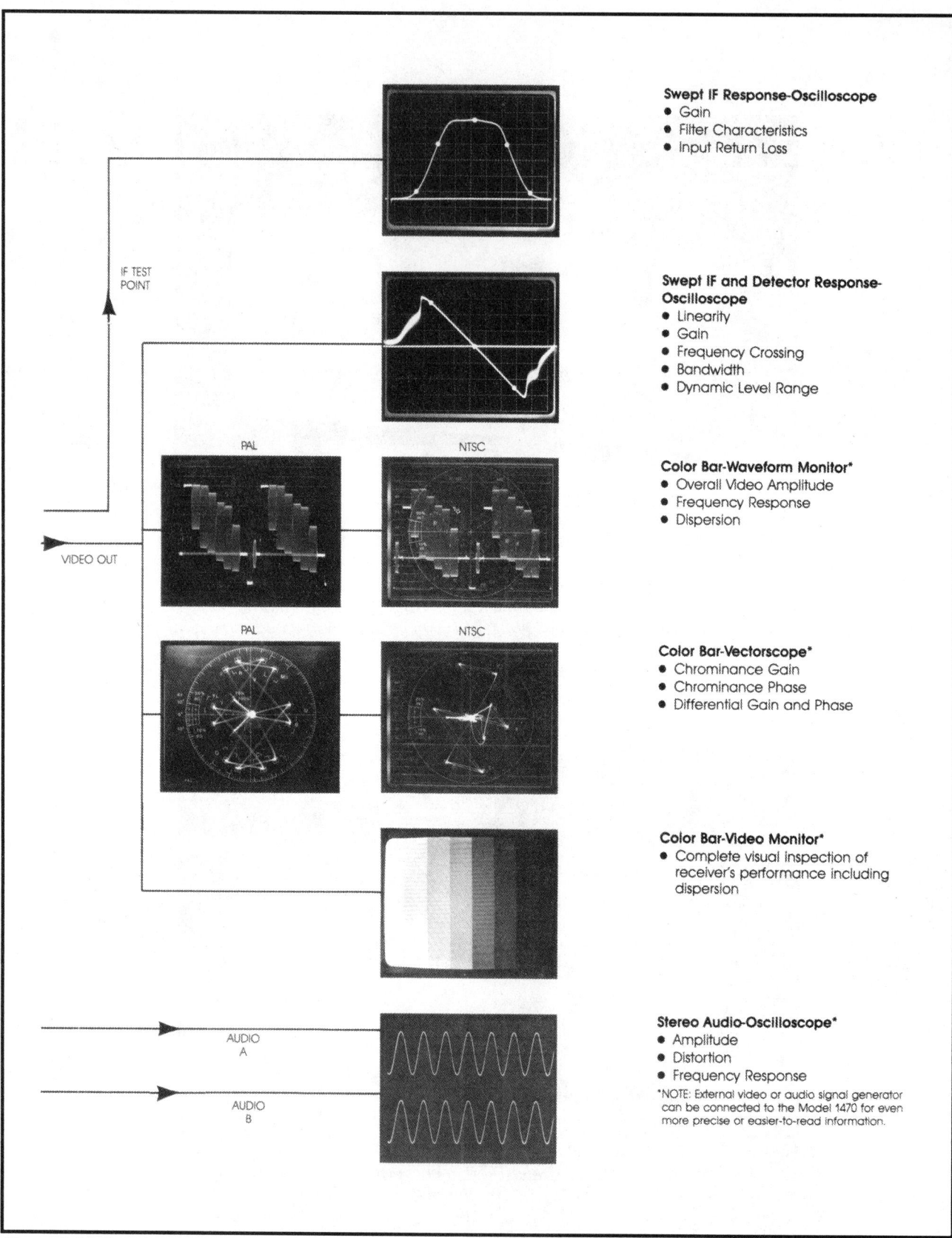

Figure 3-19. Receiver Measurements. *This diagram indicates the type of receiver measurements that can be generated when either the CT Systems model 1470 or 1471 drives a satellite receiver. The receiver IF test point, video or audio outputs are connected to the instruments indicated to measure the listed parameters. (Courtesy of CT Systems, Inc.)*

IV. SELECTING EQUIPMENT FOR KU-BAND EARTH STATIONS

The satellite TV industry has evolved at an astonishing pace during the 1980s and early 1990s. Today, selecting Ku-band equipment has become a matter of sifting through scores of antennas, receivers and other components. The growth in available Ku-band components appears to be following the course set by the C-band industry. Hundreds of brands have been marketed while just nine years ago only a few types of equipment were available and antenna actuators were virtually unknown.

A competent satellite technician will have a thorough understanding of system operation and will keep abreast of new developments and equipment. He or she should know how to design a reception system with an acceptable performance margin, should be able to anticipate the effects that factors such as reducing spacing between satellites and weather will have on picture quality and should be prepared to deal with a host of other issues.

This chapter is not intended to be an advertisement for the various brands of equipment on the market. Including the photos of products should not necessarily be taken as an endorsement. A number of excellent trade journals do provide surveys of equipment (see Appendix G). Our intent here is to help a technician or installer read and understand such journals or equipment specifications so he can then make well-informed judgments about satellite equipment.

Figure 4-1. A Fixed Mount Installation. *This "Astra" dish is fixed in place of the wall of a large commercial building. (Courtesy of WV Publications, London)*

A. EVALUATING EQUIPMENT FOR RECEIVING KU-BAND BROADCASTS

Among the criteria used in choosing components for TVROs are performance, cost, features, durability and aesthetic appeal. Each of these must be carefully weighed in selecting equipment for any customer. Some will be satisfied with a lower cost system having acceptable quality pictures on most but not all transponders and a relatively high probability of rain outage; others will pay top price for a TVRO having all the bells and whistles and will demand perfect video on all channels.

Mounts

The antenna, feedhorn and LNB are the most important components of any satellite system. If they are not properly chosen, no extra money spent on satellite receivers or other electronics known today will improve performance.

The antenna must be mounted on a stable foundation that allows accurate targeting of the satellite arc (see Figures 4-2 and 4-3). A flimsy mount will permit an antenna to sway in the wind and may even collapse under the weight of a heavy snow or gale force winds. It must be securely attached to the supporting pole and mount, and have a method for setting and maintaining declination and polar axis adjustments. The bearings that allow movement in a tracking system should be well designed to move freely and to resist excessive wear. The points of attachment to the antenna must be strong but should not warp the reflector or allow twisting or excessive movement under loads.

By far the best method to evaluate a mount is by a detailed hands-on examination of its structure. There should be no play when the antenna is pushed from its outer rim; otherwise, the various adjustments may be upset too easily. A visual comparison of mounts is usually quite instructive because there can be large differences in construction as well as in quality between various brands.

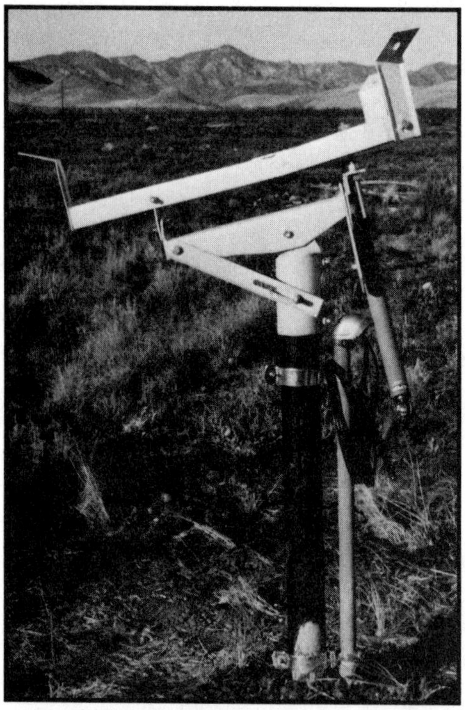

Figure 4-2. Polar Mount *This photo shows all the necessary adjustment points on a polar mount including elevation, declination and azimuth control. A linear jack tube is shown here.*

Figure 4-3. Az-El Mount. *This type of az- az-el mount supports a 4.5 metre Ku-band antenna. It is part of the Hi-Net satellite network used by the Holiday Inn hotel chain in North America. Such a large antenna is required because the system is designed for 99.9% availability and needs a high fade margin.*

SELECTING EQUIPMENT

Actuators

Actuator arms and horizon-to-horizon assemblies are also subjected to mechanical stresses. Jack tubes and motors are especially prone to problems and failures, most weather-related. However, satellite TV technology has developed over the past few years to the point where reliable units are readily available to discerning buyers (see Figures 4-4 and 4-5).

The actuator jack of choice should not have excessive play. This could cause an antenna to flop around in the wind, a particular concern with narrow-beamwidth, Ku-band TVROs. When a larger antenna is used, a longer arm is needed so that adequate leverage and force is maintained throughout its sweep across the entire arc. Seals against water entry should be effective and drain holes should be drilled at the lowest point on the motor housing. Actuator arms should be shipped with ball joints for attachment to the antenna and mount. The hardware must be properly protected against inclement weather. Galvanized, stainless or zinc-plated steel is the best for this purpose.

Actuators are controlled by counting pulses relayed from reed switches, Hall effect transistors, optical counters or by dividing voltages from potentiometers. Each type has a unique set of advantages and disadvantages depending upon the quality of its design and construction.

Actuators also differ in their mechanical tracking characteristics. Some brands may take an undue amount of time to track an antenna from one end of the arc to the other while others may be rapid but inaccurate. In general, the slower moving actuators are more finely geared and can have greater tracking accuracy. Actuators also differ in their ease of installation and programming as well as reliability. It is essential that an actuator have electrically programmable limits or some protective device such as a slip clutch in its motor, in case an over ambitious customer or malfunctioning microprocessor tries to drive it beyond the end of the arc.

Antennas

Antennas are the eyes of a satellite system. They must accurately concentrate and reflect the satellite signal while also rejecting interference and noise (see Figure 4-6). This is a tall order considering that an antenna is installed outdoors and is subjected to corrosion, wind, ice and snow loading, hail, heat, cold, children and pets. Of course, a brand that may be perfectly adequate in a warm climate may not stand the test of time where snow and winds are often excessive.

Figure 4-4. Tracker II & III Positioners. *The Tracker II SC is a digital east/west antenna positioner. All circuitry is on a single chip and shielded from external noise and voltage transients. The chip has a non-volatile memory, eliminating the need for a battery backup. The Tracker is a microprocessor-controlled, programmable positioner with 71 satellite capability. Both units are available with UHF remote, hand-held control for operation from locations visually removed from the controller. (Courtesy of Houston Tracker Systems)*

Figure 4-5. BlackJack Actuator. *The BlackJack actuator is a compact, lightweight, efficient linear drive designed to be fully compatible with Ku-band antennas. Its drive mechanism has been geared very low and it is therefore capable of positioning the antenna in smaller, more accurate increments. The counting sensor is a reed switch. The arm is available in an 46 centimetre (18 inch) stroke. (Courtesy of Houston Tracker)*

A simple, physical inspection of an antenna can be very revealing. First, when sighting along one side of the rim, does the other side line up perfectly parallel? If not, the reflector is warped. Second, does it feel smooth to the touch or is the surface rough or bumpy? If the surface has visible waves or has panels that do not line up accurately, beware. Third, how is the hub or central portion of the antenna attached to the mount? The mount must be adequately strong and be well attached to the reflector to support its full weight without allowing the antenna to sag and warp in time (see Figure 4-7). Some antennas are secured with only a few screws or bolts and may therefore twist as well as rock when under stress. This type of construction may not withstand the test of time in very windy or

SELECTING EQUIPMENT

snowy environments. Any rocking or twisting movement caused by wind will eventually produce metal fatigue. This can lead to metal failure and subsequent deterioration or even complete loss of reception.

Figure 4-6. Saturn 3.1 Metre C/Ku-Band Antenna. *This photo was taken on a test range where antenna gain and side lobe patterns were being evaluated at both C and Ku-band. Individual antenna petals are available in solid as well as perforated aluminium or steel. (Courtesy of Metal Specialists, Inc.)*

The design and method of supporting the feed assembly is also very important (see Figure 4-8). The weight hanging out in front of the antenna must be minimised. Some earlier designs have resembled a bowling ball on a bamboo pole: these certainly should be supported by guy wires attached to the outer antenna rim. In addition, the feed should block a minimum amount of antenna aperture or side lobes may be unnecessarily large. An offset fed antenna has a number of advantages in this respect. No antenna aperture is blocked by either the feed, LNB or its cover so that the feed support members can be as massive as necessary to provide the required stability.

Reflector design and material composition must be selected so that expansion and contraction caused by temperature changes will not degrade its surface accuracy or structural integrity (see Figure 4-9). If a panel is damaged it should be inexpensive and simple to replace. If the antenna is located near an ocean, the materials must be specially treated to resist the highly corrosive effects of salt spray.

Other important criteria used in selecting an antenna are its weight, appearance and ease of shipping and assembly. When a TVRO is installed on a roof mount, it makes sense to use a lightweight reflector such as one constructed from stamped or spun aluminum.

Antennas can be grouped into two major categories: mesh and solid. The solid antennas are either fibreglass, or stamped, spun or hydroformed metal. For purposes of this discussion, perforated reflectors are included in the solid category. Brands from each group can be evaluated by all the above criteria as well as by some which are specific to a given type of antenna.

Figure 4-7. Inspecting a Solidly Mounted Antenna. *This photo shows how rigid an antenna and mount should be. Antennas and mounts must withstand all the forces that mother nature will produce. This 80 kilogram (180 pound) man is in the centre of the reflector where he is checking for correct feed positioning. Standing in this position is not a recommended practice for spun aluminium or perforated antennas. Applying downward pressure on the outer rim is a less risky method to evaluate the amount of flex in both the antenna and mount.*

SELECTING EQUIPMENT

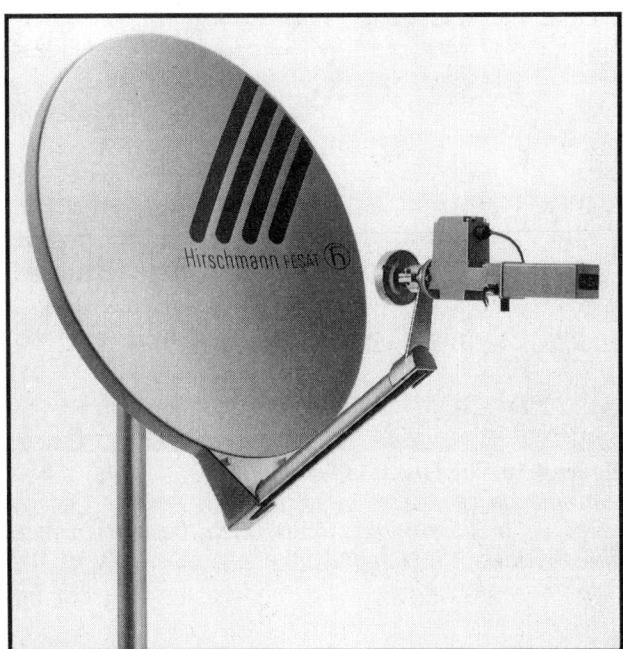

Figure 4-8. Hirschmann Offset Fed Antenna. *This system features feed, LNB and either a 55, 85 or 120 cm offset fed antenna. Gains of the 55 and 85 cm dishes are 34.75 and 38.5 dBi at 12.1 GHz. (Courtesy of Richard Hirschmann GmbH & Co.)*

In some brands of spun or hydroformed antennas, the mount is bolted to a back ring that is attached to the reflector surface. Others have mounts that are bolted directly to the back of the dish and, if not carefully assembled, can distort the antenna surface as the bolts are tightened. This is especially true if the mount does not properly match the reflector shape and if rubber washers, which act as shock absorbers, are not provided for insertion between the mount and reflector. These washers can also prevent an electrolytic action between the different metals which can result in corrosion.

The quality of a fibreglass reflector depends upon which manufacturing method was used (see Chapter II for more details). If such antennas are cured and molded properly, they can have a very accurate, smooth surface. However, improper techniques can, in time, result in warpage, shrinking, surface deterioration and even separation of the protective gel coating from the underlying structure. In the worst cases, the embedded reflective material could even delaminate and warp.

Fibreglass antennas can be very heavy and must have solid mounts attached securely at numerous points. Like other types of antennas, sections should have compound curves and rib structures to form a true parabola and these should fit snugly together. The surface should also feel smooth to the touch. However, note that this does not guarantee that the embedded reflective material lying just below the surface has maintained an accurate shape. Some antennas which have performed well at C-band frequencies have surprised installers when converted to Ku- band or dual-band reception.

Solid Antennas

Solid antennas are usually stronger and most often are more durable than mesh reflectors. In general, they can be assembled more quickly. Snow and ice tend to slide off instead of accumulating as they do with some mesh brands. But solid reflectors are usually heavier, are more subject to wind loading and can more easily sag when loaded.

Spun or hydroformed aluminum or steel antennas are constructed from one large disc of material and generally have the most accurate surfaces. If not cut into sections they can be rather cumbersome to ship and handle on the job. This variety as well as stamped or hydroformed antennas must have their surfaces protected from rusting or oxidation. Of course, this is not an issue with stainless steel varieties. But unpainted stainless steel surfaces can have problems with heat build-up caused by solar reflections.

Figure 4-9. Lip-Sighting an Antenna. *Sighting along one rim of an antenna and lining up the opposite side is a good visual method to check for warping. An antenna can also be lip-sighted at various points in its tracking range to check for sagging.*

SELECTING EQUIPMENT

Fibreglass antennas can have the advantages of being easy to assemble, resistant to corrosion and durable. Of course, the best test of any brand of fibreglass reflector is probably how it will maintain its accuracy and structural integrity over a period of time in the field.

The surface of any antenna should be finished to protect the components at the focus from damage. If not, when it points directly at the sun, it may reflect these rays so efficiently that they are focused to generate heat that is intense enough to burn up the feed and LNB (see Figure 2-10). Either dark or matte paints can be used because they diffuse light. Even rolled mesh and fibreglass surfaces are smooth enough to cause a problem unless they are properly painted.

Figure 4-10. The Effects of Solar Energy. *A highly reflective antenna can literally melt a plastic feedhorn cover, especially at solar outage times when the sun is aimed directly down the antenna boresight. (Courtesy of WV Publications, London)*

Mesh Antennas

Mesh antennas are generally made from expanded steel or aluminium (see Figure 4-11). These materials are protected by painting, powder-coating or anodizing. Judging which methods are best for the given climate depends upon where the antenna will be installed. For example, a painted mesh reflector may be perfectly durable in a desert environment but a disaster on an ocean coast where salt corrosion is a concern.

Rarely will mesh antennas have adequate surface accuracy to ensure high efficiency at Ku-band. When an existing C-band mesh reflector is retrofitted with a Ku-band feed it must have an accurate and stable surface. A flimsy mesh will probably not withstand a hail storm without dimpling, deforming or having some panels badly damaged or destroyed.

The method used to maintain the shape of a mesh antenna is crucial to its performance and is often related to the ease of assembly. As few as eight or as many as 24 supporting struts and panels of mesh may be used. Struts are usually manufactured from tubular steel or extruded or angular aluminium for structural strength in order to maintain their shape under wind or snow loading. Some mesh antennas are designed so that the actuator arm attaches to just one rib. This member must be strong in order to maintain a true shape as the actuator pushes the antenna through the arc. If not, the antenna could easily warp towards one side. Mesh antennas often have one outer circular ring con-

Figure 4-11. Winegard 1.2 Metre Perforated Antenna. *This perforated aluminium antenna, model CK-4014, has the advantages of lower wind resistance and see-through appearance while retaining an accurate reflector surface. An az/el mount is shown here. (Courtesy of Winegard Satellite Systems)*

necting the ribs to the middle hub; others have one or more middle supporting circular rings as well. Some designs and materials yield an accurate and stable surface; others are poorly designed and manufactured. Common sense judgments of these designs are usually surprisingly accurate.

Feed Supports

Prime focus parabolic antennas have two varieties of feed supports: buttonhook and tripod or quadpod (see Figure 4-12 and 4-13). Whichever type is used, the feed must be stable under wind or snow loads and should not move during tracking. Both the attachments between the antenna and support structure and the support and feedhorn must be strong. The feedhorn and LNB mounted at the antenna focus can easily stress their mounting hardware and cause the feedhorn to move from the desired "sweet-spot" at the exact phase centre of the reflector.

Some antennas have feeds which simply pop into a preset position so that no adjustment of the focal length is possible. This is both an advantage and a limitation depending upon the attention an installer wishes to give to adjusting the feed position. The adjustable varieties are more difficult to install but allow a more accurate fine tuning of picture quality. This is critical at Ku-band.

Many smaller Ku-band dishes, especially those used in the European market, have offset feeds. It is critical that these support arms hold the feed securely in place so that very little motion is possible. These feeds generally illuminate the reflector surface with carefully designed patterns and must be correctly installed so they sit in the correct spot.

Figure 4-13. Quadpod versus Tripod Feed Support. *This photo illustrates the additional feed/LNB support that can be achieved compared to a buttonhook design. Problems with blocking the off-axis feedhorn can be encountered in mounting dual-band systems with this arrangement. This can be accompanied by an increase in side lobes caused by the additional shadowing.*

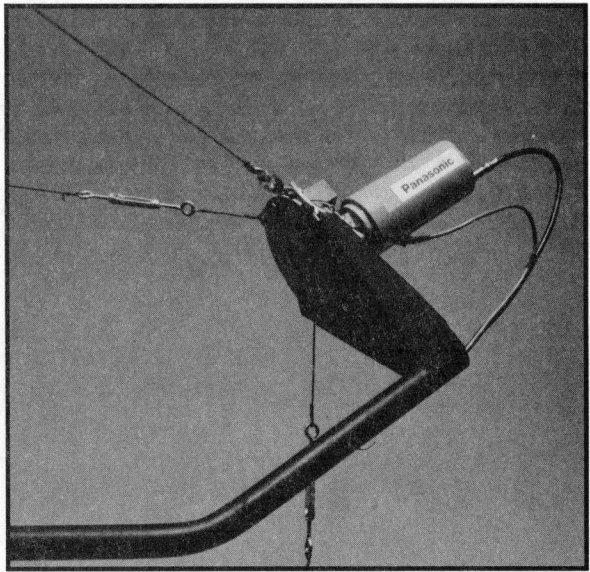

Figure 4-12. Prime Focus Buttonhook Support. *This photo shows a Ku-band feed and LNB mounted on a buttonhook. Notice the installed guy wires with turn buckles to allow flexibility when centring the feedhorn and provide extra stability in high winds. This additional support is necessary at Ku-band as the antenna tracks the satellite arc and is critical for mounting heavier dual band feeds*

Feedhorns

Feedhorns have the important job of properly illuminating an antenna and of capturing as much of the incoming signal as possible (see Figures 4-14, 4-15, 4-16 and 4-17). It is critical that the proper feed be chosen for each f/D ratio. Deeper antennas (f/D from 0.32 to 0.25) will use different feeds than more shallow ones. Offset antennas which have f/Ds ranging from 0.50 to 0.60 must have specially designed feeds for each system.

Well designed and manufactured feeds have very low insertion losses, i.e. they transmit most of the signal into the LNB. The selection of a feedhorn can be an important determinant of picture quality. The cost of a feed is a small enough portion of the entire system expense to warrant purchasing a top-of-the-line model.

Polarity selection can be accomplished by mechanical rotors, ferrite devices or pin diodes. Each has its strengths and weaknesses. For example, a ferrite device will probably perform more reliably in very cold climates where motors may seize. Pin diodes have relatively high insertion losses and do not have the capability for skew adjustment. However, this may not be a limitation for large, fixed-antenna systems. Polarities can also be selected by switching between LNBs attached to orthomode feeds.

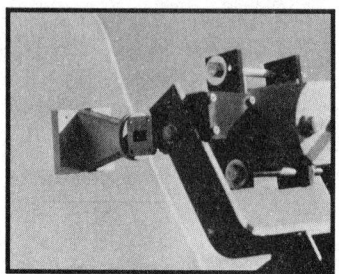

Figure 4-14. Offset Fed Dual Orthomode Feedhorn. *This Seavey orthomode feedhorn was specially designed to simultaneously detect both horizontally and vertically polarised signals on the Pico Kid antenna..*

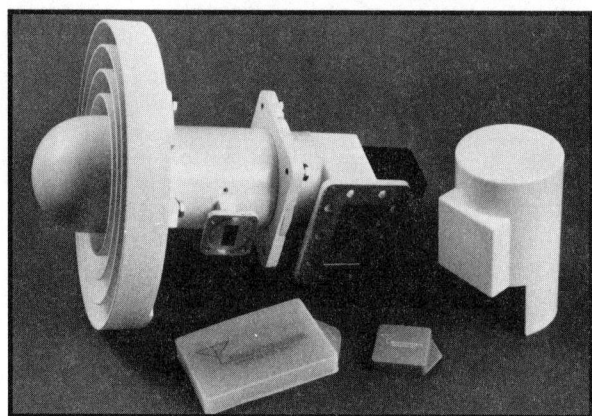

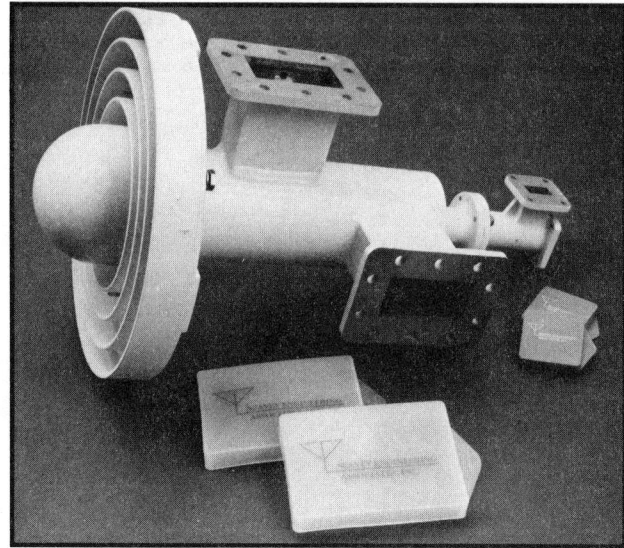

Figure 4-16. Seavey Coaxial Dual-Band Feedhorns. *The ESR 124-H prime focus, coaxial dual-band feedhorn is designed to receive C and Ku-band horizontal or vertical transmissions without any degradation of signal normally experienced in an off-axis configuration. One servo motor controls both C and Ku-band polarisation selection. This feed weighs 2 kilograms (4.5 pounds). The ESA 124-D four-port, coaxial prime focus feedhorn is designed to simultaneously receive both horizontal and vertical polarities on C and Ku-band without any loss comparable to that encountered when using an offset configuration. Its total weight without LNBs is 5.5 pounds. Courtesy of Seavey Engineering)*

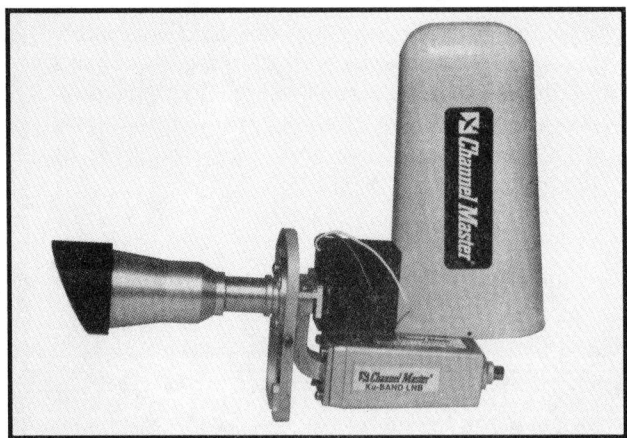

Figure 4-15. Channel Master Feed. *The line of channel master feeds include a single linear polarity offset and single circular polarity offset feed, a linear polarity orthomode transducer and a linear/circular polariser. (Courtesy of Channel Master)*

SELECTING EQUIPMENT

Figure 4-17. The Omni-Feed. *The Omni Feed™ is designed to fit the straight feedhorns supplied with many parabolic reflectors. It features the same servo motor and probe as the Polorator I-KU. (Courtesy of Chaparral Communications)*

LNBs

LNBs are the most complex but generally the most reliable satellite reception component (see Figure 4-18). Noise figures should be low enough to yield a good quality picture and LNB gain should be at least 50 dB. For a slightly higher price tag, a 60 dB gain is good insurance against excessively long cable runs. LNBs must be matched to the frequency range of the satellite broadcast. A unit designed to detect a 500 MHz wide North American transmission will be insufficient for use over the entire 750 MHz European bandwidth.

LNAs are rarely encountered in Ku-band systems. However LNBs have one minor disadvantage. Since the low noise amplifier and first downconverter are contained in one sealed box, notch or bandpass filters cannot be inserted between these components. In rare situations, this may be a problem. Readers can refer to *The Home Satellite TV Installation and Troubleshooting Manual* for more details on terrestrial interference.

Figure 4-18. Gardiner Micro Modular LNAs and LNBs. *These photos show how small and lightweight the Gardiner C-band LNA and block downconverter is compared to a typical LNA. The lower photo shown is of a Micro 12 GHz LNB that is not much bigger than a standard 12 GHz SuperFeed. This relative size is clear when compared to the background 8.5 x 11 inch piece of paper. When mounting dual band systems, it is important to keep the weight as low as possible on the feed support. The combined weight of the LNA and downconverter is 0.43 kilos (15 ounces), while the 12 GHz LNB weighs 0.11 kilos (4 ounces). (Courtesy of Gardiner Communications)*

SELECTING EQUIPMENT

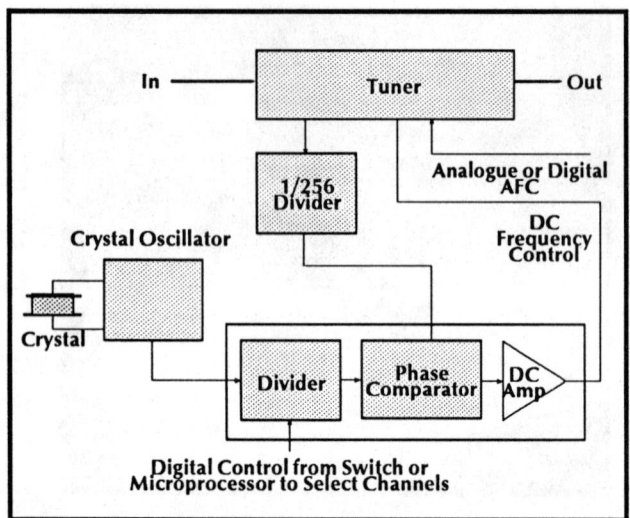

Figure 4-19. Digital Phase Locked Loop (PLL) Tuner Schematic. *Digital PLL local oscillator frequency control is becoming increasingly common in modern satellite TV receivers. The phase of the frequency divided local oscillator and the crystal-controlled oscillator signals are compared to derive the necessary dc voltage for frequency control and stabilisation.*

Satellite Receivers

Sometimes it seems that there are as many receivers available as salesmen to sell them. While each brand of receiver may have a slightly different appearance, most models draw from quite a similar set of features.

Picture quality can vary substantially between receivers. Some will produce jumpy, flickering or grainy pictures. These characteristics reflect receiver electronic design which has been examined in Chapter II. If a low threshold has been obtained by the least desirable method, reducing the bandwidth, the result is often fewer sparklies but a picture with less resolution and smeared colours. Phase lock loop (PLL) receivers normally have lower thresholds than older model coaxial delay line models. However, in extreme cases when using PLL receivers, bright colours or high contrast scenes can tear or streak. Of course, the optimal strategy in evaluating satellite receivers would be to observe the pictures played by various receivers receiving a signal from the same dish/feed/LNB at the same time.

Audio quality can also vary. The audio should be easy to tune over the full subcarrier band. There should be no buzzing or rasping heard on weaker transponders when bright colours appear on the screen. Observing the video and audio quality on transponders which are loaded with audio subcarriers is a useful worst case test. Each audio subcarrier should be clearly and crisply heard if signal strength is adequate. When narrow deviation processed audio will be received, the receiver must have a bandpass filter which can switch between wide and narrow operation.

Receivers also differ in their method of tuning between channels (see Figure 2-19). It is imperative that Ku-band satellite receivers have the flexibility to manage the range of bandwidths and channel centre frequencies encountered. While continuously variable tuning may be annoyingly inaccurate compared to synthesized channel selection, and may make determining what channel has been selected difficult, it allows the required adjustment of centre frequency. Some receivers have an even/odd polarity selection button used in addition to the conventional tuning knob or button. This prevents the polarity selection device from rapidly switching back and forth as all channels are scanned in their natural order, i.e. 1,2,3,4, etc...

The requirements for bandwidth and channel tuning capabilities differ widely between countries as well as between satellite types. These must be taken into account. For example, while American Ku-band receivers have a 500 MHz bandwidth spanning the block downconversion range of 950 to 1450 MHz, the European ECS, Astra and Intelsat spacecraft broadcast in one or more of three separate ranges, 950 to 1120 MHz, 1200 to 1450 MHz or 1450 to 1700 MHz. There are therefore three bands, each 250 MHz wide. This means that receivers designed for the entire European market must either have three separate ranges or must span the entire band. In some situations, Ku-band satellite receivers must either have an adjustable bandwidth control or have an IF "loopthrough" that allows insertion of bandpass filters so that broadcasts with variable bandwidths can be deciphered. Typical second IF loopthroughs are at 70, 130 or 134 MHz while other frequencies are not uncommon.

Detent tuned receivers for use in C/Ku-band satellite reception systems are more difficult to operate than brands which are programmable by

SELECTING EQUIPMENT

Figure 4-20. Monterey 90 Satellite Receiver. *This advanced C/Ku-band IRD features picture-in-picture, an audio/video switcher, Dolby™ surround sound, MTS and digital stereo, an adjustable bandwidth, parental l ockout, on-screen menu displays and 100 favourite channel memory.*

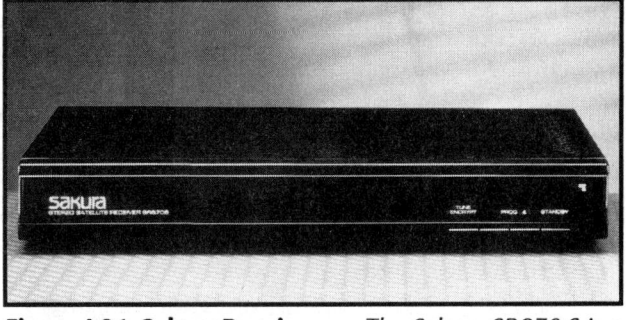

Figure 4-21. Sakura Receiver. *The Sakura SR870-S is a fully remote controlled receiver with one rear panel SCART connector. It features 70 channel memory and automatic station tuning and has received an award for the best "Astra System" from "What Satellite" magazine in 1991 (Courtesy of Sakura Electronics Ltd.)*

satellite, channel and polarity. Locating channels on the former type of receiver involves using the video invert switch and hunting for centre frequencies. However, receivers which have one continuously variable tuned potentiometer for both channel selection and fine tuning are easier to use than detent types for Ku-band.

It is essential that satellite receivers be capable of processing the local broadcast standard whether it be PAL-B, PAL-M, PAL-N or SECAM or NTSC. There are variations in the number of scan lines, the amount of pre-emphasis and in line voltages as well as colour transmission formats. North American broadcasts differ from European ones in that they do not use an energy dispersal waveform so no clamping circuit is necessary. In general, answers to questions about receiver compatibility can be obtained from either the equipment manufacturer or a local distributor who has experience with the operation of various brands of equipment. The book *World Satellite and Scrambling Methods - The Technician's Handbook* explores these design considerations in some detail.

Some minor operational points also should not be overlooked. For example, whenever a handheld remote is available, it should have the five main satellite receiver control functions: channel selection, fine tuning, audio selection, volume control and polarity selection. If any of these features were absent, a customer would have to walk over to the receiver to change the missing function, thus eliminating a good deal of the advantage of the remote

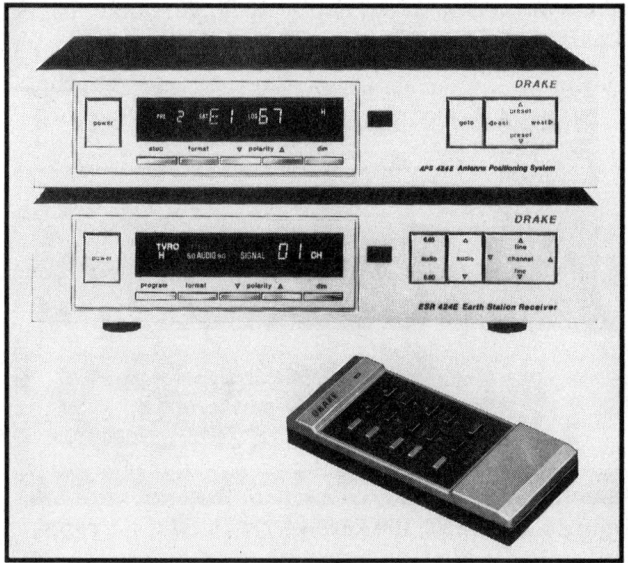

Figure 4-22. Drake ESR-424E Receiver and APS 424 Actuator Controller. *This receiver is designed for use with PAL or SECAM satellite television broadcasts. It operates on 220 Vac at 50 Hz in either the Ku or C-band. The matching APA 424E motorised programmable antenna positioner can program up to 30 antenna positions. The microprocessor memory stores satellite polarity format and skew. (Courtesy of R.L. Drake Company)*

control. Also realise that some of these functions might be automatically selected by the receiver and not be available for user control.

Figures 2-20 to 2-33 present a pictorial sample of the receivers available today. Two of the many receiver peripherals are shown in Figures 2-34 and 2-35.

SELECTING EQUIPMENT

Figure 4-23. SatTel PRK-4 Receiver. The remote control Ku-band receiver accepts a European 950 to 1750 MHz Ku-band input and features parental lockout. It can be accompanied by a matching antenna positioner.(Courtesy of Space Communications Ltd.)

Figure 4-24. General Instrument 2750R Receiver. The 2750R IR/D is C/Ku-band compatible and accepts a 950 to 1450 MHz input. Its features include digital stereo sound, parental lock, 32 satellite and 150 channel memories, favourite channel recall, built-in TI filter, on-screen display and UHF remote. (Courtesy of General Instrument)

Figure 4-25. STS SR 300 Receiver. This IRD accepts a 950 to 1450 MHz American input. Its features include on-screen display, 90 satellite and 40 favourite program recall, parental lockout, full stereo, auto positioning channel scan and audio muting when changing channels. An optional LCD "smart remote" that control up to 6 independent audio and video components is available. (Courtesy of Satellite Technology Services)

Figure 4-26. Hirschmann CSR 200-A Satellite Receiver. This remote-controlled, stereo receiver offers 50 channels for reception of Astra and Copernicus satellite broadcasts. (Courtesy of Richard Hirschmann GmbH & Company)

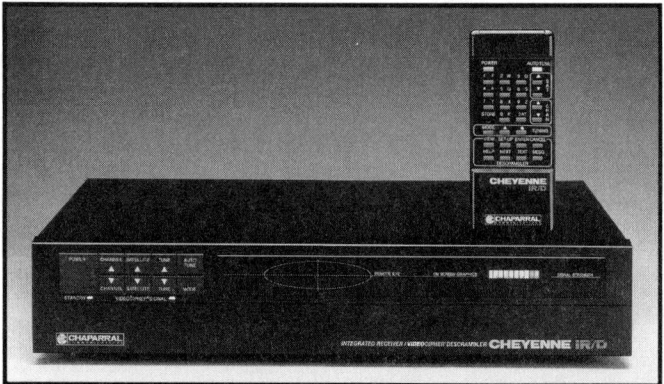

Figure 4-27. Cheyenne IRD and International Receiver. This Cheyenne receiver and it its North American counterpart, the C/Ku-band Cheyenne IRD, features computer synthesised tuning, digital stereo, complete pre-programmed tuning, on-screen graphics, parental lockout, an integrated receiver/actuator in one component and an IR remote control. It is designed to operate in either the 11 or 12 GHz bands. The Cheyenne IRD also includes a built-in VideoCipher II module. (Courtesy of Chaparral Communications)

SELECTING EQUIPMENT

Figure 4-28. Uniden UST-9000 Receiver. *The UST-9000 receiver is a full-featured, infrared remotely controlled, C/Ku-band compatible receiver with stereo and a 99 satellite programmable positioner. (Courtesy of Uniden Corporation)*

Figure 4-31. Grundig STR-201 Receiver. *The European model stereo satellite receiver has a 49 channel memory, a baseband output for D2-MAC or anther type of external decoder and an integrated PSD control. (Courtesy of Grundig)*

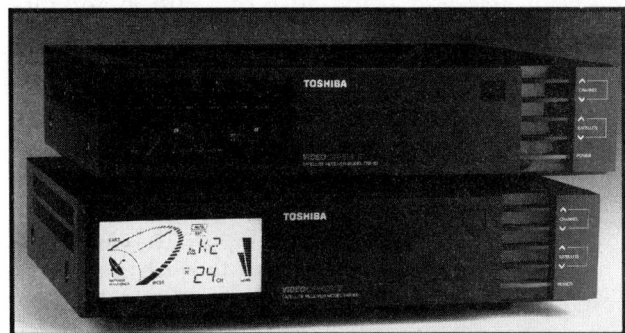

Figure 4-29. Toshiba TRX-80 and TRX-100 Receivers. *This IRD is C-Ku-band compatible, accepts a 950 to 1450 MHz input and features an on-screen display, frequency synthesized tuning, a built-in VideoCipher II descrambler and 20 channel favourite programming. (Courtesy of Toshiba)*

Figure 4-32. TechniSat ST-6000S Receiver. *This remote controlled, integrated receiver and antenna positioner features 99 pre-programmed channels, stereo, either J17 or 50 μsec de-emphasis and a 7 dB threshold. It has connections for either ferrite or mechanical polarisers. (Courtesy of TechniSat (UK) Ltd.., Horsahm, West Sussex and TechniSat Sieh Fern Product GmbH, Daun, Eifel, FRG)*

Figure 4-30. Echostar SRD-7000 Receiver. *This Videocipher II equipped receiver features 950 to 1450 MHz block downconversion and programmable video noise reduction circuitry, automatic C to Ku-band antenna positioning which includes an optional C/Ku dish controller accessory, Echostar SRD on-screen menu display, phase/locked loop synthesised video and audio with programmable fine tuning, and parental lockout. All these function are controlled via a UHF wireless remote. (Courtesy of Echosphere Corporation)*

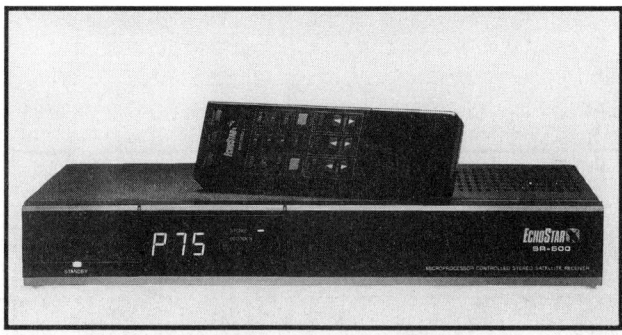

Figure 4-33. Echostar SR-500 Receiver. *This remote-controlled receiver features 79 pre-programmed video channels and 40 audio frequencies for Astra, Kopernikus, Eutelsat and Intelsat reception. (Courtesy of Echosphere Corporation).*

SELECTING EQUIPMENT

Figure 4-34. Toshiba U-Remote Adaptor Kit. *This component transforms the remote controls for the Toshiba TRX receiver series from infrared to UHF. The UHF remote has a range of 65 metres (200 feet). (Courtesy of Toshiba)*

Figure 4-35. Hidden Link. *This is one component in a complete system that allows one IR remote to control audio and video equipment in a radius of hundreds of metres. (Courtesy of Video Link)*

Stereo Processors

While stereo processors are generally built into most satellite receivers, they have been available as stand-alone units. They should be capable of handling at least the two most popular stereo formats, discrete and matrix. The separation between channels should be clear and distinct. The audio quality should be strong and clean with no audible popping or static. There are two ways to tune in audio channels: a continuous tuning circuit such as a potentiometer or, the most accurate method, voltage synthesised tuning. In the latter case, all audio channels can often be preprogrammed into a memory, selected at the push of a button and displayed on a digital read-out.

Television Monitors

Television monitors accept raw audio and video signals from a satellite receiver and bypass the intermediate modulation step. As a result, picture quality can be better than that obtained on conventional television sets. A satellite dealer may find a good quality monitor useful in order to demonstrate how clearly satellite broadcasts can be received.

Warranties

Even the best designed and manufactured equipment can fail. If the factory offers a warranty covering too short a period, does not provide a fast turnaround on service or, even worse, is out of business when the equipment fails, a dealer can incur considerable costs.

The typical warranty for satellite TV components offers full coverage for at least 90 days and limited coverage for one year but there are wide variations. A few manufacturers of electronic equipment have extended warranties that last up to two years. And some antenna manufacturers warranty their products for five years indicating either faith in their design or an unrealistic hope about the longevity of their product. In any case, responsible manufacturers should stand behind their products with prompt service. But realise that companies have been known to go out of business when high equipment failure rates overly taxed their financial resources. In these cases, their warranties were only as good as the paper on which they were written.

SELECTING EQUIPMENT

Therefore, a prudent course of action to follow should include careful selection of well-made equipment, a knowledge of the underlying warranties and an evaluation of the manufacturer's stability and reputation. Word-of-mouth information from knowledgeable sources can be invaluable in judging a manufacturer. Whenever possible obtain schematics when purchasing equipment.

Multistandard Television Receivers

Because television broadcasts are relayed in three principal formats throughout the world, PAL, NTSC and SECAM, a television set, television monitor or VCR designed for one system will not properly interpret any of the others. As the satellite television business grows, companies are beginning to manufacture components for use in countries around the world. Until a common international standard is adopted, is important to be familiar with such international standards. The technical characteristics underlying these formats have already been explored in Chapter II.

Video components that are equipped to process any of these formats are available. A single microchip acts as the translator by "reading each language," translating to a common one and then re- translating to the chosen format. Such multistandard televisions, monitors and VCRs can therefore be used anywhere in the world.

B. KU-BAND TERRESTRIAL INTERFERENCE

Ku-band reception systems are much less likely to encounter terrestrial interference (TI) than that experienced by C-band systems. However, it is the common misconception that TI is never a problem in this higher frequency range. This is not the case.

A band of frequencies known as the CARS band, which is adjacent to the upper edge of the DBS range, has been allocated to line-of-sight, terrestrial communications (see Table 4-1). As a result, out-of-band interfering signals with frequencies just above 12.7 GHz have the potential to be an annoying problem especially in dense urban areas. The TI becomes worse as power levels increase and as the separation between the CARS communication and satellite channel frequency decreases.

When appropriate, satellite TV installers may have to conduct "site-surveys" with test TVROs or sensitive spectrum analyzers to identify and overcome TI. When natural screening methods cannot be used and bandpass filters are required, they must be located between the feedhorn and LNB input. Readers are referred to *The Home Satellite TV Installation and Troubleshooting Manual* for details on identifying and combating TI.

TABLE 4-1. CARS BAND AML AND FM FREQUENCY ALLOCATIONS

CARS Group	Channel Range Number	Frequency	Bandwidth (MHz)
A	A1-A20	12700.0 - 13200.0	25.0
B	B1-B19	12712.5 - 13187.5	25.0
C	C1-C42	12700.5 - 12946.5	5.0
D	D1-D42	12759.7 - 13005.7	6.0
E	E1-E42	12952.5 - 13198.5	6.0
F	F1-F32	13012.5 - 13918.5	6.0
K	K21-K40	12700.0 - 13200.0	13.5

SELECTING EQUIPMENT

C. SCRAMBLING TECHNOLOGIES

Until recently, no satisfactory methods other than regulation of the use and sale of home satellite systems had been developed to allow "pay-TV" producers to collect revenues. The development of scramblers and matched decoders which have evolved from cable TV technologies has ushered in a new and revolutionary era in satellite television. Customers scattered anywhere over an entire continent and beyond can now be addressed and managed from one central computer. The economic potential is staggering.

A number of scrambling methods have been used and introduced in Europe. These include the FilmNet, Sky Channel, EuroCypher, D2 MAC, Teleclub Payview III and the BSB systems. In North America, two encryption systems, the Oak Orion and the M/A COM VideoCipher are employed in the home satellite TV market. The latter has captured the lion's share of the satellite broadcasting pay-TV market. IN addition, the BMAC system is used on some commercial networks (see Figure 4-36). The MAC (multiple analog component) broadcast format can also be securely scrambled and is employed to encode broadcasts such as the American Armed Forces Television and Radio System (ARFTS).

The methods and systems available for scrambling satellite TV transmissions and surveys some of the technologies are outlined here. Due to space limitations it must be brief. Much more detail on television operation, scrambling principles as well as on the various European and American encryption systems can be found in *World Satellite TV and Scrambling Methods - The Technician's Handbook*.

Scrambling Methods

Television signals are composed of audio and video components. Such broadcasts can be scrambled or rendered unusable if either or both of these signals are altered so that a television receiver cannot recognize and therefore cannot reproduce the original material.

Basic Parameters

Scrambling systems can be defined by a number of parameters including level of security, method of control, encryption and key distribution, addressing, subscriber authorization and tiering.

The level of security translates into the degree of difficulty an unauthorized person would experience in recovering the original television signal. The least secure methods can usually be defeated by a competent do-it-yourselfer; the most secure systems generally employ top-secret military technologies which are extremely sophisticated and well protected from unauthorized use.

Control information is transmitted with the television signal and determines which decoders are authorized and the identity of the programs a subscriber is permitted to receive. The control signal has no direct relationship to the entertainment content of any broadcast but is usually relayed as a component of the baseband signal.

Figure 4-36. Scientific Atlanta B-MAC 9730 Integrated Receiver Decoder. *This unit accomplishes the tasks of decoding and processing B-MAC satellite audio and video broadcasts. (Courtesy of Scientific Atlanta)*

SCRAMBLING

SELECTING EQUIPMENT

Keys are digital "words" that unlock a decoder and allow it to descramble the message relayed on a television channel. The method used for key distribution is as critical as creating a secure scrambling formula, the matching "lock." All satellite broadcast systems relay their keys within the television signal. Each decoder is programmed to recognize a unique key or set of keys which are only accessible to the broadcaster. Only then can any particular broadcast be descrambled.

Each satellite TV decoder also has a unique address, just like each telephone line has a unique telephone number. The control data addresses each unit or any combination of decoders and is capable of authorizing or deauthorizing any particular subscriber. Tiering refers to the channel or group of channels a subscriber is authorized to receive. For example, permission may be granted to view a single special event or a year's worth of ten pay-TV channels. Each of these selections is considered a particular tier.

Video Scrambling

Scrambling of the video information is a process of altering the character of the signal. System designer may choose to leave the video content unaltered but remove, suppress or shift the non-picture portions of the signal such as the sync pulses are from their normal positions. The picture information can also be altered by inverting its waveform, shifting its voltage level, time shifting or dicing individual lines or adding an interfering signal to mask the entire television signal.

Audio Scrambling

Affordable digital methods for securely scrambling audio signals are now available because the audio bandwidth is relatively more narrow and the rate of information transmission is substantially lower than that of the video signal. Earlier audio scrambling devices used simple analog techniques such as hiding the audio information on a subcarrier in an unusual position within or outside the channel bandwidth or remodulating the basic audio subcarrier also to hide the audio information. Today, however, sufficiently low-cost, high-speed, analog-to-digital converters are available to transform the analog audio input to a digital stream which can easily be encrypted to a high level of security. This digital audio signal can then be mixed with control and addressing information and embedded within the horizontal blanking interval of the video signal or elsewhere for transmission. Transforming the video information into a digital stream has not been commonly used since is quite costly as faster converters are required to handle the higher associated data rates.

A Review of Scrambling Systems.

Some of the more common scrambling systems are briefly outlined below. These are the VideoCipher, Oak Orion, Filmnet, Teleclub Payview III, VideoCrypt and various forms of MAC including B-MAC, EuroCypher and D2-MAC.

VideoCipher II and Oak Orion

The VideoCipher II system is a relative newcomer to the satellite broadcasting industry (see Figure 4-37). By contrast, the Orion (Oak Restricted Information and Operation Network) equipment has been used in over 20 major systems in North American including over two years of service on the Canadian CANCOM, the US Chamber of Commerce Biznet and the now defunct ON-TV networks, as well as on numerous private broadcast systems. Nevertheless, the VideoCipher II decoder has recently occupied centre stage in North America having been

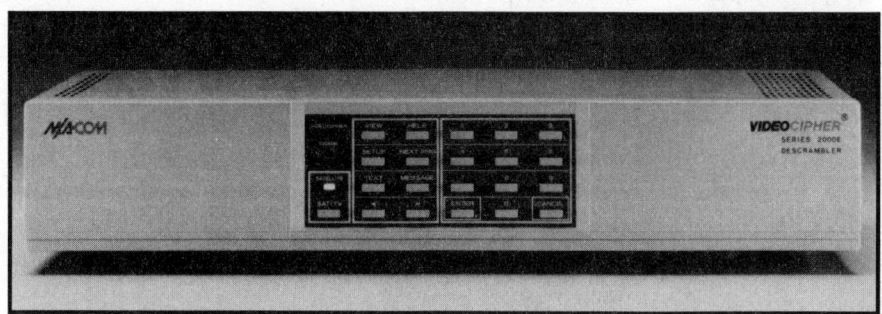

Figure 4-37. VideoCipher 2000 Decoder. *The VideoCipher 2000 stand-alone decoder is available as either baseband/RF model 2000E or the baseband-only model 2000E/B. The RF input level must fall between 55 and 0 dBmv (3 and 1 millivolt). The baseband input should be close to 1 volt. This unit has been incorporated into many North American receivers. (Courtesy of General Instruments)*

SELECTING EQUIPMENT

Figure 4-38. Sakura Receiver with Build-In SMART Card/Decoder.
This Sakura SR 890 IRD is pre-tuned to Astra and comes with a built-in Sky Movie Channel decoder. The smart card which is required for authorising the decoder is periodically mailed to subscribers. The receiver has on-screen graphics, 70 channel memory, automatic tuning and full Wegener stereo capability. (Courtesy of Sakura Electronics Ltd.)

chosen by most programmers including Disney, HBO/Showtime, Cinemax/The Movie Channel and others.

In Europe, Sky Movies has been the only satellite service to use the Oak Orion system (see Figure 4-38). The minimum security level has been used for this application as Sky has not really functioned as a pay-TV channel in the true sense. The audio scrambling facility was not used because monophonic and stereo subcarriers are transmitted with the Sky signal.

Of the two, VideoCipher II is perhaps the most important scrambling system in the history of signal security because it marked the first successful attempt to become accepted as the de facto standard for satellite television broadcasting. Although it was based on the technology of the early eighties, it has been rapidly upgraded and modified to improve system security.

Both of these satellite TV systems use relatively "soft" analog video scrambling methods and relatively secure digital audio encryption. At the VideoCipher II uplink the composite baseband video signal is scrambled by eliminating and replacing both the horizontal and vertical sync pulses with a data stream, inverting the video waveform and moving the colour burst to a non-standard voltage level. As a result, a total of thirteen video scrambling configurations are possible.

The Oak Orion System scrambles the video with horizontal/vertical sync replacement and random or sequential field or line inversion (see Figure 4-39). The usual horizontal and vertical sync pulses are removed from the scrambled video and replaced with 2.5 MHz bursts. The video on each line can be either inverted or normal polarity. A pulse situated just before the start of video in each line indicates the video polarity. Inversion can take place on a line, field or frame basis. Six possible inversion modes can be used in the system. One of four video inversion modes can be used during each field: inversion on odd lines, even lines or all lines, or no inversion. The original patent application also includes two further options. The voltage level of the digital packet in each line could be shifted or alternatively it could be varied by a sine wave gated to vary the levels in the horizontal blanking period.

The audio signal as well as all control and addressing information is digitized in both systems, then these bits are mixed up or encrypted, and are embedded into the horizontal sync pulse. A secure encryption "algorithm" or formula known as the Digital Encryption Standard (DES) is used to encode the audio signal on the VideoCipher II. Because the audio information is transmitted within the video signal, the need for a separate audio subcarrier is eliminated. Therefore, the uplink energy can be concentrated on the video signal and the transmitted signal-to-noise ratio can be increased. If so desired, the portion where audio subcarriers would have been transmitted can also be used for other purposes.

These encryption systems are controlled by one or multiple mainframe computers. Control, addressing and tiering information is continually uplinked to the whole population of decoders. The VideoCipher II and Orion systems are capable of managing 56 and 49 tiers, respectively. The computers can address any individual or group of subscribers to change tiering structure, activate decoders or terminate service within seconds. Security is enhanced because both systems have the capability of changing keys at will from the uplink control centre.

SELECTING EQUIPMENT

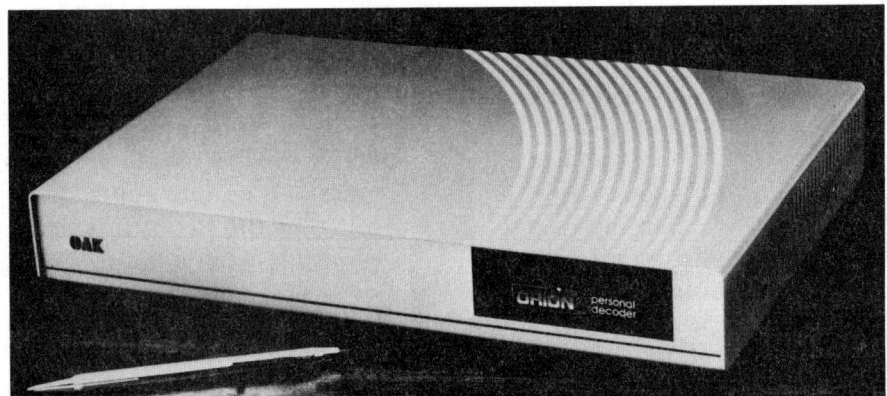

Figure 4-39. Orion-PD Decoder. *This small set-top decoder is designed for use in direct-to-home applications where low cost is essential cost is essential. It incorporates a channel 3/4 internal modulator. The features of this decoder match those of all other Orion units. LSI chips decoder the video and digital audio and data while accepting a variety of receiver video level inputs. (Courtesy of Orion Industries)*

The controlled access section of the VideoCipher system uses the DES algorithm to encrypt the keys and the programme attributes. The monthly key is encrypted with the unique key of each decoder. If a subscriber is not current, it is a simple matter to disable a decoder. The encrypted monthly key transmissions do not include a monthly key encrypted with the decoder's unique key. There are a number of unique keys in each decoder. The monthly key is decrypted by the decoder and used to decipher the programme attributes. If the decoder is enabled for the programme, it would display the descrambled signal.

The Filmnet System

The Filmnet/SATPAC/Matsushita system was first used on Filmnet in September of 1986. This channel had become one of the most popular premium film selections because the soundtrack of most of the films was in the original English. Subtitles in Dutch appeared on the video until recently.

Filmnet has been one of the most advanced channels in Europe in respect to subtitling. A viewer can select the subtitling language from a number of European languages via the teletext facility. This system is similar to the closed captioning used in the United States.

Video scrambling is accomplished by either gated sync shifting, or sequential or pseudo random line, field or frame video inversion. Because neither the vertical or horizontal scanning lines lock, the picture flashes from negative to positive image. Audio scrambling has not yet been used.

The first level of security was employed from early September 1986 to March 23, 1987. This phase was extremely simple to hack and a booming market in pirate or blackbox descramblers developed. The upgrade that occurred on March 23, 1987, inverting the video in each alternative field, dented this market. A further upgrade is expected soon and there are some rumours that Filmnet will begin scrambling the audio signal.

Teleclub Payview III

The Payview III system was first tested via satellite for the now defunct Spanish Canal 10. This system was not eventually chosen for use on this channel. Teleclub tested this system sporadically throughout 1988 and introduced it in 1989 in its basic form, alternate line inversion and no pseudo random line delay.

Teleclub Payview III, which is compatible with PAL, SECAM or NTSC like most of the other

SELECTING EQUIPMENT

scrambling systems, uses a variety of methods including sync modification, pseudo line delay, random and sequential video inversion for video scrambling. The horizontal blanking interval is raised above peak white level. This confuses the television receiver automatic gain control and clamping circuitry. When the scrambled video is displayed on a television set, the picture appears dark. The horizontal sync pulse is dithered or rapidly shifted in time. This causes a pseudo line delay effect on the scrambled video. The line video can be inverted on a sequential or random basis. Digital audio scrambling is available but has not to date been used over satellite.

The system has not proved difficult to hack in its basic forms but if random line inversion is used many of the pirate descramblers would probably be disabled. The raised horizontal blanking section can cause problems with some types of satellite receivers especially those with inadequate bandwidth since the Teleclub satellite transmission format uses a 36 MHz transponder bandwidth. Some hackers have tried to receive this signal using an 26 MHz ASTRA bandwidth receiver to no avail. Doing so causes a problem with the colour burst which appears to be suppressed or attenuated.

VideoCrypt

The VideoCrypt system was introduced on Sky Movies on February 1, 1990. Audio scrambling is not yet used. Video scrambling is accomplished by a line video cut and rotate method. Each line of video is cut at one of 256 possible points. Then the video information is rotated around this cut point. Of the 625 lines in the PAL system only 585 lines or so are used for video. The remainder are reserved for non-video information such as test signals and teletext. Thus only 585 lines must be scrambled.

The cut point on each line can be defined by one byte or eight bit word. The sequence of the cut points is pseudo random, derived from a pseudo random number generator. This chip generates a sequence of numbers that repeat if the sequence is left to run for long enough. The start point in the sequence is set by a seed transmitted over the air. This seed could be changed for every frame or field or even every few lines.

The encryption methods used to protect the keys and the over-the-air addressing data are some of the most advanced introduced to date. The encryption process is based on the Rivest Shamir Adlemann (RSA) algorithm. This algorithm uses the multiple of two large prime numbers to encrypt data. The main strength of the algorithm lies in the fact that it is computationally difficult to factor the product of the two primes and thus recover the original prime numbers.

Each descrambler is controlled by a Smart card which is mailed to the subscriber every three months. The Smart card differs from automatic teller machine (ATM) cards in that it has built in circuitry while an ATM card stores the data on a magnetic strip on its back. This built in circuitry holds the keys for decrypting the seeds and other data transmitted over the air.

Each descrambler is supplied sterile. It does not have any built-in identity number or coding. The first card that is inserted into the descrambler contains a program that implants an identity in the descrambler. Subsequent cards for this descrambler will have this identity encoded, to ensure that stolen cards will not work on other descramblers. This identity could be encoded on a magnetic strip similar to ATM cards.

The VideoCrypt system was developed by the French Thompson company. The system is one of the more modern scrambling systems that makes use of digital television technology. The primary weakness of most of the non-digital scrambling systems on the market is that they affect the sync pulses in the video. This makes the signal unrecordable in its scrambled state. It is as yet unclear if the VideoCrypt signal can be recorded in its scrambled state and then recovered.

Multiplexed Analog Components (MAC)

In the MAC system, the luminance (brightness) and chrominance (colour) components are compressed and transmitted sequentially whereas in PAL, NTSC and SECAM formats the chrominance and the luminance are simultaneously transmitted (see Section M, Chapter II). At the receiver, the components are digitally expanded and combined to give YUV or RGB outputs. This compression of the video information allows greater depth for audio and teletext services. The line sync in the PAL system is transmitted as a pulse in each line. This takes a considerable amount of the actual line time.

In contrast, in the MAC system, the line sync is derived from a six bit sync word in the data block.

The MAC system has a specific advantage compared to PAL, SECAM or NTSC formats in satellite broadcasting. In an FM system, the noise voltage rises almost linearly with frequency. In the PAL signal, the colour or chrominance information is mainly in the upper section of the baseband signal, (3.5MHz to 5.5MHz). As a result, the highest noise voltage appears in the chrominance signal thus producing a lower signal-to-noise ratio. Pre-emphasis only offers slight compensation for this effect. And de-emphasis can have one disadvantage when dealing with a near threshold PAL signal, namely, dot sparklies become streak sparklies due to the response of the de-emphasis network.

B-MAC

B-MAC was developed by Digital Video Systems, now owned by Scientific Atlanta. This system has the capability to deliver one scrambled video channel, six audio channels and a teletext channel over a single satellite transponder.

This standard has been chosen for what is currently the largest private network in Europe – the Racing Channel. British Telecom supplied the decoders, used in conjunction with modified JRC receivers, to owners of the system. Receiver bandwidth has been widened and the baseband response has been flattened to 10 MHz. An integrated receiver decoder is also available. This system is available in MAC 525 lines and MAC 625 lines. The outputs from the decoders are generally RGB but a PAL or NTSC modulator is included.

The system as used by the Racing Channel has a common video channel, a teletext service and four audio channels. The separate audio channels are for the system users: Ladbrokes, William Hill, Coral and Mecca.

Video scrambling is accomplished by pseudo random line delay. There are no usable sync points in the signal for PAL or NTSC television receivers. This generally ensures that there is no line lock and in most cases, no vertical lock. The audio is digitally encrypted by a technique technically known as adaptive delta modulation. A DES-like algorithm is used. The audio channels are transmitted using the Dolby™ Deltalink II system. The basic operation of the B-MAC system is the same as the other MAC systems with the exception of the data structure. Encrypted ASCII teletext data is transmitted during the field blanking interval, nominally 25 lines long.

The B-MAC encoder generates the encryption pattern based on a seed or key that can be varied every quarter of a second. This encrypted seed is transmitted within a data packet in the field blanking interval. It is not fixed for security reasons but can be changed by the system operator using the keyboard on the encoder. The new seed is then individually transmitted to each subscriber in individually addressed data packets.

The encryption routine used in the B-MAC system is similar to DES but not identical. This use of custom algorithms makes reversing of the algorithm more difficult. The primary reason for using a custom algorithm was that the B-MAC system is used outside the United States and DES ICs are restricted from export. Therefore an alternative was required. Since the B-MAC decoder uses custom algorithms, clearly it also uses custom ICs. In order to hack the teletext and audio, the potential hacker would have to reverse engineer and produce these ICs.

EuroCypher (D2-MAC)

EuroCypher is the European variant of VideoCipher II, albeit a more secure version because it is based on more recent technology and builds upon the technological countermeasures developed for VideoCipher. The transmission standard of the EuroCypher system is D–MAC. This means that even in its unscrambled state it is unusable with an ordinary PAL television or monitor. It requires the television to have a MAC to RGB or PAL converter circuit, generally referred to as a transcoder.

The D-MAC transmission format allows for the transmission of up to sixteen compact disc quality audio channels as well as superior quality video. The colour and the luminance components of the video signal are separated and time compressed. They are then transmitted sequentially with the digital audio data. The digital audio packet occupies the first section of the line. The digital audio packet occupies the section that would be occupied by the horizontal blanking interval in the PAL video line. The compressed chrominance or colour information packet, approximately 17 µS long, comes next followed by the The luminance packet, approximately 35 µS long.

SELECTING EQUIPMENT

The Eurocypher system uses two security facilities; double cut and rotate video scrambling and audio encryption. The double cut and rotate is applied to the video. The chrominance and the luminance packets are cut and rotated at separate and unrelated points. There are 256 cut points per packet. A 16 bit pseudo random number generator (PRNG) is used for generation of the cut point on each line.

The double cut and rotate is a necessary security precaution in the MAC system. Since digital time compression and expansion are used in the MAC process, it would be theoretically possible to work out a single cut point in the chrominance by comparing the scrambled chrominance with an expanded gray scale derived from the unscrambled luminance.

The audio is digitally encrypted. It is first digitised and then combined with a PRNG, identical to the one in the decoders. The initialising word or seed for the PRNG is encrypted using the DES algorithm. The key used for this encryption is the monthly key. This key generally does not change until the end of the month. The new key is then transmitted to the decoders but is encrypted using the decoder's unique key. This facility allows the system owners to deactivate the decoders of those users who have not paid their monthly subscription.

D2-MAC

The D2-MAC system is the variant of MAC that has found favour with most European countries. One of the main advantages of this MAC variant over D-MAC is that it fits within the cable bandwidth allowed for the PAL signal. While the D-MAC variant can transmit 16 compact disc quality audio channels, the D2-MAC variant can manage only 8.

The backwards compatibility of the D2-MAC variant is a major factor in its growing acceptance. Many countries in Europe have cablenets. Indeed in some countries, most of the larger cities and towns rely more on cable than on conventional over-the-air transmissions for their television viewing.

The D2-MAC variant supports the double cut and rotate video scrambling as used in D-MAC. It also allows the encryption of the digital audio by similar means to that used for D-MAC.

V. INSTALLING KU-BAND SATELLITE TV SYSTEMS

One of the objectives of this manual is to teach readers how to install home satellite systems that will stand the test of time. A theoretical knowledge, while academically interesting, is of little value to an installer unless the result is a system that works. Even the most sophisticated troubleshooting expertise may be overtaxed if the installation was seriously flawed or if unacceptable shortcuts were taken.

In this chapter, each installation procedure is examined in detail. For simplicity's sake, some steps are presented in a slightly different order than might occur during an actual installation. For example, the section on aligning the feed precedes the description of mounting actuator arms, although it is convenient to hold an antenna in place by installing the arm before the feed assembly. Of course, in those situation where, for example, a fixed Astra antenna is being installed, the sequence would be different compared to a full-blown installation. Nevertheless, the exact sequence of steps is often a matter of personal taste: each technician will develop a personal strategy and special methods as he or she gains experience.

Today the techniques used to install some of the newer small-dish Ku-band systems are quite different from those used in the earlier days of the home satellite industry in North America and elsewhere. DIshes ranging in size from 30 cm to 1.2 meters are often installed on south facing walls or on roofs (see Figure 5-1). Larger antennas most often have been installed in poles set in concrete on the ground. The smaller dishes are lighter, less susceptible to wind loading and are simply easier to install. Many of these antennas are fixed onto one satellite, especially in Europe, so tracking is not necessary. The techniques outlined in this chapter are intended to cover both situations.

An Overview

Tables 5-1 and 5-2 provide an overview of installing both a fixed and a tracking satellite reception system. These illustrate that the general process is rather simple, although the details might require some time and experience to master. Table 5-1 applies to a somewhat larger ground-mounted dish while Table 5-2 to a small dish system. Note that in some cases installing some systems may involve steps from each sequence.

Figure 5-1. Small Dish Installation. *This Ku-band Grundig satellite system was unobtrusively installed on the rear wall at the second storey level of this home. Cable runs are simple and neat. (Courtesy of Grundig International Ltd.)*

INSTALLATION

TABLE 5-1. THE INSTALLATION PROCESS - GROUND-MOUNTED TRACKING SYSTEM

STEP	TASK
1. Conducting Site Survey	Locate a position with a clear view of the arc and plan the overall installation.
2. Installing Support Structure	Install the supporting pole in a perfectly vertical orientation.
3. Trenching and Cable Runs	Dig cable trenches, where necessary, and lay conduit.
4. Assembling Mount	Assemble mount and lift it onto the pole.
5. Installing Actuator	Install the actuator on the mount and set elevation so that it sits. so the dish will rest in either a vertical or horizontal position.
6. Lift Antenna onto Mount	After assembling the antenna, if necessary, either lift it onto the mount so it faces upward and can be bolted to the mount or roll it into position so it can be bolted in place.
7. Assembling LNB/Feed	Bolt the LNB onto the feedhorn and then attach both structures onto the dish.
8. Wiring	Attach the coaxial cables, polariser wires and actuator wires.
9. Aligning Antenna	Set the elevation and declination angles, turn on the power and align the antenna onto the arc.
10. Fine Tuning	Complete the tracking and electrical fine tuning.
11. Waterproofing	Once the system is working faultlessly, waterproof all outdoors electrical connections.
12. Sealing Cable Runs	Bury the conduit.
13. Programming Receiver	Program the satellite receiver by following manufacturer's detailed instructions.
14. Connecting Accessories	Hook up accessories such as VCRs or stereos.

TABLE 5-2. THE INSTALLATION PROCESS - WALL OR ROOF FIXED SMALL DISH SYSTEM

STEP	TASK
1. Conducting Site Survey	Locate a position with a clear view of the chosen satellite and plan the overall installation.
2. Installing Support Structure	Install the supporting bracket.
3. Trenching and Cable Runs	Dig cable trenches, where necessary, and lay conduit.
4. Assembling Mount/Antenna	Assemble the antenna, mount and LNB/feedhorn. Lift and attach this equipment onto the mount.
5. Setting the Angles	Set the elevation and azimuth angles.
6. Wiring	Attach the coaxial cables, polariser wires and actuator wires.
7. Aligning Antenna	Set the elevation and declination angles, turn the power on and align the antenna onto the arc.
8. Fine Tuning	Complete the tracking and electrical fine tuning.
9. Waterproofing	Once the system is working faultlessly, waterproof all outdoors electrical connections.
10. Programming Receiver	Program the satellite receiver by following manufacturer's detailed instructions.
11. Connecting Accessories	Hook up accessories such as VCRs or stereos.

INSTALLATION

A. THE SITE SURVEY

Every installation must begin with a thorough site survey, one of the most critical yet often most neglected steps in a satellite TV installation. The survey is the equivalent of a doctor's examination and diagnosis prior to surgery. No surgeon would ever open up a patient without a complete understanding of what he wishes to accomplish. A competent dealer or installer would not cut corners in conducting a site survey.

Two important tasks are completed during a site survey. First, a location with a clear view of the entire arc of satellites is found. Second, the entire installation is planned. Although a check for terrestrial interference is crucial in C-band installations, TI is rarely encountered in Ku-band systems except perhaps in some large metropolitan areas.

Numerous facts must be established before beginning an installation. It is important to know which satellites and transponders will be accessed; what type of equipment should be used; where to place the antenna, how much concrete or wall fittings might be necessary; whether or not to use a long pole, roof, tower or ground mount; what length, route and type of cable is required; the distance from the antenna to the satellite receiver and a host of other questions. The planning process, carried out with the cooperation of the customer, must deal with seemingly minor concerns as well as issues with potential legal ramifications. For example, on one hand, cable paths must always be clearly marked because the installation crew might be unfamiliar with the site. On the other hand, permits which grant permission to install the system and receive the programming may need to be obtained.

Above all, the customer's wishes must always be considered so that all decisions are mutually acceptable. For example, if the customer does not want the antenna next to her home but 50 metres away, a line amplifier and 12 gauge actuator wire may be required. This may result in a substantial variance from the original bid. The customer's home and yard should be treated as if it were the installer's home. It is a wise practice to respect the saying that "the customer is always right."

Ensuring a Clear View of the Arc

An antenna must have a clear view of each satellite because any obstruction will absorb or reflect microwaves and subsequently lower the detected signal-to-noise ratio. Water is a particularly strong absorber of Ku-band microwaves. Some reflector installed in winter had perfect reception until spring when foliage returned to trees obstructing its view: then pictures either disappeared or deteriorated in quality (see Figure 5-2).

Two basic instruments are required to aim the antenna at communication satellites: an inclinometer and a compass (see Figure 5-3). Each satellite can be targeted by adjusting the antenna's azimuth and elevation angle. The azimuth is measured in degrees of rotation from true north and the elevation in degrees above the horizon.

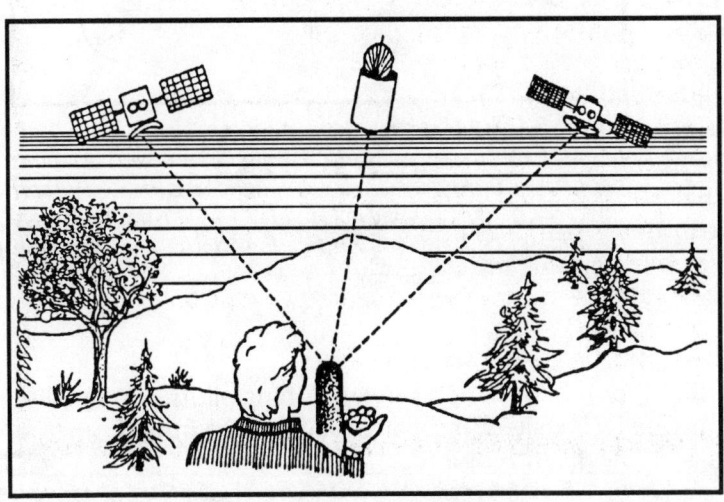

Figure 5-2. Checking for a Clear View of the Arc of Satellites. *When conducting a site survey, one of the most important steps is to be sure that all satellites to be received can be clearly targeted. Any trees, buildings, mountains or other obstructions blocking a clear line of site to any satellite will make receiving a broadcast difficult, if not impossible.*

INSTALLATION

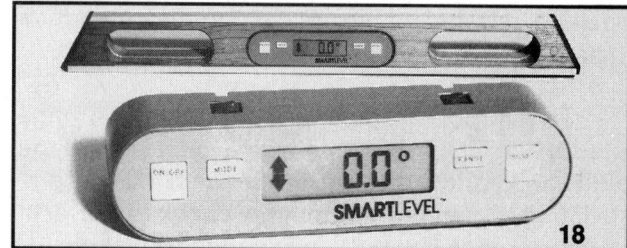

Figure 5-3. Site Survey Tools. *A compass (left photo) and an inclinometer are the two basic tools required by a satellite TV installer for locating and aiming at the geosynchronous arc of satellites. The inclinometer is used to measure the angle from the horizon to any point in the sky as well as to set the correct angles on a polar mount. This inclinometer pictured here has a digital read-out. (Top Photo Courtesy of Natropolis International)*

Determining a True North/South Bearing

A compass will point along the line between magnetic north and south. In most locations, there is a difference between the bearing to the north and south poles and the needle reading on a compass. Magnetic north, the strongest magnetic centre near the north pole, is located near Bathurst Island off the northern coast of Canada (see Figure 5-4). As a result, magnetic variation can exceed 30° in locations such as the American state of Alaska. However, magnetic variation is zero along the "agonic line." In North America this line runs just off the west coast of Florida, through Lake Michigan, over the Great Lakes, and to the magnetic north pole. In Europe, the agonic line runs through Sicily to Prague, and on through the middle of Sweden. East of this line, the north pole is east of the compass reading; west of this line it is west. Agonic lines are also found in various areas of the globe as the accompanying maps (Figures 5-5 and 5-6) illustrate.

A true north/south bearing can therefore be found by using a compass and correcting for magnetic variation (see Figure 5-7). For example, in the American city of Denver, Colorado, magnetic variation is minus 12.5°. So the arc of satellites centred on true south is found by rotating 12.5° east of the south reading on a compass. This is equivalent to a compass reading of a 167.5° (180° less 12.5°) as measured from true north. A similar procedure can be followed at any other point on the globe.

Magnetic variation at any location in the world can be found from the charts here and, in some countries, by requesting the value from a local airport information service. Airline pilots, like satellite TV installers, must also take their bearings from true north and south.

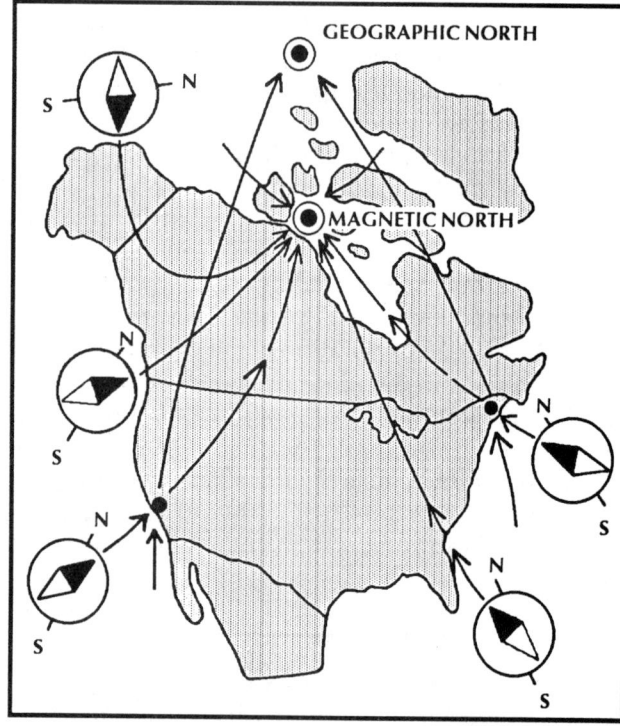

Figure 5-4. True North versus Magnetic North. *In both hemispheres magnetic north is quite far removed from the geographic north pole. In North America it is located just off the coast of Hudson Bay in Canada. This illustration shows the difference between the bearing to true and magnetic north.*

SITE SURVEY

INSTALLATION

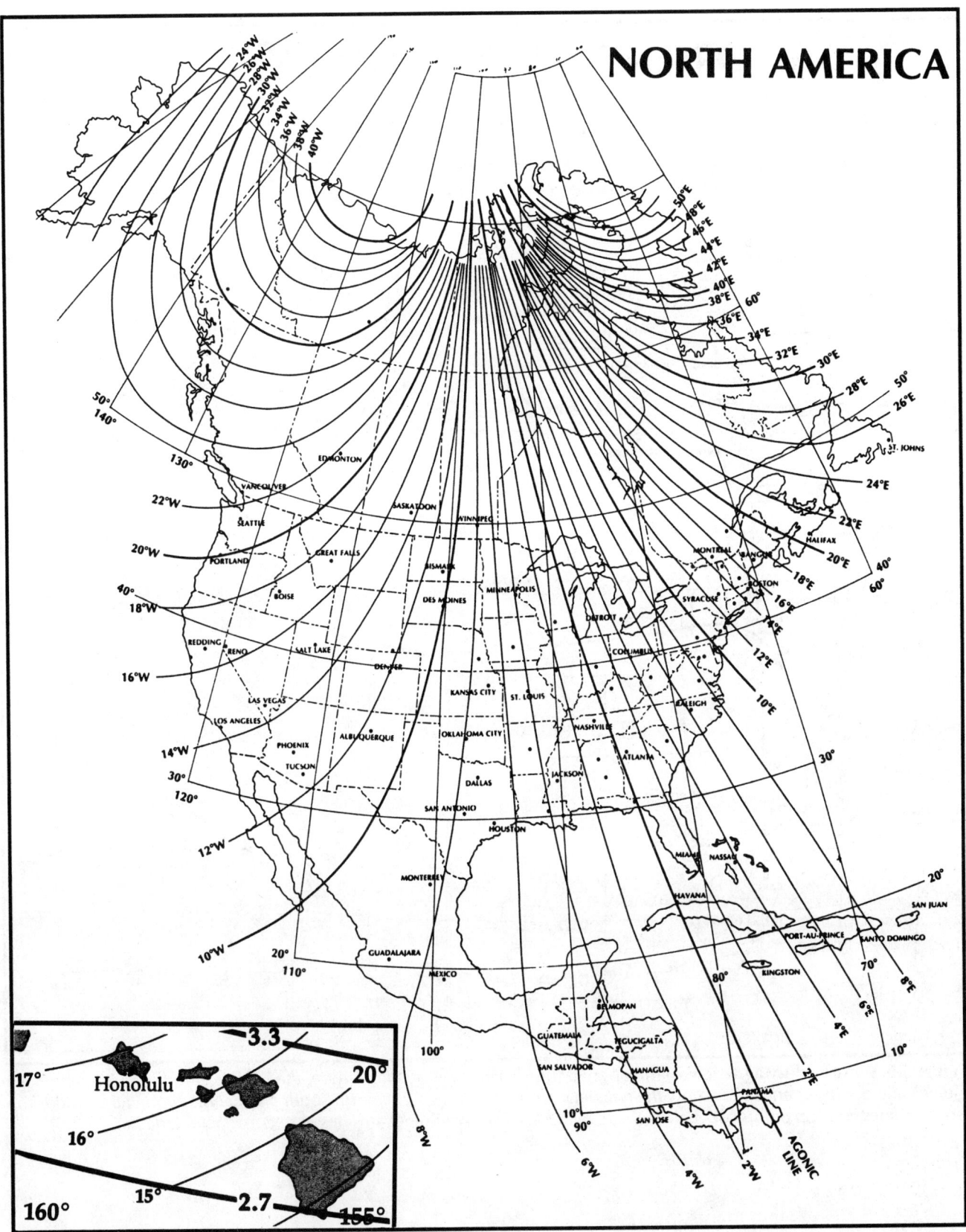

Figure 5-5. North American Magnetic Variation Map. *A compass will point to magnetic north and not true north except along the "agonic line" because of the difference in location between the magnetic and geographic north pole. North of the equator, a reading taken with a compass west of this line must be corrected by rotating east to find true south. South of the equator the opposite is true.*

INSTALLATION

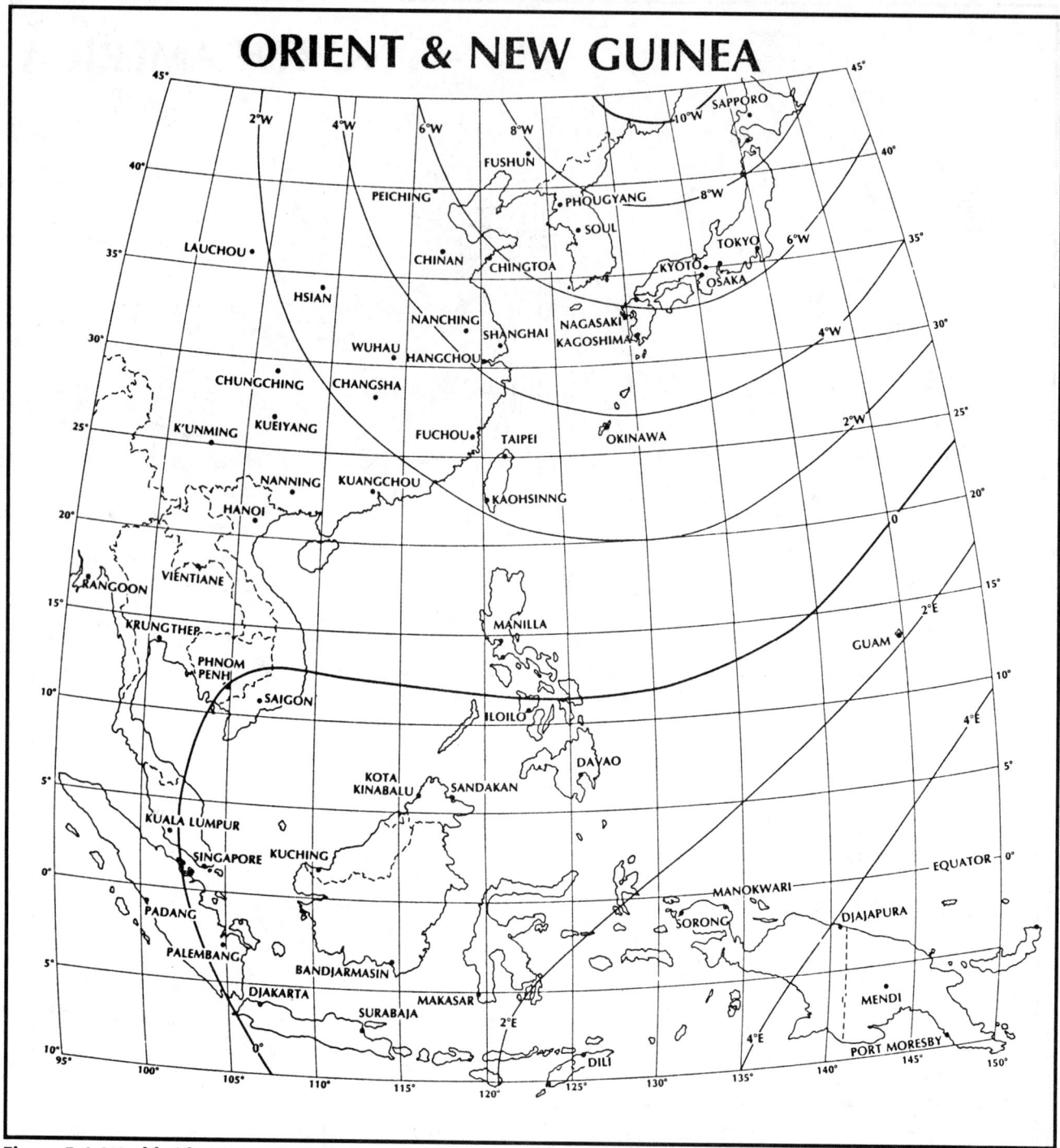

Figure 5-6. Worldwide Magnetic Variation Maps. *These maps of Australia, Europe, the Middle East, the Orient and New Guinea and South America show how the magnetic field varies from a true north/south orientation through the world. The isogonic lines connect points of equal magnetic variations. The agonic lines are lines of zero magnetic variation.*

SITE SURVEY

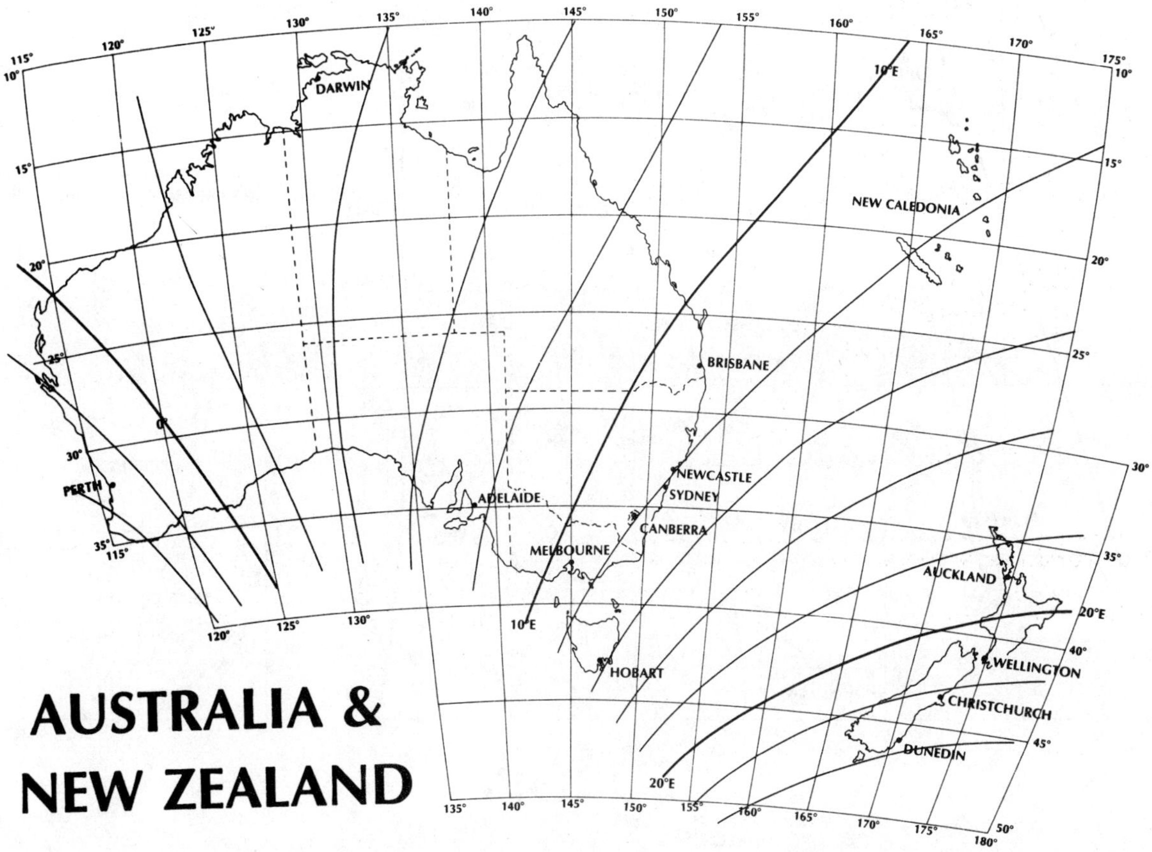

INSTALLATION

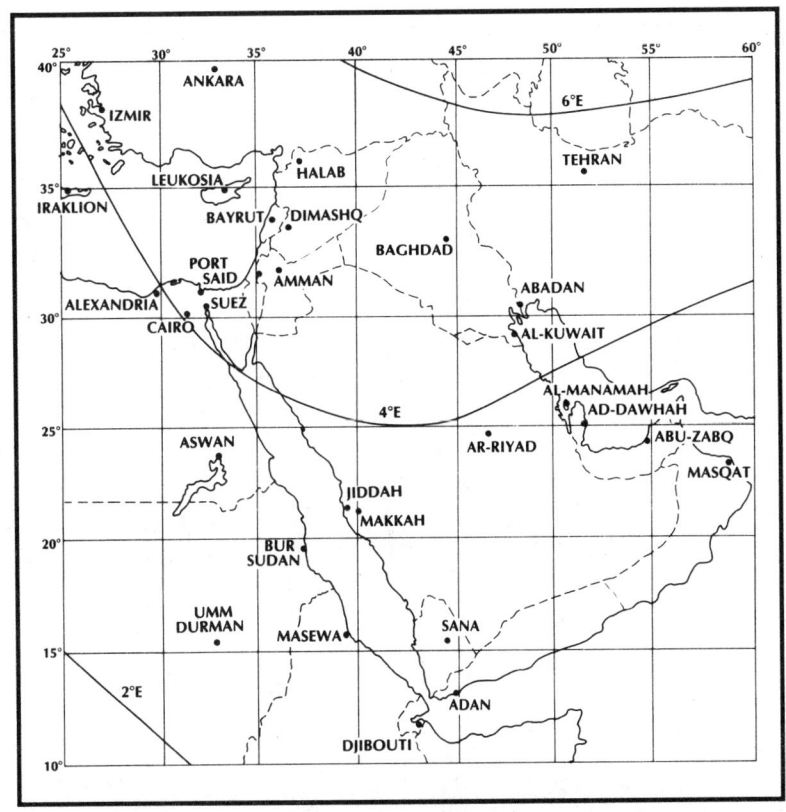

AUSTRALIA & NEW ZEALAND

SITE SURVEY

INSTALLATION

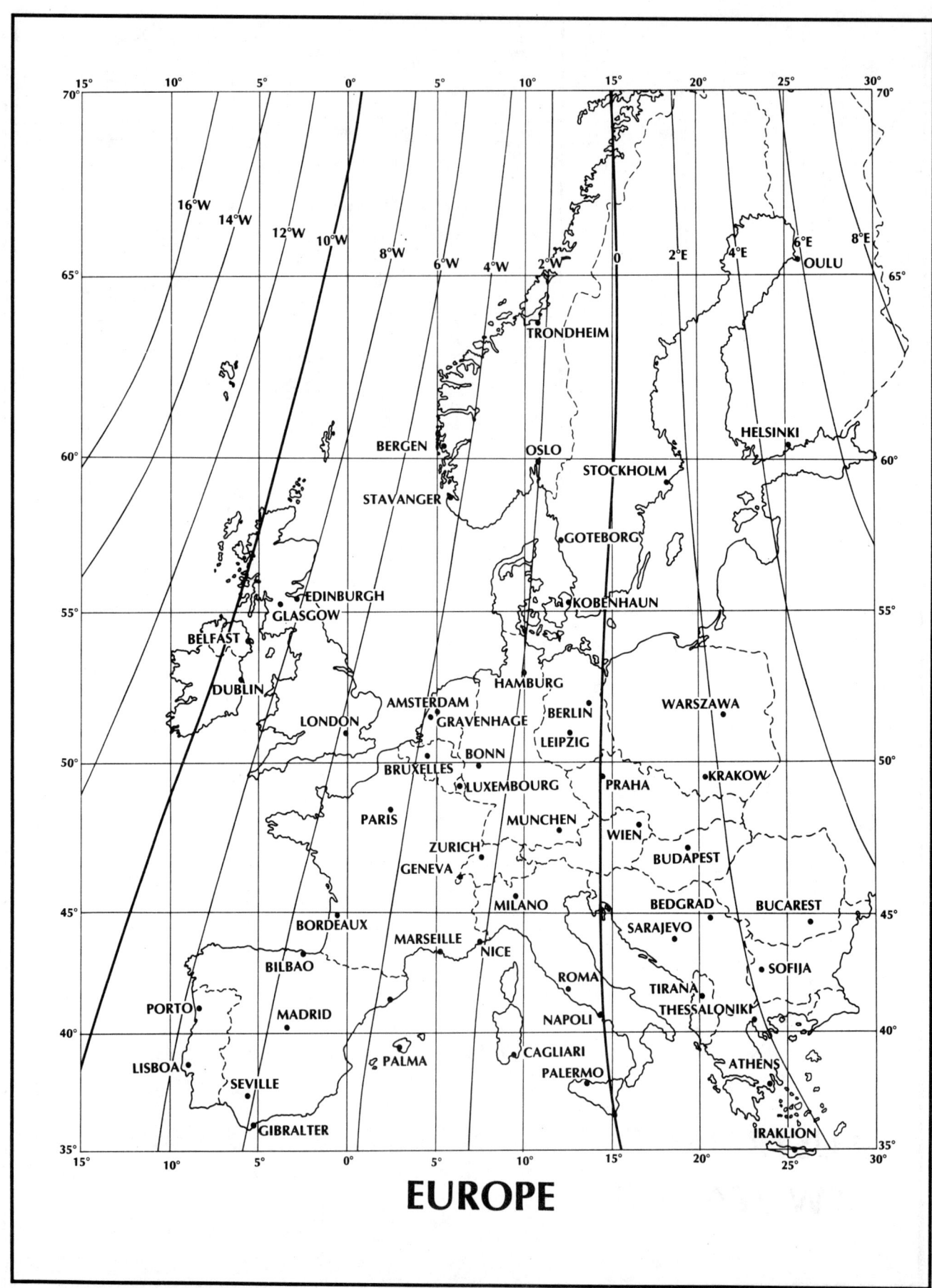

EUROPE

SITE SURVEY

INSTALLATION

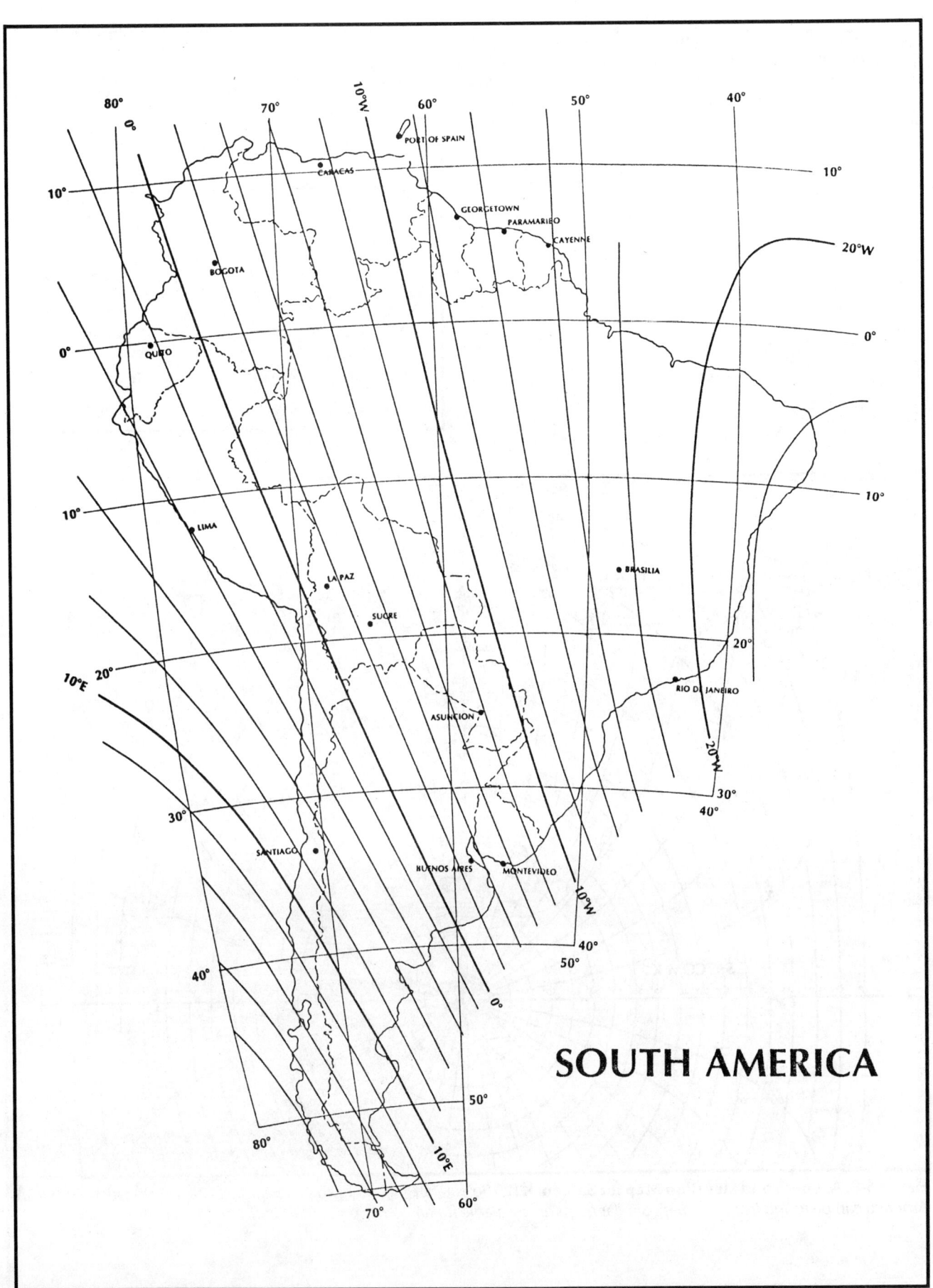

SOUTH AMERICA

SITE SURVEY

INSTALLATION

Figure 5-7. Adjusting for Magnetic Variation. *The bearing to true south can be found by aiming a compass east of magnetic south when the variation is negative and vice versa. Magnetic variation is negative when west of the zero-variation, agonic line at a site north of the equa- tor. South of the equator the settings are reversed.*

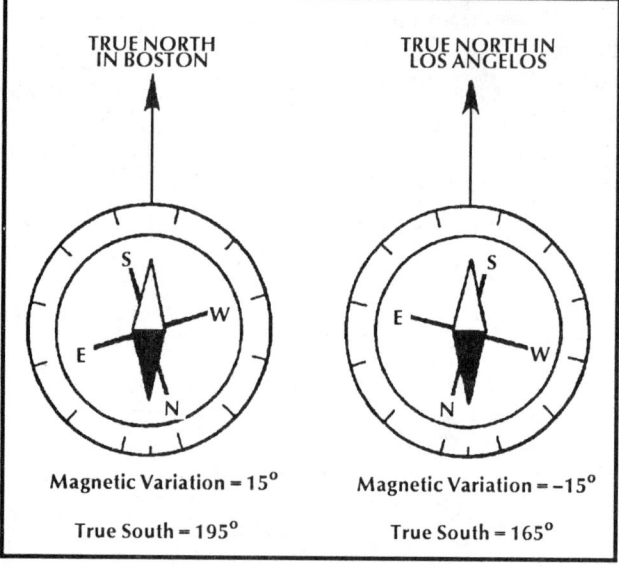

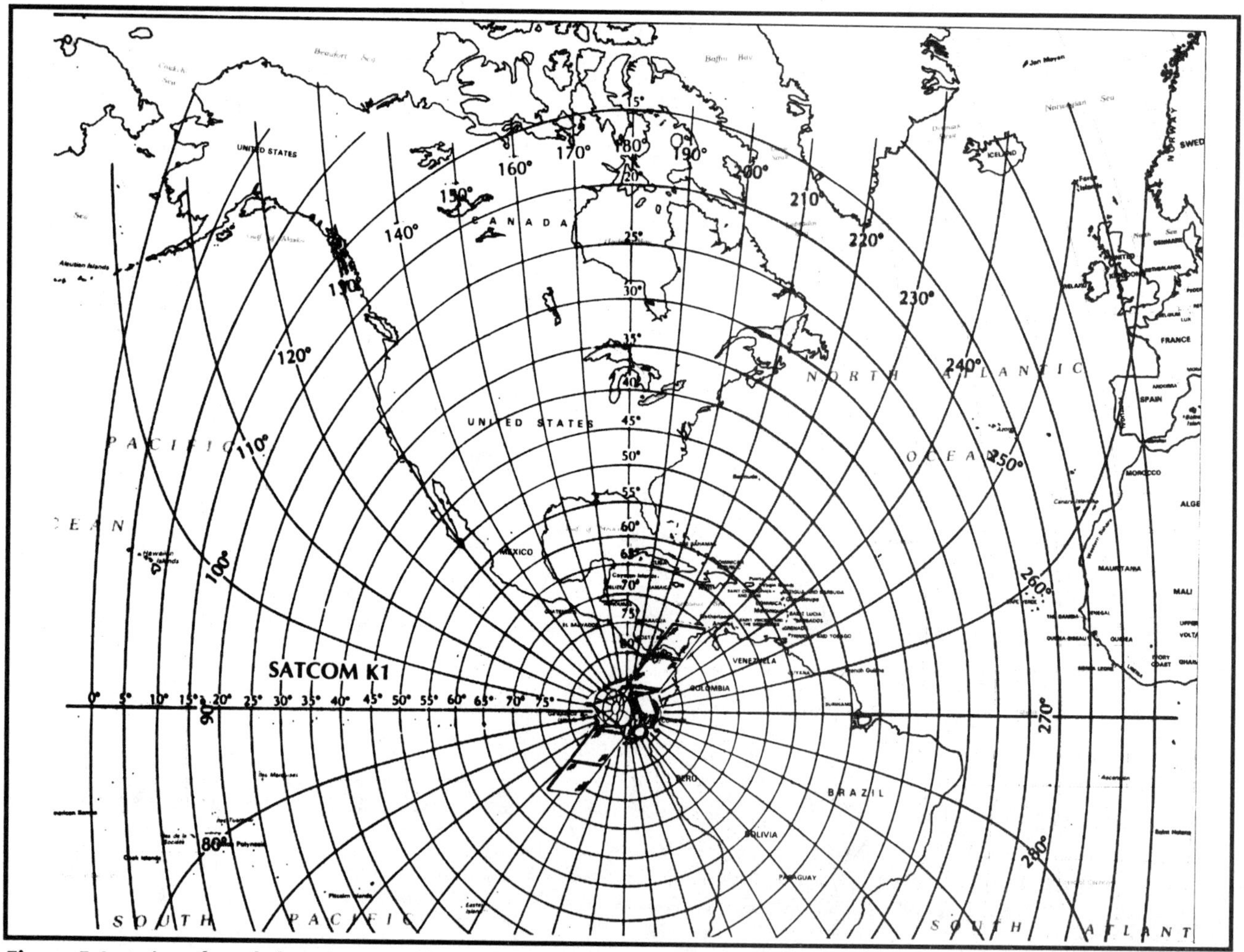

Figure 5-8. Azimuth and Elevation Map for Satcom K1. *The satellite azimuth and elevation from any location in North America can be found from this map. Similar maps are available for any other particular satellite or area in its footprint.*

SITE SURVEY

INSTALLATION

When using a compass, it is important to realise that it can yield inaccurate readings when used in the vicinity of iron-containing objects. Therefore, a bearing taken next to a large truck or any other large steel object could be many degrees off the local average magnetic north/south line.

Determining the Azimuth and Elevation Look Angles

The azimuth heading towards any chosen satellite can be found from a table, from equations based on simple geometry (see Appendix B) or from commercially available computer programs. A useful fact to realise is that the satellite whose angular location corresponds to the site longitude is found by aiming due south in the northern hemisphere and due north in the southern hemisphere. For example, GStar A2 is located at 105°W which is exactly south of Denver at 105°W longitude. Both Rome and Munich are located at approximately 13°E longitude which equals the Eutelsat 1-F4 angular position.

Similarly, the elevation to any satellite can also be found by calculation, computer programs, from tables or from satellite location charts (see Figure 5-8 and Appendix C for examples). For example, Spacenet II has an elevation angle of 30.8° and an azimuth of 138.0°W in Denver, Colorado. During the site survey it would therefore be found by rotating 29.5° east of true south and then by aiming up to 30.8°. A compass would read 125.5° when pointing at Spacenet II. This equals 138.0° less the 12.5° correction for magnetic variation. If a tree or any other obstruction were blocking the view, this proposed installation site would have to be changed.

Elevation angles are easily measured with an inclinometer. This instrument can be placed on a long ruler or any other straight edge and used in a fashion similar to sighting a gun (se Figure 5-9). The ruler is raised until the desired elevation is reached. While sighting along its length, see if there are any objects blocking a clear view to each satellite in the applicable portion of the belt. When sighting with some types of inclinometers, having a second person read the scale makes the process a little easier.

Various companies market instruments designed to make sighting the arc even simpler. Some of these products are see-through charts having satellite locations marked. These are mounted on a tripod and pointed due south for sighting the satellites. Others have telescopic view finders which are preset for any geographic location so a user can visually scan across the entire arc (see Figure 5-10)

Choosing the Best Antenna Site

In those cases where an antenna that will track the arc is to be installed, at least three satellites should be sighted, the most easterly, the most westerly and the one highest in the belt. Then a line connecting these three points should be visualised in the sky. The entire arc should be free of obstructions to a clear view of the desired satellites. If there is a question about any other satellite in the arc being partially or fully blocked, it should also be targeted. If a clear view is not possible from one location, then another site should be tested. If no appropriate location which is acceptable to the customer can be found, raising the antenna higher on a roof mount or long pole mount may be the solution. There is, however, a trade off to elevating a larger dish above the ground because the higher

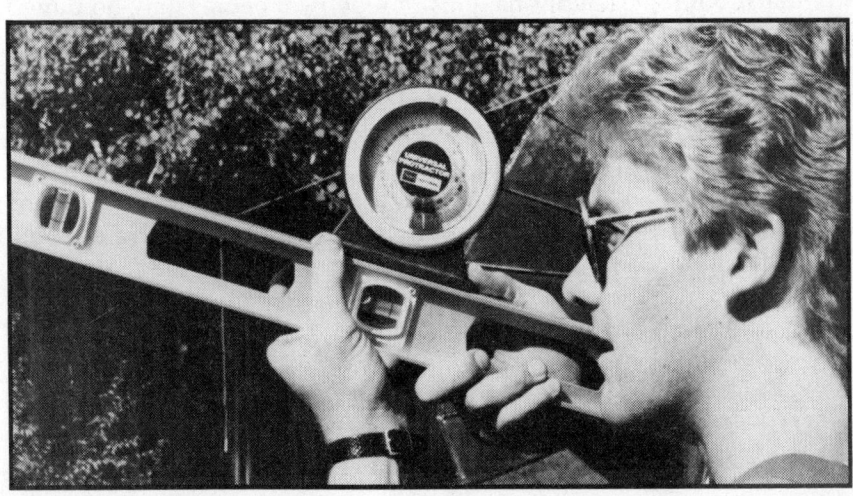

Figure 5-9. Measuring the Elevation Angle. *The elevation angle can easily be measured by placing an inclinometer on a straight edge and sighting along it in the*

INSTALLATION

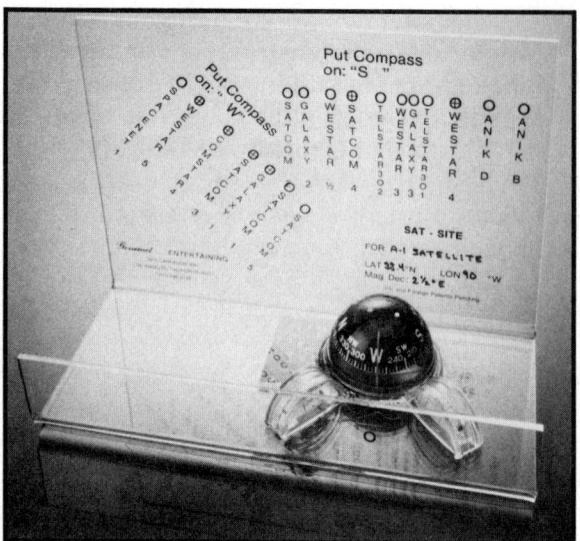 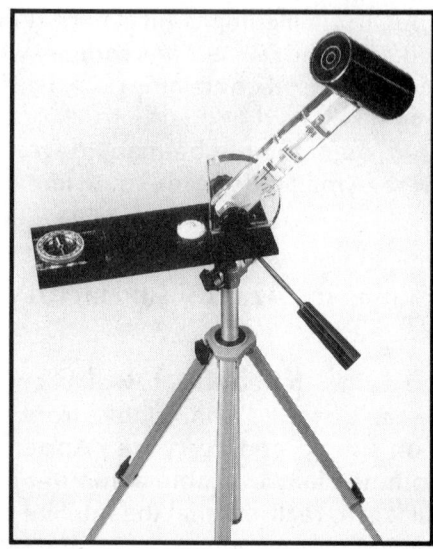

Figure 5-10. Satellite Sighting Tools. *There are many useful tools available to facilitate installing TVROs. Two varieties of these tools are shown above. The Sat-Site compass/satellite locator (on the left) has a plastic viewing screen with a scaled profile of the satellite arc plotted for a particular location within an 80 kilometre (50 mile) radius. When this screen is superimposed on a Plexiglas grid, an installer can view the satellite arc through the grid and spot any obstruction that may impede clear reception. The Viewfinder (on the right) allows an installer to set the azimuth and elevation to any satellite and then check for obstructions to its view through the eyepiece. (Courtesy of Gourmet Entertaining and Focii Manufacturing Company)*

it is installed, the more susceptible it is to wind loading and interfering signals.

Note that power or mains lines and other low frequency lines attached to telephone poles are not a concern unless they are blocking a large portion of the antenna aperture (see Figure 5-11). Such signals could however be a source of low frequency ingress interference that leaks into poorly grounded or shielded distribution equipment.

Customer participation should be encouraged at this point in the installation. He or she should understand that perfect pictures will be obtained only if the view to each satellite is unobstructed. It may be decided, if necessary, that the antenna should be installed in a location where the signal from one or more satellites will be weakened or lost. Or, if there are minor obstructions to a clear view at a preferred installation site, the customer may agree to purchase a larger antenna at a higher cost to overcome this partial blockage of signal. In some cases, the extra gain of a relatively large reflector can compensate for signal power lost to partial reflection and absorption by obstructions. In rare cases, it may be necessary to bring a small test dish to the site to ensure clear reception is possible (see Figure 5-12).

The installation plan should include informing the customer of any necessary building permits and local ordinances. Distances from fences are usually set by zoning laws and easement restrictions. For example, in Canada the minimum distance to the antenna edge is 5 metres although 4 or 6 metre clearances from the property line are not unusual. Changing an approved site may involve filing for a new permit. Customers generally have very strong opinions concerning the aesthetics of antenna location. In some European countries, local ordinances forbid installing dishes on roofs so eave mounts are a choice.

Underground utilities such as water mains, telephone lines or electrical cables may be buried at these easement boundaries to conform to local building codes. Planting the pole a few more metres inside the property line than is mandated may therefore be an intelligent strategy.

The site should also be chosen with cable runs in mind. Unexpected difficulties may be encountered. For example, tunneling under a 3 metre wide driveway may prove nearly impossible. If, as a result, the antenna must be located at a less desirable site where it is not hidden from view, it may prompt the customer to purchase a smaller, perhaps lower performance but more attractive brand.

INSTALLATION

Figure 5-11. A Possible Source of Ingress Interference. *This 3 metre antenna was installed near power lines. Although these lines do little in the way of blocking microwaves from reaching the reflective surface, in this case the 60 cycle line power might be a source of ingress interference. All precautions against ingress TI such as properly grounding equipment should be taken to eliminate this possibility.*

One additional important factor must not be ignored. An antenna needs adequate freedom of motion to track the entire arc. If, for example, it is set too close to a fence, the reflector may either be bent or may demolish the fence before the end of arc limits are set. Nothing can replace careful planning.

Planning the Installation

Clearly, a satellite TV installation should be planned as completely as possible during the site survey. The antenna location must be chosen and coordinated with the cable runs to the receiver and all televisions. How cable will enter the the building must be established. The type of cable and number of splitters and/or taps needed, as well as the necessary connectors must also be determined.

Satellite receivers can be placed anywhere, not necessarily on top of TV sets. If radio controlled actuators and receivers are used, a customer might decide to leave all the indoors electronics in a well-ventilated closet.

Figure 5-12. Portable Antennas for Site Checks. *A small portable antenna should be used if any uncertainty about obstructions blocking the vision of a TVRO or interference causing impaired reception exists.. The 90 cm antenna at the left is a Ferguson "Astra" dish. The 1.5 metre (5 foot) dish at the right is manufactured by DH Antennas. (Courtesy of WV Publications, Ferguson and DH Antennas)*

It is essential that the customer be consulted and be heard during a planning process. Too many dealers and technicians have preconceived notions and may not pay close attention to the desires of their customers. Every satellite system should be installed as if it were for personal use. That is the route to success. When an installation has been completed, always have the customer sign the site survey worksheet release (see Figures 5-13 and 5-14).

SITE SURVEY

INSTALLATION

SITE SURVEY WORKSHEET

Job Number_____
Date_____

Customer Name_____
Address_____
    City_____State_____Zip_____
Telephone (Home)_____(Work)_____
Lead Source_____Salesman_____

SITE MAP

Size and Make of Antenna and Mount_____
LNB Noise Temperature or Noise Figure_____
Style and Manufacturer of Feed_____
Receiver Manufacturer and Model_____
Length of Cable Installed_____
Length of Conduit Installed_____
Number of Wall Plates_____

Figure 5-13. Site Survey Worksheet. *By documenting everything about an installation and obtaining a customer acceptance signature, the job will be properly competed. In addition, information will be available for any future troubleshooting work that may become necessary.*

ANTENNA SUPPORT STRUCTURES

INSTALLATION

SITE SURVEY WORKSHEET (continued)

Make and Model of Actuator_____

Number of Extra TV Sets Installed_____

Description of Cable Routes_____

    In Home_____

    Out-Doors_____

Obstructions to Cable Runs_____

VCR or Stereo Hook-Up_____

Type of Mount (pole, roof, etc...)_____

Types of City Ordinances in Effect and Permit Requirement_____
_____

Type of Warranty_____

Status of Cable TV Disconnect_____

Type of Program Guide Requested_____

Type of Off-Air Antenna and Connection_____

Method of Waterproofing Used_____

Number of A/B Switches or Combiners_____

Date Installation Requested_____

TI Carriers Noted_____

Further Comments_____
_____
_____
_____
_____
_____
_____
_____
_____
_____

Customer Signature_____

Figure 5-13. Site Survey Worksheet (continued).

INSTALLATION

PRE-INSTALLATION WORKSHEET

Customer Name_____Customer Number_____

Date_____

Equipment List:

Number of Items	Item	Manufacturer	Model Number	Serial Number
	Antenna			
	Mount			
	Feed			
	LNB/Feed Cover			
	LNB			
	Coax			
	Receiver			
	Actuator			
	TI Filter			
	Combiner			
	A/B Switch			
	Line Amplifier			
	Splitter			
	Wall Plate			

                                                                                                Initials

All electronics have been checked before leaving office   _____

Back-up electronics are on hand   _____

Enough wire has been taken for installation of all TV sets   _____

Proper tools are in installation vehicle   _____

Customer contacted before leaving office   _____

All lights and signals on vehicle and trailer have been checked   _____

Correct customer invoice has been taken   _____

Pole mount has set and has been checked   _____

Connector kit has been checked   _____

                                                        Time left office_____

Figure 5-14. Pre-Installation Worksheet. *This chart can be used as a guide to ensure that all necessary equipment has been taken to the work site.*

INSTALLATION

B. ANTENNA SUPPORT STRUCTURES

Satellite TV antennas and their mount assemblies are usually supported on a pole, especially when more than one satellite will be targeted. However, there are now numerous installations in Europe where customers who have purchased systems to view exclusively one satellite, such as Astra, have fixed antennas that are mounted directly on az-el frames. These supports are often bolted onto an eave, a roof or secured to the ground by weights or concrete.

This section is organized into two main topics: installing antennas at ground level and installing antennas in locations above the ground. The information required and the techniques used in these cases are somewhat different. These differences as well as the similarities discussed below.

Pole and Pad Mounts at Ground Level

A variety of construction methods can be used to secure a pole to a ground-mounted antenna site. These include pole supports, pads, pier foundations or combinations of these types. Most require the use of concrete or another strong binding material that will withstand the test of time.

When antennas are being installed on pole supports, securing the pole is critical because an antenna and its mount can often be very heavy and may be subjected to severe winds and other environmental stresses. This is especially the case in North and South America where large antennas might be used to simultaneously receive both C- and Ku-band transmissions. If the pole is solidly set in a true vertical position, all the remaining system adjustments can proceed smoothly. The three basic rules to follow when installing a Ku-band TVRO are stability, stability and stability. Even the minor instabilities in the support structure can degrade performance.

Preparing for the Job

Before work begins on the antenna support structure, some preliminary steps should be taken.

Underground Utilities

Gas, telephone, electric and cable television companies often have underground lines in unmarked areas. In many countries these utility companies will send personnel at no charge to locate and mark the route of their cables or pipes. If a utility line is severed during the digging and trenching portions of an installation, the costs in customer dissatisfaction and time lost are high. In addition, the dealer will probably have to pay an hourly rate to the utility company to have the damage repaired.

Pre-Trenching for Conduit Placement

When a ground mount installation is chosen, once the cable runs have been mapped, a few metres of the trench should be excavated starting at the antenna mount. This will allow the metal or PVC conduit leading in to a supporting pole to be set at the same time as other concrete work such as a pad is completed. If this step is overlooked, the cables would lie exposed near the base of the pole and could be susceptible to damage from gardening utensils such as lawn mowers, weed-eaters or shovels.

The Fundamentals of Concrete

Concrete is a building material which was invented centuries ago. Its properties are well known and new varieties are constantly being developed for a host of applications. When prepared and hardened properly, it has tremendous strength and is very durable. However, incorrect mixing or setting procedures can result in a weak base which can easily crack or disintegrate. If this were to happen, the antenna base could move and it would be extremely difficult to realign the heavy block of concrete.

Concrete is like a synthetic rock and derives its strength from its ingredients - gravel, sand and cement. When water is added to a cement mixture, a chemical reaction begins. Setting of cement is not simply water drying out. The cement acts like a strong glue which powerfully binds the gravel and sand with other materials within the mold.

INSTALLATION

During the chemical reaction, heat is generated. If this heat is lost too quickly during cold weather installations, the concrete will not cure properly. Under such conditions it can lose strength and crack. If the surface of the concrete is covered with straw, blankets or other insulating materials during curing, heat will be retained and such problems can be avoided. Calcium chloride can also be added to speed up the setting process; generally 2% by weight of this salt is recommended. But be aware that too much calcium chloride can rapidly corrode metal structures such as the supporting pole.

Post type foundations can be made from ordinary Portland cement or standard premix. Slab foundations having larger areas of concrete should be made from air-entrained Portland cement and wire screen. This mixture contains additives that allow microscopic air bubbles to be entrained or trapped. These bubbles are like an internal lubricant which makes pouring and spreading easier. They also act like tiny shock absorbers which allow the larger piece of concrete to expand and contract without cracking.

Both of these types of cement can be purchased in pre-mixed bags or they can be mixed on site. Proportionate ingredients for plain and air entrained concrete are listed in Table 5-3.

TABLE 5-3. INGREDIENTS FOR VARIOUS TYPES OF CONCRETE

Type	Proportion	Ingredient
Plain Concrete	2.5	Portland Cement
	3	Sand
	5	2cm aggregate
	1.25	Water
Air Entrained Concrete	21-A	Portland Cement
	2.25	Dry Sand
	5.66	Coarse Aggregate
	1.25	Water

Other mixes of concrete or replacements for concrete having specifically designed properties are available. For example, in the United States, QR Inc. sells a special mix sold under the trade name QUIK-ROK which sets up in 15 minutes at temperatures as low as -7 °C. Its manufacturers claim that it is stronger than concrete. It expands as it sets, thus forming a tight seal; ordinary concrete tends to contract during setting. Similar characteristics are advertised by the makers of another substitute called Sat-Based Cement. Dish Set™, an alternative to concrete, is a closed-cell, expanding polyurethane foam which sets in about 15 minutes depending upon ambient temperatures (see Figure 5-15). It can be used for pole mounts since it generates force against the pole and the sides of the hole during expansion. It cannot be used in extremely sandy soils which do not support this action. When it is used, the hole must be narrow, 4 to 8 cm on either side of the pole, and deeper than normal for this expansive force to have full effect. All of these quick setting mixes do not allow much time for mistakes; be totally prepared before pouring these materials by using stakes to level the pole.

Figure 5-15. Dish-Set Concrete Substitute. *One alternative to using concrete is a two-component chemical foam, trademarked as Dish-Set, which when mixed with a high speed drill will set a pole in approximately fifteen minutes. Instructions must be carefully followed to achieve full strength. When using this concrete substitute, the hole size must be deeper than normal and must be only slightly larger than the pole diameter so that compaction can occur. Because the setting occurs so rapidly, leveling stakes must be used to secure the pole in place in advance of mixing. This product does not perform well in extremely cold weather or in loose sandy soils.*

INSTALLATION

Freezing, Underground Water and Stability

Underground water is the chief enemy of any foundation. During winter, water at the base of sidewalks, patios or fence supports may freeze and generate tremendous forces which cause the surrounding rocks and soil to shift. When a thaw occurs, the water percolates away, soil and rocks collapse inwards and the concrete can twist or settle. This heaving effect is usually not too great a bother for fence posts or other such structures. However, if a pole supporting an antenna is moved even slightly away from a vertical orientation, tracking of the geosynchronous arc can be ruined.

The frost line is the depth below which no freezing occurs. In North America, it varies from approximately 2.5 cm (one inch) in southern Texas to over 3 m (10 feet) in northern portions of the United States. In even more northerly locations such as Alaska or the Canadian Northwest Territories, special construction techniques are always required to manage the difficulties associated with the great depth of the frost line.

Pole or Post Supports

Pole or post type antenna supports are the most common variety in use today in North and South America, especially for installing larger C-band antennas. They provide strong, stable platforms (see Figures 5-16, 5-17 and 5-18). But pole mounts cannot be used in locations where the ground is too rocky to dig the necessary hole or where the water table is too high. Because water is a good conductor of heat, a high water table may mean the frost line extends much deeper than normal in a given region. A pole support should also not be used for antennas larger than 4 m in diameter since wind generated forces are simply too great. A 3-legged tower which distributes the force over its whole base should be installed.

The hole for a pole support should be 2 to 4 times the diameter of the pipe. It should extend at least 15 cm (5.9 inches) below the frost line and 2 or 3 cm (1 to 2 inches) of gravel should fill the bottom to allow for proper drainage. Usually between 100 and 300 kg (200 to 700 pounds) of concrete are needed for an average pole support.

Pole supports typically extend approximately one metre into the ground and 1.5 m (5 feet) above.

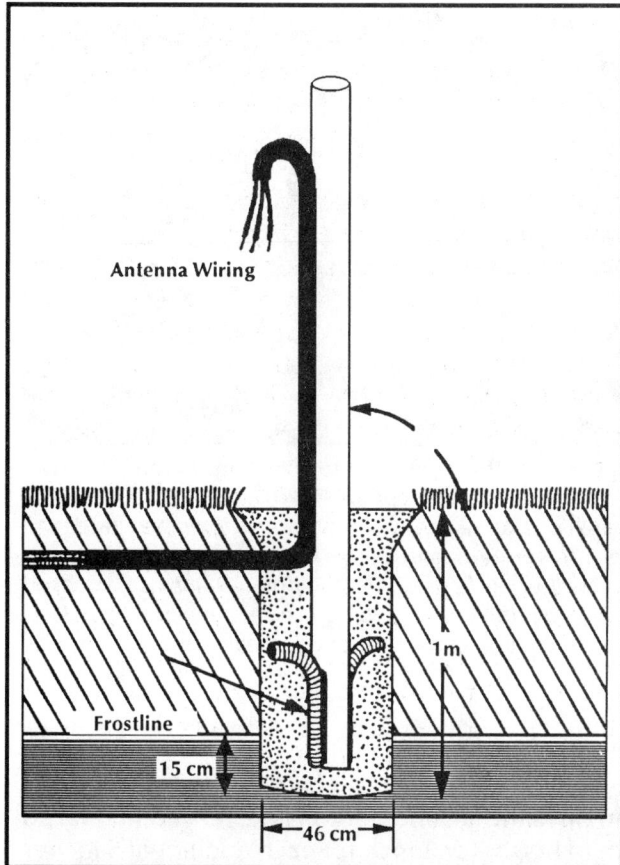

Figure 5-16. Ground Mounted Pole Support. *This drawing shows all the necessary procedures required to install a pole mount for a residential TVRO.*

Figure 5-17. Digging the Hole for a Pole Mount. *After the site survey is completed and all cable routes are mapped, digging the hole for the mounting post can begin. A few simple tools including a post hole digger, a digging bar, a shovel and a garden hoe as well as a wheel barrel are all that is necessary. All the excavated soil should be placed in a wheel barrel and not thrown on a sodded lawn. This simple step will make clean-up after installation easier.*

INSTALLATION

Figure 5-18. Pre-Trenching for Conduit. *Before the concrete is poured it is recommended that at least 2 metres of trench be excavated and that conduit be aligned inside the hole. This ensures that the conduit will be completely buried and will extend parallel to the pole. Otherwise, exposed cable near the concrete base can be damaged by pets as well as by garden tools such as weed-eaters or lawn mowers.*

A general rule of thumb is to add 10 cm of extra length in the ground portion for every additional 50 cm of height (or 1 inch for each additional 5 inches of height, see Figure 5-19). For example, a pole extending 5 m above ground should be planted an additional 70 cm, or 1.7 m deep because there are 3.5 m of extra length compared to a standard pole. Poles set in very soft or sandy soil should be either planted deeper or have a wider concrete base.

Once the pole is placed into the ground, it can be temporarily supported by a few rocks and by wooden stakes or guy wires. The pole must be perfectly vertical so that the antenna will be capable of properly tracking the entire arc of satellites. A carpenter's level at least one metre in length or a plumb bob can be used to check this orientation from three or four vertical positions around the pole. These readings should be taken periodically as the concrete is setting. Note that after the concrete has been mixed it should have a thick consistency. Too much water makes it "soupy." This will delay the curing process because all the excess water must either evaporate or drain off. The caustic ingredients in concrete will kill grass so care should be taken to avoid spills.

Some believe that concrete should also be added to partially fill the centre of the pole for even further stability in high wind areas (see Figure 5-20).

Figure 5-19. Recommended Pole Depth. *The hole for a regular pole mount should extend at least one metre deep with a diameter of four times the pole size. In any case, it should extend at least 15 centimetres below the local frost line.*

Figure 5-20. Stabilising the Pole with Concrete. *Concrete can be poured down inside the pole to give it additional support.*

ANTENNA SUPPORT STRUCTURES

INSTALLATION

The pole must have sufficient strength to withstand both bending and twisting forces. At least schedule 40 steel poles should be used. Using schedule 80 or 120 steel is even better. (Schedule 120 is like pipe drillers stem used in the oil industry). Most antennas typically require poles having 7 to 10 cm (3 to 4 inch) outer diameters.

In order to prevent an antenna from breaking free from the concrete form and rotating, a 1 cm metal rod should be inserted through the bottom part of the pole so that it extends out 7 to 15 cm (3 to 6 inches) on both sides (see Figure 5-21). Or a piece of re-bar with similar dimensions should be welded onto it. An alternative to using one continuous pole is to implant a sleeve into which the long pole can be bolted. If designed properly, this type of assembly can allow a pole to be re-leveled in case the base eventually shifts.

Most pole supports have no mechanism for making a leveling adjustment once the concrete is set. This can be a concern because the effects of wind and water can, in time, cause the concrete base and thus the pole mount to shift. This effect is encouraged because a large antenna angled towards the equatorial arc of satellites does not have its weight evenly distributed on the pole. This situation places quite a large bending force on the support. Without a leveling adjustment, the alternatives are to dig up the pole and replant it using additional concrete, install a new pole and cut out the original one, or attempt to level it with a piece of heavy machinery such as a truck or jack. None of these are attractive options.

For the dealer or installer who regularly digs holes for planting poles, a gas-powered auger similar to those used to dig holes for fence posts is probably a good investment. This tool can save many hours of installation time (see Figures 5-22 and 5-23).

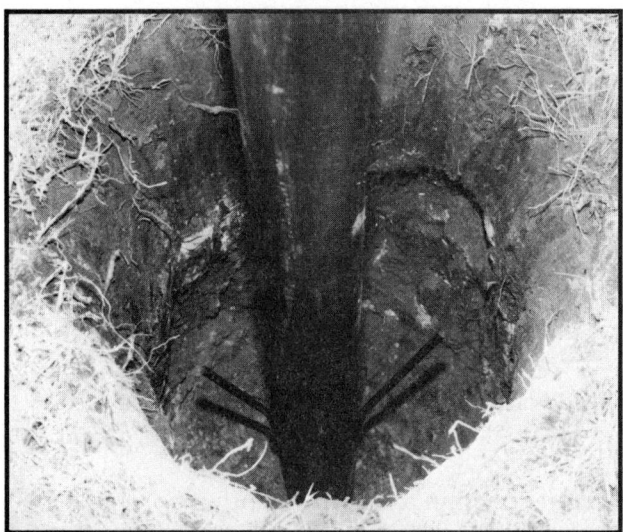

Figure 5-21. Pole-to-Ground Stabilisation. *It is recommended that construction rebar or a 15 to 20 cm (six to eight inches) bolt be welded to the bottom of the pole to keep it from twisting loose when high winds exert powerful forces on the antenna.*

Figure 5-22. A One-Man Auger. *For those installers who do not have additional help, one alternative is the use of a Singleman Power auger. (Courtesy of Ground Hog, Inc.)*

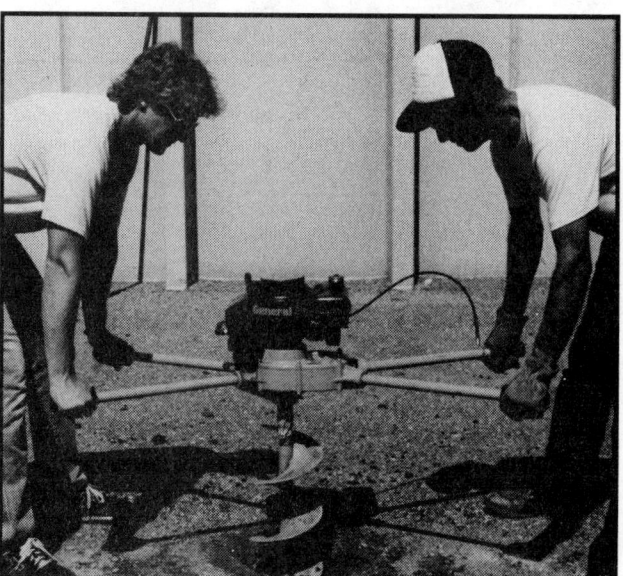

Figure 5-23. Using a Power Auger. *Digging holes with a power auger can save many hours of time and can avoid uncomfortable blisters. Safety must be stressed when using this type of equipment in the vicinity of buried power and utility lines.*

INSTALLATION

A completed pole mount is shown in Figure 5-24. The section of conduit that is set in concrete with the pole must have a weatherhead to prevent moisture from entering the conduit.

Concrete Pads

In those cases where the ground is too rocky or hard to allow digging of a narrow, deep hole, where the ground water table is too high or where an especially large reflector will be installed, a concrete pad can be poured on site.

A pad is constructed by digging a shallow trench, building a wooden form to hold the concrete, adding gravel for drainage, embedding wire mesh for strength and pouring the concrete with the pole and the conduit in place (see Figure 5-25). It is best to recess the pad into the ground so that a lawn can be easily mowed and so that the customer will not stub his or her toe during the night on a concrete obstruction. When installing a pad in a non-level location it is important to excavate the high ground and never to fill dirt into low terrain. Otherwise, the pad is likely to settle and cause the pole to tilt (see Figure 5-26 and 5-27).

When concrete has set, the wooden form can be removed. These forms should be made from 5 by 25 cm (2 x 6 inch) green lumber because dry wood will absorb water from the concrete mix and hinder the setting process. Also, water should be periodically sprayed onto the surface of newly poured concrete to improve hardening. This keeps the outside as wet as the inside and thus prevents cracking during setting.

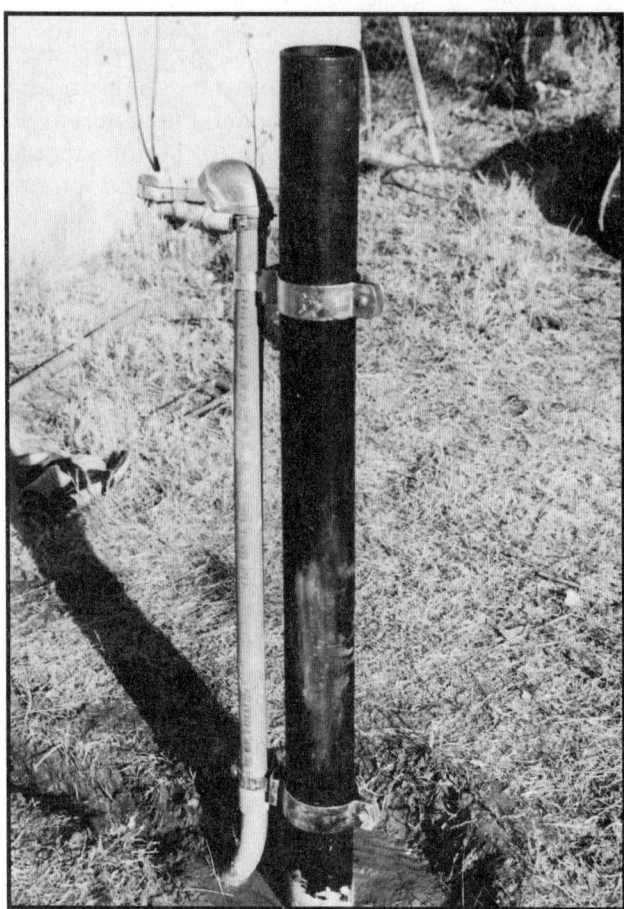

Figure 5-24. Pole Support with Conduit in Place. *This photo shows the correct installation procedure for the pole support with the conduit extending into the concrete and secured in place with clamps. Concrete should be poured to about 9 centimetres (six inches) from the top of the hole to allow grass to be replanted adjacent to the pole.*

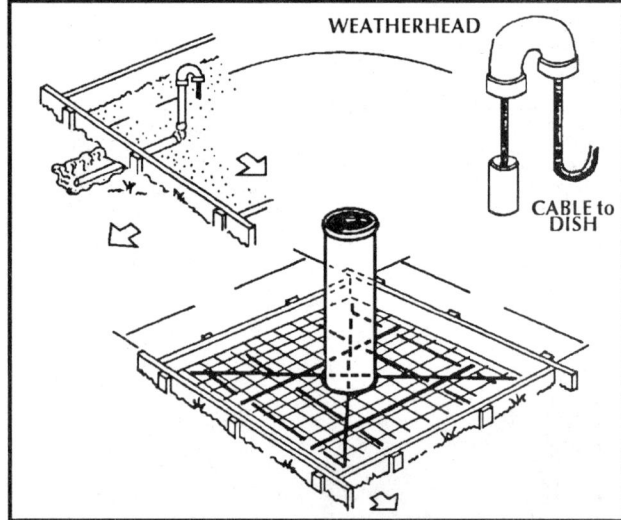

Figure 5–25. Schematic of a Pad Support. *A concrete pad support has steel rebar and wire mesh for strength and a gravel base for drainage. The pole has its weight distributed by extending rebar or channel iron from its base to near the edges of the pad.*

INSTALLATION

Figure 5-26. Pad Support on a Hill. *This pad support was poured onto a sloping hill. It is important to remember when pouring a pad mount on an uneven surface that the pad should be dug down into a high spot and never installed on a dirt-filled low spot. Doing so could cause the pad to settle over time resulting in inaccurate tracking of the arc. About 1.6 m³ (1.5 cubic yards) of concrete was used to pour this pad.*

Figure 5-27. A Unique Installation. *This pad support was poured on a granite rock outcropping in the mountains of Colorado. A jack hammer was used to drill holes to insert rebar into the granite base. Over 3 cubic metres of concrete were used to pour a flat pad and four 19 mm (3/4 inch). A redheads secured the steel mount to the pad. (Courtesy of*

A slab foundation supports an antenna by virtue of its weight and size. Rules of thumb have been developed to size pads. A pad reinforced with mesh should have concrete at least as thick as the pole diameter. Without reinforcement, increase this width by 50%. The length and width of a pad should be at least half the antenna diameter or the measurement from the pad surface to the base of the reflector. For example, a 2 metre antenna supported by a 9 cm pole should be supported by a pad measuring 1 by 1 metres and should be from 9 to 15 cm thick.

Wire mesh used for reinforcement should be at least 8 gauge and have holes spaced so that openings are no more than 40 cm on a side. It should be located at a depth equal to about 1/3 the thickness of the concrete as measured from the pad surface.

A pad mount requires adequate drainage. Sandy soils require less drainage than rocky or clay soils which are both more impervious to water. The gravel bed under the concrete should range from a minimum of 50% to 150% of the concrete thickness. For example, a reinforced pad set on solid rock using a 9 cm (3.5 inch) pole should have about 12 cm (5 inches) of gravel and 16 cm (6.5 inches) of concrete.

As in a standard pole mount, the mounting pipe must have two pieces of rebar inserted through its base at right angles. When these are set with the pole in concrete, they spread the load over the whole base of the pad for stability and prevent rotation of the pole under high wind loads. This rebar can be tied to the mesh reinforcement with baling wire for further strength.

Three Point Pads and Pier Foundations

Three point pads and pier foundations both bolt onto a tripod assembly which holds the antenna support pole. The former is a concrete pad with three or more anchor bolts either preset into the concrete or drilled in after hardening (see Figures 5-28 and 5-29). The latter uses three smaller pads called piers, each one having a small concrete foundation attached to deeply set reinforcement bars and anchor bolts (see Figure 5-30). While an az-el mounted antenna is typically mounted on a three-point support, this differs somewhat from both of these mounting structures (see Figure 5-31).

211 ANTENNA SUPPORT STRUCTURES

INSTALLATION

Figure 5-28. Northsat 1.2 M Dish on Pad. *This photo clearly shows the re-leveling bolts supporting this antenna. (Courtesy of Northsat)*

Figure 5-30. Three-Point Pad Mount. *This triangular az-el mount was installed with minimal concrete by using a three-point pad mount. The excavated hole was 1.6 m deep and 1 m in diameter.*

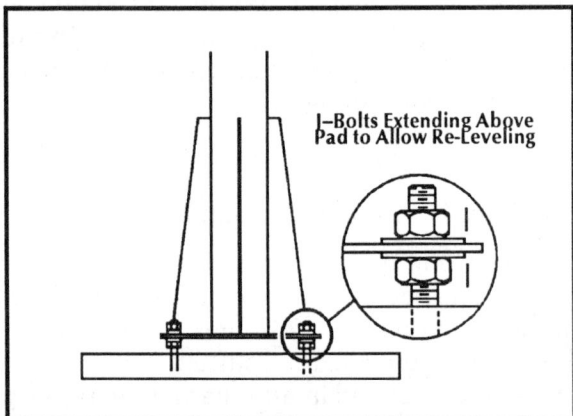

Figure 5-29. J-Bolts with Double Nuts for Mount Re-Leveling. *Allowing additional threaded rod to extend above the surface of a pad will permit leveling at any time in the future.*

Figure 5-31. Az-El Support on a Concete Pad. *About 3 cubic metres (3.3 cubic yards) of concrete were poured onto this concrete patio. The pad was secured by redheads to prevent shifting. This 5 m Scientific Atlanta aluminium antenna is supported by an az-el mount and is permanently targeted onto one satellite. Notice how the cable runs are well protected by conduit. (Courtesy of Brent Gale)*

INSTALLATION

ANTENNA SUPPORT STRUCTURES

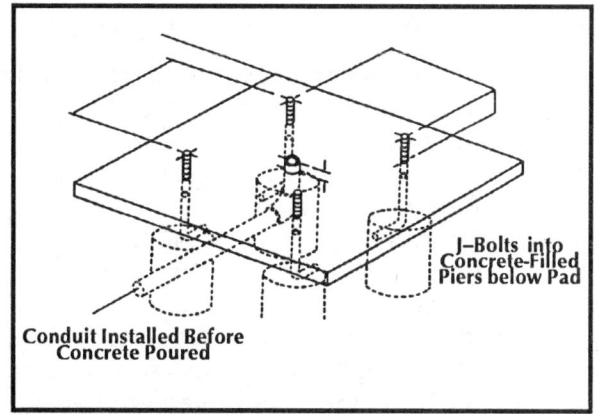

Figure 5-32. Pad Support Using J-Bolts. *Here the J-bolts are set into piers that extend into concrete piers below a pad mount. The conduit is also set in place before the concrete is poured to protect cable runs.*

Figure 5-33. Redhead Anchor Bolts. *These specially designed bolts are to be used after the holes are drilled in the concrete base or wall. They can support thousands of kilos of force.*

Figure 5-34. Drilling Holes for Anchor Bolts. *Once concrete is cured, an electric hammer drill and a masonry bit are needed to drill the holes for the appropriate mount support.*

Anchor bolts should be made of galvanized steel, 2 cm (3/4 inch) thick by 60 to 90 cm (25 to 35 inches) long. After installation they should extend from 7 to 10 cm (2.5 to 4 inches) above the surface of the concrete. This design allows re-leveling of the pole when necessary (see Figures 5-29 and 5-32, 5-33 and 5-34). If the visible portion of the bolt is greased before pouring the cement, clogging of its threads can be avoided.

A template conforming to the layout of the base of the pole should be used to place the anchor bolts, generally J-bolts, before concrete is poured or while it is still "green" or soft. The bolts should be carefully kept away from the reinforcing mesh or rebar since any shifting of this supporting structure during pouring can cause them to move out of place. The bolts should never be welded to the rebar since too much heat will cause them to become brittle. When the concrete is completely cured, 15 centimetre (6 inch) "redheads" having at least a 16 mm (1/2 inch) diameter can be inserted in a hole made by a hammer drill. Redheads, like some drywall screws or moly bolts, have a collar which folds up when the bolt is tightened, they can support at least 20,000 kilos (more than 40,000 pounds) of force.

When the antenna supporting tripod is mounted, two nuts with lock washers placed both above and below the attachment points allow a fine leveling adjustment during installation and at any time thereafter. When using either a pad or a pier to set the tripod, care should be taken to have the three supporting points as level as possible to facilitate making this vertical adjustment. It is also important to orient the legs of this tripod along the north/south axis during this procedure.

Some installers pour concrete at their shops and then truck the completed structure to a site. This technique has the advantage of lowering costs and making winter installations easier. However, slabs can settle in the spring, require special equipment to move them to a site and can be difficult to place once at the site. A 1500 kilo (over 3000 pound) trailer will crush sprinkler pipes in the ground and cannot often be moved into a customer's backyard. Nevertheless, such an approach can be very effective in retaining business activity during spells of cold weather.

INSTALLATION

Combinations of Pole and Pad Supports

A creative installer can construct any variation of pole and pad supports that a given situation warrants (see Figure 5-35). For example, if an immovable rock is encountered half a metre below the surface when digging a hole for a pole mount, a combination pole/pad support can be created. The pole would still be set into the undersized hole. But a small pad would also be formed around the top of the pole to provide further support. Thus, less concrete could be used and the labor spent in digging the preliminary hole would not be wasted. In this case, the pole would extend further out of the ground than would normally occur and lifting the antenna onto the pole might prove to be a little more difficult.

Ground Supports Not Requiring Concrete

Foundations which use a tripod support design but do not require concrete are also viable structures. Three or more rods can be driven deeply into the ground with a power auger. Then a tripod or even a four-legged assembly can be attached with weather resistant, adjustable bolts. For example, the Earthbound, Inc. system uses a tri-leg steel platform secured with 1.2 metre long anchors driven into the earth with a power auger (see Figure 5-36). It is interesting that this type of support is not considered a permanent improvement by some building codes because no concrete is used. Thus, additional taxes can sometimes be avoided.

Another technique employs an A-frame support, known in Europe as the TRAC A-Frame polar mount. This assembly can be installed in about fifteen minutes. A three point anchoring structure is fabricated as shown in the accompanying diagram. This pre-built support can be bolted to a pad or roof or set in concrete.

Figure 5-35. Combination Pad/Pole Mount. *This photo shows a concrete pad poured with J-bolts inserted while the concrete was green, namely before it had cured. A template of the antenna base should be used to make certain that the bolts are in the correct pattern. If not, it may be necessary to remove them and reinsert redheads at the cost of additional labor and time.*

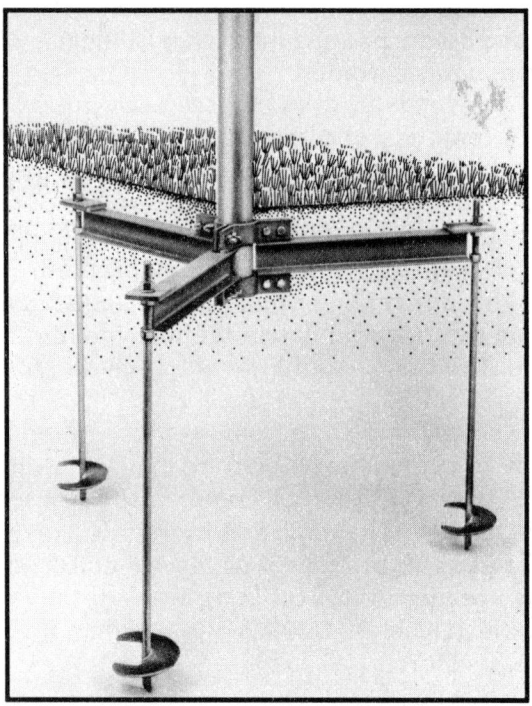

Figure 5-36. Auger Antenna Support. *The Earthbound system uses a tri-leg steel platform secured with 1.2 metre (4 foot) long anchors screwed into the earth with a manual driver. Leveling is possible at any time following installation. Although this support system is permanent, it can be removed and installed elsewhere by screwing the legs out from the ground. (Courtesy of Earthbound)*

INSTALLATION

Above-Ground Mounts

As transponder power levels from Ku-band satellites increase, LNB noise temperatures fall, and the antenna diameter required for clear reception continues to shrink, wall, roof or eave mounts, towers or long pole supports are becoming more feasible and common. This is especially the case in Europe where most antennas being installed today are less than a metre in diameter so that eave and roof mounts are the preferable alternative.

In some cases, installing an antenna at ground level is simply not possible because obstructions block a clear view to the satellite arc. In addition, in some countries and locales, real estate is a premium commodity and backyards are small or non-existent. This presents an installer with two options: choose a different site or mount the antenna at or above roof level. There have even been demanding situations where antennas have been mounted on free-standing towers like some commercial broadcast antennas.

Roof, Wall and Eave Mounts

Although some of the same concerns including reflector stability apply to installing both small and larger antennas on roofs, walls or eaves, the supporting structures can be quite different (see Figures 5-37 and 5-38).

Figure 5-37. Large Antenna Mounted on Flat Roof. *This antenna was mounted on a gravel-top roof. After the gravel was cleared away a 6 mm (1/4 inch) steel diamond plate sheet was attached to the underlying steel roof members with four 19 mm (3/4 inch) double nutted all-thread rods. Tar was poured onto the roof surface beneath the steel plate and on its top surface after all necessary tracking adjustments were completed. This installation detail thoroughly sealed the roof and prevented water leaks. A professional structural engineer was consulted before the work was started.*

Figure 5-38. An Eave Mounted Antenna. *This Astra dish installed in London was securely attached to the eave of this stone building, high above the ground. (Courtesy of WV Publications, London)*

ANTENNA SUPPORT STRUCTURES

INSTALLATION

Large Antennas

Historically, mounting C-band sized antennas has been regarded with caution by many in the satellite industry. Roofs are often not designed to withstand weights in excess of normal snow loading or the tremendous upward forces caused by winds acting on a large aperture antenna. For example, a 80 kph (50 mph) wind blowing directly into the face of a 3.5 m antenna can generate a force of more than 500 kg (over 1200 pounds). This tremendous force can rip a roof off its rafters and cause even more damage when it finally lands! An additional hazard is often not considered: if a reflector is mounted on a long pole near an area where people walk, snow or ice resting on the antenna may loosen, fall and cause injury to those below. This outcome is less likely when using offset fed antennas which are inclined more steeply towards the horizon than prime focus dishes. In any case, if damage or injury does occur, both the owner and satellite dealer can be held responsible.

Anyone considering roof mounting a relatively large dish would be wise to hire a qualified structural engineer to analyse each particular situation. If a roof mount is judged to be feasible, the engineer takes responsibility for making this recommendation. This is mandated in those countries where a permit is required to change the structural integrity of a building.

The fundamental principle to be followed in designing and installing a roof mount is to tie the support structure in to as many rafters as possible in order to spread the load (see Figures 5-39 and 40). A number of companies now manufacture welded steel assemblies with built-in leveling adjustments designed to support poles on most types of roofs (see Figure 5-41). The support pads must be attached to a length of lumber or metal inside the attic which itself is secured to as many rafters as possible. If the roof has steel beams, it is highly recommended that the mount be tied in to these members. Two or three guy wires, preferably made from a material as strong as aircraft control cable, can be used to secure the antenna for further stability. These will also prevent the reflector or mount from landing a hundred metres away if the worst should happen and it does break away from the roof.

Figure 5-39. Top and Bottom View of a Spun Aluminium Antenna on a Roof Mount. *This 2.6 metre (8.5 foot) Birdview antenna was attached by spanning five rafters with two pieces of 20 cm (8 inch) channel iron both above and below the roof. The resultant "sandwich effect" helped in distributing the antenna/mount weight over a larger surface area. Holes were caulked and then tar was poured onto the roof. If a pitch pan had been installed under the channel iron, water sealing would have been even more effective.*

INSTALLATION

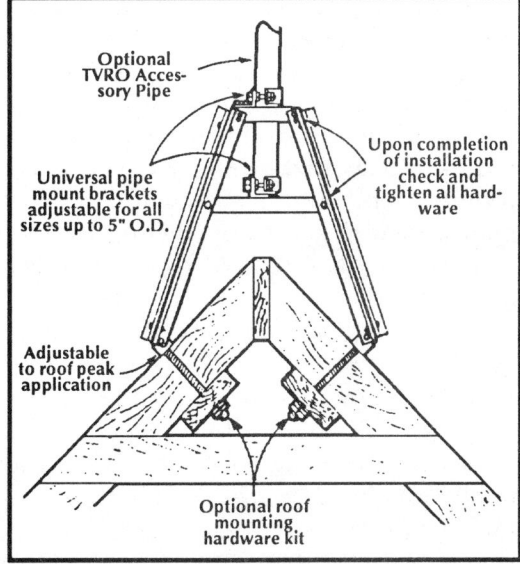

Figure 5-40. Structure of a Peak Roof Mount. One alternative to supporting antennas as large as 2.5 m on a roof is a peak mount. Before installation a professional engineer and zoning authorities should be contacted and the appropriate license should be obtained. It is critical that as many as eight beams under the pitched roof be interconnected with additional steel or lumber members. Great care should also be taken to make certain that no water can leak through the holes causing damage to the inside sheet rock and electrical wiring. (Courtesy of Rohn Tower)

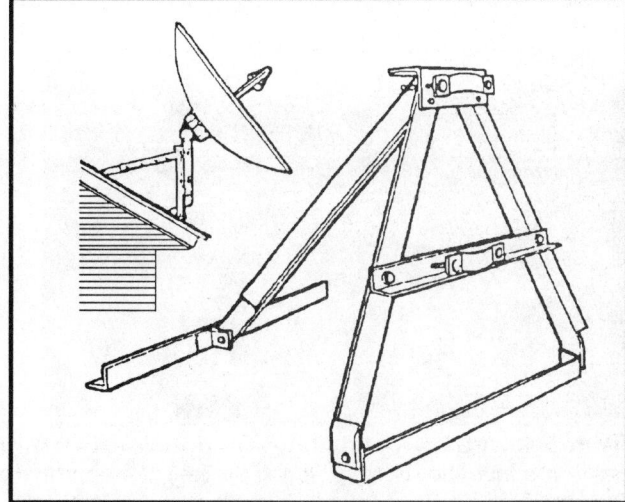

Figure 5-41. Pitched Roof Mount. When using a side roof mount, the same care should be taken as a peak roof mount in making certain that as many roof beams as possible are tied into with additional support lumber to distribute the weight of the antenna/mount. As in all installations, roof mounts must also be well grounded. (Courtesy of Rohn Tower)

ANTENNA SUPPORT STRUCTURES

INSTALLATION

Great care must be taken in breaking roof seals to prevent leaking (see Figure 5-42). It is wise to plan this step with the help of a reputable roofing contractor. When seals are broken, the original roofers warranty is voided and a dealer can become liable for water damage to the structure or contents of a building. Even small unsealed holes can allow water to penetrate a roof and ruin a ceiling. A lesson in this matter can be learned from local, competent solar installers who have faced and solved related problems of wind loading and water leakage for years.

Roof mounts do have the advantage of being easier and less expensive to assemble than most ground foundations. There are no problems associated with mixing and pouring heavy concrete, with waiting for concrete to harden in harsh winter conditions and with trenching through frozen ground to lay cable runs. In addition, there is a much lower chance of theft when the satellite equipment is mounted out of reach. However, winds are much stronger above a building than on the ground below and installing on a high or steep roof can be dangerous.

A type of roof mount that also can potential use concrete is known as a gravity mount. Concrete pads or other supports secure the antenna (see Figures 5-43 and 5-44). A gravity mount can also be used as a ground mount.

Figure 5-42. Flat Roof Pitch Pan. *Before beginning this installation a roofing contractor was hired to lay down a 1.2 x 2.4 m (4 x 8 foot) aluminium pitch pan which was filled with tar after the installation was completed. This sealed the roof and prevented leaks.*

Figure 5-43. A Gravity Mount. *One alternative to piercing a roof and causing possible water damage would be the use of gravity mounts. Such designs spread the weight over a large area spanning 3 metres or more on a flat roof. (Courtesy of Rohn Towers)*

INSTALLATION

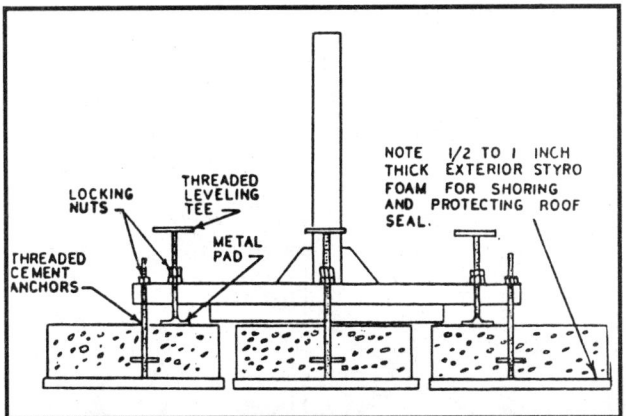

Figure 5-44. A Concrete Pad Gravity Flat Roof Mount. *This mount is bolted onto four concrete slabs which are spread over a one square metre area. "Tea handle" screws are used to allow leveling of a polar mount. No leveling is necessary for az/el type mounts. (Courtesy of Upper Midwest Satellite Supply)*

Small Antennas

Ku-band antennas are generally supported from walls, eaves or mounted on roofs in regions such as Europe where satellite power levels are relatively high and dishes of one metre or smaller diameter are common (see Figures 5-45 and 5-46.

Even when installing antennas as small as 45 cm in diameter, wind loading and stability must be seriously considered. When attaching a bracket to a wall, under an eave or onto a roof, the weight of the dish as well as forces caused by winds and other loads can act to pull the assembly away from the mounting surface, known as a tensile force, or to shear it along this surface. The considerations mandate that an adequate safety margin must be factored into the choice of materials.

The mounting assembly should generally secured to the wall or eave before a dish is attached. If this is not possible, first mark the position of the necessary holes using just the mount as a template. After drilling, the dish/mount assembly can be lifted into position and easily secured with the bolts. Although both steel or nylon wall plugs can be used on walls, nylon plugs are recommended when the antenna is subject to vibrations caused by wind loading or when it is being mounted in solid masonry (see Figure 5-47).

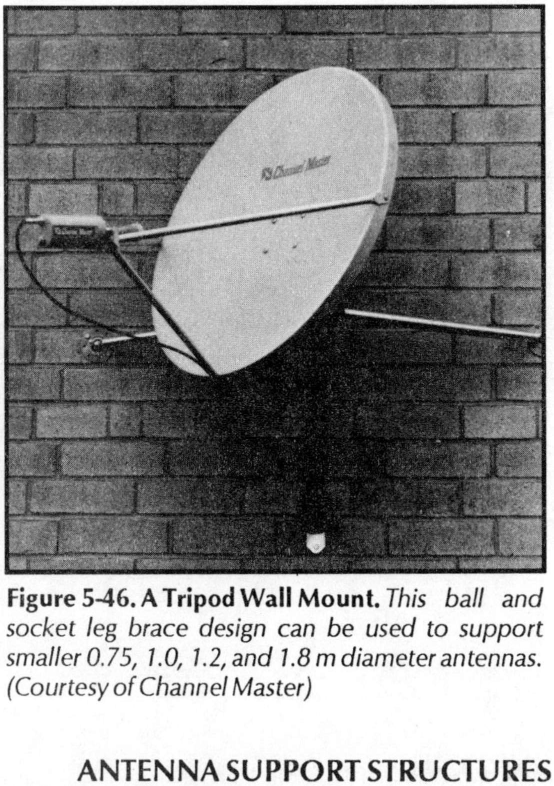

Figure 5-45. A Bracket Wall Mount. *This NEC home satellite system is securely attached to this brick wall far above ground level. (Courtesy of WV Publications, London)*

Figure 5-46. A Tripod Wall Mount. *This ball and socket leg brace design can be used to support smaller 0.75, 1.0, 1.2, and 1.8 m diameter antennas. (Courtesy of Channel Master)*

ANTENNA SUPPORT STRUCTURES

INSTALLATION

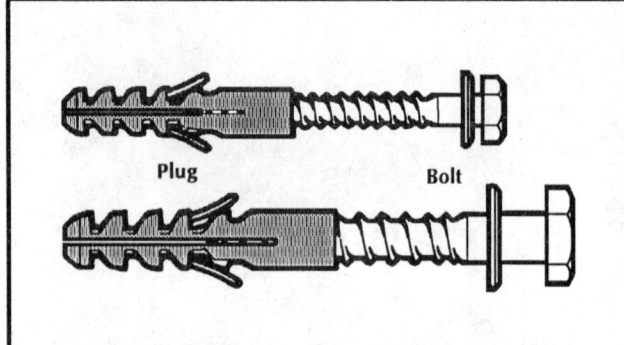

Figure 5-47. Wall Plugs and Bolts. *Wall plugs are used to attach a mount to a brick, concrete or concrete block wall. The plug inserts into a snug pre-drilled hole. When the bolt is inserted, the sides are forced to expand and make tight contact with the wall.*

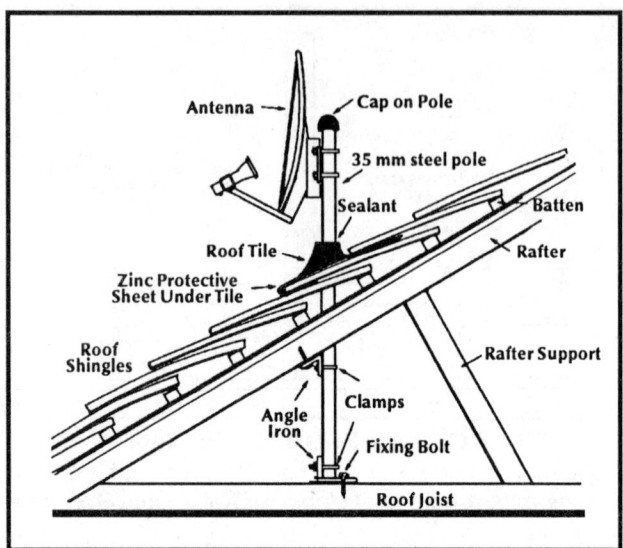

Figure 5-48. Antenna Mounted on Mast. *This roof mount system uses a mast that is secured by channel iron just under the roof and at its base. The channel iron spans at least three roof joists and rafters for stability. The mast passes through a rubber form developed by Cranleigh Aerials in the U.K.*

Walls plugs can exert tremendous expansive forces that ultimately can crack concrete or brickwork. The bolts should be tightened to only the recommended torque. When walls plugs are being installed in bricks the bolts should be secured in the brick not the binding mortar which is much weaker. In situations where a weaker material such as aerated concrete serves as the foundation, expansion type fixings should be used. The sides of this type of device detach from its body and expand into a beveled hole when the bolt is tightened.

Wall plugs and expansion bolts designed for masonry cannot be used with other building materials such as wood or shingles. In these cases brackets have to be secured directly into the underlying wooden frame or onto a plate that rests inside the wall. Techniques similar to those described for mounting large antennas can be used.

Small antennas can also be installed directly onto roofs, when building codes allow. An example of one such method is presented in Figure 5-48. Another option, securing an antenna directly onto a standard aerial mast, is shown in Figure 5-49. In this case, the mast must be securely attached the the structure underlying the shingles.

Figure 5-49. Roof Mount on Antenna Mast. *This Herschmann 80cm dish is mounted on an array of lower frequency antennas. A roof mount becomes easier as antenna size decreases. (Courtesy of Richard Herschmann GmbH & Co.)*

ANTENNA SUPPORT STRUCTURES

INSTALLATION

Towers and Long Pole Supports

An alternative to a direct roof mount is to support an antenna on a long pole (see Figure 5-50). This type of support is generally not used when installing smaller dishes. In some cases, a long pole can also be somewhat easier to install because the pole is secured in to the ground adjacent to a building and is more easily accessible. In addition, building permits are generally not required because no structural alterations other than attaching the pole to the adjacent wall are required. There should be at least one attachment point where the pole meets the top of a wall, the point of greatest torque (see Figure 5-51, 5-52, 5-53 and 5-54). It is better to weld tabs to the pole instead of using collars which may slip when high winds may twist the reflector. The structure of the wall must also be considered in attaching these tabs. Weaker parts of the wall could be damaged when the antenna and pole twists or bends under forces of wind or snow.

When installing large dishes on long poles, for every 10 cm (4 in) addition to a typical 1.5 m ground-mounted pole length, an extra 50 cm (20 in) should be added to the normal 1 m below ground segment. For example, if an 8 m length were required above ground level, 130 cm (10 cm for each 50 cm of the extra 6.5 metres) should be added to the 1 metre normally below ground. Thus 2.3 metres would be underground and 8 metres would be above ground. Of course, this is a rule of

Figure 5-50. Free-Standing Long Pole Installation. *For free-standing long-pole installations it is recommended that the post diameter be double the normal size. A pole reducer must then be welded to the top of the pole where the mount attaches. These types of support structure should not extend over 9 metres for antennas having diameters greater than 3 metres. Doing so would cause excessive unacceptable pointing inaccuracies at Ku-band.*

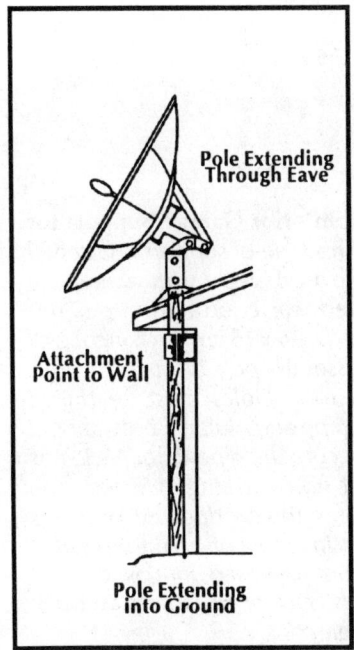  

Figure 5-51. Exterior Wall Long-Pole Wall Mount. *Two methods to attach a long pole to an exterior wall, bolting the pole directly against the building and extending the pole through a hole in a pitched roof, are shown here. Both poles were secured at the base in a concrete filled hole. Extreme care should be taken when attaching brackets because high wind forces can damage or detach some of the weaker types of wall materials. In order to determine how closely a support may approach a wall, a small hole should be dug at its base because some building footings may extend as far as two feet past the edge of the building. In these cases, a plumb bob can be extended from the roof to the centre of the hole in order to determine the required length of brackets to secure the pipe. (Courtesy of Upper Midwest Satellite Supply)*

ANTENNA SUPPORT STRUCTURES

INSTALLATION

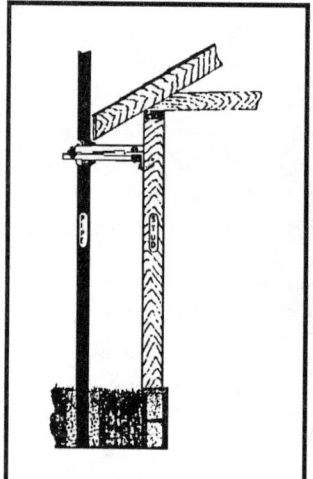

Figure 5-52. Adjustable Exterior Wall Mount. *This mount is designed so that its position can be adjusted beyond the eave of a building. It should be secured in at least two places as well as at its base in a hole filled with concrete. (Courtesy of Upper Midwest Satellite Supply)*

thumb that does not have to be adhered to rigorously.

In addition to securely attaching the pole to one or two points on the adjacent building, it should be oversized in diameter by 5 to 15 centimetres (2 to 6 inches) for extra stability against twisting and bending. In general, this will require use of a reduction coupling at the top of the pole. Guy wires should be attached to the base of the mount and to nearby support points. Note that a lesson can be learned from the designs of billboard installations which also use long poles. These supports begin with larger diameter poles at the bottom and step down to smaller diameter poles near the top.

Figure 5-53. Swivel Bracket to Support Long Pole. *This type of swivel bracket can be used for a flat or pitched roof to secure the top of the pipe where the mount attaches. (Courtesy of Upper Midwest Satellite Supply)*

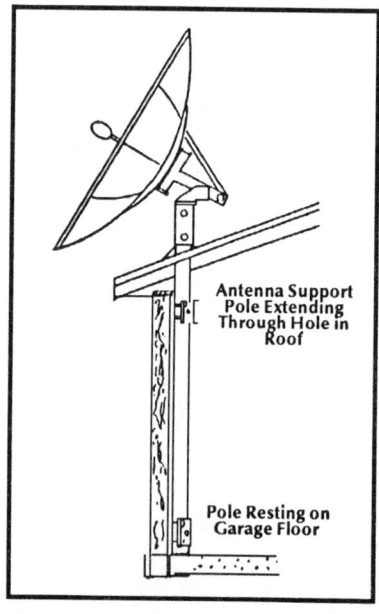

Figure 5-54. Interior Garage Support for Long Pole. *This type of support is fastened to the inside wall studs with a multiple header located near its top and bottom. This connects it to five 5 x 15 cm (2 x 6 inch) wall studs and passes the pole through a hole in the pitched roof. Holes must be drilled through the pipe and multiple bolts should be used to secure the pipe to the V-clamps to prevent it from twisting in winds. The manufacturer of this product recommends that antennas in excess of 1.8 metre diameter should not be used in this type of installation. (Courtesy of Upper Midwest Satellite Supply)*

ANTENNA SUPPORT STRUCTURES

C. CABLE RUNS AND TRENCHING

Cable runs between the antenna and indoor equipment should be safely protected and neatly installed. A variety of techniques can be used depending upon the type of cable required, the length of the run, antenna location, and, when applicable, ground conditions. For example, trenching is not necessary in conjunction with a roof or some types of long pole supports.

Digging the Trench for Ground Mounts

When a trench is necessary, before digging begins the first step is to determine where all the water, electrical, gas and telephone lines are located. In many countries, utility companies will be more than glad to send a crew out to find these lines and even to help dig to ensure that they are not severed. Customer owned underground systems such as lawn sprinklers can be more difficult to pinpoint. Often a homeowner will not be sure where the pipes are located. It makes good sense to proceed with caution. In those regions where sprinklers are common, having parts on hand for repairing pipe can be a great time-saver.

The depth of the trench depends upon many factors (see Figure 5-55). If rigid conduit is used there is less risk that an avid gardener will slice a shovel through the cable. If the cable is running through an unused portion of the yard, it should be buried at least 15 to 20 cm (6 to 9 inches) and preferably 30 to 45 cm (12 to 18 inches) under the surface. Electrical codes in many countries specify required burial depths.

It makes good sense to keep a record of the length, route and gauge of buried cables. This information on a map of the site can be invaluable if an update or repair is required at some future date. The customer will also find a copy of this record useful if landscaping changes are made.

Leaving unburied or unsecured cables lying around is an invitation to an animal or someone with a lawn mower or a destructive curiosity to cause damage.

Whenever a trench is being cut across a lawn, care should be taken with the sod (see Figure 5-56). It should be cut with a sharp knife or a flat shovel to a depth of at least 7 or 10 cm (3 to 4 inches) on both sides of the trench. The sod should then be carefully rolled in sections and laid next to the trench. All loose earth should be placed in a wheelbarrow since it is often difficult to clean loose fill from the surrounding sod with a rake after the trench is refilled. Any sod left above ground for periods longer than a day must be watered to avoid damage because it can easily dry out and wither.

A trench must not be refilled until the installation is completed and it is ascertained that there are no defects in the cable. This precaution will prevent unnecessary re-excavation of a deep trench. Refilling the trench is one of the last steps in a successful installation.

Figure 5-55. A Power Trencher. *This machine is capable of digging a 1, 2, or 3 inch (2.5, 5.1 and 7.6 cm) wide trench from 8 to 12 inches (20 to 30 cm) deep at a rate of 20 to 25 feet per minute (6 to 7.5 metres per minute). It can be a great time saver in certain situations but can damage the sod on a well-kept lawn. (Courtesy of T.H. Riley Manufacturing)*

INSTALLATION

Figure 5-56. Soil Spreader and Cable Tamper. *These two tools separate sod and soil to a maximum depth of 8 inches (20 cm) and bury cables. Note that conduit cannot be buried by this method. If such a method is used it is advisable to first be sure that the installation is complete and working correctly. (Courtesy of M.B. Sales)*

The Use of Conduit

Many direct burial, multi-conductor cables do not necessarily require encasement in a rigid, protective conduit. However, conduit does provide extra insurance against damage (see Figure 5-57). In certain situations using conduit is strongly recommended. For example, it provides the necessary protection for runs under driveways or roads. If rodents are common, conduit can prevent them from chewing through cables. In areas with excessive ground water, conduits can protect against water damage. Be aware that there have been cases where improperly installed conduit has done just the opposite and has become a water trap. As a result, the only solution was to dig up the conduit to drain the excess water.

Conduit should also be used in many above-ground installations (see Figure 5-58). While PVC conduit is fine for underground locations, it emits noxious fumes when burning. Metal conduit is therefore a better choice for locations on or within a building.

Local building codes in some countries may require the use of conduits when cables will carry power above a predetermined level. In these cases, white PVC water pipe or black rolled flexible tubing is not permitted. Only rigid varieties like gray PVC or aluminum are permissible.

Figure 5-57. The Use of Conduit. *Using conduit will protect cable from any damage that could be caused by rodents as well as from moisture where the cable enters and leaves the ground. A weatherhead must be installed to prevent moisture from entering the conduit.*

CABLE RUNS & TRENCHING

INSTALLATION

Conduit should be a least twice the diameter of the enclosed cables. Installing sweeps instead of right angles at corners makes pulling cables into and out from conduits easier. Since sweeps are not available in standard white PVC used for plumbing lines, either gray PVC or standard metal electricians conduit should be used. Some types of conduit are designed to be bent into shape by applying heat when they are wrapped with a material similar to heat tape. It is important to realize that if more than four sweeps forming a full $360°$ turn are used throughout the cable path, it would be impossible to push or pull cables into or out of the conduit. If this happens, one major advantage of using conduit is eliminated. The proper installation of conduit allows removal and replacement of cables years after burial without the need for excavation.

Precautions should be taken to prevent water ingress, especially at the antenna. Sections of conduit can be tightly glued together. Care must be taken when using PVC conduit, to avoid dripping solvent or cement onto cable insulation. Some types can be dissolved by these solvents possibly resulting in damaged coax. Weatherheads or $180°$ sweeps attached to vertical conduit should have removable screw fittings and should be used at all exposed ends (see Figure 5-57). If the cable ever has to be removed from the conduit and replaced, these fittings can then be undone. The use of weatherheads as well as correct bonding at joints is very important since any water build-up can ruin cables and corrode connectors. In those cases where the conduit runs downhill from the antenna to a building, improper installation could cause water to be channeled directly into its interior (see Figure 5-59).

Figure 5-58. Metal Conduit on Wall of Building. *Metal conduit i s an excellent protection for cables. PVC plastic should not be installed inside buildings because in the event of a fire it will emit poisonous gases.*

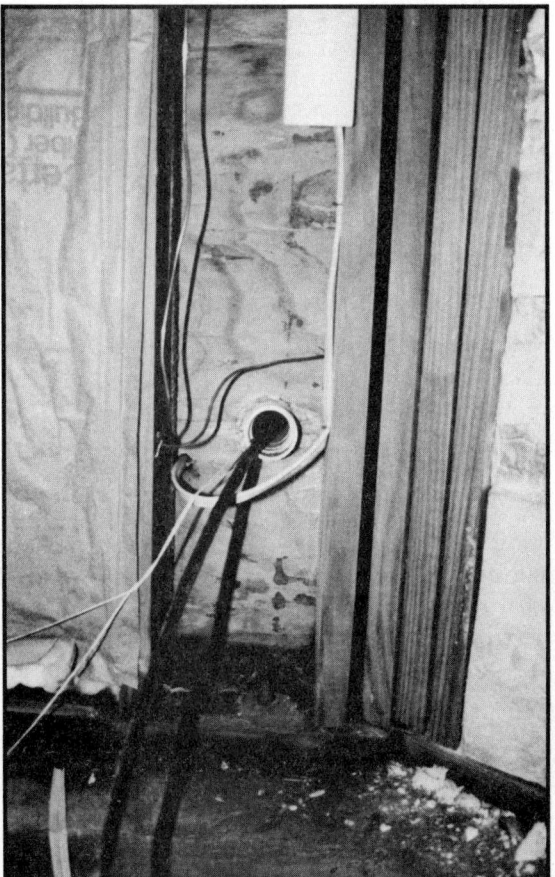

Figure 5-59. Conduit Entering Building. *This conduit was installed in the basement wall of this home so that no exposed cable would be visible. Care was taken to keep water from following the cable route to the interior of the building.*

CABLE RUNS & TRENCHING

INSTALLATION

Figure 5-60. A Drip Loop. *A drip loop allows any water build-up caused by melting snow, rain or condensation to drain away from components such as actuators or LNBs. Notice how the cable here has its lowest point below the base of the actuator motor.*

All the cables should be attached to the supporting pole with wire ties or hose clamps. Drip loops can also be formed where the cable terminates to channel water away from the actuator motor (see Figure 5-60). A drip loop is simply a loop in the cable which has its lowest point below connectors, a hole in the wall or an entrance in the bottom of a waterproof box. Any water from rain or melting snow and ice tends to collect at this low point and drip onto the ground.

When pulling cable through conduit, do not lay the cable on the ground, as it can pick up debris and make the job more difficult. A "fish tape," which is a piece of rigid wire, can also be used to pull cables through conduits. It is inserted at one end and the cable is attached to it at the other end. Cables can also be lubricated with a liquid soap to allow them to slide through the conduit more easily. Of course, this soap must be cleaned off where the cable enters a building. If the cable is too short to span the entire length between the antenna and receiver, never splice two pieces together. Start again. Run a complete and sufficiently long cable. Splicing would not only increase resistance to current flow, but would also present a perfect opportunity for cable breaks to occur in the future or for water to enter and short out the system. The splice is also likely to pull apart when cable is being fed through a conduit.

All of the cable used for both pole and roof mounts running from the antenna to the building entry should be placed in electrical conduit. This will prevent the cable from being damaged by ultraviolet light and by other enemies such as birds and squirrels. Many building codes also mandate the use of conduit. Using well-installed conduit also results in a neater and more professional installation.

Types of Cables for Satellite TV

Cables for satellite TV systems can be purchased individually as needed or in a form known as direct burial lines. Direct burial lines typically have two coaxial cables, a triple-wire feedhorn polarity control run and a 5-wire actuator power/control cable all bound together (see Figure 5-61). Installing an additional 3 or 4 conductors should be considered if coaxial relays, which are electronic switches, are to be used in dual-band systems. Cables are generally swept tested by the manufacturer to ensure that there are no breaks or discontinuities. In this test, signals spanning a range

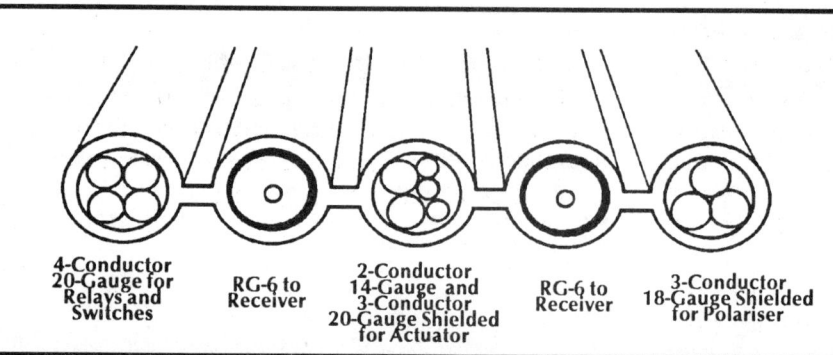

Figure 5-61. Multi-Run Cables. *Commonly used cables for home satellite television consist of single or dual run of RG-6 coax and a two-conductor 14 gauge, a 3-conductor 20 gauge shielded, a 3-conductor 18 gauge shielded and a 4 conductor 20 gauge wire. This particular composition of multi-run cable permits all the possible wiring combinations necessary for Ku-band as well as for C/Ku dual-band installations.*

CABLE RUNS & TRENCHING

INSTALLATION

TABLE 5-4. WIRE SIZES RECOMMENDED for POLARISERS

Maximum Cable Length (metres)	Wire Gauge (shielded)
25	20
50	18
Longer Runs (to 100 metres)	16

TABLE 5-5. WIRE SIZES RECOMMENDED for ACTUATOR CONTROL CABLES

Maximum Cable Length (metres)	Wire Gauge Motor	Shielded Sensor
25	16	20
50	14	20
100	12	20

TABLE 5-6. RECOMMENDED COAXIAL CABLE SIZES.

Maximum Usable Frequency (without amplification)	Cable Lengths (Metres) 25	50	100
70 MHz	RG-59	RG-59	RG-6
950 MHz	RG-6	RG-6	RG-11
1,450 MHz	RG-6	RG-6	RG-11

of frequencies are fed into one end of the cable and the output is measured at the far end. In either case, to optimise performance of a satellite system, cables should adhere to the minimum standards outlined in Tables 5-4, 5-5 and 5-6.

The cable run from the antenna to the indoors electronics nearly always carries frequencies in the 950 to 1700 MHz range. Therefore, RG-59 should only be used in lower frequency 70 MHz distribution systems after the output of a headend launch amplifier.

Polarity control wires should have three shielded wires of the minimum gauge as outlined in Table 5-6. In those dual-band installations where two polarisers are wired in parallel, increase the recommended diameter by two gauge numbers (e.g. 20 to 18 gauge).

The actuator line consists of two #14 gauge 36 Vdc motor wires and three #20 gauge sensor lines including one for a ground connection. Such cables are usually adequate for runs up to about 60 metres. For longer distances, #12 gauge should be used to power the motor.

Minimising cable lengths has a number of advantages. The main benefits are lower cost, improved signal-to-noise ratio and, for ground mounted antennas, less trenching. When runs exceed 100 metres, heavier gauge lines must be used to prevent problems such as servo hunting or incorrect actuator counting. These phenomena are explored in detail in Chapter IX. In those cases when the signal-to-noise ratio generated by an antenna/feed/LNB is inadequate for transmission over a long cable run, a line amplifier may be required to boost the IF signal power.

Most direct burial cable manufactured today has an extra coax line left over from the C-band days when LNB power and the IF signal were each transmitted on a separate line. Additional wires may be unused if an actuator is not installed. When present, these extra lines should always be used so that an extra component like an actuator can be added in the future without the need for additional trenching. Also, if ground loop problems arise, having an extra line to make a common ground connection can prove very useful (see Chapter IX for more details).

Working with cables can be difficult in very cold weather. Some manufacturers supply an especially flexible "cold-weather" type of cable for these situations. More details on cold weather installations are given later in this chapter.

INSTALLATION

The Entry into the Home

Most customers prefer to have all cables hidden from sight. Holes drilled in both exterior or interior walls should be as small as possible and well caulked after the installation is completed. Multi-run cable can be rolled up to minimise the size of the necessary hole (see Figure 5-62). If the cable entry is through a crawl space, holes may be drilled below the ground level only if they are well sealed. In general, above-ground cable entries are recommended. Always have sets of masonry and wood bits on hand. Many older homes have very thick walls so that extra long wood bits (up to 45 cm or 18 inches) should be available during an installation. Holes should be drilled at least 30 cm (12 inches) above or below and away from any electrical outlets to avoid hitting power lines. Exercise caution when drilling near bathrooms, kitchens or other places where plumbing may be close and where substantial damage can be done.

Drill from the inside to the outside of a wall, not the other way around, and only after all measurements have been taken and the path of the bit is known. When possible, use wallplates similar to those on electrical outlets (see Figure 5-63). A wide variety of brands are available at most electrical parts distributors. Cables fed either into an attic or crawl space can usually be fished up or down into a wall to make an unobtrusive entry. Difficulties can be encountered if the walls have cross-supports between studs.

When carpet must be disturbed, cut a small "X" with a razor blade or an "exacto" knife through both the carpet and padding so later if the system or cable is moved the rug can be easily repaired. A drill can often catch on the rug pile and tear a wide area open or ball up padding. Never run cables under a rug. The tacks used to attach the rug to the pad and floor can also easily short cable out. Cables when under rugs cause bumps and make tripping all too easy.

When cables are routed through either crawl spaces or attics, they should be attached to rafters or floor joists. Special staple guns are made to fit onto the cable surface to avoid damaging the conductor. For example, the Arrow T-25 staple gun used with 14 mm (9/16th inch) round-top staples, which are made especially for RG-59 cables, makes this job easy. The T-15 staple gun can be used with 18 or 20 gauge wires and cable. Small plastic clips which are designed to be nailed onto a wall are another cosmetically attractive option.

On final note: never run cables parallel to electrical lines. Doing so can result in interference such as 60-cycle hum.

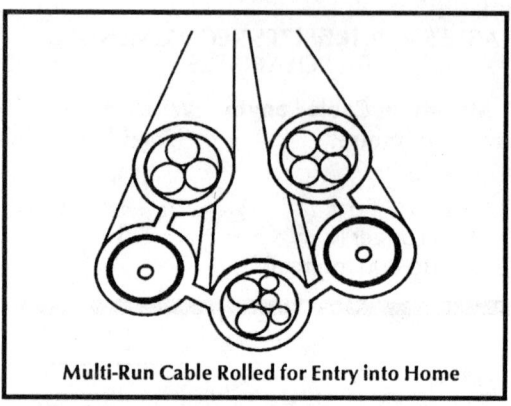

Figure 5-62. Multi-Run Cables. *Commonly used cables for home satellite television consist of single or dual run of RG-6 coax and a two-conductor 14 gauge, a 3-conductor 20 gauge shielded, a 3-conductor 18 gauge shielded and a 4 conductor 20 gauge wire. This particular composition of multi-run cable permits all the possible wiring combinations necessary for Ku-band as well as for C/Ku dual-band installations.*

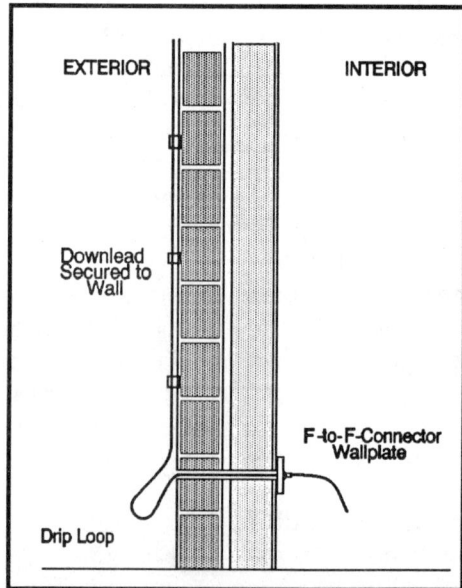

Figure 5-63. Wall Plate and Drip Loop. *Some fixed satellite systems relay the polariser pulses down the coaxial cable. In this case, just a single cable is required. A drip loop should be installed on the exterior wall and a wallplate F-to-F- connector socket on th inside wall. This is a good opportunity to make the transition from exterior black cable to another colour that matches the interior decor.*

CABLE RUNS & TRENCHING

D. ASSEMBLING THE ANTENNA, FEEDHORN & LNB

The antenna, feedhorn and LNB are the most critical components of any satellite system. If they are not selected and installed properly, using the best and most expensive satellite receiver and modulator in the world will not improve picture quality.

Assembling the Antenna

Antennas generally are accompanied with a set of assembly instructions. These instructions should be carefully read at least twice before beginning an installation to determine if any special tools or manpower will be required or if there are any unanswered questions. Usually, most of these questions can be solved by simply beginning the assembly, but occasionally a call to the distributor or manufacturer is necessary. Be sure that all parts are accounted for before proceeding to the job site.

Small Antennas

Small antennas usually have a one-piece reflector and all that is required is to bolt the mount to its surface. The mounts of off-set antennas usually have an integral feed support that holds the LNB.

Large Antennas

Larger antennas are generally not as simple too assemble as smaller off-set models. Each variety and type of antenna has some unique aspects to its assembly (see Figure 5-64). Spun or stamped reflectors are most often delivered in one piece ready to be bolted to their mounts. In cases where holes have to be drilled to attach a mount, it is critical that they be centred. Do so by crossing two strings each tied to two opposite corners of the mount. Lining up their intersection with the antenna centre will properly align the mount. Spun reflectors have been formed in a lathe and therefore always have a hole located precisely at their centre. Stamped antennas often have a hole for the buttonhook or an identifiable centre point remaining from the manufacturing process. It is good practice to sequentially tighten all bolts or screws on any antenna only after all are in place to avoid warping the reflector. It is just as important not to over-tighten these bolts. This could cause a dimpling or indentation and a substantial loss of gain. Rubber spacer washers or grommets are sometimes included to act as shock absorbers and should be used when provided. If the antenna is aluminum and the mount steel, it is even more important to insert these rubber grommets to prevent electrolytic action between these dissimilar metals. In time, this type of corrosion would weaken or even destroy the points of attachment.

Figure 5-64. Assembling a Mesh Multi-Petal Antenna. *These photos are an example of the step-by-step procedures required to assemble a sectional antenna. These illustrations show (1) and (2) inserting ribs into the centre hub, (3) tightening the centre hub, (4) inserting perforated metal, (5) installing endcaps and (6) assembled antenna on mount. (Courtesy of Brent Gale)*

INSTALLATION

Fiberglass antennas should be assembled on a flat surface so that all sections fit together with the rim lying perfectly flat (see Figure 5-65). Again, do not tighten bolts until the assembly is completed. This procedure allows each section to be lined up to smoothly interface with the adjacent panels. Some fiberglass reflectors can be assembled on the mount. In this and all other cases, the bolts should be tightened progressively from the centre to the outside to make the antenna expand outwards from the centre and maintain a true parabolic shape.

Mesh reflectors can be assembled on any surface but completing the work on grass or a soft material such as the cardboard boxes used for shipping avoids scratching the paint. It may also be convenient to rest the central hub on a stand such as a small table or a garbage can to make attaching all the ribs, panels and bolts or screws easier. If wind clips are provided they should be used, especially in areas having potentially strong and gusty winds. There have been cases of panels popping out partially or completely in a wind storm. Some installers run a small wire through each section near the rim to provide further protection against this type of damage.

There are two accurate and very simple techniques used to check that antennas are not warped either during manufacturing, shipping or assembly. Lip-sighting an antenna can be accomplished by sighting along one rim so that the other edge appears in the same plane. Both rims should line up as two straight lines one on top of the other. If not, the antenna is warped. The crossed string method involves using two or three strings taped or otherwise fastened over the outer edge of an antenna. If these strings either mesh together or are separated, the reflector surface is warped. They should

Figure 5-65. Fiberglass Antenna Assembly. *This segmented fiberglass antenna is being assembled on a cardboard box for centre support.*

just lightly touch one another (see Figure 5-66). Using three crossing strings makes it easier to locate the centre point. If there is some warpage, loosening and re-tightening the bolts which hold the antenna structure can often be an effective method to realign the surface. A warped antenna will not perform up to par.

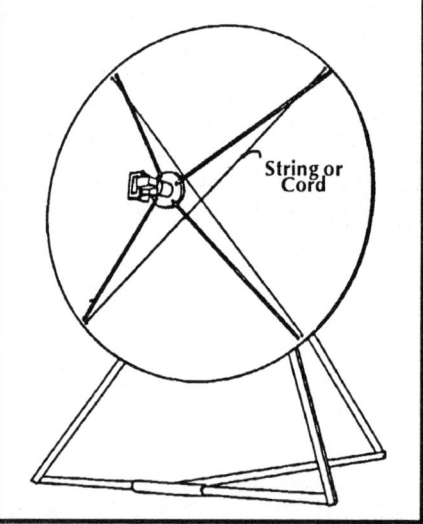

Figure 5-66. Stringing an Antenna. *Running two or three strings between opposite sides of the outer rim across an antenna face, like lip sighting a reflector, is an effective method to determine if the surface is warped. If these strings are meshed tightly together or are separated, the surface is warped. Sighting from behind the antenna through a hole in its centre should line up the feedhorn hole exactly with the string intersection like the cross hair sights on a rifle scope.*

Mounting an Antenna on the Pole

Except for bulky fiberglass antennas, most can be easily lifted by two people (but see Figure 5-67). Large fiberglass antennas in excess of about 4 m (13 feet) in diameter usually require four or more people to share the lifting. In some cases, a crane may even be required to complete the job. This extra weight has prompted some manufacturers of fiberglass antennas to design their products so they can be assembled directly on the mount after it has been set on a pole.

Other types of antennas can also be installed by first setting the mount on the pole. Then the completely assembled antenna is lifted into the correct position where the attaching bolts can be inserted. Since the reflective surfaces of spun or stamped antennas are often easily deformed and can be dented or scratched these should be lifted into their final position after their mount has been attached to the pole. This will avoid bending the rim at the points of lifting. Damage can also occur before the actuator arm is attached since the antenna can often move freely and strike other objects.

Figure 5-67. Lifting a 5 Metre Mesh Antenna onto a Pole. *Five men were required to lift this 5 m antenna into position so it could be properly bolted to its mount. (Courtesy of Brent Gale)*

Assembling the Feedhorn/LNB Structure

The feedhorn and LNB should be bolted together before being installed on the buttonhook or tripod support. Two important rules should be observed. First, never touch or bend the probes on either component. They are finely tuned at just the correct location and any tampering will increase the VSWR and result in a lower signal-to-noise ratio. Second, always use the gaskets provided for insertion between the flanges. These protect against water ingress. But never use any sealant between these gaskets because there must be metal-to-metal contact in these joints. The distance between these joints should not be increased as this would upset the impedance match between the waveguides. The bolts around these flanges should also be evenly spaced. They should be first tightened by hand and then with a spanner.

Correctly Aligning Mechanical Polarisers

When using mechanical servo motor feeds, align them so that their 270° range of rotational motion is correctly centred. Note that some older mechanical polarisers had only 180° range of motion so that more care had to be taken in the alignment process (see Figure 5-68). For linearly polarised signals, the center must be halfway between the directions of the two polarities. For circularly polarised signals, the feedhorn probe must be able to move at least 45° on either side of the dielectric slab as described below.

The range of motion of a servo motor can be observed by hooking it to a test receiver when limits are set. When properly installed, the motor should have the freedom to rotate 90° between positions of horizontally and vertically polarised signals and still have extra room to move past either end of this range. This will permit the necessary skew adjustment when tracking between satellites transmitting different orientations of linear polarisation. Most importantly, it will prevent the motor from running up against a limit and burning out either the motor or the timer in the receiver. This installation procedure also guarantees that the probe motor is functioning properly. Signs that a feed does not have an adequate range of motion are snowy pictures or floating lines (hum bars) that appear on the television screen (see Chapter IX for more details).

INSTALLATION

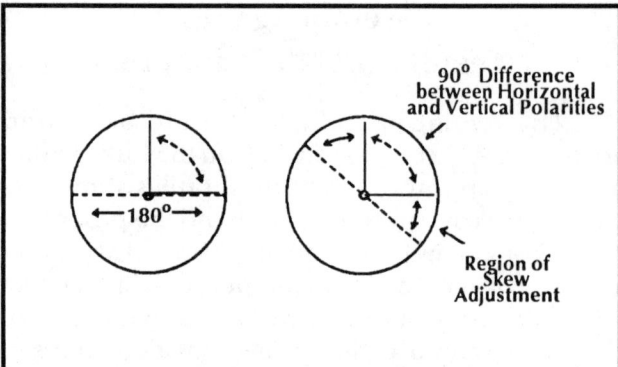

Figure 5-68. Servo Motor Mechanical Limits in Older Feeds. *Most older servo motor feeds have a 180° range of motion. If the mechanical limits do not lie past the orientation of both vertically and horizontally polarized signals, the probe will not be able to be properly aligned and the motor might be burned out in the attempt to move the probe to the correct position. Modern servo motors now have a 270° range of motion and the alignment problem is less critical.*

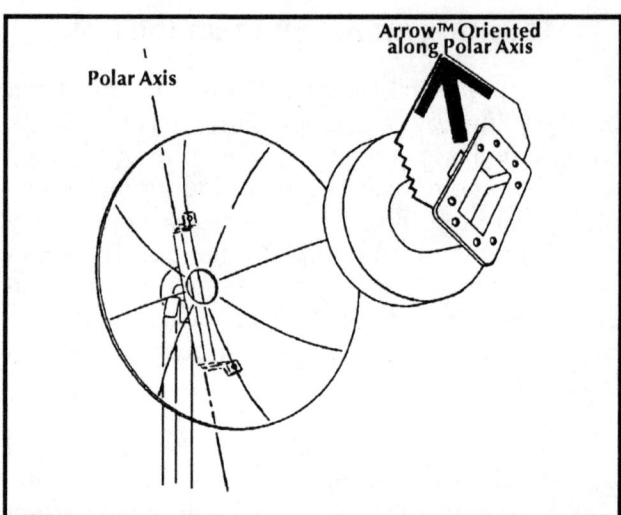

Figure 5-69. Orienting a Mechanical Feedhorn with The Arrow. *Chaparral Communications makes a useful tool, the Arrow™, which aids in properly aligning a ferrite or servo motor feed. The point of the arrow is lined up parallel with the polar axis on the mount and guarantees that the motor will not be driven against a mechanical limit and damaged. (Courtesy of Chaparral Communications)*

Polarisation Offset

The probe in a waveguide must have the same orientation as that of the signal polarisation. Therefore, if the downlink antenna relays vertically polarised signals, the probe must be in a vertical orientation. Two factors makes aligning the feed somewhat more difficult. First, when the receiving antenna is pointed toward either horizon the polarisation is offset because the view towards that satellite is somewhat "twisted." This "polarisation offset" increases the further away from south that the dish is aimed. Second, not all the satellites transmit vertically or horizontally polarised signals in a perfectly vertical or horizontal orientation, respectively.

The best strategy when aligning a mechanical polariser is to first target a southerly spacecraft. Then set the probe halfway between the two mechanical limits. Finally rotate the feed until the probe is halfway between vertical and horizontal orientations. Some manufacturers provide assistance in this alignment process (see Figure 5-69).

Retrofitting for Detection of Circularly Polarised Signals

Feedhorns can be retrofitted to discriminate between left and right-handed circularly polarised (LHCP and RHCP) signals by inserting a dielectric slab such as the teflon insert described in Chapter II. The insert converts circularly polarised signals to linearly polarised ones which are then detected as usual by the feedhorn. If properly installed, losses incurred when detecting of the linearly polarised signals can be kept to an acceptable minimum.

In order to maximise reception of both linear and circular polarisations by the same feedhorn, the probe position must first be set for best reception of either linear polarisation from a low power transponder. The slab is then inserted parallel to the direction of the feedhorn probe to guarantee that linearly polarised signals are attenuated as little as possible. LHCP and RHCP satellite transmissions can then be received when the probe is rotated to a position 45° from either side of the dielectric slab. It is important that the feedhorn be oriented to allow the probe movement to 45° from both sides of the slab as well as to both horizontal and vertical polarity positions. This requires aligning the feedhorn so that the dielectric slab is set about 15° or 20° closer to what normally would be the centre of the probe's range of motion.

INSTALLATION

Setting the Focal Distance and Centring the Feed

Installing the feed precisely at the phase centre of either a prime focus or offset fed antenna is critical to system performance. This involves correctly setting the focal length and, in the case of prime focus dishes, aligning the feed along the antenna axis. In order to maximise the performance of the entire system, great care should be taken to ensure that this step is properly completed. This is especially critical with Ku-band broadcasts where the wavelength is one third that of C-band microwaves. Small misalignments are three times as critical at Ku-band and cause three times the reduction in signal strength compared to that at C-band.

If the feed is not correctly located, gain will be lower and additional noise will potentially be detected because the feed would be illuminating more than its intended target, the dish surface.

Always follow the manufacturer's installation instructions for measuring the correct focal length. Each antenna has a unique f/D ratio and focal length. To illustrate, if a 1m reflector had a 30 cm focal length, the distance from the centre surface of the antenna to the front edge of the feedhorn must be precisely 30 cm. If the antenna were warped this could change the position of the focal length and cause a marked impairment in reception. To avoid this outcome, the procedures for assembling and testing the antenna reflective surface discussed earlier should be carefully practised.

Centring the feedhorn is just as critical as setting the focal distance. At least three methods can be used:

1. Measure from the outer lip of the reflector to the throat of the feedhorn from four different points at 3, 6, 9 and 12 o'clock positions. The resulting measurements should be very similar in each case. If not, the feed should be adjusted until the measurements are nearly identical (see Figure 5-70).

2. Begin by crossing the antenna using two pieces of string at right angles to each other over its surface. This is similar to the method outlined above to check for warpage. Spun aluminum, hydroformed or other reflectors having a hole in their centre can then be tested for feed alignment by stringing two additional pieces of

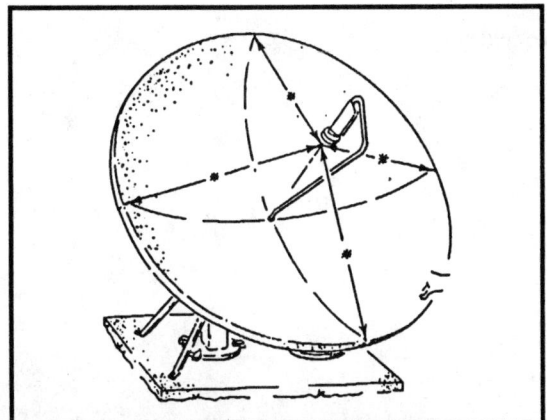

Figure 5-70. Centring a Feed by Measuring. *A feed can easily be centred by making sure that three or four measurements from points around the antenna rim to the feed are equidistant.*

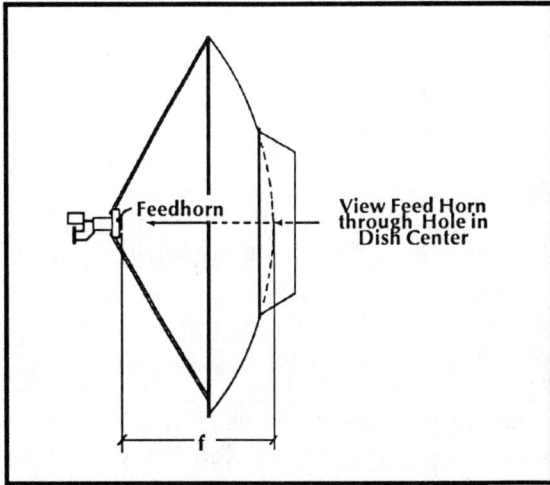

Figure 5-71. Sighting a Stringed Dish. *When looking through the hole in the center of many dishes to the point of crossed strings, the feedhorn should be located directly on center. This is similar to sighting through cross hairs on a gun. It is crucial in this process that the strings span the full diameter of the antenna so they will cross in its center.*

twine behind the antenna. They should cross over this hole. Sighting along both cross-points should target the centre of a properly aligned feedhorn (see Figure 5-71).

3. Use an instrument called a focal finder. One end inserts into the mouth of a feed and the other end has a telescoping rod that should strike the antenna at its centre (see Figure 5-72). If the tip of the focal finder is not correctly aligned, it is a relatively simple matter to readjust the feed position until it does.

ANTENNA/FEEDHORN/LNB ASSEMBLY

INSTALLATION

Figure 5-72. Focal Finders. *These focal finders are inserted into the mouths of these two scalar feedhorns. The telescoping rods are fully retracted in this photo. They extend to when a feed is being aligned. (Courtesy of Natropolis International)*

Centring a feed can be simplified when using a buttonhook feed support. Three small cables with S-hooks can be attached to the feedhorn at the 12, 4 and 8 o'clock positions for stability. The other ends of these cables are then attached to the antenna with metal clips that slide onto the outer lip of the reflector. This "wind kit" makes centring the feed by tightening or loosening the different threaded eye hooks easier.

All these efforts could be in vain if the correct feed had not been chosen. Remember that every feedhorn is designed to function optimally at a rated f/D ratio. For proper performance, the feedhorn must be mated to antenna design and construction. Tests have shown that a mismatch between feedhorn and antenna could cause in excess of a 1 dB drop in performance. That extra gain can be the difference between having a few sparklies or no sparklies in a poor footprint area or when using a small aperture antenna.

When all these adjustments are completed the plastic cover that fits over the throats of most feedhorns must be installed (see Figure 5-73). These covers which have small holes that allow any condensed water to escape, protect the throat of a feed. These covers should always be used. Occasionally, when a feedhorn has been left uncovered, wasps or other creatures have built nests inside. Needless to say, such an obstruction would absorb microwaves and cause complete loss of reception.

It is wise to use an LNB/feedhorn cover when the installation and all waterproofing has been completed. This protection against the elements provides extra insurance against water intrusion (see Figure 5-74).

Figure 5-73. Capping a Feedhorn. *It is important that the plastic feedhorn cover which is transparent to microwaves be used to seal the front opening of C or Ku-band feedhorns. This prevents foreign objects like insects or small birds from lodging in its throat and interfering with reception of the satellite signals.*

Figure 5-74. Installing an LNB/feedhorn Cover. *The necessary water sealing and attachments should be completed before installing the top portion of the feedhorn/LNB cover.*

ANTENNA/FEEDHORN/LNB ASSEMBLY

E. INSTALLING ACTUATORS

Step number one in installing actuators is reading the instructions. No one knows better than the manufacturer how many problems a poorly installed actuator can have. The instructions clearly spell out what mistakes to avoid as well as correct installation methods.

Mechanical Assembly

Once the antenna is mounted on its pole, the actuator arm can be attached. Both ends of the arm must be bolted to the reflector and mount pivot points with ball joints in order to allow for lateral movement and to avoid binding (see Figure 5-75). This is crucial because lateral forces on the inner tube can make it bend and seize up. These forces can also cause deterioration and potential failure of the internal O-ring seal which protects against water and dirt entry between the inner and outer tubes.

The arm should also be attached so it forms a minimum of at least a 30° angle with the rear surface of the antenna at all points in its sweep. Having a smaller angle causes the motor to work harder because the arm has less leverage. Actuator arms vary in length from 30 cm (12 inches) and less to over 120 cm (48 inches). The larger and heavier the antenna, the longer the arm required to provide adequate leverage to easily move the antenna. This arm should be installed as parallel to the plane of rotation as possible to minimise forces which could cause it to bend.

The collar on the outer arm should be positioned so that the antenna is aimed just past its most westerly target (or easterly target depending upon which side the actuator is mounted) when the inner tube is fully retracted. This can be accomplished by allowing the arm to move in its collar while the antenna is rotated to this position. Then the collar bolts can be tightened.

Most ball jack actuators can exert 700 kg (1500 pounds) or more of force. Acme arms can generate about 350 kg (750 pounds) of force. This force applied via the reflector outer rim could easily cause damage to surrounding objects. Be sure that the antenna has a full range of movement without coming into contact with any obstructions such as trees or fences. In addition, the actuator must not bind on any part of the mount or reflector over its full range of movement, even past the preset electrically programmed east and west limits. Enough slack cable should be left on the actuator so that the antenna can move through the entire arc without pulling this cable and so that a drip loop can be created (see Figure 5-76).

Preventing Water Damage

Water is the chief enemy of actuators (see Figure 5-77). If it manages to accumulate in the motor housing or in the gear box it can cause corrosion. More immediate damage results if this water freezes so that motors and gears seize.

Actuators should be mounted so that the motor is facing upwards and the drain holes under the housing are facing down. If there are no drain holes, drill one or two 5 mm openings into the housing at the lowest point. This will allow water which enters by condensation to escape before it accumulates.

Figure 5-75. Correctly Mounted Actuator Arm. *An actuator arm attaches to the mount at one end and to the dish frame at the other. The angle between the arm and the back surface of the antenna should be at least 30° to give it the leverage necessary to transfer adequate force for tracking satellites. This arm must be mounted with the motor facing upwards and the drain holes pointing downwards to allow proper drainage. Ball joints should be used at both points of attachment so that any potential binding is avoided.*

INSTALLATION

Figure 5-76. Cable Protection and Waterproofing. *This photo illustrates proper techniques for installing an actuator. This includes using water sealing boots, protecting cable runs from the actuator motor to the weatherhead with a harness and securing all cables in a neat bundle with tie wraps.*

Another place where water can enter is the opening between the outer and inner sleeves. Most manufacturers provide a rubber fitting, called a shaft wipe, to protect this opening. Further protection can be realised with a neoprene accordion sleeve (see Figure 5-78). Because neoprene is transparent to sunlight, it permits any moisture to evaporate. There should be a vent hole at both ends of this sleeve to allow water and air passage so that before any water accumulates it can evaporate and escape. If a rubber boot is used to cover the motor housing, make sure that it also has large drain holes. If not, it can do more damage than good.

Remember that actuators are the sole piece of equipment responsible for a great deal of mechanical work. If not properly installed, failure will probably result. An estimated 75% of all service calls on C-band systems have been for actuator related problems, especially in cold climates! These statistics are expected to be closely matched in tracking Ku-band home satellite systems.

Wiring the Actuator

Most direct burial, multi-run cables provide the necessary five-wire conductor needed to connect most actuators (see Figure 5-79). This includes two #14 gauge wires carrying typically 36 Vdc to power the motor. If the east button is pressed on the internal control box and the antenna moves west, or vice versa, reversing these wires will correct the situation. Three shielded #20 gauge wires are used to send the counting pulses between the

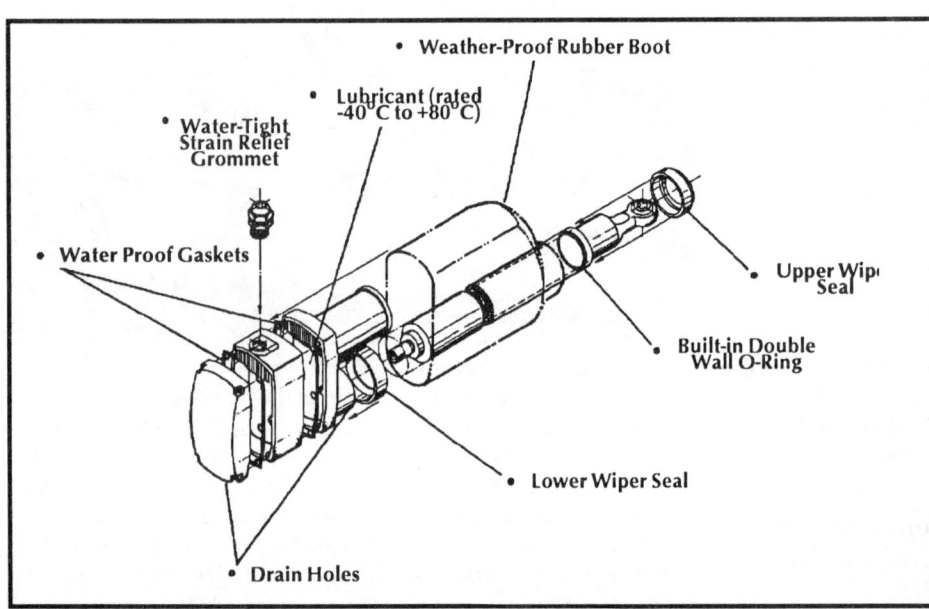

Figure 5-77. Waterproofing an Actuator Arm. *This drawing shows the points where an actuator can be protected from water entry. Drain holes must be installed and weatherproof covers as well as shaft wipes are necessary. (Courtesy of Pro Brand International, Inc.)*

INSTALLING ACTUATORS

INSTALLATION

actuator sensor and indoor control box. These wires are shielded and grounded in order to prevent spurious voltage spikes from being mistaken for counting pulses when using controllers that expect similar pulses from Hall effect transistors or reed switches.

It is important to use the seals provided to protect the points of cable entry against water intrusion. These should be waterproofed with a material such as coax seal.

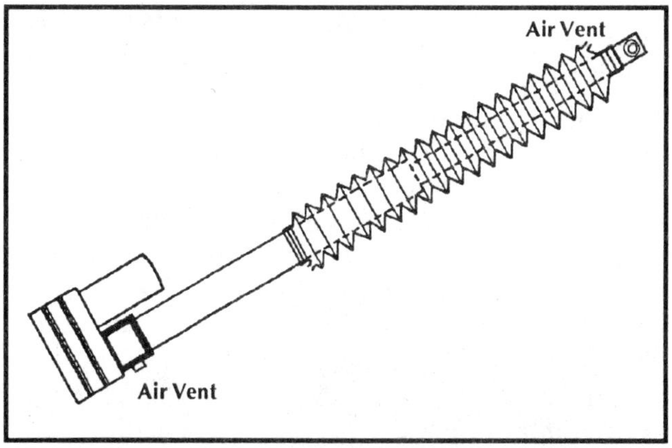

Safety Considerations

Installing and repairing actuators is a potentially dangerous activity. Actuators operate with the most current and voltage of any TVRO component, from 1.5 to 6 amps of current and typically either 24 or 36 Vdc. If a live circuit is touched by an installer who, for example, has wet feet and happens to be at ground potential, the shock will certainly be uncomfortable and possibly deadly. It takes only one milliamp of current to override the heart's electrical pulses.

The key to avoiding electrical shocks is to take steps to not be at ground potential. Always carrying a rubber mat to work on is a good habit to develop. If the old rule "always keep one hand in your back pocket when working on live circuits" is followed, the chances of receiving a fatal shock will be reduced to near zero. Never work on the electrical components of a system if an electrical storm happens to be nearby.

Using an ac power strip with a built-in circuit breaker known as a ground fault interrupter is an additional safety precaution. The satellite receiver, actuator and TV should all be connected via this outlet. If a ground problem such as a short circuit through a person's body develops, the breaker will open before any harm is done to either the person or the electronic equipment.

Actuators also have the potential of being a fire hazard because they draw a relatively large amount of current. This danger is recognized in some countries like Canada where actuator power supplies must be enclosed in a separate container to meet government safety standards. It is generally a wise decision to install actuators as well as satellite receivers in a non-flammable environment having adequate ventilation. Actuators have been known to overheat and cause melting of carpet fibres.

Figure 5-78. An Actuator Boot. *A neoprene actuator boot should have air vents at top and bottom and should attach securely to both the inner and outer actuator arms with hose clamps. This will prevent any dirt which can be drawn into the jack tube as it is extended and retracted thousands of times during its lifetime.*

Figure 5-79. Electrical Connections to an Actuator. *Five wires are used to interconnect this actuator motor to the indoors positioner. The two heavier gauge wires on the right are for motor power. If they are interchanged on the positioner terminals, east-west movements will be reversed. The ground and the two lighter gauge wires on the left side are used for control voltages or pulses. Notice how the incoming cable is sealed against water entry by a tightly fitting cable strain relief gasket and by the use of a drip loop.*

INSTALLING ACTUATORS

INSTALLATION

F. ELECTRICAL CONNECTIONS

Poor wiring techniques or improper connections can ruin an installation. A poorly grounded cable or a leaky connector may be sufficient to ruin reception of a satellite broadcast. Installers should be familiar with the correct techniques for installing all connectors that may be encountered. Fortunately, only a few types of connectors and cables have become standard in the satellite TV industry, so acquiring the necessary tools and knowledge can be rather simple.

Connector Types and Techniques

The most familiar types of connectors encountered include F-connectors, N-connectors, RCA or phono jacks, moly plugs, the solderless lug terminal, Skotch Locks and SCART connectors. Other varieties often encountered in video accessory systems include K10, K14, BNC and DIN connectors. Each of these has certain uses and may require special tools to be properly installed.

F-Connectors

F-connectors are the industry standard used for attaching coaxial leads to the LNB, satellite receiver and television sets (see Figure 5-80 and 5-81). There are a variety of types used for both RG-6 and RG-59 coax. Some come with the crimp ring as a separate part. It is imperative that these connectors be correctly installed particularly when working with higher frequency signals which are all too easily attenuated.

F-connectors are attached by first stripping off about 2 cm of the outer insulating layer. The ground braid is next trimmed to a length of about 3 mm and folded back over the outside insulation past the 2 cm point. Then about 10 mm of the white centre insulating jacket is removed from the inner wire. Use a knife and scrape the inner conductor wire clean to guarantee that a good electrical contact will be made. The crimp ring, if a separate part, is twisted on before the body of the connector. The connector should be twisted onto the coax so that the inner white dielectric jacket is flush with the inner shoulder of the F-connector. Finally, a crimp tool is used to bind the crimp ring onto the ground wire and outer sheath.

When installing an F-connector, it is important that a crimp tool and not simply a pliers be used for the final step. This special tool compresses the connector onto the jacket equally on all sides. A pliers would distort the shape of the inner insulating jacket, alter the electrical characteristics of the cable and result in signal losses due to impedance mismatches.

The use of push-on F-fittings should be avoided. They do not provide as tight an RF connection as threaded connectors and can potentially let radiation leak into or out from the attachment point. They are also more easier to pull off accidentally than the threaded type, for example, when furniture is rearranged during house cleaning.

Figure 5-80. F-Connector. *These connectors are used with coaxial cables such as RG-59, RG-6 or RG-11 which are rated for operation at frequencies below 1500 MHz. (Courtesy of MACOM Industries)*

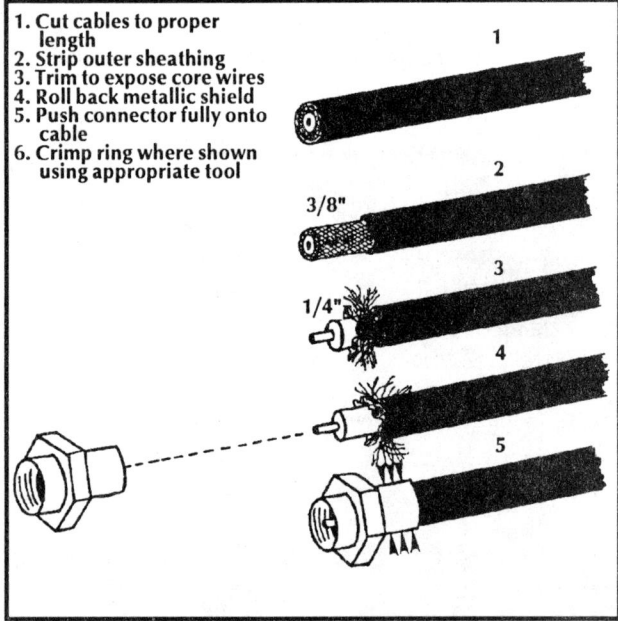

Figure 5-81. Installing F-Connectors. *This illustrates a step-by-step method to correctly install F-connectors. During this process, it is important not to score the central conductor wire because this would affect the current flow at the high frequencies used in satellite television.*

INSTALLATION

F-connectors can be joined together with couplers called F-81 barrels which have a female lead at either end. This is not a recommended practice except when absolutely necessary as it causes some loss of signal at the junction and creates another possible route for moisture entry.

Tools Required: Scalpel, Wire Strippers, Cutters, Crimping Tool

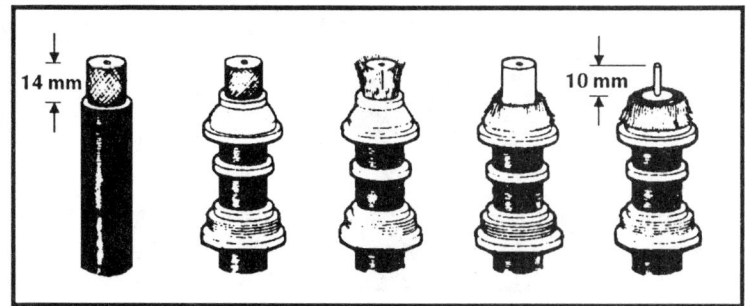

Figure 5-82. Installing N-Connectors. *Cut the outer insulating jacket back about 14 mm (9/16 inch) with a sharp knife. Then slide the pressure nut, rubber grommet and the retaining ferrel onto the cable. Peel back the insulating braid, remove about 10 mm (3/8 inch) of dielectric from the inner conductor making certain not to score the wire and solder the pin onto the centre conductor. Care should be taken not to use too much solder in the hole. Place the N- connector over the completed assembly and screw in the pressure nut from the back. Be very careful not to turn the N-connector while tightening the rear pressure nut because the inner braid can easily be broken. The centre pin should not extend past the end of the connector because equipment can be damaged if the pin is to long.*

N-Connectors

N-connectors are similar to BNC fittings. These are heavy-duty connectors that are weatherproof when properly installed. The centre pins must be soldered to the centre wire of the coax (see Figure 5-82).

N-connectors were used in earlier LNA systems and, in Europe, in LNBs until about 1987. However, some earlier versions of brands of equipment sold on the European continent such as the NEC and certain MASPRO receivers as well as some commercial systems still employ this fitting.

Tools Required: Scalpel, Wire Strippers, Cutters, Soldering Iron, Solder.

Phono Connectors or RCA Jacks

RCA jacks are often required for relaying audio and video outputs from receivers to stereo processors or to external modulators. They should never be used for higher frequency signals because they are not shielded as well as F-connectors. Most phono connectors need to be soldered onto a dual run wire or coax. A simple way to avoid the difficulty associated with soldering is to use phono-to-F-connector adapters. Varying lengths of cables having pre-attached phono connectors can also be purchased at audio or electrical supply stores.

Tools Required: Scalpel, Wire Strippers, Cutters, Soldering Iron, Solder.

Nylon Block Plugs

Nylon block plugs (the Moly plug is a familiar variety) are occasionally required in installing satellite receivers or actuators. These have internal pins which are inserted after the wires have been attached. These pins easily fit into the housing but are nearly impossible to remove without an extraction tool specifically designed for this task. When Moly plugs are installed, be careful to seal them against water entry.

Tools Required: Stripper, Cutter, Crimper and Extraction Tool

Scotch Locks

Scotch locks are another similar type of connector for joining two wires together. These crimp-on connectors are filled with a flooding compound which is squeezed out during the crimping procedure. A good mechanical and well-protected electrical connection results (see Figure 5-83). These connectors are ideal for use in attaching polariser wires.

Tools Required: Stripper and Crimper

ELECTRICAL CONNECTIONS

INSTALLATION

Figure 5-83. Scotch Loks. *Skotch Loks connect two unstripped wires. Both wires are inserted through the connector bottom which is then securely crimped. This causes conducting pins to puncture the wires and an internal flooding compound to ooze out and seal the junction against water entry. Scotch Locks are commonly used by telephone companies.*

The Solderless Lug

Solderless lugs are designed to be used with screw-on terminal strips which are often used as rear panel outputs on satellite receivers. The wire fits in one end and is either crimped down with a standard tool or soldered in place (see Figure 5-83). Such lugs are recommended for attaching wires to screw-on terminals. Bare wires can fray and, in time, even short out to adjacent terminals. As a minimum, if lugs are not used, wires should be twisted and tinned before being screwed onto a terminal.

Tools Required: Wire stripper/cutters

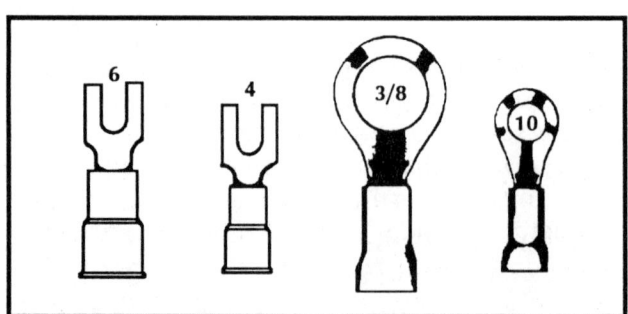

Figure 5-84. Solderless Lugs. *This lug should be used on any equipment that has a screw-type barrier strip. Attaching conductors to this connector will prevent stray wires from touching an adjacent terminal and shorting out.*

SCART Connectors

The SCART or Euroconnector ("Peritel" in France) is used on virtually all new European television receivers, other video equipment and some satellite receivers, especially those manufactured in Europe. The pin assignment is uniform among manufacturers, although in some cases not all pins are used (see Figure 5-85).

The SCART fitting can be used to interface between decoders and televisions. The baseband video from the TV set is routed through the decoder and back into the television. The operation of this connector has been examined in more detail in Chapter II.

Tools Required: Stripper, Soldering Iron, Solder

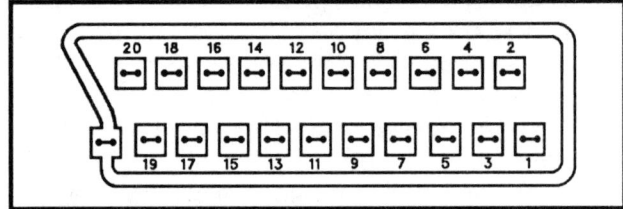

Figure 5-85. The SCART Connector. *Virtually all modern European video equipment is fitted with a SCART connector. Although the pin assignment is uniform among manufacturers, not all pins are used in many cases.*

Other Connectors - DIN, BNC, PL259

DIN connectors, soldered to very thin wires on a multi-wire cable, are standard audio/video fittings. BNC connectors are commonly used with video recorders, generally as video in and video out. PL259 connectors have been used on older brands of video recorders.

Tools Required: Scalpel, Wire Strippers, Cutters, Soldering Iron, Solder.

Additional Pointers on Installing Connectors

When stripping insulation or ground braid from cables and wires, be careful not to score the internal conductor. This would weaken it and could cause a break at some later time. Also, be careful, especially when peeling the outer insulation off coaxial cables, that the tiny pieces of wire do not fall into a vent in the actuator controller or satellite receiver. Such wires could easily short an internal component and burn out the unit.

INSTALLATION

As a general rule, never splice cables. However, some cables can be extended with the appropriate adapters joined to fittings on both pieces of cable. If the point of union is exposed to weather, use both coaxial sealant and a rubber boot or shrink fit tubing.

Adapters are also available to join different types of connectors. For example, an RCA to F-connector adapter has a male F-fitting at one end and a standard phono jack at the other. This particular adapter is often needed when attaching stereo processors or external modulators. It is much easier to make a cable with F-fittings at either end and use an adapter than to install a phono jack at one end and a F-connector at the other.

Whenever cables are terminated in connectors or adaptors care should be taken to eliminate any forces between the cables. This can be accomplished by either leaving slack cables on both side of the unions or by using devices such as wire ties to hold the cables and counteract these forces (see Figure 5-86).

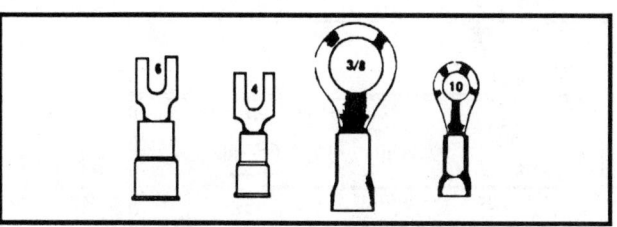

Figure 5-86. Cable Ties. *Cable ties should be used to relieve strain on cables or when interconnecting cable pairs.*

The Final Wiring

The wiring sequence to be followed depends upon how the final adjustments of antenna position will be made as well as the type of system being installed.

Small Dishes

When installing small antennas ranging in size from 35 cm to 1 m, the aiming angles can be set once the unit is in place. The beamwidths of such dishes are large enough that signals from a chosen satellite can usually be detected quite easily. If this is the case, the wiring to the indoors equipment should be completed with no intermediate step as outlined below for larger antennas. Then final tracking can be easily done either using the television or an inexpensive test meter. Since much of this type of work occurs above ground from the vantage of a ladder, there is a strong motivation to complete the electrical connections and tuning as quickly as possible.

The process is further simplified because many small dishes are fixed in position and the pulses to control ferrite polarisers as well as to deliver the signals are relayed down just a single run of coax.

Large Dishes

Larger antennas are usually installed in easily accessible positions on the ground. However, the greater the diameter of the antenna the more narrow the beamwidth so more tracking accuracy is required and the process is more demanding.

When the antenna will be adjusted either using the indoor satellite receiver or from a test setup at the dish, the LNB and polariser should be connected before it is raised into position to track the arc. Note that the electronics at the focus are more easily reached with the antenna angled down to either horizon.

If a test setup at the antenna is used for tracking adjustments, temporary wiring is required. If the antenna will be aligned using walkie-talkies to communicate from the indoors receiver to a crew member outdoors, the final wiring is completed without an intermediate step.

The three polariser leads should be twisted tightly together and joined with Scotch locks. If not, they should be soldered and then covered with wire nuts or other types of crimp-on connectors. These are then covered with "shrink spaghetti" which, like shrink tubing, contracts when heated by a small flame or hair dryer. The F-connector output from the LNB should be waterproofed with a mate-

INSTALLATION

rial similar to coax seal. Finally, the five-wire actuator lead should be hooked up and double checked to ensure that a proper connection has been made. All weatherproofing should be done only after the installation is complete.

The satellite receiver has an F-connector input for the IF signal which falls in the range of 950 to 1700 MHz (see Figure 5-87). Note that some earlier European LNBs have an N-connector output which must be connected to an N-to-F adapter and subsequently to a similar connection at the receiver. Power and RF signal are relayed between the LNB and satellite receiver on a signal coaxial cable. The rear panel of receivers also usually has a terminal strip for polarity power and control. When mechanical polariser are used, this strip has three connections: +5 Vdc, pulse control and ground. Actuator motor wires are attached to either a separate control box or directly onto those receivers having built-in actuators. In those cases where a unit such as the Houston Tracker V which controls the receiver functions is used, the polariser leads are wired onto the actuator rear panel.

As a final check, every coaxial cable connector should be carefully examined prior to mating to ensure that the centre pin has not been broken off and that it is centred and properly extended. There will be no electrical contact if the pin does not extend far enough, and damage may occur if the pin is not centred or extended too far. Usually LNBs damaged by off-centre pins are not covered under warranties. It can be very difficult to locate a poor connector once an installation is completed.

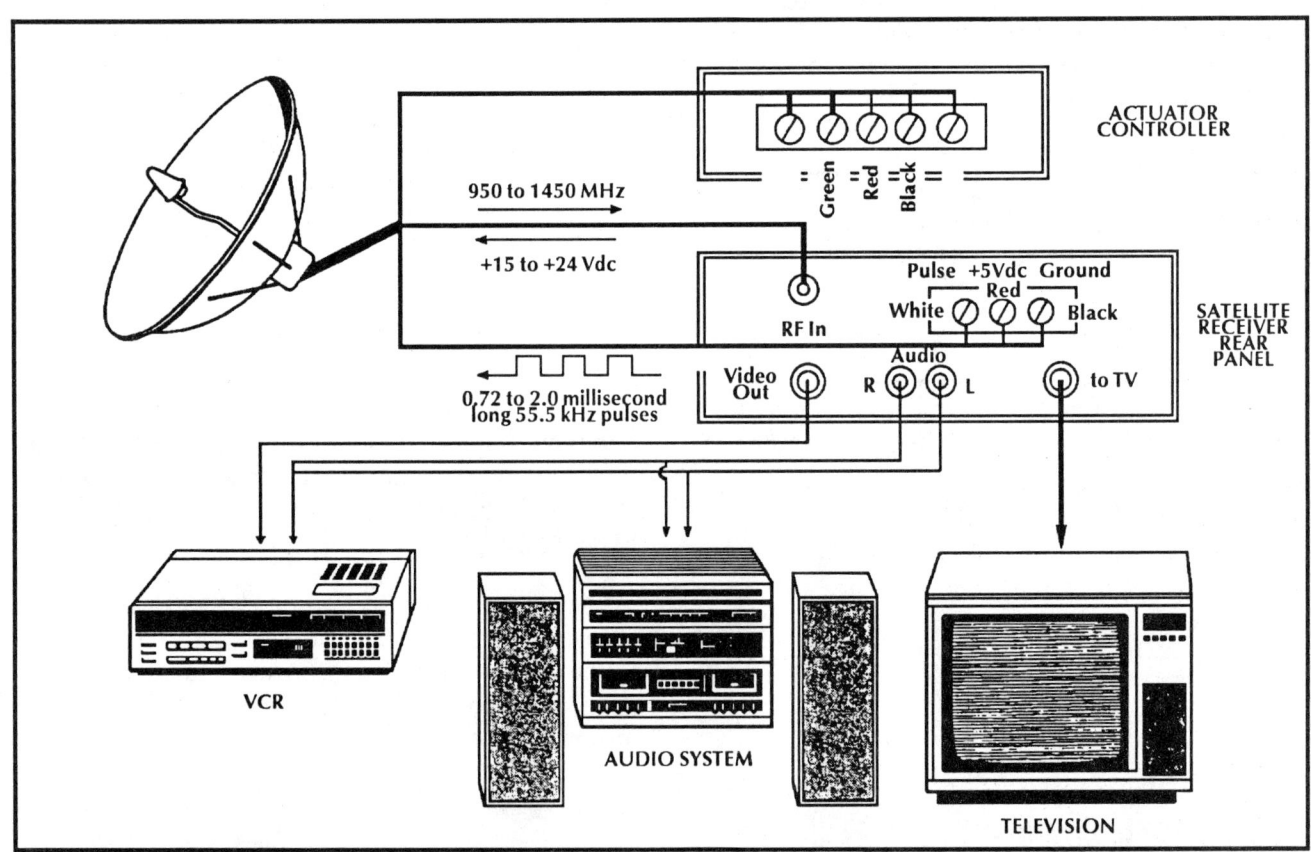

Figure 5-87. Typical Satellite TV System Wiring. *A satellite receiver can be connected to a complete audio/video home entertainment center. This illustrates how a system is wired. The receiver relays voltage to the LNB and receives RF signals on a single coaxial cable. Three wires send 5 Vdc pulses every 18 milliseconds (which equals 55.5 kHz hence the name 555 timer chip) to the polarizer. The position of the servo motor which moves the probe is controlled by the width of these pulses which vary from 0.72 to 2 milliseconds. The actuator controller transmits 6 Vdc to and receives pulses on the three sensor wires. It also sends 36 Vdc via two motor wires to control movement of the actuator. The receiver here has a built-in stereo processor which hooks directly into the home audio receiver via its output audio terminals. Alternatively, the baseband output could be used to drive a stereo processor. The VCR can be connected to either the TV set or video output ports.*

ELECTRICAL CONNECTIONS

INSTALLATION

Safety Considerations

Satellite receivers typically draw up to a maximum of one ampere of current to provide a range of regulated power outputs. The polariser usually requires +5 Vdc while internal circuitry most generally requires a regulated +12 Vdc, +18 Vdc and often +23 Vdc. Other voltage outputs are not uncommon.

While it is rare that these voltages are sufficient to cause serious harm, the same precautions taken when installing actuators should be followed. Better be safe than sorry. Many receivers often have built-in actuator controllers and, therefore, are potentially dangerous.

Protection Against Power Surges and Voltage Fluctuations

A direct lightning hit is a rare occurrence and can not only destroy the electronics in a satellite system but can also set a house on fire. However, most hits are indirect and their effect is seen as power surges (see Figure 5-88). Standard home power provided to many rural regions is often not very "clean" and produces similar but much smaller voltage transients which can cause a receiver or actuator microprocessor to lose its memory.

Power or voltage fluctuations are common in rural areas where the majority of satellite receivers are used. These can range from typically 10 to 20 Vac but can be as high as 30 to 40 Vac during storms or times of peak summer heat. These variations in line voltage will be more prevalent if a system is located near the end of a power line or if there are high-current loads between the installation site and the power source.

Whatever the source of voltage spikes or fluctuations, it is better to take precautions during installation than to be forced to make a time-consuming service call at some later date. Most satellite receivers and actuators have built-in protection in the form of voltage regulators and filters. These will be examined in more detail in Chapter IX. Although component warranties do not cover "abnormal conditions of operation" such as a direct hit by lightning, or other "acts of God", a technician can take some very effective preventative measures during installation.

Ground Rods

Ground rods are constructed from long narrow poles of copper-clad steel which are driven into the ground to provide an escape route for lightning (see Figures 5-89 and 5-90). The effectiveness of such "lightning rods" depends upon the ability of the soil to pass current, the soil resistance, and details of their connection to the antenna and mount. A 2 m ground rod will usually be sufficient to handle most lightning strikes.

Most soils have sufficient conductivity so that a well-made ground connection will effectively shunt voltage transients caused by nearby lightening strikes. Soils which have little clay or loam but are mainly rocky may need to be treated with materials like rock salt, copper sulphate or magnesium sulphate to increase their ability to pass current.

A 2 to 2.5 m (6 to 8 feet) lightning rod should be driven into the ground about 30 cm (12 inches) or so away from the supporting pole foundation. A 6 mm (1/4 inch) copper or 9 mm (3/8 inch) minimum diameter aluminum wire is bolted onto the

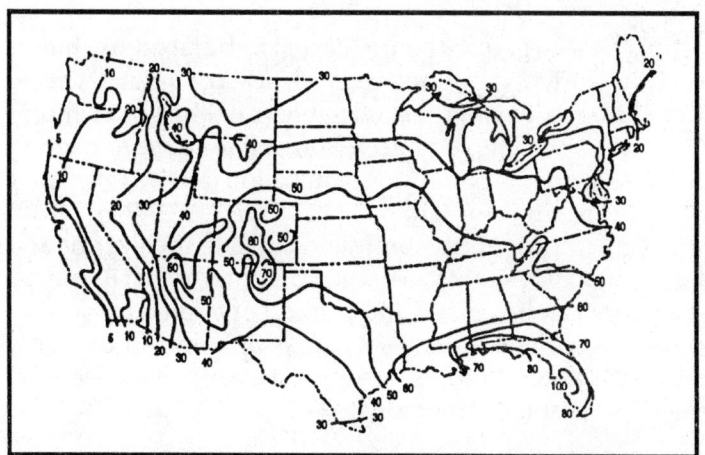

Figure 5-88. Probably of Thunder Storms in the United States. *This map indicates the number of days thunder is heard at specific locations throughout the United States. It is recommended all satellite systems be grounded with a separate ground rod installed at the antenna.*

243 ELECTRICAL CONNECTIONS

INSTALLATION

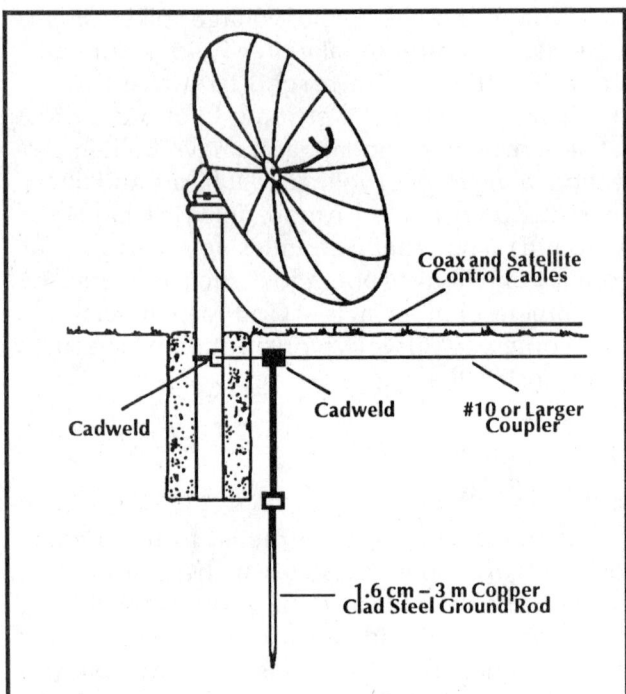

Figure 5-89. Ground Rod Configuration. *Proper grounding technique for a satellite reception system is illustrated here. The rod should be at least 2 m (6 feet) in length and should be driven just below the ground surface. This will prevent injury to an unsuspecting person who might stumble over a protruding rod. Even though pole mounts are installed directly into the ground, they should also be electrically grounded because concrete is an insulator not a conductor.*

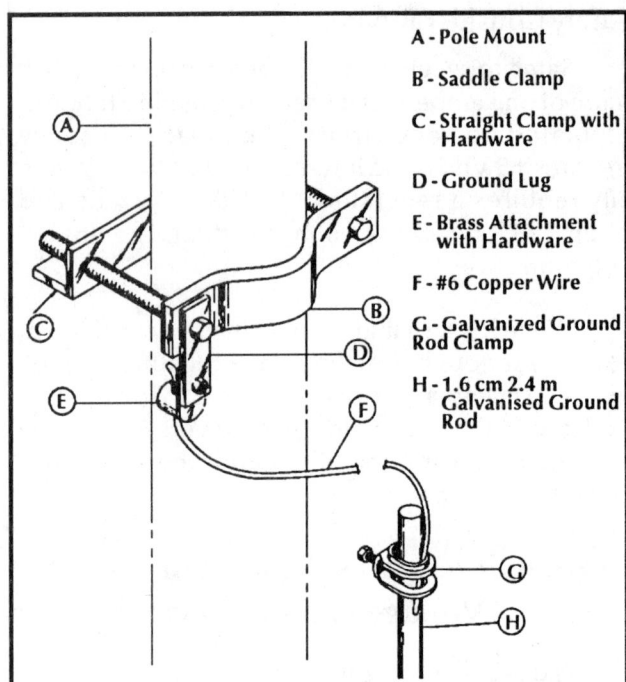

Figure 5-90. A Grounding Kit. *Grounding hardware can be purchased at many electrical supply houses. Number four or six gauge solid copper conductor is recommended. It is attached with brass hardware to the ground rod.*

ground pole. A tight clamp, often provided with the rod, is used for making contact at the top end. This wire should follow a smooth curve; any sharp bends will impede current flow. It is interesting to realize that for the same reason, if knots are tied in regular power cords, voltage transients will have a harder time finding their way into electronic components.

An effective method of planting a grounding rod is to use a pipe with a cap on one end as a driving tool. If the soil is difficult to penetrate, water can be added to the hole as the rod is driven. The top of the rod should be planted at least 15 cm (6 inches) below the ground for best results. This will also protect the customer's lawnmower as well as the customer's feet from damage. A driver sleeve is also available for use with a standard jackhammer to plant the rod. This is a required tool in many cases.

Probably the best alternative to using a ground rod is to run a heavy wire, #8 gauge or larger, from the pole or antenna to the location where the building's electrical circuits are grounded. This should effectively shunt any excess currents to ground. Paint must be removed from these points of contact prior to attaching the ground strap so that a good electrical contact is made.

Surge Protectors

Surge protectors can also be used to shield electrical components from voltage spikes. There are many varieties available ranging from simple units which plug into a power line to more expensive ones requiring the installation talent of an experienced electrician (see **Figure 5-91**). Any microprocessor controlled satellite receiver or actuator is similar to a home computer in that its memory can be erased, interrupted or altered by power surges. A surge might cause channels to change, the antenna to move unpredictably or a carefully programmed memory to fail.

ELECTRICAL CONNECTIONS

INSTALLATION

A surge protector consists of a combination of varistors and filters. A varistor allows voltages up to a pre-defined value to pass without attenuation but transfers large spikes directly to ground through the third prong on the wall outlet. Some brands can handle up to 50,000 volts and are able to respond in microseconds. Electrical filters employ capacitors to shunt any rapid changes in voltage to ground but do not effect the passage of direct currents. Good quality surge protectors incorporate varistors and filters rated to respond rapidly to large surges. However, the least expensive brands may lose their ability to operate after only a few large spikes.

All satellite reception components should be joined together to one common ground as further protection against voltage spikes. Unless a separate ground line connects all components to a common point at the antenna, the ground on a three-prong power cord must always be used. In those cases where only two-prong outlets are available, a ground relief adapter can be used. Paying good attention to ground procedures will also minimise the possibility of an annoying type of interference known as hums bars (see Chapter IX for more details). Further protection is built into some brands of LNBs having "grounded probes" that can withstand short duration power surges of up to 1000 amps.

Constant Voltage Transformers

When a receiver often overheats and shuts down, a constantly high or low line voltage may be the cause. It may be necessary to install a constant voltage transformer to counteract this problem. This component can also be useful in improving the performance of TVs or monitors.

Figure 5-91. A Surge Protector. *Surge protectors should be used when connecting any microprocessor-type equipment to a power source. These filter out any voltage surges by shunting excessive currents to ground and therefore protect sensitive equipment. Some types have internal circuit breakers that can be reset; other types will self-destruct if an excessively large voltage transient occurs.*

Protection Against Overheating

Actuators and satellite receivers generate enough heat to cause problems when they are not properly vented. At the very least, the electronics may not function at peak performance or could shut down or blow a fuse. In the worst situation, a fire may result. These components should always be placed on their feet above a flat surface, and a 7.5 cm (3 inch) air space should be provided above and to their sides. Care should be taken never to cover vent openings or install the receiver or actuator in a non-ventilated wall location or directly on a thick pile carpet.

INSTALLATION

G. DECODER INTERFACING

Hooking up a descrambler or a decoder to a receiver is not in itself a difficult process. However, the interface between decoders and receivers can sometimes be problematic. In Europe difficulties can arise with some of the older non-ASTRA receivers when baseband output signal levels do not fall within specified limits. A similar situation occurs in North America when the outputs of some earlier receivers are fed into the VideoCipher II decoder.

Clamped Or Unclamped Input?

The majority of descramblers need an unclamped composite baseband input signal to operate properly. Most receivers have such an output accessible on their back panel.

The need for an unclamped video input stems from the fact that the receiver clamp circuit can distort a scrambled signal. For example, in European encoders such as the MAAST/SAVE that operate by virtue of sine wave interference, the sine wave in the scrambled video can flattened out as a result of the clamp operation. Problems with interference can arise when the descrambler uses a properly formed sine wave in an attempt to cancel out the flattened sine wave.

Also, for example, in a decoder such as the Filmnet that uses a sync shifted signal, the satellite receiver clamp slightly distorts the vertical sync pulse and the video. The Filmnet system also uses alternate field video inversion to scramble the signal. Clamping circuitry in a satellite receiver could cause a slight offset between fields in the descrambled picture. This distortion would show up as a flicker on the screen, not unlike that effect when feeding an unclamped, unscrambled picture into a television set.

In most receivers, the clamp/unclamp switch is located on the back panel. In others, the clamp can be selected from a front panel switch or via the remote control when programming the channel parameters.

Baseband Or Composite Video?

Most descramblers can decode the unclamped baseband or unclamped composite video. However, in systems such as Filmnet, it can be better to use the baseband output of the receiver. This is primarily due to the fact that Filmnet transmits a composite, vertical and horizontal, sync on a FM subcarrier at 7.56 MHz. Some decoders demodulate this subcarrier to reassemble the sync and thus descramble the picture. Many of the newer models do not rely on this carrier to descramble the signal and therefore the use of unclamped composite video suffices.

The baseband signal is unfiltered and not de-emphasised. If a descrambler did not have any filtering and de-emphasis circuitry, the colours and definition of the descrambled picture signal would be non-linear and therefore the picture would be somewhat too sharp in appearance.

It is important to note that the SCART connector employed on many ASTRA receivers features a composite video, not a baseband output. The baseband output when needed is generally available from a separate connector on the receiver back panel.

The MAC baseband output from some satellite receivers is fed through a separate amplifier with a flat output response up to about 10 MHz. This buffered signal is the output, in some cases, from demodulator circuitry. Sometimes a filter circuit is required to even out the response. This output should only be used for MAC decoders though it can also be used for ordinary descramblers, if necessary when there is no regular baseband output.

Modulators And VCRs

Some of the more expensive pirate descramblers incorporate modulators. These are generally of poor quality and can produce large amounts of unwanted interference. In these cases, it would be better to use a direct video input to a video cassette recorder (VCR) and use the VCR modulator to feed a television receiver.

DECODER INTERFACING

INSTALLATION

The majority of VCRs now have direct composite video input for cameras and other accessories. A user can connect such a direct video source and take the direct video output of the VCR to a television monitor for higher quality pictures. Decoders can also benefit from this ability. In many cases when installing descramblers, the use of a VCR as a modulator is necessary because some older model descramblers do not have internal modulators. Direct video input into television receivers is a more recent trend. It is not unusual to find a customer with the latest model VCR and an old television set.

It is not difficult to use a VCR as a modulator. Such a procedure is generally covered in the VCR manual. Fortunately, there is a good deal of uniformity in this procedure among different VCRs.

Documentation

If a number of different brands of receivers are installed in the course of operating a satellite business, it is wise to assemble a database about the connector types used with each receiver. Since many satellite television installers also have a video and television section in their businesses, it would be useful to keep a list of the video and audio connector types available on VCRs used with direct input televisions. A sample chart that could be employed is shown in the Tables 5-7, 5-8 and 5-9.

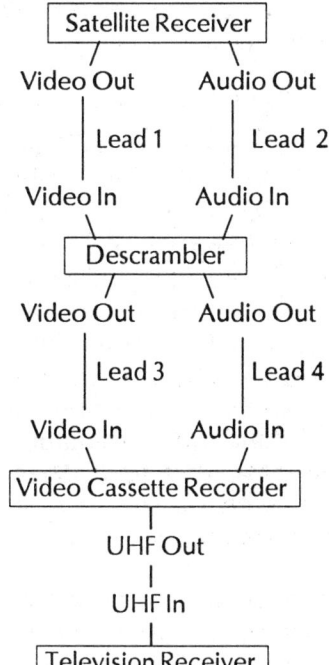

TABLE 5-8. A TREE STRUCTURE FOR INSTALLING VIDEO/AUDIO DESCRAMBLERS

Satellite Receiver Type: ....
Descrambler Type: ....
VCR Type: ....
Lead 1 ....
Lead 2 ....
Lead 3 ....
Lead 4 ....
Receiver Modulator Channel: ....
   (Is inter-channel separation sufficient?)
VCR Modulator Channel: ....
Off-air Antenna Connection: ....
   (connect to receiver or VCR?)

Descrambler Setup Procedure
Potential Problems:

Does the customer understand descrambler operation? (if not autoswitching descrambler)

If descrambler selected is a pirate descrambler, is the customer aware of the legalities?

If the descrambler is legal, is the customer aware of the subscription and contractual obligations?

TABLE 5-7. RECEIVER, DECODER AND VCR CONNECTORS

Satellite Receiver:
   Video-Out Connector_____
   Audio-Out Connector_____
Descrambler:
   Video-In Connector_____
   Audio-In Connector_____
   Video-Out Connector_____
   Audio-Out Connector_____
Video Cassette Recorder:
   Video-In Connector_____
   Audio-In Connector_____

INSTALLATION

A tree structure procedure can then be used for the descrambler installation. As each connection is completed, the relevant lead is ticked off on the chart. This provides a record of the installation details which could be quite useful if there is ever a service call-back.

The use of the connector database saves time and effort when installing descramblers. If an installer has this information on hand at the office or the shop, then he or she can prepare the necessary connector kit before setting out to an installation site. The alternative is to carry a video and audio connector interface kit to each installation location.

It should also be noted that the satellite television receiver and the VCR modulator channels are included on these charts. Some sources recommend two-channel spacing. This might be somewhat too close when installing modulators other than the crystal controlled variety. In practice, the frequency separation should be greater than two channels when driving some older televisions with inexpensive modulators.

Note that when home computers and game consoles that output onto European channel E36 in the UHF band are used in other areas of the house they can actually cause interference problems with VCRs and satellite modulators. Some of the popular makes of such equipment contain noisy UHF modulators. This should be checked when installing a satellite system. Otherwise, it could result in a service call.

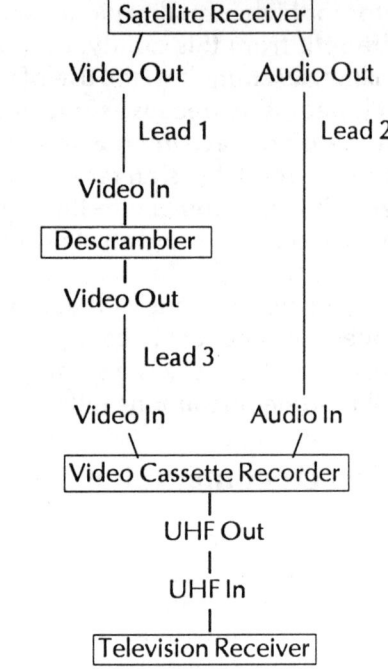

TABLE 5-9. A TREE STRUCTURE FOR VIDEO-ONLY DESCRAMBLERS

Satellite Receiver Type: ....
Descrambler Type: ....
VCR Type: ....
Lead 1 ....
Lead 2 ....
Lead 3 ....
Receiver Modulator Channel: ....
    (Is channel separation sufficient?)
VCR Modulator Channel: ....
Off-air Antenna Connection: ....
    (to satellite receiver or VCR?)

Descrambler Setup Procedure
Potential Problems:

Does the customer understand descrambler operation? (if not use an autoswitching descrambler)

If descrambler selected is a pirate descrambler, is customer aware of the legalities?

If the descrambler is legal, is the customer aware of the subscription and contractual obligations?

DECODER INTERFACING

H. ALIGNING THE ANTENNA ONTO THE SATELLITE ARC

Aligning an antenna and actually seeing television from Ku-band satellites is by far the most exciting installation step. The adjustments to prepare for receiving signals can be set from one of two locations: indoors or outdoors. Once the final wiring is finished, a set of walkie-talkies can be used to communicate between one person at the receiver/positioner and another at the antenna or a complete test setup can be wired at the antenna. The second method is recommended if weather permits. There is less potential for miscommunication and, if necessary, just one person can complete the installation. Note that aligning a large antenna Ku-band system requires more precision than positioning a C-band dish because the beamwidth is narrower. However, the very small dishes now being used in European countries are quite easy to align because of their wide beamwidths.

Fixed Antennas

In many installations an antenna is aimed at just one satellite and a positioner is not required. This is particularly the case today in Europe where tuning to just a single powerful Ku-band satellite such as Astra or BSB can be sufficient to provide a complete range of television entertainment. In these cases, a satellite can often be easily targeted knowing just its elevation and azimuth bearing and establishing a vertical plumb line. The elevation and azimuth angles to any satellite from any location in can be easily calculated from the equations presented in Appendix B. In particular, the elevation angles to TDF1, ASTRA 1-A and Eutelsat 1-F4 from any location in Europe can be determined from the chart in Appendix C.

Nevertheless, there are two reasons why it is important for a technician to be familiar with installing systems that are capable of tracking. First, there may be situations where locating true south may prove to be difficult and some of the techniques outlined in the sections below may prove quite useful. Second, as more satellites are launched and as the curious customer learns of the diversity available, those having fixed systems may very well want to install tracking systems at some point in the future. Please note that many of the techniques and pointers outlined in the sections below apply to both fixed and tracking systems.

In either case, the initial test setup should include a satellite receiver and a small television set or monitor. An actuator is required when installing a tracking system and a signal strength meter may be quite useful when fine tuning the alignment. Note that the signal strength meter built into many brands of satellite receivers can be adequate during the initial alignment procedures. However, a portable spectrum analyser or another lower cost signal detection instrument can also be used at this stage and is a valuable tool later when fine tuning the sys-

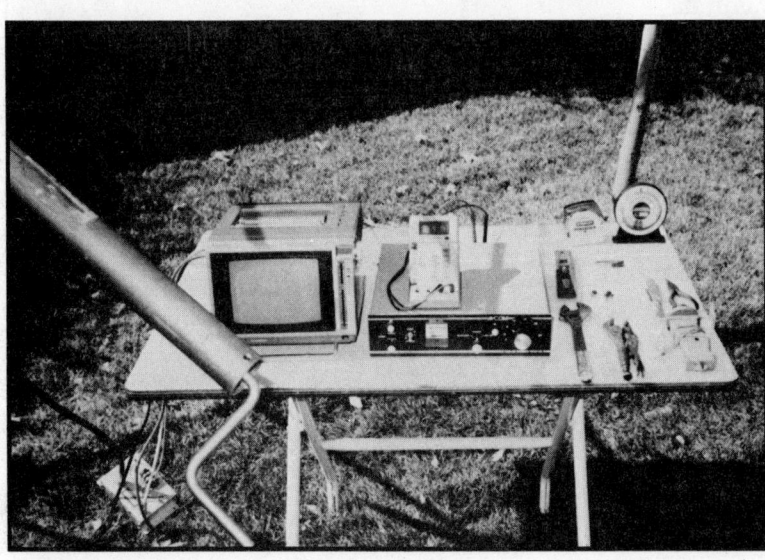

Figure 5-92. Equipment for Aligning a Dish.
This photo shows the satellite receiver, television set, and an external meter test setup at an antenna ready to aid an installer in tracking the polar arc.

INSTALLATION

tem as described below. An alternative to hooking up the antenna to an actuator during dish alignment is to remove the motor and its housing from the actuator arm and then to use a small hand crank to scan between satellites. It is quite useful to have a small monitor or TV as a standard part of an installer's test equipment. This avoids the need to disconnect a customer's set and lug it out to the site. Using a familiar TV known to be working eliminates the possibility of a malfunction occurring with unfamiliar equipment. It is also important to have a supply of 75 to 300 ohm transformers on hand. These will be necessary if the customer's TV set is also an older model and does not have F-connector attachments.

The antenna and feed can be aligned using the receiver and actuator that are being installed. But if difficulties with any electronic component arise, it is best to have a spare, shop-tested receiver and actuator on hand for use as substitutes.

Testing Components Before Setting Angles

By far the easiest method to test if equipment is working is to check out each electronic component on a working system back at the shop. Taking the necessary half hour of time to do so in a familiar environment can eliminate many installation headaches. It is strongly recommended.

A technician also can use pre-alignment tests to determine if the equipment being installed is working as expected. First be sure that the television set is tuned to the same channel as the modulator output. If necessary, the health of the TV can be checked by viewing a local over-the-air channel. Next, attach a coaxial cable between the TV set input and the RF output on the satellite receiver. Set the audio mode selection to the mono position if there is a choice between this and stereo.

If the cable from the LNB to the receiver input is disconnected and if the receiver and TV are on, the screen should be black or solid white. If not, the receiver has a short or the cable between it and the television is open. It is important always to turn the power off and unplug a receiver before disconnecting or hooking up any connectors since damage can be caused by shorting a centre lead to ground.

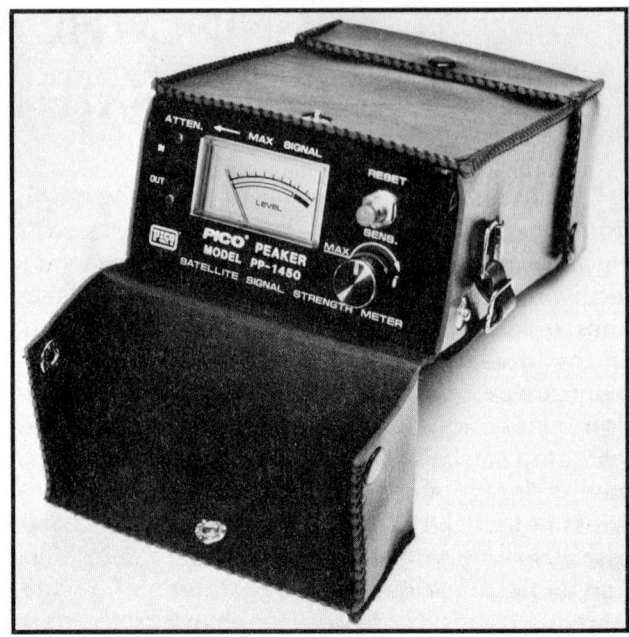

Figure 5-93. An Inexpensive Signal Strength Meter. *This line-powered signal level metre is used for aiming, focusing and peaking C/Ku-band TVROs with block receivers. It has an adjustable sensitivity and can be installed at the LNB output and is powered from the indoors receiver therefore eliminating the need to carry equipment out to the antenna when troubleshooting. (Courtesy of Pico Products)*

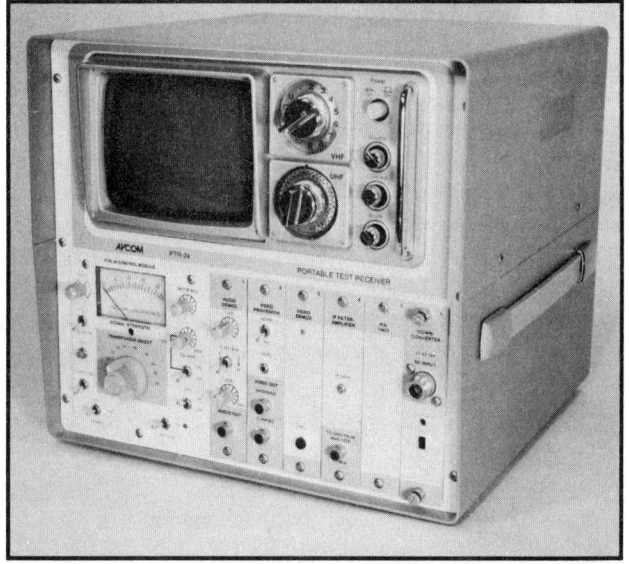

Figure 5-94. Avcom PTR-24 Portable Test Receiver. *The PTR-24 provides the necessary power and video outputs to peak or align a dish. This eliminates the need to use a customer's satellite receiver outdoors and avoids any possible damage due to mishandling during installation. (Courtesy of Avcom of Virginia)*

ALIGNING THE ANTENNA

INSTALLATION

Finally, hook up the LNB. The screen should display white noise and a loud hiss should be heard. If not, the LNB is either defective or is not being powered by the receiver.

Another option is to use a voltmeter to check for the necessary 15 to 24 Vdc at the receiver output and LNB input. If the proper reading is indicated, the components are probably good. Next check that the voltage at the LNB input is adequate. This will require installing a 2-way, dc-passive power splitter that allows current to pass in both directions. Alternatively, disconnect the cable and measure the power between its ground and centre conductor. If power is not getting to the LNB, the cable or one or both connectors are faulty. These and other troubleshooting procedures are outlined in much more detail in Chapter IX.

Aiming a Tracking Antenna

Three adjustments are required on most polar mounted antennas to ensure proper tracking of the arc of satellites: north/south heading, declination offset angle and polar axis angle. It is usually simpler to set these parameters when the antenna is at its highest position and is aimed towards the centre of the satellite arc. This can be accomplished by turning on only the actuator and moving it to this position with the east or west buttons, or by using a hand crank or slide arm bar.

North/South Orientation

A polar mount must have its axis aligned with the north/south axis of the earth in order to be able to detect all satellites in the viewable arc. This is easily understood by visualising an antenna at the equator. If it rotates in an axis aligned with the centre of the earth, it will correctly scan the circle of satellites in the sky directly above from horizon to horizon.

Most antennas have a flat plane on the mount which can be used as a sighting reference. A handheld compass with cross-hairs is the most effective type for lining up with the north/south plane. Remember that a correction for magnetic variation is necessary. West of the line of zero variation, the agonic line, rotate the antenna and mount east of magnetic south by an amount equal to the variation. East of this line, rotate it west to correct for vari-

Figure 5-95. Setting the North/South Orientation. *A true north/south orientation can be obtained by using either a compass or the north star for alignment. When using a compass, stand well back from the satellite antenna to avoid pointing errors caused by iron in the mount. Sight along the centre line of the polar axis or along a flat portion of the mount parallel to this axis. The antenna should face south in the northern hemisphere and vice versa. Remember that the correction for magnetic variation which can be found in the accompaning maps is necessary.*

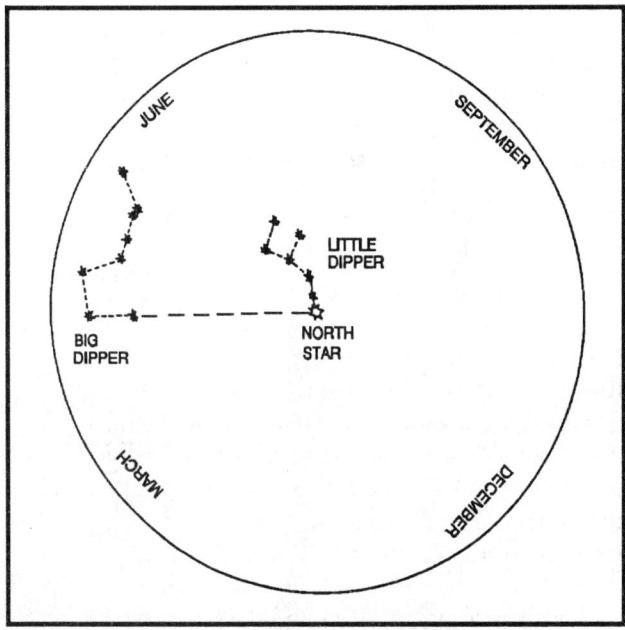

Figure 5-96. Using the North Star for Antenna Alignment. *A second rather unusual option to line up the polar axis with a true north/south bearing is to use the north star. It is the small star at the end of the constellation called the Little Dipper and, of course, can only be seen at night*

ation. During nighttime installations, the north star can also be used for alignment.

ALIGNING THE ANTENNA

INSTALLATION

The Polar Axis Angle

The polar axis angle is set equal to the site latitude. This targets the antenna directly out into space along a plane parallel to one passing through the equator as described in Chapter II. Most antennas have one or two long threaded rods which are used to adjust polar axis angle. An inclinometer resting on the axis bar or back part of the mount is used to set this angle (see Figure 5-97).

Declination Offset Angle

The declination offset adjustment lowers the antenna's sight from a plane parallel to the equatorial plane down to the arc of satellites. The declination angle is greater in locations farther away from the equator. Table 5-10 shows how declination varies with latitude. These values can be used to set declination in either the northern or southern hemisphere.

Declination offset angle is measured with an inclinometer. The difference between two readings, one on the main part of the mount, the axis bar, and one on a flat spot on the back of the antenna, should equal the declination offset angle. The easiest way to set the declination offset angle is with an inclinometer placed on a back surface which is parallel with the face of the antenna or a flat board placed in a vertical direction spanning the antenna rims (see Figures 5-98 and 5-99). This reading should be set equal to the sum of the site lati-

TABLE 5-10. POLAR MOUNT ELEVATION AND DECLINATION

Site Latitude	Elevation	Declination	Site Latitude	Elevation	Declination
0	0	0	34	34	5.510
1	1	0.178	35	35	5.641
2	2	0.355	36	36	5.770
3	3	0.533	37	37	5.897
4	4	0.710	38	38	5.669
5	5	0.887	39	39	6.142
6	6	1.063	40	40	6.260
7	7	1.239	41	41	6.376
8	8	1.415	42	42	6.489
9	9	1.589	43	43	6.600
10	10	1.763	44	44	6.708
11	11	1.936	45	45	6.813
12	12	2.108	46	46	6.799
13	13	2.279	47	47	7.015
14	14	2.449	48	48	7.112
15	15	2.618	49	49	7.205
16	16	2.786	50	50	7.296
17	17	2.952	51	51	7.385
18	18	3.117	52	52	7.470
19	19	3.280	53	53	7.552
20	20	3.442	54	54	7.632
21	21	3.603	56	56	7.782
22	22	3.761	58	58	7.792
23	23	3.918	60	60	8.047
24	24	4.073	62	62	8.162
25	25	4.226	64	64	8.265
26	26	4.377	66	66	8.357
27	27	4.526	68	68	8.437
28	28	4.674	70	70	8.505
29	29	4.819	72	72	8.562
30	30	4.961	74	74	8.608
31	31	5.102	76	76	8.643
32	32	5.241	78	78	8.666
33	33	5.377	80	80	8.678

INSTALLATION

tude plus the offset angle. For example, in the American city of Denver, the polar axis angle is set equal to its 40° latitude. The declination offset angle is 6.26 degrees so the antenna should be set at a 46.26° elevation. Similarly, at a latitude of 35°, the declination offset angle is 5.64°, so the final elevation is 40.64°.

When the modified polar mount geometry is used, as explained in Chapter II, these declination offset angles as well as the elevation angle must be slightly altered (see Table 5-11). This adjustment can be made by directly measuring the required angles or by finely tuning antenna alignment by the methods described below.

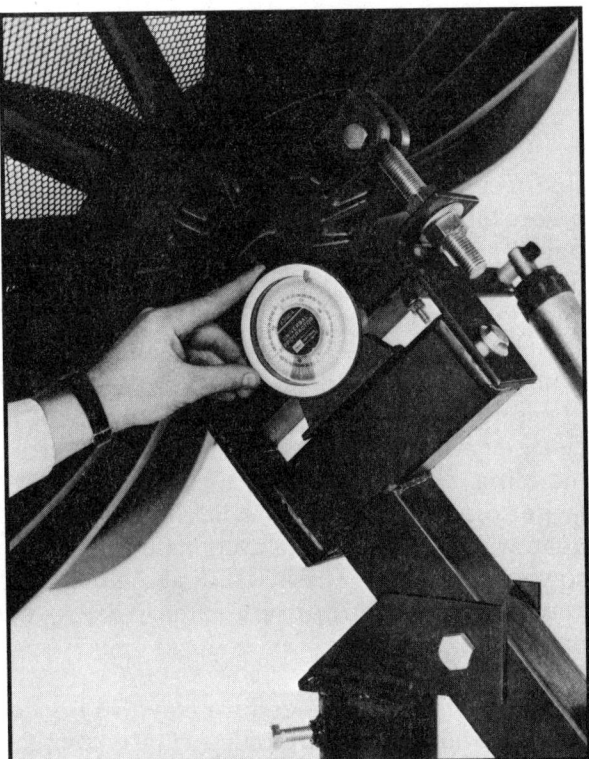

Figure 5-97. Setting the Polar Axis Angle. *To set the polar axis angle place an inclinometer on the polar axis bar and raise or lower the elevation to equal the site latitude.*

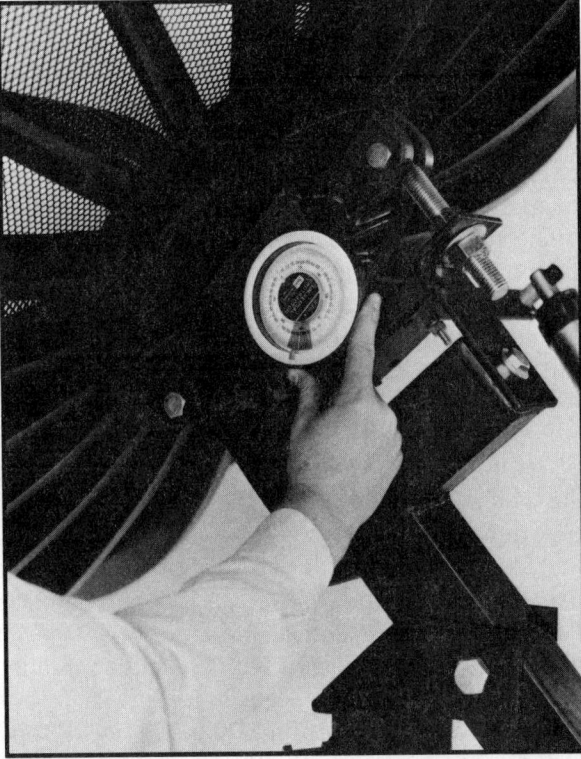

Figure 5-98. Setting the Declination Offset Angle. *To set the declination offset angle, place the inclinometer on a flat surface of the reflector and adjust the long threaded rod until the reading equals the sum of the declination plus the site latitude angle.*

TABLE 5-11. MODIFIED POLAR MOUNT TRACKING ANGLES

Latitude	Polar Axis Angle	Declination Angle
5	5.12	0.77
10	10.23	1.54
15	15.33	2.29
20	20.43	3.02
25	25.51	3.73
30	30.57	4.40
31	31.59	4.53
32	32.60	4.66
33	33.60	4.79
34	34.61	4.91
35	35.62	5.04
36	36.63	5.16
37	37.63	5.28
38	38.64	5.40
39	39.64	5.51
40	40.65	5.63
41	41.65	5.74
42	42.65	5.85
43	43.65	5.96
44	44.66	6.07
45	45.66	6.18
46	46.65	6.28
47	47.65	6.38
48	48.65	6.48
49	49.65	6.58
50	50.64	6.67
55	55.61	7.11
60	60.56	7.51
65	65.49	8.84
70	70.41	8.11
75	75.32	8.33
80	80.22	8.48
85	85.11	8.57

ALIGNING THE ANTENNA

INSTALLATION

Most antennas have continuously adjustable declination offset angles that are changed by moving the assembly on all-thread bolts. However, some antennas manufactured for C-band operation had a preset declination offset angle allowing no adjustment. This type cannot be used when tracking Ku-band satellites which require a finely tunable adjustment.

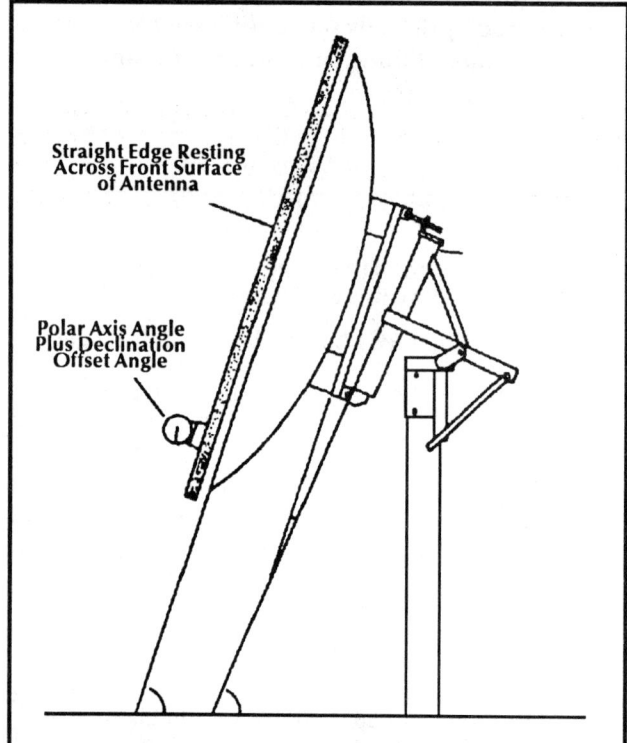

Figure 5-99. Setting the Declination Angle on the Antenna Face. *If a flat surface cannot be found on the rear of the antenna, as the case with some spun aluminium antennas, a flat board can be placed across the face of the antenna and the declination may be set by resting the inclinometer on this surface.*

Powering On and Aligning Onto the Arc

The time has come to watch satellite television. Check all connections and leads one final time before turning the equipment on. In some cases, especially when using smaller dishes, a satellite will happen to be targeted by the initial adjustments and clear pictures will immediately appear on the TV. However, while locating a C-band satellite by simply scanning across the arc may be rather easy at this point if all angles had been carefully set, precisely targeting Ku-band spacecraft can be more demanding. This is especially the case when using larger diameter reflectors having very narrow beamwidths, because the antenna must be aimed so very accurately. The azimuth and elevation angles to a satellite can be useful in the initial targeting procedure.

The alignment procedure can also be aided with a portable spectrum analyser that is connected to the LNB output. Its sensitivity should be set to maximum and its range should be set for 950 to 1700 MHz in Europe or 950 to 1450 MHz in North America. As the antenna is tracked across the arc, any signal received would be seen as a clear blip on the screen. If more than one transponder is active and if a satellite is on target, more than one peak would be visible. The centre frequency of these peaks can be determined and they can be compared to those of the broadcast channels expected from a given satellite. When a Ku-band system is being retrofitted onto a C-band TVRO, the arc can be more easily found by first aligning onto a C-band spacecraft. Recall when targeting Ku-band satellites that many but not all of these spacecraft transmit only horizontally polarised signals in North America.

Often a good starting point is to target a satellite which has many active transponders, preferably one near due south. For example, in North America, Satcom K2, GSTAR I or Anik C3 are in frequent use, while in Europe Intelsat VA F11 and Eutelsat 1-F4 have numerous active transponders. When the alignment process begins with an active satellite located near due south, it is easier to arrive at an accurate polar axis angle setting. For example, in Denver having a longitude of $105°W$, GSTAR I at $103°W$ is only $2°$ off from due south. During this entire procedure, it is necessary to have either a TV guide or an excellent memory so that those transponders which are active during the installation can be located and identified.

Tuners on many brands of C and dual-band receivers can be set in the scan mode. This causes all channels to appear on the screen in rapid succession and is a very useful feature for locating active transponders. As a reflector is slowly tracked across the arc, all channels on each satellite targeted for the preset polarity will flash across the screen. This mode is useful for receiving Ku-band broadcasts, but when the scan is turned off their channel centre frequencies may be far from those preset for C-band systems.

Whether or not the scan mode is used, attempt to find a picture by tuning through all active transponders. As soon as anything other than white noise appears, the process can begin. All that is needed to start the alignment fine adjustments is just a semblance of a picture. If the receiver does not have a scan feature, check the TV guide and pick an active channel on one of the satellites which is broadcasting at the time of the installation. Then aim the antenna as closely as possible to this satellite and begin the hunt.

At this point, be aware that a poor quality Ku-band antenna might have side lobes which could have good enough "vision" on their own to actually detect broadcasts. There have been cases of installers being able to receive signals from one satellite at three locations across the arc, one location for the main lobe and two for the side lobes. Needless to say, the alignment should be performed only on the signal received by the antenna main lobe.

In those rare cases when a receiver signal strength meter is "maxed out," it can usually be set to a lower sensitivity by either a front panel button or a back panel or internal potentiometer setting. Occasionally, these adjustments may not be sufficient to lower the meter sensitivity when testing outdoors at the antenna since cable runs are very short and signal-to-noise levels are higher than usual. Most meters built into receivers are set to compensate for losses in the typical 30 m run of coaxial cable. Inserting a coil of 30 m of cable at the antenna may reduce the signal strength enough so that the meter returns to the readable range.

Once a rough picture has been detected, try to improve it by moving the antenna ever so slightly east and west and by adjusting the polarity and

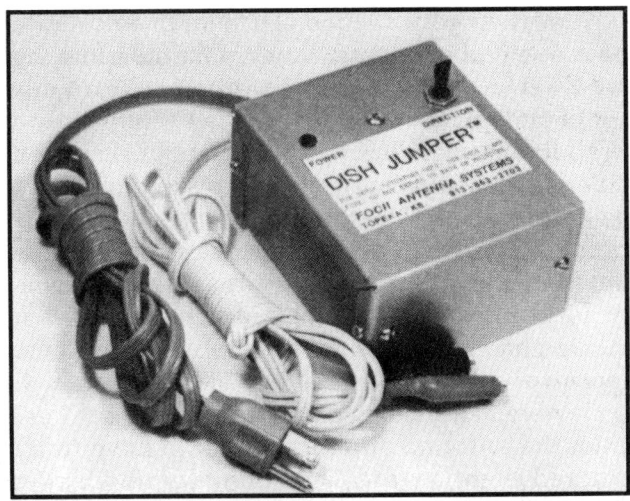

Figure 5-100. A Low-Cost Actuator. *The Dish Jumper™ is a simple, low-cost antenna actuator that could be part of an tool kit for installing bulkier antennas. A bi-directional switch drives the dish in either direction. (Courtesy of Focii Manufacturing Company)*

video fine tune (see Figure 5-100). If a side lobe is being targeted by mistake, no adjustments will improve the picture quality. It may be necessary to move the antenna until the picture disappears and then reappears more clearly. During this procedure, be certain that the receiver's automatic frequency control (AFC) is off. Also, make sure that when a polariser is used, the probe is free to move so that the skew can be adjusted. The probe must also easily move between even and odd channel positions.

Once the clearest possible picture has been obtained by using the standard mechanical adjustments, gently push the antenna up and down from its front bottom rim. Take care not to bend the reflector. If the picture improves, adjust the elevation in that direction until the best picture is received or until the signal strength meter has the highest possible reading.

At this point, the final feedhorn adjustments should be made. A more detailed discussion of these methods is presented in the following section. It is important that once the feed has been peaked, it remain in this optimal position so that the antenna can be aligned as closely to the arc as possible.

INSTALLATION

Next, target a second active Ku-band satellite near one end of the arc. Either make sure that the receiver is set to an active transponder on this spacecraft or use the receiver scan setting. It will probably be necessary to readjust the skew and video fine tune as well as the polar axis angle once a picture is seen. Note the movement needed to improve pictures. If it is necessary to raise or lower the antenna, the north/south axis usually needs some minor correction. For example, if pictures at a site in Europe had been peaked on Intelsat VA F11, the most westerly satellite, but the reflector needed to be lowered after being swept east to Eutelsat 1-F4, then the antenna and mount would have to be rotated slightly west. The geometry underlying these adjustments is outlined in the Figure 5-101.

North/south readjustments should be made in very small increments, on the order of 1 or 2 mm (1/16 inch) as measured by the movement of the mount relative to the support post. This fine tuning can most easily be accomplished by partially loosening the mount-to-pole bolts and by pushing on the outer rim of the antenna where leverage is greatest. Pushing from the mount is more difficult and can cause jerky, unpredictable movements. It is also wise to use a piece of chalk to mark the starting point so that if too much movement occurs, the process need not be started over from scratch. An antenna can often be quite heavy and should not be allowed to pull the mount forward on its supporting pole. The bolts should be only slightly loosened. Even so, the fine adjustments are often made with

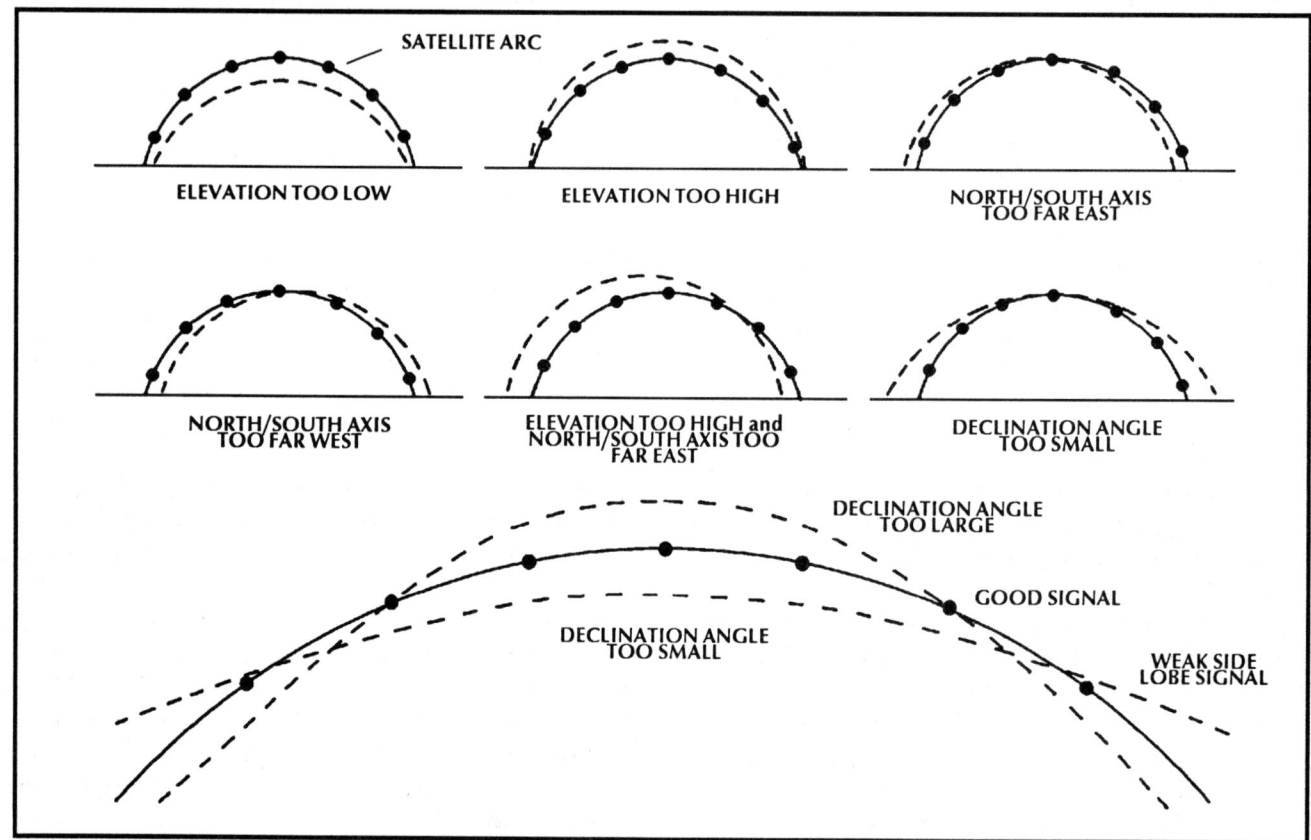

Figure 5-101. Common Antenna Tracking Problems and Solutions. *Most tracking problems are associated with an incorrect north/south orientation. However, if the declination angle has not been correctly set, tracking will also be incorrect. Lining up an antenna with the geosynchronous arc of satellites is simply a matter of lining up two half circles, the one in which the satellites are located and the one scanned by the polar mount.*

ALIGNING THE ANTENNA

the mount tilted slightly forward so that when the bolts are tightened, the mount will pull back and the antenna will be pulled slightly off target. Be aware that this can happen and that in most cases a slight decrease in elevation will correct the error. Do not make more than one adjustment at a time; it generally simply causes confusion.

After the antenna is tracking the arc reasonably closely, the initial fine tuning should begin. Accurate readings are now taken from the signal strength meter. It is important during these steps that the scale chosen on the meter be left unchanged so that comparisons are possible between these readings before and after changing any settings. For example, assume that the meter read 6.23 on the American Ku-band satellite, Satcom K2, transponder 20. Then the antenna was moved to the Canadian Anik C3 and fine adjustments in north/south position and polar axis angle were made. When the antenna was returned to the same channel on Satcom K2, the new reading should have been compared with the old one on the same transponder. After readjusting to obtain the best signal, this reading should have been at least as high as the old one. For the same reason, it is also important to keep the skew adjustment knob unchanged when satellites are transmitting linearly polarised broadcasts. This is possible if transponders on satellites having the same polarity format are being compared.

When the antenna is accurately aimed, maximum signal strength readings should be observed on all satellites in the arc. Generally this can be accomplished by adjusting just the north/south position and the polar axis angle. However, the declination offset angle will occasionally need some small readjustment. The accuracy of this iterative fine tuning procedure is better if satellites at far ends of the arc are used. If a stronger signal can be obtained on any satellite by pushing either up or down on the antenna, the arc is still not perfectly targeted.

For further clarity, this procedure is outlined step by step in Table 5-12. If this table is understood, accurate tracking can be easily accomplished.

Six common tracking problems are illustrated in the Figure 5-101 for locations north of the equator. South of the equator reverse the sense of these drawings. The best way to visualise these is to picture the arc of satellites as one half circle and the tracking movement of the antenna as another. Both of these circles must be aligned for perfect tracking. It really is that simple. It is important to understand that none of these procedures will work if the pole is not perfectly plumb or if the mount is not sitting vertically on the pole. If problems are encountered, check these mechanical settings. Be perfectly sure that the antenna does not have the opportunity to be driven into a table, toolbox or even an installer's vehicle by the actuator while it is being repeatedly scanned across the arc.

This fine tuning process is completed by repeating the necessary adjustments at both ends of the arc. Satellites at the centre of the arc should also be targeted to ascertain that all the fine adjustments are correct. Tighten all bolts and nuts and then recheck to make sure that doing so has not slightly altered the elevation alignment.

During this procedure, familiarity with the programming relayed by each active transponder on every satellite is invaluable. By far the easiest way to learn this information is by having experience with a working satellite TV system. It becomes clear what to expect in a relatively short time. Become familiar with those satellites and transponders having test patterns with the satellite name and transponder number. These are invaluable landmarks in the search.

INSTALLATION

TABLE 5-12. THE BASIC ANTENNA ALIGNMENT PROCEDURE.

1. Carefully check that the feed system is centred and located at the correct focal distance. It is important that the feed be peaked before tracking is started.

2. Set the two tracking angles, polar axis and declination offset angles, and align the antenna polar axis in a north/south direction as well as possible. It is easiest to adjust the polar axis and declination offset angles on a satellite high in the arc, i.e. at the centre of the arc, and to adjust the north/south orientation during the following step on a satellite lowest or at either end of the arc. The simple rule is elevation adjustment on a high satellite; north/south adjustment on a low satellite.

3. Check all electrical connections and turn the power on.

4. Aim the antenna at a preset position or slowly track across the arc to find even the faintest picture. It may be necessary to use instruments outlined in the text to find a signal. Attempt to target a satellite located as close to due south as possible so that the polar axis angle can be accurately adjusted.

5. Once even a semblance of a picture has been detected, move the antenna slightly east and west and adjust the polarity and video fine tune controls on the satellite receiver to get the best possible reception. Then make any adjustments necessary in the polar axis angle to further improve reception.

6. Re-peak the feed system to maximise the signal strength reading. Once this has been accomplished, do not change the feed position again.

7. Move to another satellite, preferably at one far end of the arc, and attempt to tune in a second channel. During this and procedure #3, having a receiver with a scan setting may be useful. Otherwise, set the receiver tuner, both frequency and polarisation, on a channel known to be active on the satellite being targeted. If signals from this transponder cannot be located on this satellite, move to another closer to the first successful target. If three or four satellites in a cluster but none others further away in the arc can be found, the north/south alignment is probably substantially off position and needs correction.

8. Tune in this second satellite as well as possible with the video and polarity fine tune controls and east/west scanning.

9. Next, determine if the antenna has to be pulled up or pushed down to improve reception. If neither, move to another satellite at the other end of the arc and try again. In those rare cases when all satellites are targeted accurately, jump to step number 12. If adjustment is necessary, consult the common tracking error figure to determine which direction to rotate the north/south axis.

10. Make a small movement of the north/south axis in the correct direction. Make sure to mark the original position just in case the mount moves too much.

11. Return to the original satellite and repeat this procedure to further zero in on the arc. It makes sense to do this for one satellite at the far eastern and far western end as well as one in the centre of the arc for best results.

12. Next, if possible, chose one satellite at the middle and one at each end of the arc all three of which have the same polarisation format and a common active transponder. Adjust the skew and video fine tune controls one last time to maximise the reading on the signal strength meter on one satellite. Do not touch these controls until this procedure is completed. Repeat steps 4 through 9 again to get maximum readings on all three satellites. At this point, some adjustment to the declination offset angle could possibly be necessary. It will probably be necessary to move back and forth between these three transponders on the three satellites until all three readings are consistently at their maximum value. The accuracy of this procedure can be improved by using a more accurate measuring instrument.

13. As a final test, tune to the lowest power transponders in the arc. If excellent pictures can be obtained on these as well as on satellites all across the arc, the adjustments are optimised and the arc is corrected targeted.

14. Tighten all bolts carefully watching to be sure that the signal strength meter maintains its maximum readings.

ALIGNING THE ANTENNA

Aligning a Polar Mount Without a Compass

After some experience in the field, many installers have developed useful shortcuts to align a polar mounted antenna with the arc of satellites. One such method that allows a complete installation to be accomplished using just an inclinometer is described below.

First, two satellites that can be used as reference points are located. A satellite east and another west of due south are chosen. Note that due south always corresponds to a satellite located at an orbital slot equal to the longitude of the TVRO. So, for example, in Denver, Colorado, due south would correspond to an orbital slot of 103°W, while in London, England a satellite at 0° would be seen by aiming due south.

The antenna is aimed in a fashion similar to aligning an az-el mount. The elevation angle of the first reference satellite is set. Then the antenna is rotated around the azimuth axis, i.e. the mount is rotated on the supporting pole, until a signal from this satellite is detected. The most common technique is to slowly rotate the dish and watch the television screen for even a flicker of a picture. This method can be preferable to using a spectrum analyser because it allows visual confirmation of picture content to verify that the satellite observed is the actual target. Then this first point is scribed on the pole with a marker or tape.

Next, a second satellite is located by the same procedure. Ideally the two satellites should be at equal distances from due south so they have equal elevation angles. This makes this aiming procedure quite easy. Once the second satellite is found, its position is also marked on the pole.

The calculation of the heading for south is now simple. The position of the two reference satellites have are already been calculated. The circumference of the base pole is measured. This distance divided by 360° gives the linear distance per degree. Since the azimuth angles of the reference satellites are known, the linear distance of due south from each of the reference points can easily be calculated. Once this is done, the mount axis can be rotated to a true south heading as measured from the previously marked bearings from either of the two reference satellites.

This technique is simple to implement. It is also quite well suited to cold weather installations since most of the effort and time spent in tweaking a polar mount is focused on finding a due south heading. In contrast, adjusting the elevation and declination angles can be accurately and rapidly accomplished using an inclinometer.

The azimuth and elevation bearings to the two reference satellites should be calculated before traveling to the installation site. This can be done using the equations in Appendix B or via a computer program such as the one listed in the reference materials.

An Example

The example of this simple technique in Table 5-11 uses the satellites ECS 1-F4 at 13°E and Intelsat VA-F11 at 27.5°W. This same procedure would obviously be applicable in any location on the globe. Tables 5-13, 14 and 15 can be a useful guide in determining the information required in this procedure.

TABLE 5-13. OUTSIDE DIAMETER, CIRCUMFERENCES AND DISTANCES

Outside Diameter		Circumference	mm/degree
Inches	mm	mm	
2.0	50.8	159.6	0.4433
2.5	63.5	199.5	0.5541
3.0	76.2	239.4	0.6650
3.5	88.9	279.3	0.7758
4.0	101.6	319.2	0.8866
4.5	114.3	359.1	0.9975
5.0	127.0	399.0	1.1083
5.5	139.7	438.9	1.2191
6.0	152.4	478.8	1.3300

INSTALLATION

TABLE 5-14. INFORMATION REQUIRED AND METHOD TO DERIVE TRUE SOUTH BEARING WITHOUT A COMPASS

	Satellite 1 (East of south)	Satellite 2 (West of south)
Name		
Azimuth		
Elevation		
Channel & Polarity		

Elevation to Satellite at True South: _____
Declination Angle: _____
Modified Polar Adjustment Angle: _____
Site Coordinates: Latitude _____
                  Longitude _____
Mount Base Pole Outside Diameter:
    Inches _____ Millimetres _____

METHOD

A_1 = Angular distance of satellite 1 from true south
  = 180° - Satellite 1 azimuth

A_2 = Angular distance of satellite 2 from true south
  = Satellite 2 azimuth - 180°

C_P = Pole outside circumference
  = 3.142 * outside diameter

D_{deg} = Angular distance per degree
  = Pole outside circumference/360

Distance of true south from satellite 1
  = $D_{deg} * A_1$

Distance of true south from satellite 2
  = $D_{deg} * A_2$

The parameters used in this example are:

	SATELLITE 1 ECS1-F4 (13°E)	SATELLITE 2 Intelsat VA-F11 (27.5°W)
Azimuth	155.254°	205.290°
Elevation	27.414°	27.285°

Site Location: Latitude - 52.233° N
               Longitude - 7.020° W

Mount base pole outside diameter:
    76.2 mm (3 inches)

These measurements can then be used to calculate the following:

Distance from satellite 1 to true south:
    180 - 155.25 = 24.75° from West

Distance from satellite 2 to true south:
    205.29 - 180 = 25.29° from East

Mount pole outside circumference:
    3.1415 * 76.2 = 239.4 mm

Distance per degree:
    239.4/360 = 0.67 mm per degree

Distance of true south from satellite 1:
    0.67 * 24.75 = 16.46 mm to West

Distance of true south from satellite 2:
    0.67 * 25.29 = 16.82 mm to East

Elevation of satellite at true south: 30.26°

Final Feedhorn Centring Adjustments

Final feedhorn centring adjustments should be made once a clear picture has been observed. If the feedhorn is precisely at the "phase centre" where all the signals are focused, the picture quality will be at its best. Moving this assembly away from its correct position will degrade picture quality by lowering gain and increasing the amount noise detected. If the reflector is warped, the optimal reception point may be centimetres away from the recommended focal point. It is critical that this problem be corrected, preferably early in the installation.

Moving the feed assembly closer to the antenna decreases the signal power relative to the amount of noise detected via the side lobes. This shows up as a decrease in the carrier-to-noise ratio and ultimately in the signal-to-noise ratio. Moving the feed away from the antenna past the prime focus will also decrease the signal-to-noise ratio as the signal strength decreases and as more noise from the surrounding ground is detected. If both the TV picture and signal strength meter are carefully monitored at the same time, knowing these effects, the feedhorn can be located precisely at the phase centre. The payoff in improved picture quality is more than worth the trouble.

TABLE 5-15. A RAPID POLAR MOUNT INSTALLATION TECHNIQUE

1. As part of the pre-installation check list, before traveling to the site the quantities listed in Table 5-14 should be collected.

2. Calculate true south using the simplified true south derivation procedure as outlined in this section. Set the mount facing true south.

3. Set the dish on the mount so that it is parallel with the east-west line. Then adjust the polar axis angle equal to the site latitude angle. Add the modified polar mount adjustment angle, if further accuracy is desired. Finally, set the declination angle of the mount.

   Note that the elevation of a satellite whose longitude equals the site longitude is the highest point on the geostationary arc and due south.

4. Look up or calculate the declination angle for the site location. Some mount manufacturers provide a chart of declination angles versus declination bar length. This data makes an installation simpler because it is often easier to measure a linear distance rather than an angle. Set the declination angle.

5. The resultant elevation of the dish should equal the calculated value of elevation angle for the satellite due south. If the elevation angle of the dish is not equal to this value, recheck the angles. In situations where the declination angle is given on a declination angle versus declination bar length chart, the problem is likely to be the polar axis angle. In situations where the declination angle has to be measured, the declination angle should be first to be checked for accuracy.

   Note polar mounts having fixed declination angles should be avoided at all costs. No fine tuning is possible.

6. If these steps have been correctly implemented, the mount should be properly adjusted. A quick scan across the geostationary arc should verify this tracking.

   Note that with smaller diameter base poles, the mount may be off true south by a degree or so. Given that the polar axis and declination angles were correctly adjusted the southern orientation of the mount probably still must be fine tuned.

   The procedure is designed to reduce the initial errors to a controllable level. However, the steps outlined above in the basic antenna alignment procedure and the fine tuning procedure with test instruments can be combined to develop a very skilled and effective alignment strategy.

If the feedhorn is improperly aligned, there can be substantial degradation in system performance. For example, if 2 dB of signal is lost, this is equivalent to reducing the performance of a 3 m antenna to that of a 2.5 m model. In addition to centring the feedhorn and correctly setting the focal length, the body of the feed should not be twisted relative to the face of the reflector. If this does occur, additional gain can be lost from shadowing and misalignment.

Some manufacturers of earlier C-band dishes set the focal length so that no adjustment was possible. While this design detail simplified and speeded installation, the ability to fine tune the feed system was lost. In general, having this fine tuning capability is an important feature needed for Ku-band antennas.

Alignment with Test Instruments

While a satellite reception system can be aligned by simply watching the picture on a portable television set, aiming a Ku-band dish can be improved by using a more sensitive test instrument. There are at least two reasons to use a portable carrier level meter. First, many smaller antennas are installed above ground from a position on a ladder. Therefore lugging a satellite receiver and portable TV monitor up to the dish may prove impracticable. Second, none but the most trained technician would notice a change in picture quality on a monitor as the carrier-to-noise ratio level increases from, for example, 9 to 12 dB. But this extra margin of 3 dB may be crucial during a rain storm when the C/N may be substantially reduced.

Carrier level meters fall into a number of rather broad classes: simple wide band detectors, narrow band carrier monitors, spectrum analysers and various combinations of these. Use of such tools allows for more efficient use of the available equipment so that even finer adjustments can be made to the feed position, north/south orientation, polar axis angle and declination offset angle. If the antenna mount has been designed to allow sufficiently fine adjustments in the two angles, the system can be aligned to incorporate the modified polar mount geometry settings and tracking can be extremely accurate.

Signal power is generally measured at the LNB output. It is unrealistic to expect to measure power levels at the LNB input since instruments to do so are very expensive.

INSTALLATION

RF Heads – Simple Wideband Detectors

The first device available to measure power levels from C-band downconverters was a carrier level detector known as an "RF head" that plugged into an inexpensive voltmeter so it could read 70 MHz signals. (The model "82RF" from John Fluke Manufacturing was the pacesetter). Neither a voltmeter, which is designed to detect much lower frequency signals, nor a field strength meter, which monitors AM signals on a cable TV system, was capable of detecting 70 MHz signals from C-band downconverters. Pioneer RF heads were not able to measure signals having frequencies much in excess of 70 MHz.

RF heads that monitor the full range of LNB output frequencies up to 1700 MHz can be built from simple passive circuit components. These are not particularly sensitive to the small changes in signal level that occur when a dish is aligned close to its optimum position.

Narrow Band Carrier Meters

Narrow band carrier level meters that monitor just one video channel at a time and thus eliminate much of the broadband noise are also available. These are in essence the front end of a satellite receiver that can be tuned to a particular channel (see Figures 5-102, 5-103 and 5-104). These are generally more complex and costly than simple RF heads. In addition, they can be somewhat bulky and therefore less convenient for use on ladders. Like RF heads, they usually powered from either internal batteries or via the RF coax from a satellite receiver.

Spectrum Analysers

A spectrum analyser displays the relative levels of signals being detected plotted across the entire satellite frequency band. It is used in, for example, commercial installations where alignment procedure generally requires the use of excellent electronic test gear. This type of instrument while quite accurate tends to be rather bulky and relatively expensive compared to simpler carrier level detectors. Nevertheless, relatively high quality but reasonably priced analysers having fully adjustable frequency ranges and signal sensitivities are available today.

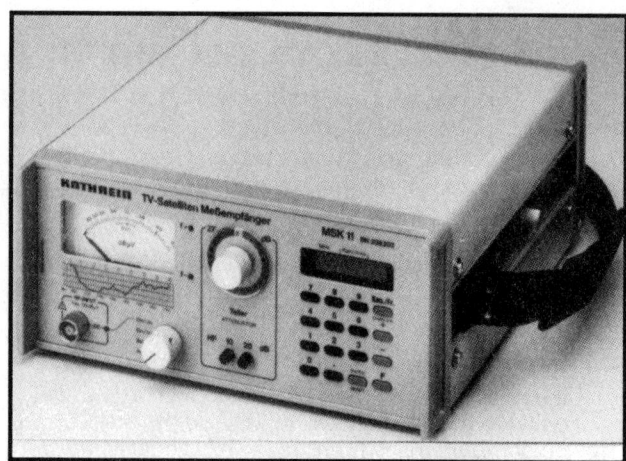

Figure 5-102. Kathrein MSK-11 Signal Level Meter. *This battery operated meter measures levels from 461 to 113 dBmV. It features a direct channel or frequency input, video output for image control, an audio tone for detection and control subcarriers between 5 and 8 MHz. (Courtesy of Katherin-Werke KG)*

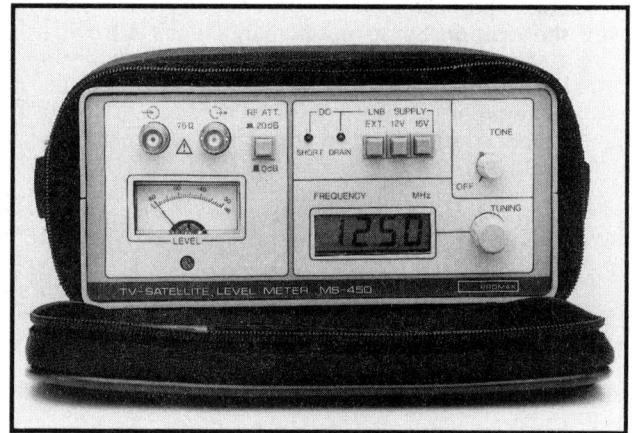

Figure 5-103. The Promax MS 450 Meter. *This portable installation aid intended for satellite TV installations has an audio output, digital frequency meter with 1 MHz resolution and a 70 dBm linear range on its two scale level indicator. It accepts an input frequency of 950 to 1750 MHz. (Courtesy of Instrumentacion Electronica Promax)*

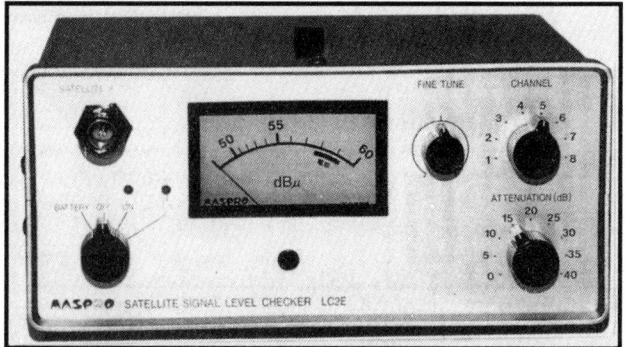

Figure 5-104. Maspro Signal Level Meter. *This battery powered meter accepts 975 to 1275 MHz inputs, has variable attenuation, indicates a level of 50 to 100 dBmv and has rotary channel selection to locate individual satellite transponder levels.*

ALIGNING THE ANTENNA

INSTALLATION

An analyser is a very useful tool that can be set to display the entire frequency band or can be used to zero in on the structure of a message transmitted on a single carrier. Even the smallest microwave signals can be observed. It can be used to determine if an LNB is functioning properly. When the LNB is connected, its noise floor should increase if all is well. It can even be used to troubleshoot hand-held remotes, which also relay RF signals.

A spectrum analyser such as can be connected via a 2-way power passing splitter to the LNB output. Its internal battery can provide power to the LNB. The second port of the splitter can be connected to a carrier level detector that offers both an audible tone and digital read-out of signal strength. Another option is to connect the second port directly to a satellite receiver. Then its signal strength meter can be used in conjunction with the analyser for an accurate determination of satellite position. In this latter case, the receiver, not the spectrum analyser, would power the LNB. More details on this instrument are presented in Chapter IX.

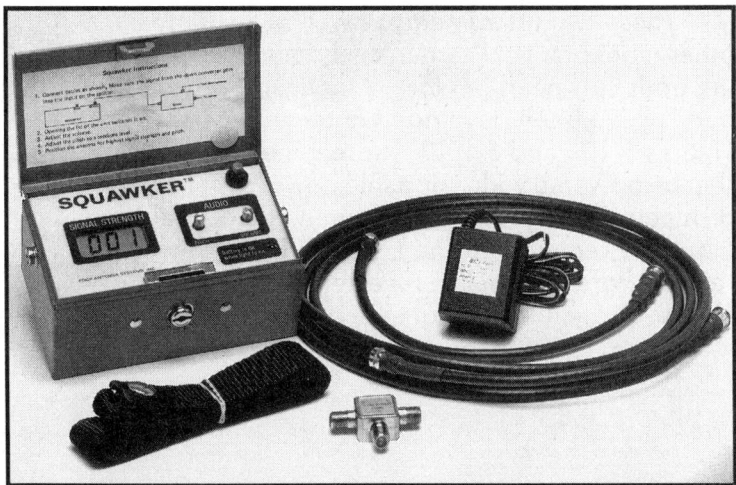

Figure 5-105. The Squawker. *This instrument has a battery powered digital signal strength metre as well as an audible tone which increases its frequency as signal power increases. This can be a very useful feature to aid in peaking an antenna. (Courtesy of Focii Antenna Systems)*

More Practical Meters

Two practical carrier level detectors designed to monitor signal power levels at the LNB output were originally developed for C-band use in North America. These were attractive alternatives to RF heads or signal strength meters built into receivers that are usually relatively crude instruments not really suited to squeezing out that last dB or two. The Pico "Peaker" one of the earlier C-band models, has a sensitivity adjustable from +30 to –80 dBm. Another, the "Squawker" has a variable sensitivity digital scale to handle from –7 to –48 dBmv but also has an audible tone output that increases in frequency as signal strengths rise (see Figure 5-105). The digital scale allows greater accuracy in reading powers than an analog scale. Both these instruments can also be used with signals up to 1500 MHz in frequency.

Such instruments must be protected from the power being relayed from the satellite receiver to the LNB. A directional coupler which passes dc between two ports but isolates the third test port should be used in those lines also carrying dc

Figure 5-106. The SAM Meter. *This simple alignment instrument is powered from the coax line, has a built-in power splitter and an audio tone for sensitive peaking. It also features a logarithmic meter to indicate carrier level and detects the presence of dc power from the satellite receiver. It accepts an input frequency range of 50 to 2000 MHz. (Courtesy of TranSat)*

power. While the Squawker is not protected, the Pico Peaker has a built-in coupler so LNB output can be connected via a short jumper cable into one port of the test device while the cable from the satellite receiver is connected into the other port. dc protection is provided internally.

263   ALIGNING THE ANTENNA

INSTALLATION

These two instruments as well as a variety of other more practical carrier level detectors spanning both European or North American frequency bands have been developed in recent years (see Figures 5-106 and 5-107). These generally can either be powered with a portable power supply so a dish could be aimed before running the RF cable from the receiver to the LNB or via the power relayed from the satellite receiver. They also often feature an audible tone for ease of use.

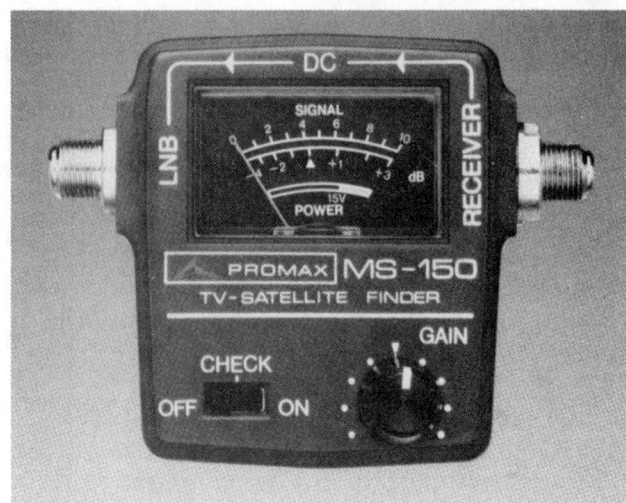

Figure 5-107. Promax MS-150 Signal Level Meter. *The MS-150 meter detects signals from 950 to 1750 MHz over a -40 to -10 dBm range. It features both an audio tone and an indicator. It inserts in the line between the satellite receiver and LNB and receives power from the coax. (Courtesy of Instrumentacion Electronica Promax)*

FINE TUNING THE SATELLITE RECEIVER

Once the antenna has been accurately aimed onto the arc of satellites, the satellite receiver can be moved indoors to its final installation location, if this has not already been done, and final fine tuning can be completed. First, make sure that each active satellite in the entire arc can be tracked from the actuator controls. Then the receiver IF gain should be adjusted to receive the optimal picture across all transponders (see Figure 5-108). In those installations where linear polarities are being received, some receivers have back panel or internal potentiometers to adjust the skew control so that the knob is near the centre of its range of movement. Also, when a mechanical polariser is being used, make sure the probe is not sitting at the end of its range of movement by observing its position or by feeling the motor for vibration when the skew knob is set at its end positions. Some brands of actuators have timers which shut the motor off in 5 seconds so this test requires two people, one at the receiver and one at the antenna. Both the IF gain and skew fine tuning are usually accomplished by rear or bottom panel adjustments.

Read the manufacturer's instructions carefully to ascertain that all that can be done has been done. These instructions are invaluable during the next step, programming the receiver and actuator.

Figure 5-108. Fine Tuning a Satellite Receiver. *This photo was taken while aligning the satellite receiver after installation was completed at Arthur C. Clarke's residence in Sri Lanka.*

At this point, if pictures are not as perfect as expected, check that all connections to the satellite receiver are secure. Another source of difficulty could be a customer's television set which will oc-

INSTALLATION

casionally require fine tuning. Its channel frequency should precisely match the satellite receiver's modulator output. This television channel is rarely used for over-the-air programming but may have been used if the customer was a cable TV subscriber or used a VCR. In addition, to guarantee that the customer's TV set is in good health, tune to a strong over-the-air broadcast or use a VCR or a portable signal simulator which generates test patterns on the modulated channel. If the television is in good shape, excellent pictures should be observed.

J. PROGRAMMING THE RECEIVER AND ACTUATOR

Each brand of programmable receiver and actuator is different and the manufacturer's instructions should be carefully followed in programming these units.. Be aware that some remote controls require that standard 9-volt batteries be added by the installer. Always have a supply on hand.

Making the System "User Friendly"

It is generally a good policy to ensure that a customer or user either fully understands any complex programming procedures for a system or perceives the system as quite user-friendly. This latter approach might mean fixing the dealer's telephone number to a satellite receiver and charging a fee for an extended warranty. Both are strategies designed to accomplish the same objective, that of making a system "user friendly."

Technicians must also be familiar and comfortable with any components that are installed as they may require service in the future. For those who install numerous systems using a common receiver type, it is wise practice to maintain a consistency in programming this particular component. For example, if possible, ensure the preset channels are identical in all receivers or that satellites are stored with similar abbreviations or mnemonics on all actuator controllers. If such continuity is maintained, a set of simple instructions for in-field work can be created.

A copy of the chart for each receiver along with a detailed setup procedure could then be issued to each installer. The installer's copy should have a diagram of the remote control. It also should be sufficiently detailed to enable the technician to talk the user through the procedure over the telephone. If it might be necessary to travel long distances in order to re-program a receiver, the cost of this should be built into the price of the system.

Many receivers have a hardware switch on the back panel that enables the programming and re-programming procedure. If this switch is set in the locked position, it prevents the receiver from being inadvertently re-programmed by the remote control. This switch should always be in the locked position after a receiver is programmed.

Two sample forms for both customers and technicians are illustrated in Tables 5-16a and b. This actuator set-up procedure would include those parameters used in programming a channel into memory. These generally would be receiver specific. The installer chart should also include the battery type used in the remote control for the particular receiver.

TABLE 5-16a. SAMPLE CUSTOMER CHART

Satellite Name:
Satellite Abbreviation:

	Channel Number	Service
Channel 01	....	....
Channel 02	....	....
| |	|	|
Channel nn	....	....

Satellite Name:
Satellite Abbreviation:

	Channel Number	Service
Channel 01	....	....
Channel 02	....	....
| |	|	|
Channel nn	....	....

PROGRAMMING THE RECEIVER

INSTALLATION

> **TABLE 5-16b. SAMPLE INSTALLER CHART**
>
> Satellite Receiver Type:
>
> Positioner Type:
>
> Satellite:
>
> Satellite Abbreviation:
>
>   Channel | Frequency | Service | Bandwidth | Audio | AFC
>
> Channel 01 ....
>
> Channel 02 ....
> | | | |
>
> Channel nn..
>
> Detailed Receiver Setup Procedure:
>
> Battery type in remote control:
>
> Detailed Positioner Setup Procedure:
>
> Parameters used in programming a channel into memory:

K. WATERPROOFING

Water in the form of vapor, rain, ice or snow is the chief enemy of a satellite reception system. It can corrode connectors, accumulate in feedhorns or weigh down antennas. Once the system has been tested and is working perfectly, the waterproofing team gets on the job. This task should be saved until the end when fine tuning is completed. If a component must be replaced, water seals would not have to be broken and then redone.

All connectors and junctions where water could enter should be sealed. Some types of caulks are made with ammonia or lactic acid which can corrode metals; others harden, become brittle and crack. A sealant which stays soft like Coaxial-Seal™ (Universal Electronics, Inc.), a hand-molded, non-corrosive rubbery material, is recommended.

The equipment at the antenna focus should be protected by an LNB cover, a small expense and excellent insurance. In addition, drip loops should be used whenever applicable.

When a ground mount is used, in climates with snowy winters, the antenna should be installed high enough so that it will not be pushing snow as it tracks the arc. In addition, if at all possible, leave adequate space under the bottom rim so a lawn mower will not run into and damage the reflector. Clear away extra space around the dish if necessary.

Be sure that water cannot accumulate anywhere on the mechanical system. There have been cases where water has collected at the bottom of mounts which were open at the top but closed at the bottom: the water froze, split the tubing and weakened the mount.

In the case of ground mounts, the cable trench can be covered up and the sod replaced after waterproofing is completed. At this point, the installation is finished.

L. CONNECTING STEREO PROCESSORS AND OTHER ACCESSORIES

Every installation has slightly different requirements for in-house equipment. Extra TVs, stereo processors, modulators, or even large screen or projection TVs may be needed. But installing peripheral equipment is usually a fairly straightforward task if the design has been carefully planned.

Extra Television Sets

One satellite receiver can power two or three nearby television sets without amplification or any number of televisions with insertion of the necessary line amplifiers. The simplest way to accomplish this is by using splitters. These divide the video signal into 2, 3, 4 or even more equal signals. The placement and type of splitters used depends upon the location of each television. For example, assume that there is one set adjacent to the satellite receiver and three at the far end of the house. A 2-way splitter would be used at the receiver output to drive the nearby television and a trunk line having a 10 to 20 dB line amplifier would be connected to a location near the other sets. There, a 3-way splitter would divide the signal. This would certainly be a much better strategy than using a 4-way splitter at the receiver and running three long cables.

The type of cable selected depends upon the distance traveled and number of televisions connected. For short distances, RG-59 is adequate. RG-6 can be used for lines up to 100 m. RG-11 can be used for longer trunk lines. However, for example, if 30 televisions all in close proximity to the receiver were driven by this receiver, it would still be necessary to use lower loss RG-6 or RG-11 since the original signal would be split or tapped so many times. The rule is common sense and minimising cable runs.

Line amplifiers which are easily inserted via F-connectors into a trunk line system may be required for extensive distribution systems. Note that larger distribution systems require a highly quality "front-end", i.e. antenna/feed/LNB and receiver, so the S/N ratio is sufficiently high to properly drive the system.

If a multiple receiver system is being installed, splitters or taps rated for use at higher frequencies are required if the coax from the LNB is divided before entering each receiver. This could be accomplished in a variety of ways. More details about such configurations are presented in Chapter VII.

Stereo Processors

Most modern satellite receivers have built-in stereo processors. All that is necessary to hook them into a home stereo system are two leads connected by RCA jacks, one for the left and one for the right sound channel.

Stand-alone stereo processors are driven by the composite baseband output from a satellite receiver. This output is most often via an RCA jack in North America and via a SCART connector in Europe. The receiver instruction manual should clearly specify which output is dedicated to a stereo processor.

Modulators

Occasionally it will be necessary to modulate the receiver composite audio and video signal onto channels other than the standard choice. Using an American system as an example, three receivers may be used to send three different channels along one trunk line to many televisions. In order not to interfere with over-the-air channels, it may be necessary to modulate onto UHF channels 20, 22 and 24 or onto VHF channels 5, 11 and 13. In general, it is best to modulate onto VHF channels since cabling losses are much greater in the higher frequency UHF range.

Modulators ranging in quality from those built into satellite receivers to commercial brands providing 30 to 60 dB of amplification or more and excellent channel isolation are available. They generally require both an audio and video input and have one RF output. Such modulators also usually have both audio and video level adjustments used to fine tune a well-designed distribution system.

INSTALLATION

M. TOOLS REQUIRED FOR SATELLITE TV INSTALLATIONS

The old boy scout adage "be prepared" applies to a satellite TV installer. Having to make an extra round trip of 60 km for a F-connector coupling or a 75 to 300 ohm transformer is certainly a waste of valuable time and money.

Essential installation tools include a compass and inclinometer for a site survey, a crimp tool for fitting F-connectors, and other standard items including screw drivers, a drill, spanners, a soldering iron and power extension cords. These should be supplemented by an assortment of connectors. A complete list of tools is presented in Table 5-17.

Tools of the trade can be carried in a small pickup or a thoroughly equipped lorry. The type of equipment needed depends upon what type of antennas are being installed, the climate and how a dealer wants to organize his installations. For example when installing a larger ground-mounted dish, one crew can be used to plant the pole, assemble the antenna and run the cable and then another can do all the electrical and alignment work. Or a single crew can complete the job in one visit.

Heavy Machinery

Heavy machinery includes vehicles needed to transport poles and antennas and all the equipment required for excavation and concrete work. Although some 3 m mesh antennas can be packaged to fit into a small car, spun or stamped reflectors come in one piece and require at least a small pickup truck for transportation. At very least, a small pickup truck or a single or tandem axle trailer which can be pulled behind a car should be owned by any serious dealer. Of course, the situation is entirely different when installing smaller 45 to 90 cm antennas.

Excavating tools include a wheelbarrow or a portable cement mixer, a pick ax, breaker bar, hoe, rake, hole digger and shovel. Others may be required in different situations.

Major Tools

This category includes all those tools required for every installation. This includes an inclinometer and compass, extension cord, wire strippers and carpenter's level.

Occasionally Needed Tools

Tools on this list are needed occasionally in special situations. For example, if no power is available at the site, a portable generator may be required. For long pole or roof mounts, an extension ladder may be needed.

Connectors, Cables and Other Accessories

This category includes all those small items that are most often required for an installation. These should always be on hand for both installing and troubleshooting. The lack of a small but essential item could be a source of great frustration.

The installer's equipment checklist in Table 5-17 is provided for convenience. Modifications to this list can be made to suit a dealer's tastes. However, using a checklist before leaving the shop to install or troubleshoot a system is a highly recommended practice.

TABLE 5-17. INSTALLER'S EQUIPMENT CHECKLIST

Heavy Machinery for Ground Mounts

___Wheel Barrow or Cement Mixer
___Post Hole Digger or Power Auger
___Shovel
___Pick Ax
___Rake and Hoe
___Bucket and Hose
___Breaker Bar
___Trencher

SATELLITE TV TOOLS

INSTALLATION

Principle Tools

___Inclinometer & Compass
___Carrier Detector Meter
___30 m Extension Cord with 3 or 4-Way Adapter
___Extra Television Set or Monitor
___Multimeter (volts, ohms, etc.)
___Hex Crimp Tool for RG-6 and RG-59 Connectors
___Wire Strippers /Exacto Knife
___Tape Measure (at least 30 m); or Ultrasonic "Tape Measure"
___Soldering Gun (20 and 40 watt) and Solder; or Gas Powered Iron
___Carpenter's Level
___Flashlight and/or Fluorescent Light
___12 mm (1/2 inch) Heavy Duty Drill, Hammer Operation Switchable
___10 mm (3/8 inch) Variable Speed Drill and Bits
___1 kg (2 pound) Hammer
___Hack Saw and Blades
___Large Diagonal Pliers
___Large and Small Long Nose Pliers
___Vise-Grips
___Assorted Screw Drivers
___10 to 30 mm (3/8 to 1-1/8 inch) Set of Opened Ended Spanners
___Set of Socket Spanners
___Small Set of Metric or SAE Sockets
___Large and Small Crescent Wrenches (30 to 40 cm, 12 to 15 inches)
___Awl or Center Punch (for Aligning Holes in Antenna Petals)
___Set of Allen Wrenches
___Extension Ladder
___String and Tape
___4 to 5 cm (1/5 to 2 inch) Electrical Conduit, Solvent and Sweeps
___Metal File and Assorted Sandpaper
___16 mm by 45 cm (5/8 by 18 inch) Wood Drill Bit
___16 mm by 30 to 45 cm (5/8 by 12 to 18 inch) Masonry Bit

Often-Needed Tools

___Generator
___Fish Tape (for feeding cable inside walls)

Connectors, Cables and Other Accessories

___F-Connectors and F-to-F Bullets (F-81 Barrels)
___Right Angle and Straight F and N-Connector Adapters
___RCA Jacks, SCART Connectors, Moly Plugs, Lugs, Scotch Locks and Assorted Adapters
___Video Connectors Kit with Cables
___Rolls of Direct Burial, RG-6 and RG-59 Cable
___75 to 300 Ohm Transformers
___A/B Switches
___2-, 3-, and 4-Way Splitters
___75 Ohm Terminators
___Spare Fuses and Voltage Regulators
___Electrical Tape
___Box of Assorted Wire Nuts
___Box of Romex Staples
___Can of Touch Up Spray Paint
___Coaxial Sealant
___Non-Conducting and Non-Corrosive Sealant
___A Spare Polarotor Servo Motor
___Surge Protector with Multiple Outlets
___10 and 20 dB Bullet Line Amplifiers
___Ground Rods and Heavy Gauge Grounding Wire
___T-25 Arrow Roundtop Staple Gun

N. COLD WEATHER INSTALLATIONS

Installing a satellite system in very cold weather can be uncomfortable. However, there are many factors to consider other than discomfort (see Figures 5-109 to 5-114). Things that are taken for granted do not necessarily happen as expected. For example, electrical tape which normally sticks perfectly well becomes brittle and useless. It is necessary to buy a good brand and keep it in a warm place or a pocket until it is needed. Low power soldering guns cannot produce enough heat to melt solder. A very large gun is needed at −30 °C. In these cases, crimp-on connectors, not solder, may be chosen.

When it is sub-zero, fiberglass becomes brittle and can shatter. Many fiberglass antennas have been damaged in these conditions by over-tightening the bolts that hold the reflector to the mount. The answer to this problem is either to be very careful or, if possible, to assemble antennas indoors in cold weather. If this option is not possible, let the unassembled pieces sit in the warm sun under a plastic covering.

Concrete will not set properly in very cold weather unless the right precautions are taken. Two percent calcium chloride must be added to prevent freezing and the surface must be covered with a blanket, straw or another insulating material until it sets. Other binding materials dependent upon chemical reactions, like Dish Set™, set more slowly when cold. A simple practice to aid the hardening of either this material or a fast setting cement is to place the pole indoors or under the exhaust of a running vehicle to warm it up just prior to planting it.

Some brands of silicone sealants will not harden. Make sure that a silicone designed specifically for cold weather is used. Other products behave in surprising ways when cold. For example, one of the authors purchased two thousand 30 cm wire ties one summer expecting that they would

Figure 5-109. Assembling a Fiberglass Antenna. *The shipping container was taken apart and laid down on the snow to keep snow from freezing between the panels as it was tightened. It was important to guarantee that the panels were lined up without any gaps.*

Figure 5-110. Concrete Pad and Cold Weather. *This concrete pad with all the necessary internal rebar and screen wire was poured in a heated warehouse. The pad was then hauled out to the installation site with a trailer. The pad was leveled and the arc was tracked. In the spring after the snow melted, the pad was re-leveled, the antenna was re-tracked and cable was buried.*

INSTALLATION

Figure 5-111. Preparing the Ground in Cold Weather.
This photo shows the hard work necessary to prepare for a ground pole installation when heavy snow is on the ground. A snow cover does have an advantage in acting as an insulator and keeping the ground from freezing. In contrast, those areas of the country that have extreme cold with no snow can have frost penetrations of over one metre. In such cases, a chemical called "thaw power" which generates heat when poured onto the ground can make digging possible. Another alternative is to use a barrel filled with burning wood or charcoals which will thaw the ground in a matter of hours.

Figure 5-112. Use of a Power Auger in Cold Satellite.
This two-man power auger is a very useful tool in allowing rapid excavation of holes in cold climates.

have a long service life. When winter came, the ties were so brittle that they snapped when used. When warmed up they were fine, but at below freezing temperatures they were useless. Another brand manufactured specifically for cold weather had to be ordered. Similarly, most non-colored nylon wire ties will fail after only one year's exposure to sunlight. Black, blue, brown or specially UV- stablilised brands will, however, last many years.

Tools can become very brittle in the cold, especially at temperatures below −25 °C. For example, ends of screw drivers snap off and crescent wrenches break. A solution is to use oversized wrenches which have greater leverage and need less torque, such as 50 cm instead of 30 cm crescents.

Plastic components are especially susceptible to damage in such extreme conditions. If television sets are treated roughly in the cold, their plastic containers easily break. Knobs also snap off receivers and TVs. In addition, some brands of satellite receivers do not work properly below a certain temperature. A recommended method is to keep the receiver and television in a warm car or truck during testing. Many types of cable will not unbend when cold and should be kept warm until needed. Some cables are made especially for such use. Centre conductors of some coaxial cables will snap in two when cold. Therefore, care should be taken when uncoiling coax in cold weather.

Cold metal can be dangerous. For example, one dealer was working on a system when the chill

Figure 5-113. Penetrating the Frost Line. *This photo shows the use of a digging bar to penetrate the frost line so that a power auger can be used.*

271   **COLD WEATHER INSTALLATION**

INSTALLATION

factor was −50 °C. He went inside to check a picture and ran back to raise the elevation adjustment without his gloves. A crescent wrench which had been outside froze to his hand and could only be removed by running water over it. It is just as easy to pull skin off a hand by tightly grasping a very cold mount. Always wear a good pair of gloves.

Many problems can be traced to freezing of water which enters components either by condensation or leakage. Water left inside a waveguide can easily freeze and crack the housing. Even small amounts of condensation can cause problems in polariser motors in extremely cold climates. Ferrite devices having no moving parts are recommended in such situations instead of mechanically rotated probes. There have been cases of water collecting in mounts or actuator motors and cracking the surrounding metal. Using drain holes, effective sealants, LNB/feedhorn covers and other protective measures is especially recommended in such climates.

A properly equipped dealer should be able to install in any weather, be it rain, snow or blizzard. For example, in very cold weather prefab concrete bases poured in a shop can be trucked to an installation site. But there are times when it is just too cold or wet or snowy to warrant the effort. Use this time to assemble antennas and mounts in the shop or to pour slabs for future use.

Figure 5-114. Use of Leveling Stakes. *This photo was taken shortly after the concrete was poured. The mount was installed on the pole only after curing is complete.*

Figure 5-116. Installation in Deep Snow. *Even though winter work can be difficult, antennas can be installed even in deep snows.*

Figure 5-115. Assembling an Antenna on a Mount. *With deep snow present the antenna had to be assembled on the mount causing somewhat more than normal installation difficulty.*

Figure 5-117. Weather Sealing. *After the installation is complete, all weather sealing can be accomplished using weatherproof boxes, drip loops and coax seal.*

COLD WEATHER INSTALLATION

O. CUSTOMER RELATIONS

Customer relations are common sense behaviour, most of the time. However, sometimes common sense must be learned, often at a high price. While customer care is simply a matter of treating people with consideration and respect, it is useful to invest some time and energy in learning how to turn a prospective customer into an advocate for your business. Referrals are the most effective and least expensive form of advertising.

First impression are important in creating an environment in which customers develop confidence and trust. It is quite important to dress neatly, to be punctual and conscious of the attitude, both spoken and unspoken, with which a customer is treated. One who is truly pleased to help resolve customers' problems, who is prepared to meet customers' need and be innovative, who is proud of their contributions and who is committed to their customers' welfare will certainly be appreciated.

Each contact with a customer can be useful in discovering information about their needs, in receiving valuable feedback and in learning to anticipate and correcting problems before they occur.

Some customers have an amazing amount of technical know-how and can teach most any technician interesting tricks and methods. Often a dealer who knows how to spot such people will make a sale where others have failed. In addition, such people can often have innovative suggestions about your installation methods or product line.

Some customers may want to participate in an installation. He or she may help assemble the antenna, plant the pole or run cable lines and might actually wish to complete the whole job except for the final wiring and fine tuning. Ninety percent of the time they can do an excellent job with the proper guidance and become a source of many referrals. Other customers are pleased to do the entire installation knowing that a competent dealer is ready waiting in the wings when needed. Such installations are often an example of fine workmanship.

While every installation should be be treated as if it were occurring in the dealer's own home, it is important to consult the customer whenever possible or necessary. Permission should be asked for and granted whenever furniture is moved, holes are drilled through walls, or facilities will be used. A good installation should be a source of pride and referrals and should last for years.

Before leaving a site, a professional will take one last look around the premises in order to see if anything else could have been done more perfectly.

Contracts and Warranties

Never make a sale or do an installation without a signed contract. Customers should be informed of all warranties and special conditions of this legal document. Each section of the contract should be explained so that there are no hidden expectations or unwanted surprises. Contracts protect both the dealer and the customer. Whenever a situation deteriorates to the point that a lawyer must be called in, both parties usually have already lost.

P. DOCUMENTATION

It is important that records be kept of each installation in case any problems arise in the future. Documentation for both the customer's and installer's records should include the following:

- Date of Installation
- Original Site Map
- Antenna Type
- Mount Type
- Pad or Pole Support
- LNB Temperature and Make
- Receiver Make
- Actuator Make
- Type and Length of Cable
- Number of TV Sets
- Serial Numbers of All Components
- Any Special Installation Details

During the installation, a table outlining the measurements taken during fine tuning should be created (see Table 5-18). This will prove very useful if performance comparisons are required at a later time. Remember, however, that as satellites age their transponder power will decrease and these numbers will slowly fall. Signal strengths can be measured using carrier level meters.

A final, post-installation worksheet as illustrated by the example in Table 5-19, is a useful final installation check and can serve as reference if problems do arise in the future.

TABLE 5-18. SIGNAL STRENGTH COMPARISON TABLE

EUROPE

Date: April 5, 1990
Location: Reading, England

Satellite	Frequency (GHz)	Signal Strength
Astra	11.235	9.7
Intelsat VA-F11	11.675	4.6
Eutelsat 1-F5	11.170	5.7
Intelsat V-F1	11.175	4.9

NORTH AMERICA

Date: March 10, 1990
Location: R.F. Jones, St. Laurent, Quebec.

Satellite	Transponder Number	Signal Strength
Satcom K2	14	5.0
Anik C3	12	5.5
GSTAR 1	19	4.7

INSTALLATION

TABLE 5-19. POST INSTALLATION WORKSHEET

Customer Name_____

Installation Date_____

                                                                                       Initials

1. Approximate distance between satellite receiver and dish = _____metres.   _____

2. Wires buried or in conduit.   _____

3. Area around ground mount cleaned up.   _____

4. All wires waterproofed and all holes into building sealed.   _____

5. Serial numbers and type of back-up equipment, if used?   _____

6. Customer signed all paperwork and funds collected, if necessary   _____

7. Splices in cables needed? _____ number of splices   _____

8. Signs of terrestrial interference, blockage by buildings or foliage? ___yes ___no   _____

9. If slab used, is ground subject to settling or an area of high moisture ___yes ___no   _____

10. If conduit used, were weatherheads installed? ___yes ___no   _____

11. Is cable protected at both the dish and entry to building ___yes ___no   _____

12. Picture quality on scale of 1 to 10 ____   _____

13. Protective cap on polarotor ___yes ___no   _____

14. Is map of installation including cable run completed? ___yes ___no   _____

15. How many wall plates installed? ____   _____

16. Comments:

TO BE COMPLETED BY INVENTORY MANAGER

Time and date job completed _____

All back-up equipment returned_____

INSTALLATION

VI. RETROFITTING C-BAND SYSTEMS FOR KU-BAND RECEPTION

Many C-band home satellite systems have been installed, especially in North America. As Ku-band broadcasts are gaining in popularity, upgrading an existing C-band system to receive Ku-band transmissions is often a favoured alternative to installing an entirely separate Ku-band system. When feasible, retrofitting for Ku-band reception is the lowest cost option to obtain dual-band reception.

A. PERFORMANCE REQUIREMENTS FOR KU-BAND COMPONENTS

Antennas, feeds and other components designed for Ku-band systems are subject to more rigourous performance criteria than are their C-band counterparts, in large part because the wavelength of Ku-band signals is approximately one third of that for C-band microwaves. Therefore, when considering a C-band retrofit, it is important to examine each component to determine if compatibility problems might arise. For example, in some cases it may be necessary to purchase a second antenna or receiver, or to alter the structure of a mount or actuator drive.

Antennas

Most C-band systems were installed long before the possibility of retrofitting for Ku-band was considered. Unfortunately, many of these antennas do not have sufficiently high efficiencies for effective operation at Ku-band, either because they were not initially designed to have the necessary surface accuracy or because they have suffered the rigours of time and have not aged well.

One piece spun or hydroformed reflectors can perform well in both frequency bands. However, as the number of panels increases, deviations from the designed parabolic shape are more likely to occur, even if the surface of each panel is perfectly smooth. Antenna efficiency is three times as sensitive to surface imperfections or roughness at Ku-band than at C-band. Figure 6-1 shows, for example, that signal loss is 2 dB if ripples of 25 mm occur with a regularity of 25 mm across the surface of a Ku-band dish. While the average deviations from a perfect parabola must be maintained at less than 4 mm (0.15 inches) at C-band for good performance, variations must be limited to one third of this value or 1.3 mm at Ku-band. In particular, expanded wire mesh reflectors often depart quite dramatically from the ideal shape. Expanded and perforated metal antennas can have additional losses because the surface holes are often well in excess of one tenth of the Ku-band wavelength (see Figure 6-2). As a result, reflection efficiencies decrease and higher amounts of ground noise originating behind the antenna leak into the feed. This is seen as an increase in side lobes. Some types of fi-

RETROFITTING

berglass reflectors may be perfectly smooth to the touch but may also have embedded mesh reflectors which, at worst, can delaminate from their rigid backing and cause poor reflector accuracy.

Retrofits of systems with larger antennas are more likely to be successful. For example, a 3 m dish has nine times the gain of a comparable 1 m model because gain increases with the reflector surface area. However, if the 3 m antenna has an efficiency as low as 10%, it will perform only slightly better than a 1 m model with a 70% efficiency.

Antennas often have more accurate surfaces at their centres than in regions closer to their outer rim. While this inaccuracy may have negligible effects at C-band, Ku-band performance could suffer. In order to optimise dual-band reception, the feed should be adjusted to give maximum signal-to-noise at Ku-band, not at C-band.

Mounts and Actuators

Since antenna beamwidths are more narrow at Ku-band than at lower frequencies, the reflector surface as well as the feed must be rigidly secured. This is a critical detail. Even slight wobbling of the support in the wind or play in an actuator arm may throw an antenna far enough off target to ruin reception. In addition, a satellite system that seems quite solidly mounted may still not perform up to par if the antenna weighs enough to cause sagging when it tracks between satellites. For example, a heavy fiberglass antenna which may be accurately aimed at the centre of the arc could twist the support enough when moved to either end to cause inaccuracies. This could develop into even more of a problem under wind or snow loading. In a few instances, a pole might even have to be replaced with a three-legged tower for the required additional support.

Some brands of actuators are not geared finely enough to allow the reflector to be accurately and securely aimed at a satellite. Those having too few sensor counts per cm of linear motion would probably be unable to aim the antenna at a Ku-band satel-

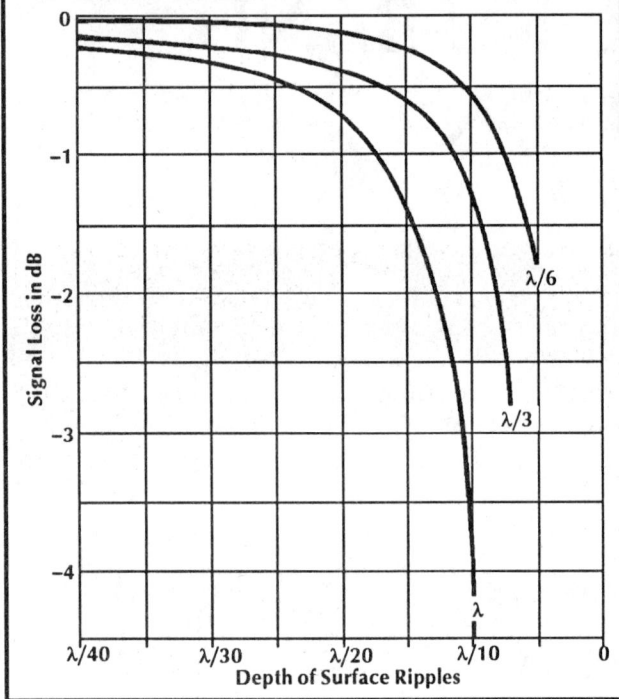

Figure 6-1. Signal Losses as Function of Wavelength and Surface Deformities. *This graph shows how antenna gain decreases as surface "ripples" increase. The bottom axis measures the depth of these ripples in units of the microwave wavelength. The three curves show the regularity with which the ripples occur. For example, one ripple in the reflector surface when receiving 2.5 cm Ku-band microwaves occurs every 2.5 cm on the bottom curve and every 2.5/3 cm on the middle curve. If the depth of these irregularities is one twentieth of the wavelength, nearly 0.75 dB of losses occur. Losses are greater at Ku-band because the wavelength is one third of that at C-band and the same relative amount of surface inaccuracy occurs in one third the distance.*

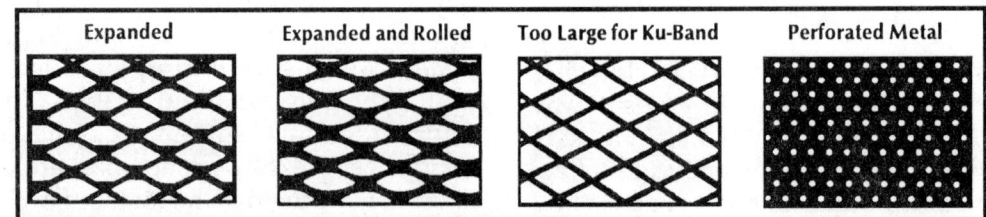

Figure 6-2. Antenna Mesh Types. *These four examples of mesh are commonly encountered in satellite antennas. The variety with the best performance at Ku-band is perforated and has the smallest diameter holes, preferably less than one tenth the wavelength of Ku-band signals, i.e. less than 25 mm (1/10 inch).*

RETROFITTING

Figure 6-3. A Horizon-to-Horizon Mount. *An attractive option for use in a dedicated Ku-band TVRO is a horizon-to-horizon gear driven mount supporting a perforated hydroformed antenna.*

Offset Feeds

In those rare cases when two separate feeds are used, it is best to offset the C-band, not the Ku-band assembly. However, this may not always be essential. For example, if a large, accurate reflector is providing more than enough gain, offsetting the Ku-band feed may be feasible. However, in general, the safest route to follow is to centre the Ku-band feed. The theory and experimental evidence backing this choice has been understood for years because the designs of multi-focus C-band antennas have been based upon multiple offset-feed configurations.

Figure 6-4. Dual Band Offset Feedhorn. *The shifter plate on this earlier Chaparral dual band feedhorn is located so that the Ku-band feed could be mounted at the antenna focus. (Courtesy of Chaparral Communications)*

lite within the necessary fractions of a degree. In this case, either a second antenna will be required or the actuator arm or both the arm and controller may have to be replaced. In general, many of the recently introduced linear actuators and nearly all horizon-to-horizon actuators are sufficient for this job (see Figure 6-3). In both types of actuators, there should be no excessive gear play or backlash.

Ku-band system mounts must have a continuous declination adjustment. Some C-band systems were designed and manufactured with a discrete set of declination settings. The simplest had factory preset positions for a range of geographic markets. Except for a stroke of good fortune, these will simply not provide the accuracy necessary for properly tracking the arc of satellites.

Feeds and LNBs

There are two options available in retrofitting a C-band system for reception of Ku-band broadcasts, since feeds and LNBs for both bands are not interchangeable. A second feed can either be offset from the first (see Figure 6-4 and 6-5) or a feed which simultaneously detects both bands, a "prime-prime" or coaxial feed, can be installed. Since both Ku-band and C-band feeds perform optimally when located at the focal point or phase centre of a microwave antenna, the second option is recommended.

Figure 6-5. Specially Made Dual Band Feedhorn. *This 4 GHz feedhorn was machined so that a Chaparral Pioneer 12 GHz feedhorn could be installed with only a 7.5 cm (3 inch) separation between the two feeds.*

RETROFITTING

There is one central reason why offsetting the Ku-band feed would be much more detrimental to system performance than offsetting the C-band feed. The loss of gain is directly dependent upon the distance that the feed is removed from the antenna phase centre. C and Ku-band feeds can realistically be positioned a minimum of about 7.5 cm apart, equal to about one C-band wavelength or three Ku-band wavelengths. Therefore, the Ku-band feed would be effectively three times further removed from the focal point than would be the case if the C-band had been offset.

Loss of antenna gain is plotted versus the number of half power or 3 dB beamwidths that the feed is positioned away from the focal point in Figure 6-6. The gain loss is substantially greater at higher frequencies because the wavelength and beamwidth of an antenna at Ku- band are one third those at C-band. For example, assume that a 3 m antenna having an f/D of 3.15 is used. The calculated C and Ku-band half power beamwidths are 1.8° and 0.6°, and if either feed would be offset by 7.5 cm, equal to an angle of 4.1°, this translates to about 2.3° and 6.8° half power beamwidths, respectively. The graph therefore shows that offsetting a Ku-band feed by 7.5 cm (about 3 inches) would result in a gain loss of a full 10 dB, while doing the same with the C-band feed would cause a much smaller but still noticeable loss of 1.5 dB.

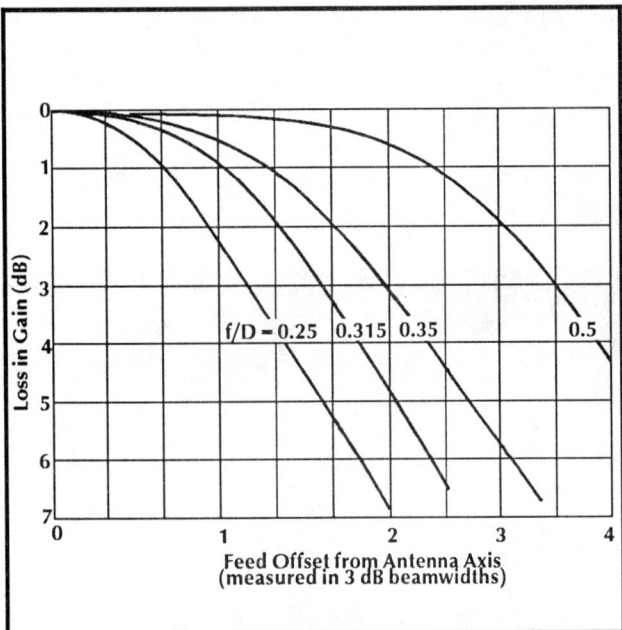

Figure 6-6. Gain Loss Due to Offset Focusing. *Gain losses that occur when offsetting either C or Ku-band feeds increase with the distance at which the feed is installed from the focal point and decrease with the system f/D ratio. In this case, the distance is measured in units of 3 dB beamwidths. Losses increase faster at Ku-band because half power beamwidths are one third compared to those at C-band.*

The second feed must be offset along the arc of satellites to ensure optimal performance. To accomplish this, the line connecting the two feeds must be set perpendicular to the polar axis of the antenna during installation. In addition, the feed should be offset on the side of the focal point opposite the location of the satellite which causes the actuator to be most extended. For example, on the

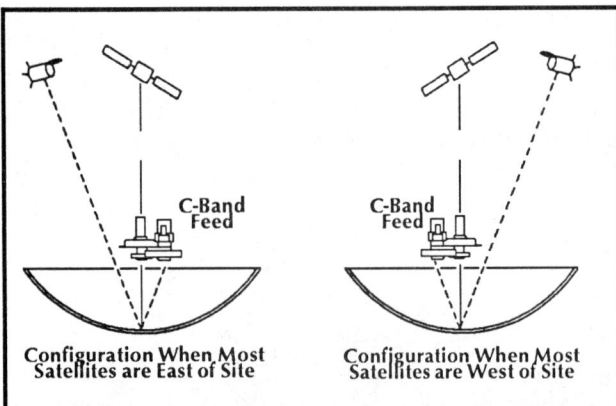

Figure 6-7. East Coast versus West Coast C-band Offset Position. *These drawings illustrate how the C-band feedhorn should be oriented when installing dual-band configurations throughout North America. A similar concept can be used in any region of the world wherever two feeds are installed and satellites are located over the entire arc.*

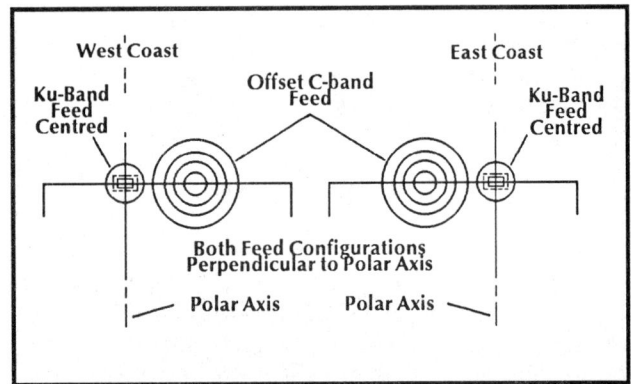

Figure 6-8. Centring a Ku-band Feed. *This drawing shows the arrangement for offsetting a C-band feedhorn in North America. The principle behind this placement is to minimise the overall antenna tracking distance. In both cases, the Ku-band must be at the prime focus and the line between the C and Ku-band feeds must be perpendicular with that of the mount's polar axis.*

KU-BAND PERFORMANCE

RETROFITTING

west coast of North America, most broadcast satellites are located in the eastern sky. So the C-band feed would be offset to the right of the Ku-band feed at the focal point as viewed from behind the antenna (see Figure 6-7 and 6-8). If the offset C-band feed were installed on the other side, the actuator would have had to extend an extra 15 cm (6 inches).

Prime-prime or Coaxial Feeds

A number of manufacturers have developed prime-prime or coaxial feeds that simultaneously detect both C and Ku-bands (see Figure 6-9, 6-10 and 6-11). Most brands have a single probe so that a polariser can be connected in a standard way to a satellite receiver. Those varieties that have acceptable performance at a reasonable price and that are mechanically adaptable to C-band home satellite systems are certainly a more desirable choice than offset feeds.

Feed Supports

Whether a coaxial or an offset feed configuration is chosen, the entire assembly must be securely supported. This is critical. Any swaying can easily move the throat of the feed away from the "sweet spot" at the antenna phase centre. While the area that the beam must hit is the size of a tennis ball for C-band detection, this shrinks to the size of a large marble for a Ku-band TVRO. Since two LNBs and the necessary feed can weigh enough to simulate the "bowling ball on a bamboo stick" phenomena, guy wires attached by S-hooks or other non-obtrusive means should almost always be used, especially with buttonhook feed supports.

Linear versus Circular Polarity

Most C-band feeds in the field are designed to select between horizontal and vertical polarities. Ku-band feeds which do the same are also available. Such scalar feeds can be converted to detect

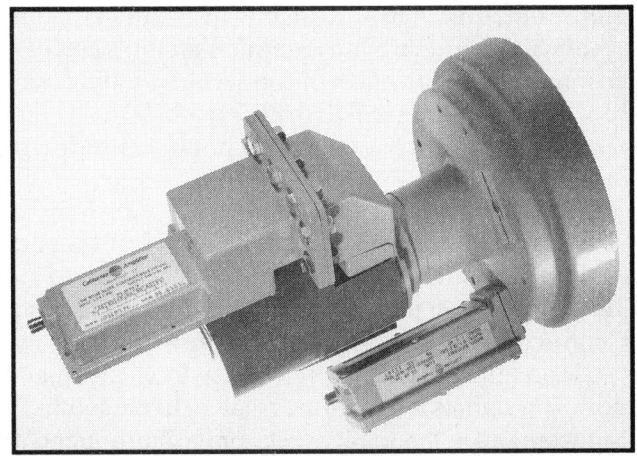

Figure 6-10. CalAmp Dual Band Feed. *This prime-prime feed uses a single servo motor to move probes for both C and Ku-band polarity selection. C and Ku-band HEMT LNBs are installed on the output flanges. (Courtesy of California Amplifier)*

Figure 6-9. Dual-Band Coaxial Feedhorn. *This coaxial Seavey feedhorn has an LNB and an LNA plus downconverter offset 180° from each other.*

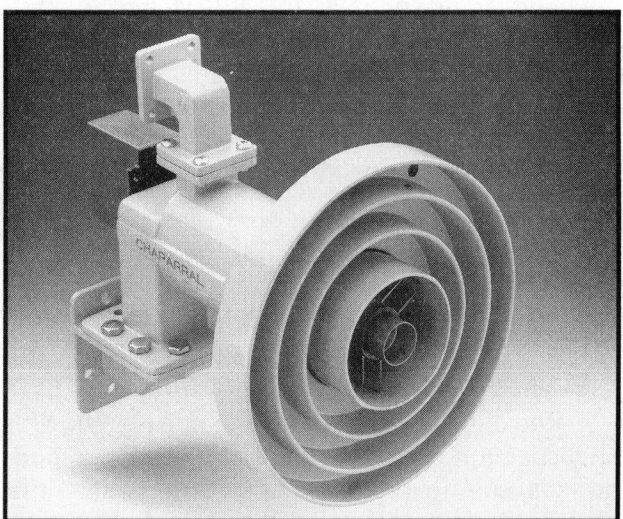

Figure 6-11. The Corotor II. *The Corotor II™ prime-prime feed has a maximum VSWR of 1.4:1, cross-polarity isolation greater than 30 dB and an f/D adjustable from 0.28 to 0.42. (Courtesy of Chaparral Communications, Inc.)*

KU-BAND PERFORMANCE

RETROFITTING

circular polarities by inserting a teflon dielectric slab. These procedures have been described in some detail in Chapters II and V. However, most coaxial dual-band feeds cannot be retrofitted in this fashion. New designs must evolve to allow simultaneous dual-band and dual-polarity operation.

Satellite Receivers

Satellite receivers used in retrofits must be capable of detecting and processing both C and Ku-band signals. The Ku-band LNB downconverts the signal to the same frequency range as the original LNB so that the existing receiver can process both signals. This is typically 950 to 1450 MHz for North and South American and Australian C-band satellite systems. C-band satellite reception systems are less common in other parts of the world, so different block frequencies such as the 950 to 1700 MHz European standard generally need not be considered in such retrofits.

A satellite receiver that processes both C and Ku-band broadcasts must have the capability to invert the video signal. This stems from the fact that C-band downconverters generally use high-side injection while Ku-band systems use low-side injection. The signals are inverted relative to each other. For example, if the local oscillator (LO) produced a mixing frequency of 5.15 GHz from which the 3.7 to 4.2 GHz C-band were subtracted, the resulting frequencies would range from 1450 to 950 MHz. If a Ku-band receiver LO injected 10.75 GHz which was subtracted from 11.7 to 12.2 GHz, the resulting band would span 950 to 1450 MHz. The same band of frequencies is also produced in this case, but it is reversed so the channel order as well as the video signal are effectively flipped. If the same channel format had been used on Ku-band broadcasts, C-band channel 1 would become Ku-band channel 23, channel 2 would become channel 24, and so on. This confusion is further complicated because different channel formats are used on different Ku-band satellites.

Dual-band receivers must also have a fine tuning capability. While C-band channels are evenly spaced at 20 MHz intervals across the allocated bandwidth, no standard has evolved for the frequency allocation of Ku-band channels. Their centre frequencies can range as far as 10 MHz away from those for the preset C-band channels. Since each Ku-band spacecraft of interest could conceivably have a different frequency plan, receivers that have a programmable memory capability are a great convenience. Every channel, both C-band and Ku-band, can then be programmed into memory for easy selection. Some Ku-band satellites also operate with more than the C-band standard of 24 channels. For example, Canada's Anik C satellites have 32 transponders. Therefore, a fully versatile satellite receiver would be required to have this extra capability.

Variable channel bandwidths are also common on Ku-band satellites. This is rare at C-band except for half transponder formats encountered on some Intelsat spacecraft. For example, 54 MHz bandwidth is occasionally used. However, this may be subdivided into two 23 MHz channels each having 2 MHz-wide protection regions on both edges. Variable tuning would be used to locate the centre frequency and then a 23 MHz bandpass filter would be required to restrict the input to just one channel. When this is done, the EIRP drops by 6 dB since half of the transponder power is allocated to each channel (a 3 dB reduction), and then the uplink power is backed off another 3 dB to eliminate the possibility of a type of adjacent channel interference known as intermodulation distortion.

B. INSTALLATION PROCEDURES

Retrofitting a C-band TVRO must begin with an inspection of the existing system. Each component should be examined to determine if all the above criteria can be met. From this information, a plan of action can be formulated and the necessary equipment can be selected.

Mechanical Requirements

All retrofits require purchase of a second Ku-band LNB and feed. If the offset configuration is selected, a special flange cut from a piece of rigid sheet metal such as hardened aluminium must be created to securely support both feeds. This is usu-

RETROFITTING

ally a rather simple piece of engineering work. Extra precautions must be taken to hold such a feed securely in place at the antenna focus because of its extra weight (see Figures 6-12 nd 6-13). If a coaxial dual-band feed is used, the dimensions should be obtained from the manufacturer or distributor to ensure that adequate room and interfaces are possible at the existing buttonhook or tripod support.

In some cases, the actuator arm may have to be replaced with a more finely geared or more stable brand. Similarly, any points on the supporting structure or mount where sagging or rocking can occur must also be properly treated. A few brands of antennas are not adequately secured in place where the declination or elevation adjustments are made. These may not be of sufficient quality for retrofit, and a second smaller, more accurate antenna may have to be purchased exclusively for Ku-band reception. A similar course of action may be necessary if a relatively small and inaccurate antenna had been originally installed. Even if the C-band reflector was oversized so that sufficient gain were available, the extra weight may prove to be a disadvantage because of associated problems with stability and sagging.

Cassegrain or other backfire designed antennas do not permit the retrofit of an offset feed. In these cases, it may again be necessary to use a second reflector for Ku-band reception.

Many existing C-band satellite receivers simply not meet the Ku-band retrofit test. These will either have to be replaced or a second Ku-band compatible or dual-band receiver will have to be purchased.

Wiring

The details of wiring a retrofit are dependent upon particulars of the installation. The signal from the second LNB may be channeled down a separate length of RG-6 coaxial cable to a second dual-band receiver or may be fed into a switch which leads to a single coax either at the antenna or indoors. A mechanical switch that passes power only to one LNB at a time generally provides more than adequate isolation between the two broadcast sources. However, in those cases where an electrically operated pin diode switch is used, it is necessary to route the dc power around the switch with a pair of dc blocks so that both LNBs will operate. Then an additional pair of control wires must be hooked up. Because signals from both bands occupy the same block of frequencies, there is a possibility of interference. If these signals are balanced, there should be at least 25 to 30 dB of isolation built into the switch. However, if the signals are unbalanced so that one source is much more powerful than the other, even more isolation will be required. For example, if the Ku-band signal is 10 dB more powerful than that from the C-band electronics, the isolation should be at least 35 to 45 dB.

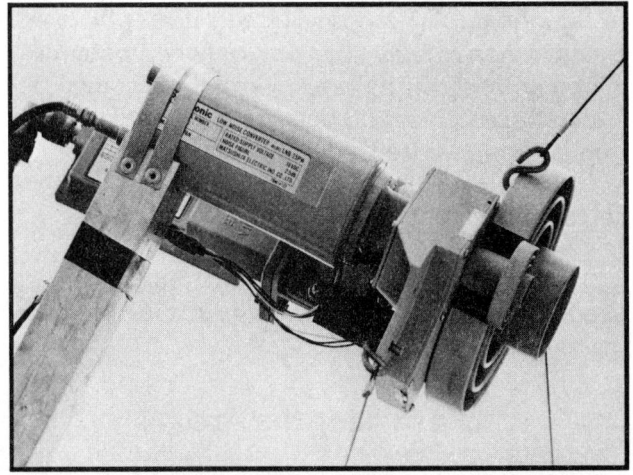

Figure 6-12. Dual Band Retrofit. *This is a close-up of a dual-band retrofit after all adjustments have been completed and the wind kit wires have been tightened to stabilise the feed.*

Figure 6-13. STS Dual Band Feed Support. *When using a heavy feed system as is the case with this dual-band prime focus feedhorn, it is very important that guy wires be installed to maintain the feed at the antenna focus during tracking of the arc.*

RETROFITTING

When a separate mechanical polariser is used for each feed, both may be wired in parallel, i.e. the input wires of each can be connected together with the triple-wire cable from the satellite receiver. Care must be taken to ensure that both probes are oriented so that neither encounters a mechanical limit over the full range of its operation. To accomplish this both must move in parallel. In addition, each polariser may draw from 0.5 to 0.75 amps of current from the 5 Vdc output terminals on a satellite receiver. Therefore, this receiver must be capable of generating up to 1.5 amps, a sizable amount of current. A 25 Vdc capacitor ranging from 470 to 2000 microfarads can be wired between the hot wire and ground to provide an extra burst of energy so that the full 1.5 amps will rarely if ever be used, except when rapidly changing between polarities. When parallel wiring is used and the distance between the receiver and polarisers is relatively long, thicker gauge wire than is normally chosen may have to be used. A wire that is too thin could conceivably overheat and become a fire hazard.

If two separate antennas and actuator arms are used, two separate positioners and associated wiring must be installed.

Tracking the Arc

Once the necessary mechanical connections have been completed, the easiest approach is to first track the full arc of C-band satellites by following those methods detailed in Chapter V. Locating active transponders on Ku-band spacecraft is relatively more difficult because the large diameter antennas used in C-band systems often have low gains and more narrow beamwidths at Ku-band frequencies. In addition, reflectors with poorer performance may also have side lobes at unexpected locations.

When the C-band arc is properly tracked, final adjustments to both the feed position and mount must always be completed by aiming at Ku-band satellites. When this is accomplished, C-band broadcasts will automatically be targeted. Be aware that it is not uncommon to discover that Ku-band reception is substantially improved when the feed is positioned as much as a few centimetres away from the C-band focal point. This occurs when the reflector surface does not accurately conform to the original parabolic design. It is quite important to ensure that the antenna is not significantly warped and, if it is, to take the appropriate countermeasures such as loosening and re-tightening bolts. Once this is accomplished, the feed will probably have to be re-peaked.

When the initial tracking is completed, the overall accuracy in aligning onto the arc of satellites can be improved by using a spectrum analyser or a carrier level meter. Of course, geostationary spacecraft of any band, C, Ku, S or any other, all fall exactly on the same arc.

In those dual-band installations where a separate antenna is used to receive Ku-band broadcasts, the luxury of easily tracking the C-band spacecraft is lost. Readers are again referred to Chapter V for all the necessary procedural details.

The Effect of a Dual-Band Retrofit on C-band Reception

The emphasis in this chapter has been to optimise dual-band operation. Using an offset fed configuration inevitably results in a deterioration in C-band performance. If the existing system was installed with an adequate safety margin, the visible effect would be minimal. However, if a marginal system is in place or if terrestrial interference (TI) exists at levels just below detection, offsetting the C-band feed could cause problems such as increasing the detected levels of TI.

If the system was just on the margin before retrofit, the only solution may be to use a lower noise temperature LNB, install a larger antenna or use a second antenna dedicated to Ku-band reception. The latter two changes would also be effective if TI became a problem. Another alternative to counteract TI would also be to either screen the antenna or to use filters.

Tuning onto Ku-Band Transponders

Confusion can arise when tuning onto channels because bandwidths and centre frequencies are so variable. For example, a programmer might refer to channel 5 on a Ku-band satellite having 54 MHz wide transponders. This video transmission might be located near the channel 9 position on a receiver tuning dial which had been designed for C-band operation. In general, it is easier to located Ku-band channels by knowing their centre frequencies. Using a programmable receiver does take some of the guess work and required operational familiarity out of the tuning process. But expect to have to consult the various charts in this book and various programming guides to locate audio and video services.

One solution to simplify locating Ku-band channels has been proposed. If satellite receivers were designed with a digital frequency read-out and sliding scale, similar to those used with SCPC scanners, as well as a memory each channel could be located and identified. Or instead of having continuous frequency adjustment capability, this sliding scale could be subdivided into perhaps 2 MHz centre frequency assignments, spanning as many as 375 tunable channels in the whole 950 to 1700 MHz band.

Chaparral Communications has presented a channel and frequency designation for Ku-band transponders (see Figure 6-14). They began by assembling a complete list of North American Ku-band transponder frequencies and their corresponding channel designations. It was noted that all Ku-band satellite transponder have the capability of being split into two channels. For non-split transponders, the centre frequencies are standard. If transponders are split to accommodate more programming, the A-side transponder should be preset and the B-side is then shifted from the standard centre-tune position. All 32 transponder formats then follow standard vertical/horizontal alternating positions.

The basis of this proposed organisation allows for both centre-tune and split transponder frequencies. As an example, transponder 1A would be designated channel 1 and then the centre frequency of transponder 1 would be channel 3. Also transponder 2A would be channel 2 and the transponder 2 centre frequency would become channel 4.

One of the advantages here is the ease by which the user can adapt to changing Ku-band satellite formats. When the centre-tune transponder format is being used, such as on Satcom K2, the transponder 1 centre frequency is Channel 3 at 11.729 GHz and transponder 2 centre frequency is channel 4 at 11.7585 GHz. When these transponders change to "split" format, the centre frequency shifts up 12 MHz so, for example, transponder 1, channel 3 shifts to 11.741 GHz. The receiver is then up-tuned by 12 MHz. Transponders 1A and 2A, channels 1 and 2, remain on their shifted frequencies and need not be changed.

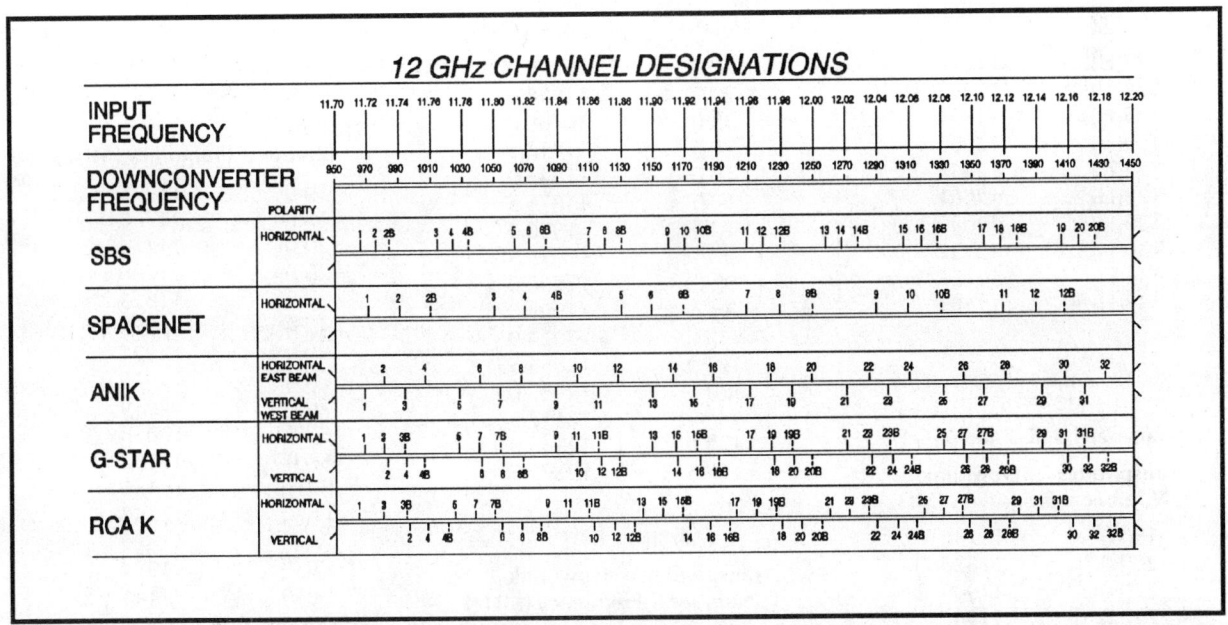

Figure 6-14. *Ku-band Channel Reference Scheme.* The 500 MHz C-band bandwidth that contains 24 channels can transmit up to 32 channels on some Ku-band satellites by simply narrowing the bandwidth on each transponder. Some K-band satellites also utilise double-width transponders that relay two adjacent channels having the same polarity. (Courtesy of Chaparral Communications)

RETROFITTING

Note that not all Ku-band satellites shift their split transponders by the same set frequency, as is the case on Satcom K1, according to this proposed format. However, the customer always is assured of receiving something on a centre-tune or split channel whether or not the satellite in question is broadcasting on either format.

Transponders numbers, polarities as well as uplink and downlink frequencies for North American Ku-band satellites are listed in Table 6-1.

TABLE 6-1. NORTH AMERICAN KU-BAND FREQUENCY ASSIGNMENTS

Anik C1/C2/C3 & E2

Transponder Number	Downlink Frequency (MHz)
T1(V)	11717
T2(V)	11743
T3(V)	11778
T4(V)	11804
T5(V)	11839
T6(V)	11865
T7(V)	11900
T8(V)	11926
T9(V)	11961
T10(V)	11987
T11(V)	12022
T12(V)	12048
T13(V)	12083
T14(V)	12109
T15(V)	12144
T16(V)	12170
T17(H)	11730
T18(H)	11756
T19(H)	11791
T20(H)	11817
T21(H)	11852
T22(H)	11878
T23(H)	11913
T24(H)	11939
T25(H)	11974
T26(H)	12000
T27(H)	12035
T28(H)	12061
T29(H)	12096
T30(H)	12122
T31(H)	12157
T32(H)	12183

Spacenet 1,2, & 3 and ASC 1

Transponder Number	Downlink Frequency (MHz)
19(H)	11740
20(H)	11820
21(H)	11900
22(H)	11980
23(H)	12060
24(H)	12140

GStar 1,2,3 &4

Transponder Number	Downlink Frequency (MHz)
1(H)	11730
2(H)	11791
3(H)	11852
4(H)	11913
5(H)	11974
6(H)	12035
7(H)	12096
8(H)	12157
9(V)	11744
10(V)	11805
11(V)	11866
12(V)	11927
13(V)	11988
14(V)	12049
15(V)	12110
16(V)	12171

GE K1 & K2

Transponder Number	Downlink Frequency (MHz)
1(H)	11729.0
2(V)	11958.5
3(H)	11788.0
4(V)	11817.5
5(H)	11847.0
6(V)	11876.5
7(H)	11906.0
8(V)	11935.5
9(H)	11906.0
10(V)	11994.5
11(H)	12024.0
12(V)	12053.5
13(H)	12083.0
14(V)	12112.5
15(H)	12142.0
16(V)	12171.5

Morelos 1 & 2

Transponder Number	Downlink Frequency (MHz)
1K(H)	11764
2K(H)	11888
3K(H)	12012
4K(H)	12136

SBS 6

Transponder Number	Downlink Frequency (MHz)
1(H)	11725.0
2(V)	11749.5
3(H)	11774.0
4(V)	11798.5
5(H)	11823.0
6(V)	11847.5
7(H)	11872.0
8(V)	11896.5
9(H)	11921.0
10(V)	11945.5
11(H)	11970.0
12(V)	11994.5
13(H)	12019.0
14(V)	12043.5
15(H)	12068.0
16(V)	12092.5
17(H)	12117.0
18(V)	12141.5
19(H)	12166.0

SBS 2,3,4 & 5

	Transponder Number	Downlink Frequency (MHz)
	1(H)	11725
	1(H)	11730
	2(H)	11774
*	2(H)	11780
	3(H)	11823
	4(H)	11872
	5(H)	11921
	6(H)	11970
	7(H)	12019
	8(H)	12068
	9(H)	12117
	10(H)	12166
**	11(V)	11748
**	12(V)	11896
**	13(V)	11994
**	14(V)	12141

* SBS 4/5 - small uplink
** SBS 5 single chanel video only

INSTALLATION

VII. MULTIPLE RECEIVER SATELLITE TV AND DISTRIBUTION SYSTEMS

Satellite television transmissions can be relayed in combination with signals from over-the-air broadcasts, videotape recorders (VCRs), video game equipment, closed circuit television security systems or any other video and audio sources to any number of video monitors or television sets. There is really no limit to the design possibilities once a sufficiently powerful satellite signal has been detected and processed.

Even the most complex channel combination and distribution systems are assembled from simple-to-understand components. When the different pieces of this jigsaw puzzle are independently studied, creating a virtually unlimited number of different designs becomes quite straightforward.

This chapter is organised into three sections. First, the basic components required to assemble a multiple receiver "headend" and a distribution system are examined. Second, the design parameters of headends and distribution networks are studied. Third, examples of some basic system designs are explained and diagrammed. The book *Satellite, Off-Air and SMATV - The Multi-Unit Handbook* which explores this subject in much greater detail.

A. BASIC COMPONENTS OF DISTRIBUTION SYSTEMS

The most complex configurations of satellite reception and distribution systems are pieced together from basic components including one or more antennas, feedhorns, LNBs, coaxial cables, satellite receivers, modulators and televisions or monitors. These can be combined with A/B switches, splitters, combiners, line amplifiers, attenuation pads, terminators, dc power blocks, coaxial relays, taps, and various types of signal processors to create any possible design.

Television Signal Requirements

The ultimate target of broadcast signals that are processed by a headend and distributed by a cable network is the television set and audio system. Televisions function best with an input signal level of 0 to +3 dBmv, although the optimal level can vary among various brands. The designation dBmv stands for decibels relative to 1mV. 0 dBmv therefore equals 1 millivolt; 3 dBmv equals 2 mv. When signal levels exceed approximately +3 dBmv, some televisions can be overdriven and pictures will be distorted. However, most sets have AGCs (automatic gain circuits) which compensate for excessively strong signals. This allows inputs exceeding 3 dBmv to be managed, to a point. But the 0 to 3 dBmv is a good target range which should be achieved for best results at television inputs throughout the distribution system.

It is much easier to manage signals that are too powerful than too weak. If the input signal-to-noise power ratio has been allowed to drop too low or is

MULTIPLE RECEIVER/TV SYSTEMS

"in the mud," no level of amplification will improve the situation because, at this point, as much noise as signal will be amplified. When signal power levels exceed the upper limits, attenuation pads can always be used. Therefore, too much power is never really a problem but allowing signal strength to deteriorate can be a cause for concern.

Many older televisions often have a 300 ohm, twin-screw VHF input where the familiar flat TV leads can be attached. Signals from satellite receiving equipment as well as from other video equipment are transmitted via 75 ohm coax and require F-connector inputs. Therefore, a 75-to-300 ohm transformer, known as a balun (derived from balanced-unbalanced), must often be used in the transition from the coax to the television VHF terminals. This transformer matches the impedance between the coaxial cable and the television receiver input to allow maximum signal power to be transferred between these two devices (see Figure 7-1).

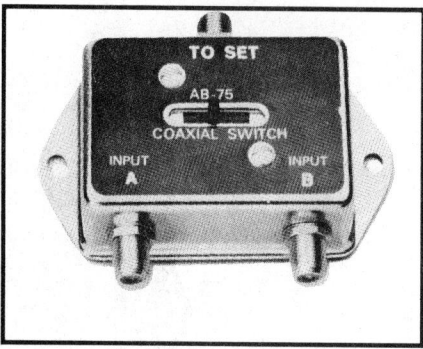

Figure 7-2. Coaxial A/B Switch. *Coax switches come in a variety of models with isolations ranging from 30 to 90 dB. They can be used to switch between inputs of either horizontally or vertically polarised satellite signals and an over-the-air broadcast.*

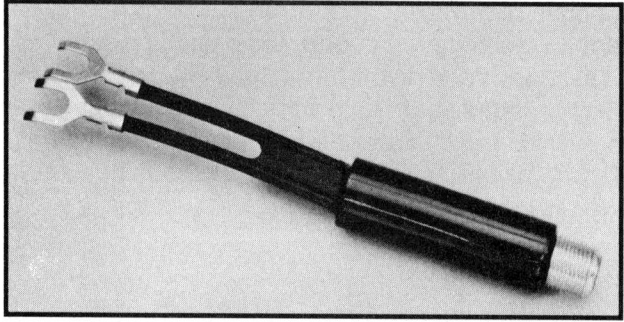

Figure 7-1. Matching Transformer. *This device is used with televisions that require 300 ohm twin-lead cable inputs. Today most sets are manufactured with 75 ohm F-connector VHF inputs.*

A/B Switches and Combiners

An A/B switch is used to select between either one of two input signals (see Figure 7-2). A high quality A/B switch has at least 40 or 50 dB isolation between its two ports. This means that if a 10 dBmv signal is present on the input port not in use, it would be attenuated by 40 or 50 dB so that its level at the output terminal would range from -30 to -40 dBmv (10 dBmv less 40 or 50 dB). Many satellite receivers already have built-in electronic A/B switches which allow selection of either satellite or over-the-air broadcasts. When power is turned off the conventional broadcast is automatically selected.

A signal combiner is used to combine and balance the powers of satellite and over-the-air signals. It can be used in place of an A/B switch (see Figure 7-3). For example, a simple unit might take an input from the satellite receiver channel 3 modulator and another line from an over-the-air antenna to create a combined signal.

If two adjacent channels are being fed onto a single coax, there is always the possibility that they will interfere with each other when a poor quality signal combiner is used. In any case, it is always important to balance the power levels of both the satellite and over-the-air signals before they enter the combiner. Another step that may be necessary to minimise the possibility of interference between channels is to eliminate the lower sideband of the higher frequency channel with a bandpass filter. This action restricts the video input to a selected narrow range of frequencies, for example 6 MHz for NTSC signals. More details about equalising and eliminating interference between adjacent channels are discussed below.

Cables, Connectors and Splitters

Coaxial cables, connectors and splitters are the conduits for carrying signals to any final destination (see Figure 7-4). Each component has characteristic power attenuation which depends upon the transmitted frequency and which must be accounted for when designing any distribution system.

MULTIPLE RECEIVER/TV SYSTEMS

Figure 7-3. Inexpensive Signal Combiner. *This channel combiner consists of a high quality, six-stage bandpass/bandstop filter with adjustable balancing attenuators. It allows 3, 4, or 6 output signals from a videotape recorder, MDS receiver, TVRO or video game to be combined on a distribution system having adjacent channels. Properly balancing signals levels in an MATV system eliminates the need for an A/B switch.*

which blend in with surrounding walls and floors are useful in improving installation cosmetics.

Splitters do exactly what their name implies, they split or divide a signal into two or more branches (Figure 7-5). In certain situations, these devices may also be used in reverse to combine signals. Each type of splitter is rated to handle a specified frequency range. For example, when dividing the 950 to 1700 MHz output of an LNB, a splitter rated up to 1700 MHz must be used or losses will be excessive. Some brands of splitters also have built-in bandpass filters so that frequencies below and sometimes above the designated range are sharply attenuated. Limiting the bandwidth in this manner can also protect cables from ingress interference. Typical insertion losses for two, three, four, eight and sixteen-way splitters are listed in Table 7-1.

Characteristic signal attenuations in standard 75 ohm cables such as RG-6, RG-11 and RG-59 have been detailed earlier in the manual. A good ballpark figure for RG-59, one of the standard cables used in video distribution networks, is 6.6 dB per 100 m at 50 MHz and 13.1 dB per 100 m at 200 MHz. Note that 50 MHz falls on the frequency range of channel 2 and 200 MHz is just below channel 13 in standard NTSC television systems (see Tables 2-8, 2-9 and 2-10 and Figure 2-92).

Coax can be purchased wound on rolls up to 300 or 400 m in length. Different colours such as black, beige or white are available. Those colours

TABLE 7-1. SPLITTER LOSS PER OUTPUT LEG	
Type of Splitter	Loss (dB)
2-way	3.5
3-way	3.5 and 7
4-way	7.0
8-way	10.5
6-way	14

  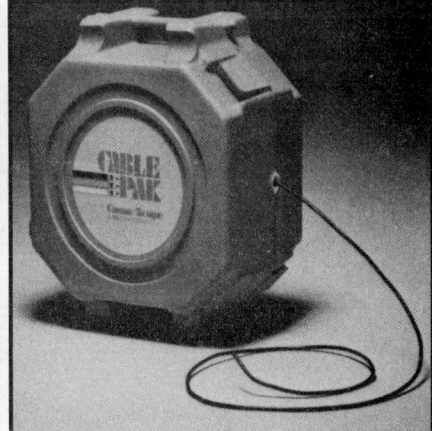

Figure 7-4. Drop Cables. *Drop cables are packaged in a variety of easy-to-use containers. These can easily be dispensed from "easy-out" boxes or reusable plastic cases to facilitate pre-wiring. (Courtesy of Comm/Scope MA/COM)*

BASIC COMPONENTS

MULTIPLE RECEIVER/TV SYSTEMS

 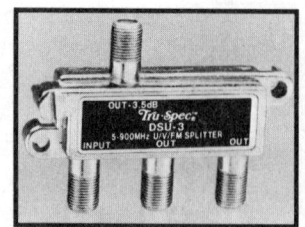  

Figure 7-5. MATV Splitters. *2, 3, 4 and 8-way splitters are most commonly employed when dividing signals in MATV applications.. The usable frequency range of this brand ranges from 5 to 900 MHz which extends past the UHF channel frequency band.*

Figure 7-6. Wall-Plate Cover. *A wall plate with a F-81 barrel connector can be used whenever single coax cable is fed through the interior of a wall to a television set. This installation technique is neat and attractive.*

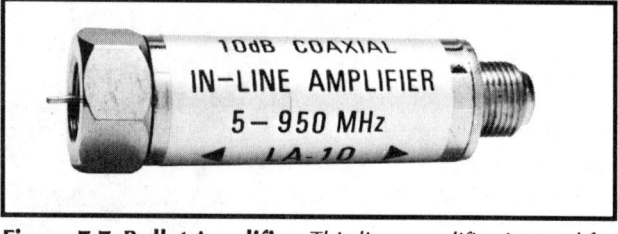

Figure 7-7. Bullet Amplifier. *This line amplifier is rated for frequencies from 5 to 950 MHz and is normally used in SMATV distribution systems or in conjunction with an LNB having a 400 to 950 MHz output. It can be inserted as a line amplifier between a pre-amp output and a channel processor input in long cable runs, but must receive line power to operate.*

Figure 7-8. 20 dB Line Amplifier. *This small 20 dB line amplifier is distribution system operating in the 950 to 1450 MHz range. It has cable slope compensation and counteracts high frequency roll-off by providing more amplification at higher frequencies. This guarantees that all channels arrive at the output of the satellite distribution system with equal amplitudes. (Courtesy of Channel Master)*

A two-way splitter cuts the signal a little more than in half (because a 3 dB reduction is a halving of power or voltage). A 3-way splitter has one -3.5 dB and two -7 dB output ports; a 4-way splitter has four -7 dB ports. For example, if a signal of 3 dBmv enters a 4-way splitter, each output leg would produce a signal of 3 dBmv reduced by 7 dB which equals -4 dBmv. This voltage level would not be enough to produce a studio quality picture because it is well below acceptable levels, and noise would begin to overpower the signal. If this signal were also relayed down 100 metres of RG-59 cable, additional losses of 13.1 dB at 200 MHz would result in a final signal of -17.1 dBmv at the television input. It is clear that 20 dB amplification would be required to restore a 2.9 dBmv level. Line amplifiers should be used before cable losses cause a situation where noise power becomes too large compared to signal power. In this case, the amplifier would be installed before the splitter.

F-connectors and at times even splitters are occasionally mounted onto wall plates so components can be hooked into a distribution system attractively (see Figure 7-6).

Line Amplifiers and Tilt

Line amplifiers, which are inserted via F-connectors directly into a coax line, are available for boosting signals spanning either a single channel or a range of channels (see Figures 7-7 and 7-8). Less-expensive, simpler units are powered from the dc voltage relayed down the coax. Most of these inexpensive amplifiers which operate in the 900 to 1700 MHz range are installed between the LNB

MULTIPLE RECEIVER/TV SYSTEMS

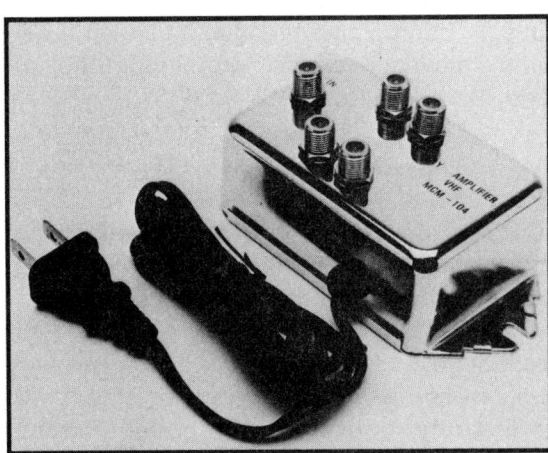

Figure 7-9. 4-Way VHF Amplifier. *This amplifier has a 4-way splitter installed. It can be used after a satellite receiver modulator and feeds a signal with sufficient power to drive four separate televisions in a home distribution system. (Courtesy of Pico Products)*

output and receiver input. Similar amplifiers operating in the 5 to 950 MHz range in an over-the-air distribution system are installed between the pre-amp output and the channel processor. More costly MATV (master antenna TV) commercial amplifiers draw current from a regular wall outlet and convert it to dc power via an internal power supply (see Figures 7-9 and 7-10). This is similar to the way pre-amps designed for distributing over-the-air broadcasts over cable systems operate.

Line amplifiers for MATV systems can have either fixed or adjustable gains. Some brands designed for longer cable runs have gains which increase with frequency to compensate for the increased attenuation of signals at these higher frequencies. The variation in attenuation with frequency is called tilt. For example, a tilt of about 6.5 dB per 100 m between 50 and 200 MHz (the range spanned by North American channels 2 and 13) is necessary for cable attenuation compensation. This is based on the difference between the 13.1 and 6.6 dB per 100 m attenuation that this type of coax contributes at these two frequencies. For example, if a cable run of 100 m were used, a line amplifier that delivered 20 dB amplification at 50 MHz and increasing to 26.5 dB at 200 MHz would properly compensate for the 6.5 dB difference in cable losses between these frequencies. Then the output at the end of the 100 metre span of cable would be flat across the entire range of frequencies. This would reduce the chance of adjacent channel interference.

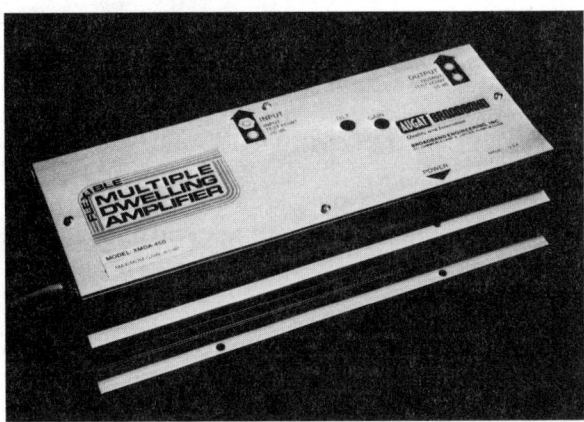

Figure 7-10. Distribution Amplifier. *This commercial MATV distribution amplifier is used in SMATV distribution systems. It operates in a frequency range of 50 to 450 MHz and has a gain of 20 to 50 dB. It is available with variable gain and tilt controls and can be tailored for operation in locations where a line voltage of 220 Vac is standard. (Courtesy of Broadband Engineering)*

Attenuation Pads

An attenuation pad, more simply known as a pad, is used to reduce the strength of an excessively powerful signal (see Figure 7-11). Pads are small, inexpensive devices which insert directly into a coax line via F- connector fittings. They are available with either fixed or variable rated attenuations and occasionally with built-in tilt compensation. Most pads are not designed to pass dc power so they cannot be used in the line between receivers and LNBs.

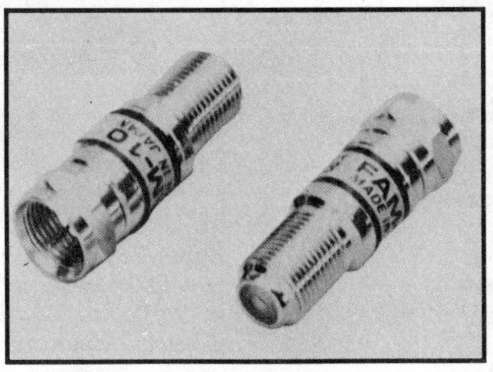

Figure 7-11. Attenuator Pads. *Pads are manufactured with one male and one female port and can easily be inserted into a coax line using F-connectors. These models have a variety of rated attenuations including 3, 6, 8, 10, 12, 16 and 20 dB. Their bandwidth ranges from dc to 1000 MHz. They are generally used in MATV systems to decrease signal levels when balancing channels. This pad does not pass power and therefore cannot be used in a satellite block distribution system.*

BASIC COMPONENTS

MULTIPLE RECEIVER/TV SYSTEMS

Figure 7-12. A Terminator. *Terminators are installed on an unused splitter port or at the end of a distribution line in order to provide a 75 ohm impedance match. Doing so eliminates both signal reflections and ingress interference.*

Terminators

Any output port on a video or audio distribution network must end in an appropriate device such as a television set, a satellite receiver or a terminator (see Figure 7-12). If not, interference could leak into the unconnected port or the resulting impedance mismatch could cause signals to be reflected back into the system as "ghosts." A terminator is simply an encased resistor which screws onto a F or N-connector fitting. It has a matching impedance, 75 ohms in the case of RG-59 or RG-6, and therefore the cable does not "see" an opened end or discontinuity at the unused port but behaves as if the cable had no end.

dc Power Blocks

A dc power block is a simple device which allows higher frequency television signals to pass unattenuated, but which blocks the passage of any direct current (see Figure 7-13). The purpose of a dc block is to isolate various components in a distribution system from dc voltages. This device is nothing but a simple circuit element called a capacitor encased with a male and female F-connector fitting at either end.

Input Selectors

Input selectors allow the user to select between two or more coaxial cable inputs, often using the polariser interface on the receiver (see Figures 7-14 and 7-15). These are electrically operated A/B switches that pass signals in only one direction at a time, while isolating the unused port by at least 40 dB from the output port. This component is commonly used in smaller scale SMATV operations where the distances are short enough to enable the transmission of the block IFs via cable. In dual-band systems, selectors are often used to switch between two input cables. Input selectors are occasionally used in conjunction with dc power inserters (see Figure 7-16).

There are basically two types of selectors; those based on relays and those on PIN diodes. A coaxial relay, the more reliable of the two types, is an electronically operated switch consisting of a coil, a solenoid and a switch, which generally has two inputs and one output. This device is commonly referred to as a single-pole-double-throw or SPDT switch. When current flows through the coil, it forces the solenoid from either input A to B, or vice versa.

In some UHF applications, the loss through an air filled coaxial relay can be significant. In these cases, a more expensive nitrogen filled relay having lower losses is used.

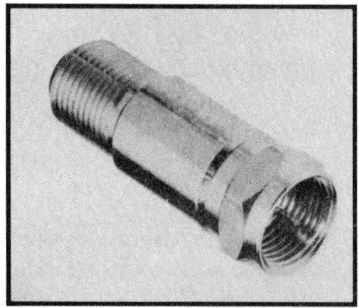

Figure 7-13. Voltage Blocking Coupler. *VBCs are used to block any dc or low frequency ac power which might be present on a coaxial cable. They can be used to protect equipment such as televisions or some types of satellite distribution system relays from damaging power surges.*

Figure 7-14. Global Magic Switch. *This 4 x 4 switch allows up to four antennas to be connected to four or more separate receivers. It is driven by dc pulses generated by most receivers.*

BASIC COMPONENTS

MULTIPLE RECEIVER/TV SYSTEMS

The PIN diode is an inexpensive selector that uses a diode as an electronic switch. The resistance of the PIN diode can be varied by the application of voltage so it can be used to pass or block RF signals. When forward biased, it passes signals; reverse biased, it acts in a fashion similar to an open circuit. The actual space occupied by the diodes is small and therefore this kind of selector is becoming common in the lower end of the market.

There are two types of receiver-selector interfaces. One switches using either a 0 Vdc or -12 Vdc level for one sense and a +12 Vdc for the other. This is a common technique for switching a coaxial relay selector. The less expensive PIN diode selector generally uses the rear panel receiver polariser interface. This type of selector has a circuit not unlike that found in polarisers. It interprets pulses from the interface to switch between vertical and horizontal inputs.

Problems with input selectors generally occur in the PIN diode versions. Some of the cheaper models are extremely sensitive to interference from noise spikes. In these cases, shielded cable should be used between the unit and receiver. The use of such cable is even more critical when a satellite receiver is microprocessor based. Coaxial relays do not pass dc power and therefore must be protected from excessive currents by dc blocks..

Taps

A tap, also known as a directional coupler or directional tap, extracts a specified portion of an incoming signal while allowing most power to pass through to its output (see Figure 7-17). For example, feeding a 30 dBmv signal into a 24 dB tap siphons off 6 dBmv and passes the remaining portion, less a small insertion loss of about 0.5 dB, through its output port. As seen in Table 7-2, the lower the rated tap value, the more signal is extracted and the higher the insertion loss. Exact insertion losses for any particular brand of tap can be obtained from the manufacturer.

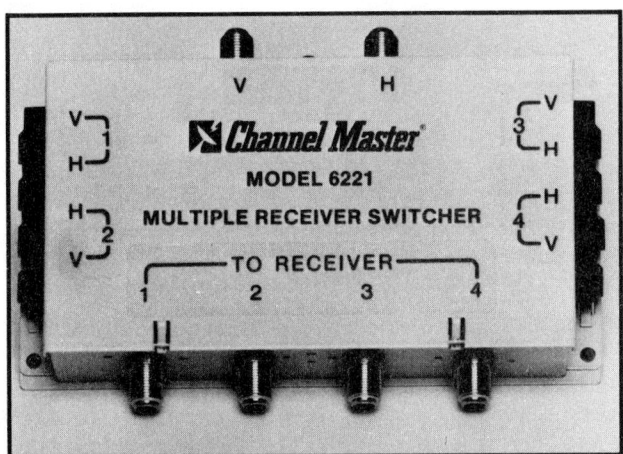

Figure 7-15. Channel Master Multiple Receiver Switcher. *This switcher/divider is designed to take a horizontal and vertical LNB input, divide this signal and then select a path to any of four satellite receivers giving each independent polarity control. (Courtesy of Channel Master)*

Figure 7-16. dc Power Inserter. *This component is used to insert power following coax relays in satellite block distribution systems. A dc power inserter used in conjunction with a power supply continuously powers LNBs whether only one or all the receivers are switched on. It can also be used to power in-line amplifiers such as a bullet amplifier. (Courtesy of Pico Products)*

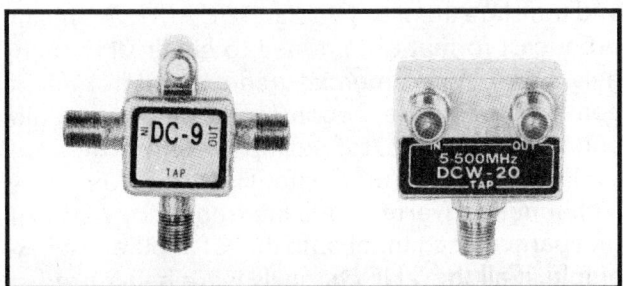

Figure 7-17. Directional Taps. *These indoor taps are used in SMATV distribution systems to tap the necessary signal from the feeder line and to pass the remainder of the signal on down the line. This concept is similar to dividing water pressure in a sprinkler system.*

BASIC COMPONENTS

MULTIPLE RECEIVER/TV SYSTEMS

TABLE 7-2. TAPS AND THEIR TYPICAL INSERTION LOSSES

Rated Tap Value (dB)	Insertion Loss (dB)
30	0.5
27	0.5
24	0.5
20	0.5
16	0.8
12	1.0
9	1.5
6	2.2

Figure 7-18 Bandpass Filter. This inexpensive bandpass filter is designed to be inserted via F-connectors into a coax line. This particular model eliminates the lower sideband on VHF channel 6 which can interfere with channel 5.

To illustrate, a 40 dBmv signal is reduced to 39.5 dBmv after passing through a 24 dB tap (which has an insertion loss of 0.5 dB) but to 37.8 dBmv following a 6 dB tap (which has an insertion loss of 2.2 dB). The former tap extracts 16 dBmv of signal; the latter 34 dBmv. A tap differs from a splitter which divides a signal equally into two or more output legs (a 3-way splitter, the only exception, has two equal output ports and one with twice the power). But both types of signal allocation devices are used to accomplish the same function. Taps are usually installed to pull signals off a main feeder line since the throughput losses are much lower than splitters. Following this point, splitters are used for distribution on local cable branches to individual TV sets.

Modulators, Processors and Bandpass Filters

The purpose of a modulator is to "rebroadcast" a television signal from one frequency range or format to another. For example, those modulators that are built into a satellite receiver take the composite baseband audio and video satellite signals and translate them onto a standard AM over-the-air broadcast format often tuned to either VHF channels 3 or 4. A commercial grade modulator might convert the satellite TV composite baseband audio and video signal to, for example, North American VHF channel 13. A modulator-like device known as a channel converter shifts the frequency range of one particular channel onto that of another. For example, if all the VHF channels were being used, a converter might accept VHF channel 7 and translate it onto UHF channel 14. The frequency selection circuits in such converters are usually crystal controlled to prevent frequency drift.

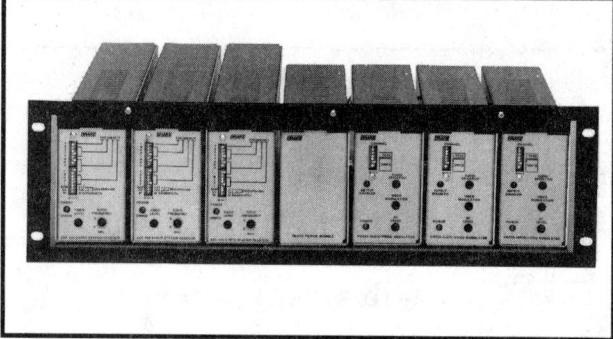

Figure 7-19. Satellite Receiver/Modulator Combination. The three frequency agile modulators on the right span PAL VHF channels 2 to 13 and is designed for the international market. It features a synthesised video carrier and selectable audio intercarrier of 5.5 MHz or 6.0 MHz. The output level is adjustable across a 14 dB range from the front panel. (Courtesy of R.L. Drake)

Figure 7-20. Channel Plus Multiplex Modulators. The unit is manufactured to transmit any channel between 14 and 60 in the UHF frequency band. A front panel LED displays the selected channel number. A ROM inside the unit's logic module contains the specifications for video and audio carrier selections for each UHF channel. A second model is also available which spans the entire NTSC hyperband spectrum and which can be configured for PAL broadcasts. (Courtesy of Multiplex Technology)

BASIC COMPONENTS

Modulators fall into three broad classes. "Home style" modulators which have evolved along with home VCR technology usually modulate signals onto channels 3 or 4 and have low output voltages, typically on the order of 0 to 3 dBmv but always lower than 10 dBmv. This variety does not have built-in bandpass filters to eliminate the lower sideband (see Figure 5-18). Commercial grade modulators have much higher output powers, with gains typically greater than 30 dB (see Figures 5-19 and 5-20). A distinction is drawn between those brands of commercial modulators with and without built-in bandpass or SAW filters (surface acoustical wave filters which have a very sharp frequency cutoff). Those not incorporating filters can generate a signal which may interfere with an upper or lower adjacent channel.

Less expensive crystal stabilised modulators typically are permanently set on one output channel. Many modulators used in home satellite receivers are crystal stabilised. Home satellite TV modulators in North America can usually be switched between channels 3 and 4, although a different pair of output channels is occasionally chosen. Those modulators which are LC (an abbreviation for inductor/capacitor) driven can be tuned by a small set screw across a range of channels. When this adjustment is made a non-metallic screwdriver must be used because metal affects the circuit elements and can make fine tuning onto the selected channel difficult. The set usually needs to be fine tuned after this to match the modulator and TV frequencies.

Commercial modulators have adjustable audio and video output levels as well as RF output controls. The audio and video settings control the "percentage of modulation" or how much of these signals are added to the carrier. Higher video outputs make for a brighter television picture but too high a level causes a buzz in the audio and washed out pictures. The RF adjustment sets the output voltage level.

Channel processors also have outputs which can be combined with signals from modulators. However, processors take already modulated signals, usually from over-the-air broadcasts, and "clean up" the signal. They do this by demodulating down to baseband and amplifying and filtering both the input and output signal to eliminate the unwanted sideband. Then the output is remodulated onto the same channel as the input signal (see Figure 7-21)

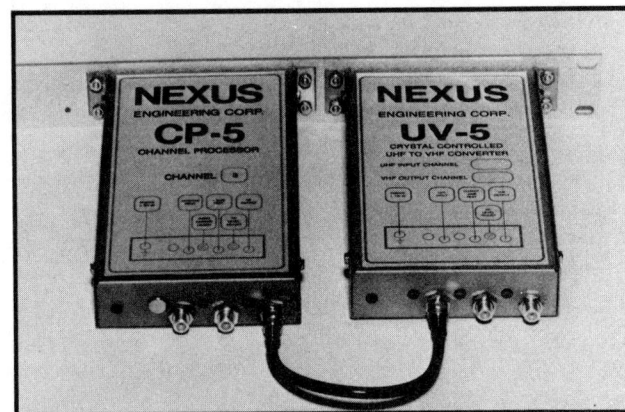

Figure 7-21. Nexus Channel Processor and Converter.
The Nexus UV-5 is a low cost crystal controlled UHF-to-VHF converter used to translate UHF over-the-air television signals to the VHF band. Its output is then fed into a channel processor which filters and amplifies the signal to be combined into the mixing system for distribution. (Courtesy of Nexus Engineering)

Multi-Purpose Components

Some manufacturers have introduced multi-purpose components in an effort to simplify installing common system designs. For example, a 4-way coaxial relay might be integrated with four dc power blocks and an output line amplifier all in one box. Some examples of such multipurpose components appear in some of the examples which follow at the end of this chapter.

B. HEADENDS AND DISTRIBUTION NETWORKS

The objective in designing either a satellite master antenna system (SMATV) or a smaller scale home satellite system is to provide each television and audio receiver on the network with one or more channels that have sufficient power and do not interfere with each other.

The Headend

Video and audio broadcasts from a variety of sources are captured and combined at the headend. Signals may originate from only satellite broadcasts or from a combination of sources (see Figures 7-22, 7-23, 7-24 and 7-25). Before entering the distribution system, signals must be modulated onto an appropriate channel and amplified. Video outputs must have nearly equal powers so that interference between channels is minimised. In general, separate audio levels should be introduced into the distribution system at about 15 dB below the video signal.

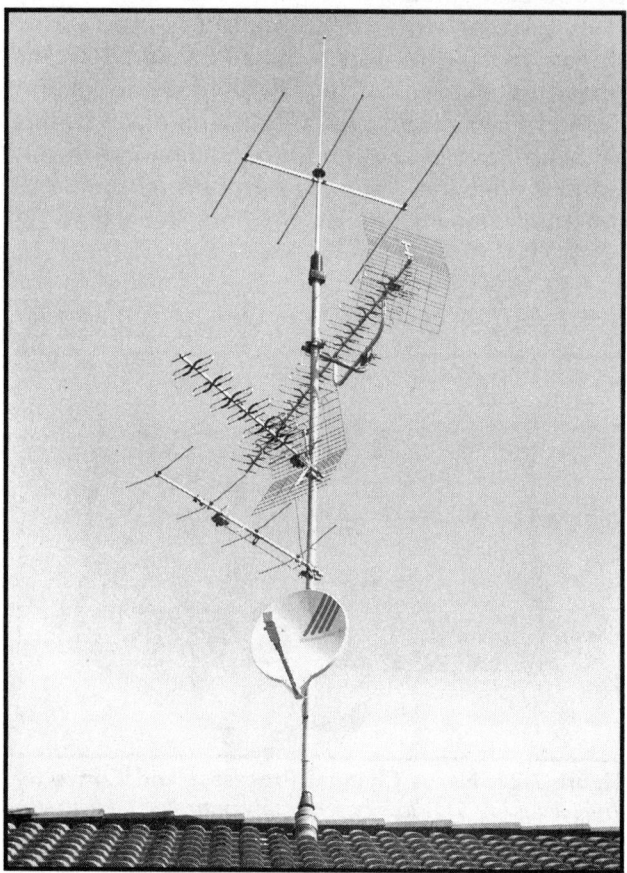

Figure 7-22. SMATV Antenna. *These antennae are designed to receive a range of off-air and satellite signals. (Courtesy of Richard Herschmann & Co)*

Signal Sources

All those rules which apply to each video or audio source used in isolation must be applied to feeding signals into an SMATV headend. Satellite broadcasts must be sufficiently above threshold in order to maintain picture quality. Margins for rain fading and the desired availability must be considered. Cable TV networks having higher quality standards than home reception systems must have an extra margin of 5 to 10 dB of S/N in order to improve picture quality and to allow for higher expected availability. Only short duration "down times" are acceptable to cable TV and larger SMATV facilities.

Other input signals from sources such as a home video camera, FM broadcasts, condominium security system, over-the-air television antenna or VCR are also prime candidates for inclusion. Commercial systems generally process and amplify each component signal before combining onto a distribution network. Home multiplexer units which add either composite baseband or over-the-air inputs are now available to create a "professional" headend in a low-cost system.

Figure 7-23. Miralite Commercial Antenna. *The Miralite C/Ku-band antenna has been tested for uplink operation at frequencies of up to 14.5 GHz. (Courtesy of Miralite)*

MULTIPLE RECEIVER/TV SYSTEMS

Television Channel Layout and Balancing Signals

Whenever signals are modulated onto two channels occupying adjacent frequency bands, there is always the possibility that interference will occur. Government agencies such as the Federal Communications Commission (FCC) in the United States wisely allocated sets of non-adjacent, over-the-air broadcast channels in each region or metropolitan area of their respective countries (see Tables 7-3 and 7-4). For example, in the United States, the city of Denver has channels 2,4,6,7 and 9 in use while Philadelphia area TV stations broadcast over channels 3,6,8,10 and 12. An explanation of why channels 6 and 7 are not adjacent is discussed below. For simplicity in this discussion, examples are drawn from North American television systems.

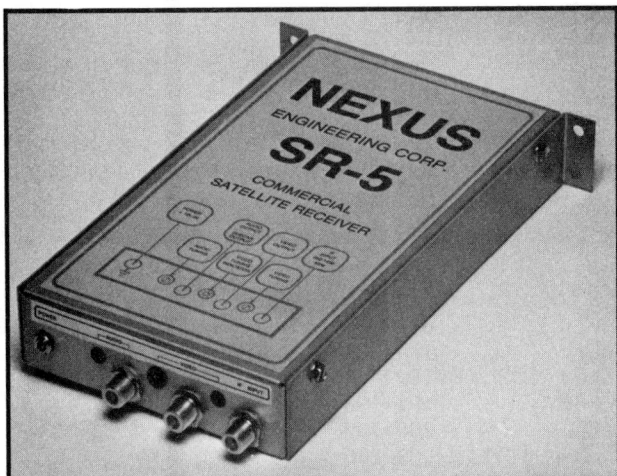

Figure 7-24. SR-5 Satellite Receiver. *This commercial satellite receiver has a low video threshold combined with high quality audio and video performance. It is available in either European PAL or NTSC models. (Courtesy of Nexus Engineering)*

TABLE 7-3. AMERICAN VHF TV CHANNEL ALLOCATIONS

Channel	Frequency (MHz)	Upper Adjacent	Lower Adjacent
2	54-60	3	none
3	60-66	4	2
4	66-72	none	3
5	76-82	6	none
6	82-88	none	5
FM BAND	88-108		
7	174-180	8	none
8	180-186	9	7
9	186-192	10	8
10	192-198	11	9
11	198-204	12	10
12	204-210	13	11
13	210-216	none	12

Figure 7-25. Televés Satellite Receivers. *These commercial completely agile SMATV satellite receivers accept an input of 950 to 1750 MHz at a -60 to -20 dBm level. The measured threshold is a C/R of 9 dB. The power supply and four receivers are resting on their one-piece mounting plate in this photo. Each unit has a built-in modulator. (Courtesy of Televés of Spain)*

Today, SMATV systems may be designed to relay, for example, five over-the-air channels along with four satellite TV broadcasts onto the VHF television band. Headends therefore often process and transmit adjacent channels and must be designed so that each one is received without interference from all others. The UHF band is also sometimes chosen as a target for modulation since more channels are available but cabling losses are substantially higher than in the lower frequency VHF band. Therefore, extensive distribution systems will most often be designed to use the VHF range. In those cases where the lower frequency channels are already occupied, cable television systems use super and hyperband channels.

HEADENDS & DISTRIBUTION

MULTIPLE RECEIVER/TV SYSTEMS

TABLE 7-4. WORLDWIDE VHF CHANNEL ALLOCATIONS

Country	Standard	Bandwidth (MHz)	Channel(s)	Allocated Band (MHz)
Britain	B	5	1	41.25 - 46.25
			2 - 5	48 - 68
			6 - 14	176 - 221
Australia	B	7	0	45 - 52
			1, 2	56 - 70
			3	85 - 92
			4, 5	94 - 108
			5A	137 - 144
			6 - 9	174 - 202
			10, 11	208 - 222
Europe	B, C	7	2 - 4	47 - 68
			5 - 12	174 - 230
Europe	B, C	7	1 - 10	104 - 174
			11 - 20	230 - 300
Italy	B	7	A	52.5 - 59.5
			B	61 - 68
			C	81 - 88
			D	174 - 181
			E	182.5 - 189.5
			F	191 - 198
			G	200 - 207
			H, H(1), H(2)	209 - 230
Morocco	B	7	4	162 - 169
			5	170 - 177
			6	178 - 185
			7	186 - 193
			8	194 - 201
			9	202 - 209
			10	210 - 217
New Zealand	B	7	1	44 - 51
			2, 3	54 - 68
			4 - 9	174 - 216
China	D	8	1 - 3	48.5 - 56.5
			4 - 5	76 - 92
			6 - 12	167 - 223
OIRT	D	8	RI	48.5 - 56.5
			RII	58 - 66
			RIII-RV	76 - 100
			RVI-RXII	174 - 230
France	E	13.15	F2, F4	41 - 67.30
			F5	162.25 - 175.40
			F6	162.00 - 175.15
			F7	175.40 - 188.55
			F8	175.15 - 188.30
			F9	188.50 - 201.70
			F10	188.30 - 201.45
			F11	201.70 - 214.85
			F12	201.45 - 214.60
Ireland	I	8	1A-1C	44.5 - 68.5
			1D-1J	174 - 222
South Africa	I	8	4 - 13	174 - 254
France	L	8	A - C	41 - 65
			C1	52.75 - 61.75
			1 - 6	174.75 - 222.75
Japan	M	6	J1 - J3	90 - 108
			J4 - J12	170 - 222
USA	M	6	2 - 4	54 - 72
			5, 6	76 - 88
			7 - 13	174 - 216

Each video broadcast has its video, colour and audio information organised as shown in the Figure 7-26. Tables 2-15 and 7-4 outline the formats used in other systems around the globe. If the audio subcarrier power is too high, it can bleed into the upper adjacent channel video and cause a cross-hatch or pattern over its picture. If the video bandwidth has not been filtered so that a signal below the lower channel edge has not been eliminated, it can bleed into the lower adjacent channel and cause scratching sounds and "cross-talk" on its audio output. In light of these considerations, whenever adjacent channel modulation is used, two rules must be followed. First, the signal must be restricted by a bandpass or SAW filter to protect the lower adjacent channel. Second, the audio level must be maintained at 15 dB below the video carrier.

Channels having adjacent numbers are not necessarily adjacent in frequency. Table 7-3 outlines channel frequency allocations in North America to demonstrate which channels are really adjacent. Since there is a large space between the VHF low band (2 through 6) and the VHF hi-band channels (7 through 13), channel 6 does not have an upper adjacent and channel 7 does not have a lower adjacent signal. Fortunately for system designers, there is also a frequency space between channels 4 and 5 which is a 4 MHz guard band occupied by marine radio communications.

Table 7-3 presents very important information for the headend designer. For example, assume that a small motel wants to feed two satellite channels to each room along with its over-the-air broadcasts. If channels 2, 4, 7 and 9 are occupied with over-the-air broadcasts, satellite signals could be modulated onto channels 11 and 13 with no trouble. But what if a conventional television station is already relaying a broadcast on channel 11. A good choice would be to modulate the satellite signal onto channel 6 which has no upper adjacent channel as well as onto the free channel 13. This sensible choice would save time and money in correctly designing a headend for a small distribution system. This would allow use of less expensive modulators and signal balancing techniques than would be used if channels 5 and 6 were chosen.

Efforts should be made to equalise signal levels being fed into a distribution system. However, precise balancing is not nearly as important when using non-adjacent channels. Therefore, systems should be designed, if possible, with non-adjacent

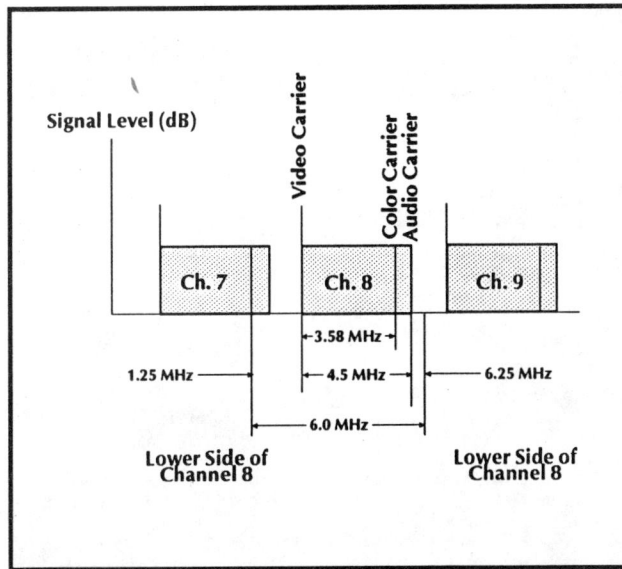

Figure 7-26. NTSC Audio and Video Channel Format. *Each NTSC television channel is organised to have a video carrier centred at 1.25 MHz above the lower side of the channel, a colour subcarrier centered at 3.58 MHz above the video and an audio signal of 0.25 MHz centered below the upper edge of the channel. When mixing or combining adjacent channels it is very important to select modulators and have adjustable audio and video gain controls that filter out the lower sideband which can interfere with reception of the lower adjacent channel.*

channels as targets for modulation. When adjacent channels are used, however, high quality processors and modulators which "clean up" the audio and video and which provide filtering and amplification are necessary.

Headends for more complex systems are usually housed in one temperature controlled location. Any substantial temperature or humidity variations can affect system performance. For example, modulator gains can increase by over 10 dB when temperatures drop from those of a hot summer day to a cold winter night. Or an excessive amount of water vapour will cause electrical components to degrade and have a shorter than normal life span.

MULTIPLE RECEIVER/TV SYSTEMS

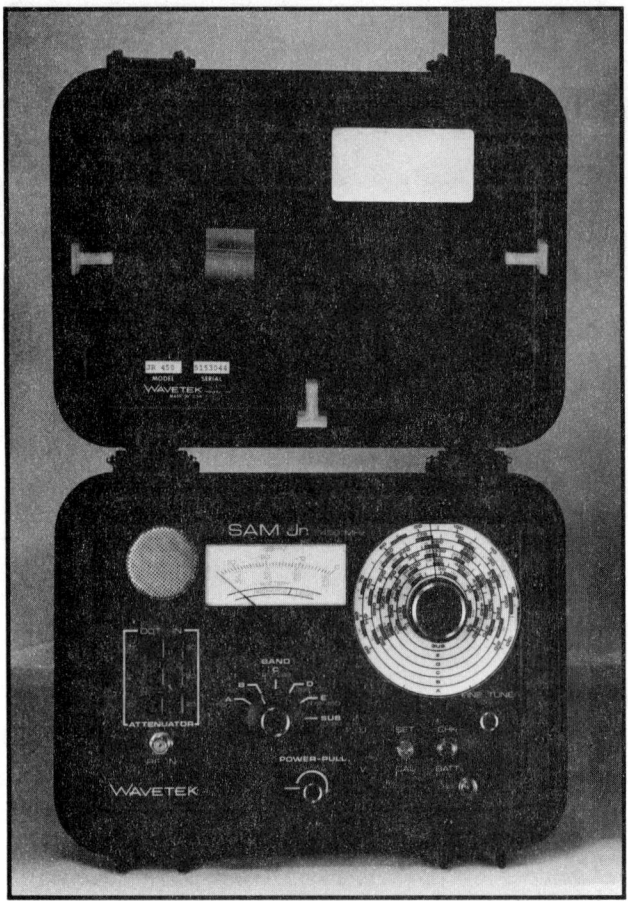

Figure 7-27. Wavetek Sam Junior. *This signal level meter can be used to balance a headend, monitor trunk and distribution amplifier signal levels and measure distribution tap outputs reaching customers' sets. One model operates in a frequency range from 10 to 300 MHz, another spans a 50 to 450 MHz band. Both models have an amplitude sensitivity from -35 to 60 dBmv with an accuracy of 0.75 dB. It is available for use with both PAL and NTSC channel formats. (Courtesy of Wavetek)*

Combining and Distributing the Signal

Signals from each modulator can be combined on a main distribution cable once power levels have been equalised. This requires appropriate adjustment of modulator output controls, bandpass filtering, and occasionally the use of signal processors. Power levels can be calculated to a fair degree of accuracy before installation. For example, if a modulator with 40 dB gain is drawing a -3 dBmv baseband signal from a satellite receiver, it will input 37 dBmv to the distribution system. If an adjacent channel is being relayed at 32 dBmv, balancing could be achieved by either inserting a 5 dB pad in the output line or by readjusting the modulator RF output level. However, no calculation will replace a signal strength meter as an essential tool for installing and aligning headend equipment. A SMATV installer would have a difficult time doing a professional job without this instrument (see Figures 7-27 and 7-28).

At the headend, output signals can be combined using either splitters or taps in reverse. Thus, for example, a 3-way splitter could take three separate signals and combine them into one stream. Or a string of taps in series could feed modulator outputs onto one trunk line. Realise that insertion losses will be the same when using both devices in either forward or reverse directions. A similar but more expensive component, a combiner, may also be a good choice to meet this design requirement (see Figure 7-29).

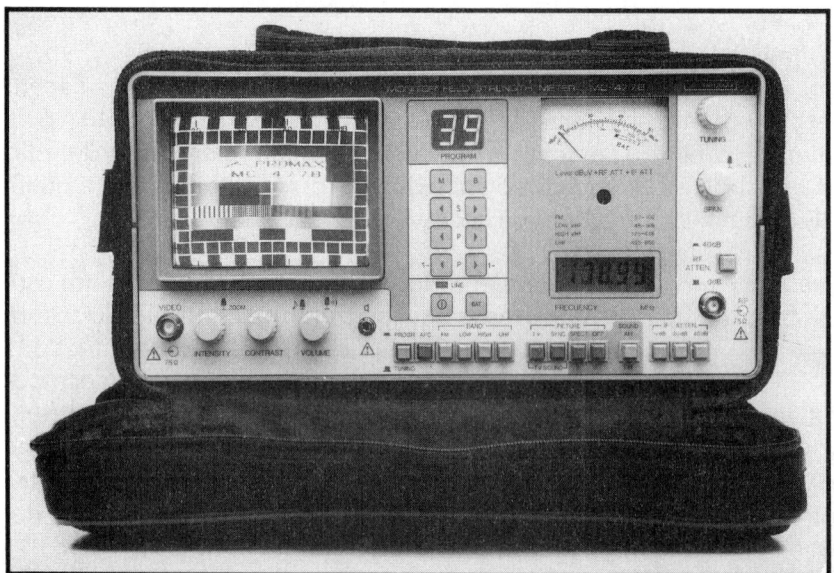

Figure 7-28. Promax MC-477B Field Strength Meter. *The MC-477B, designed for setting up and servicing TV and FM antennas and distribution systems, has a built-in monitor for audio and picture adjustments. Its digital display monitors the frequency as it tunes across the low VHF band from 48 to 169 MHz, the FM band from 87 to 109 MHz, the high VHF band from 455 to 856 MHz and the UHF band from 455 to 856 MHz. This unit also can function as a spectrum analyser. (Courtesy of Instrumentacio Electronica Promax)*

Two types of distribution systems can be designed using either taps or splitters. Splitter are generally used in smaller systems. Designs ranging from larger SMATV systems to city-wide cable networks use a combination of taps feeding the signal to a customer and then use splitters to feed additional television receivers. A main trunk line is never tapped but uses only special hybrid splitters/amplifiers so that the distribution system will be well protected from ingress interference. Then signals are tapped to individual customers from feeder lines. For example, if a hotel having two 5-room buildings were being cabled, a 2-way splitter might be used to feed each building. Then a network of either taps or splitters could be used in each building to feed all five rooms.

The objective in designing distribution systems is to provide a signal ranging from 0 to 3 dBmv to each television. Once the signal voltage leaving the headend is known, it is simply a matter of subtraction to calculate the losses at each tap, splitter or cable run to ascertain that an adequate signal reaches each set.

Figures 7-30 and 7-31 show two finished headends, a compact 12-channel headend manufactured by Nexus Engineering and a larger, more conventional headend using Scientific Atlanta components.

Figure 7-29. Active Combiner. *This active combiner is used to mix up to 12 channels together on one output. It has a 20 dB RF level control, an input frequency ranging from 50 to 300 MHz and a port-to-port isolation of 25 dB. (Courtesy of Nexus Engineering)*

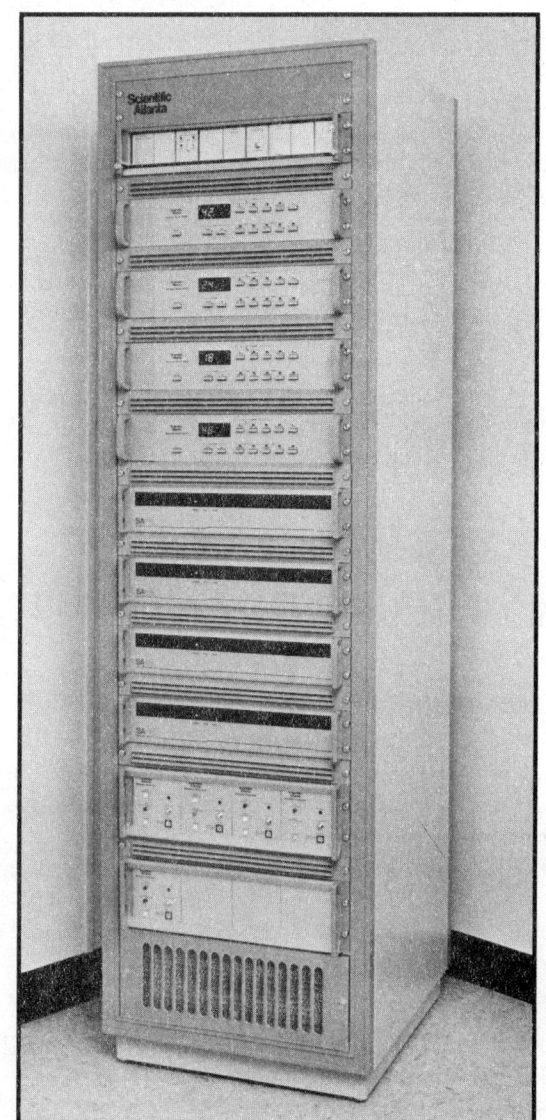

Figure 7-30. Scientific Atlanta Headend. *Four satellite receivers, four B-MAC decoders, four channel modulators and one spare comprise this headend. It received GSTAR 1 broadcasts for the Holiday Inn HighNet system. (Courtesy of Scientific Atlanta)*

MULTIPLE RECEIVER/TV SYSTEMS

Figure 7-31. Compact 12-Channel Headend. *The prepackaged headend of 12 channels include six satellite receivers, six modulators, and six channel processors. It requires a total of only 12 by 19 inches (30 by 48 cm) of space. (Courtesy of Nexus Engineering)*

C. HEADEND AND DISTRIBUTION SYSTEM DESIGNS

Each of the systems described below is diagrammed in the accompanying figures. These systems are taken as representative examples but are by no means the only possibilities. More complex or different systems can easily be designed by using the fundamentals learned in studying these examples.

The Basic Satellite TV Headend

The satellite TV headend most commonly encountered in the home market consists of a single antenna and satellite receiver whose output is combined with over-the-air broadcasts. Either an A/B switch or a channel combiner is incorporated to avoid interference between satellite and over-the-air signals. Many advanced satellite receivers also have a built-in A/B switch. Or as an alternative, a VCR can be used in place of an A/B switch. When its controls are set on "tuner" position, the satellite signal is blocked. In the "camera" or "audio/video" position the composite baseband signal feeds into the VCR. When the satellite audio/video signal feeds directly into the VCR and bypasses the receiver's modulator, recordings will be of higher quality than those from conventional broadcasts which must be demodulated, remodulated and then demodulated for viewing.

One line of RG-6 or RG-59 coax carries the LNB power and RF signal from the antenna to the satellite receiver. A power supply for a VHF or UHF preamplifier is required in the regular over-the-air television line if the signal is relayed from a distant source or if there is a long cable run from the conventional antenna. This amplifier will boost voltage to at least an acceptable 0 to 3 dBmv range as input to the A/B switch or channel combiner.

SYSTEM DESIGN

MULTIPLE RECEIVER/TV SYSTEMS

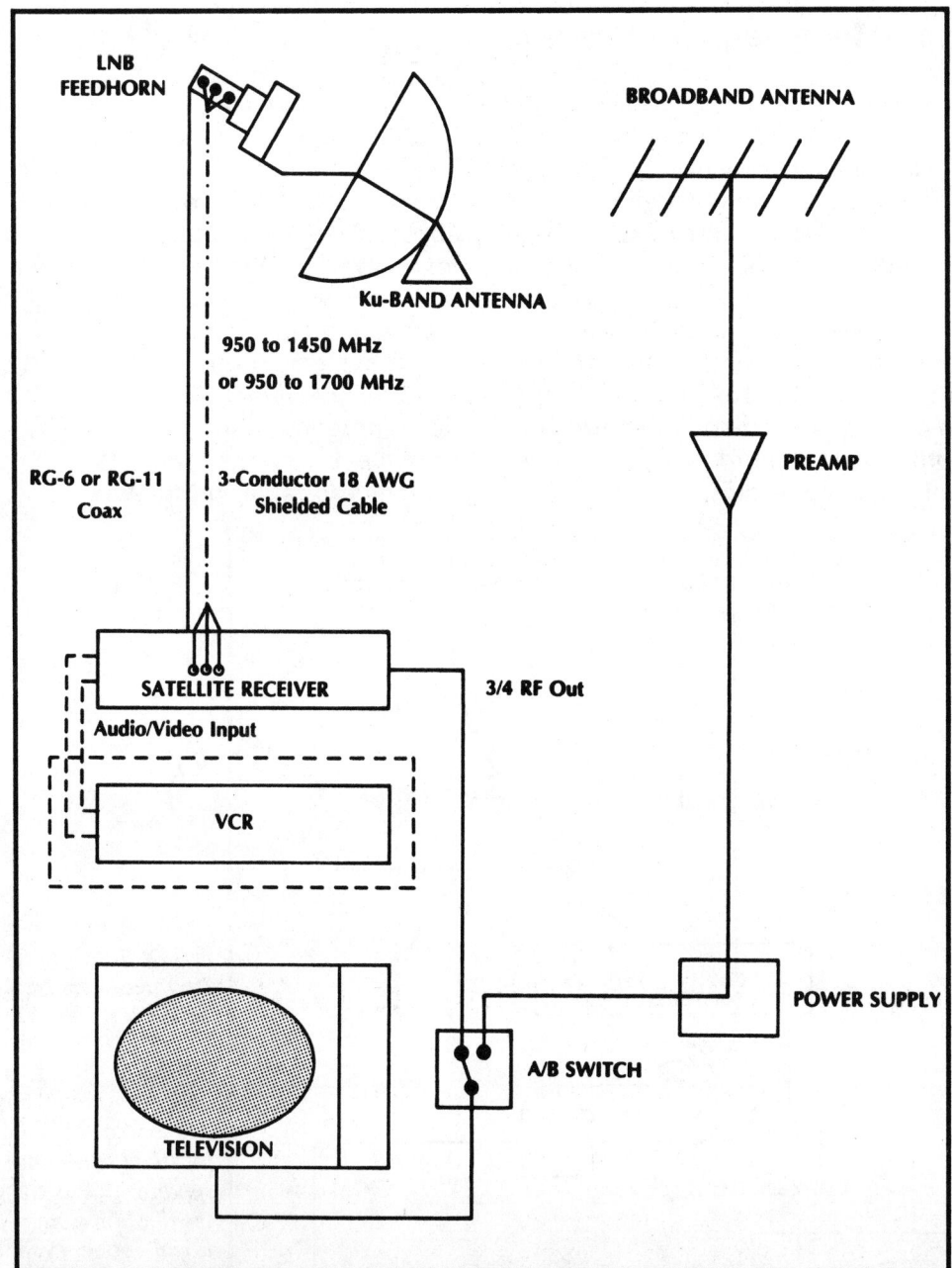

Figure 7-32. Basic Block Downconversion System Wiring Schematic. *The 950 to 1700 MHz (950 to 1450 MHz in North America) input signal is fed from an LNB output through RG-6 cable to a satellite receiver. A three-conductor, 18 gauge shielded wire carries the necessary pulses to control the polariser. The output of the satellite receiver can be fed directly into a TV set through an A/B switch or signal combiner to mix with over-the-air channels. An optional VCR can receive audio and video outputs from the satellite receiver to allow the highest quality tape recordings.*

SYSTEM DESIGN

MULTIPLE RECEIVER/TV SYSTEMS

Single Receiver, Two TVs and Extra Remote Control Headend

This design is similar to the basic system except that an extra TV has been added. The main television is located near the satellite receiver which is assumed to have an infrared, line-of-sight remote control. The satellite receiver can also be controlled from the second infrared transmitter, which is located in another room some distance away usually in the same building, by use of an extra remote control which communicates via the coax to the receiver. Using such a "remote- remote" or an "extra-link" can be very convenient for controlling the receiver from a bedroom or family room. This can be used from up to four rooms.

Some brands of receivers such as the Houston Tracker System VI feature a UHF radio frequency, hand-held remote. The satellite receiver can then be controlled from any location within a radius of about 50 metres (about 160 feet) without the need for extra wires. Even though this remote operates on UHF frequencies in the vicinity of automatic garage door opener range, it is rare that these two different systems interfere with each other. When extra range is needed on such a UHF remote, additional antennas can be installed by running coax from the antenna output port to remote locations via standard cable and splitters.

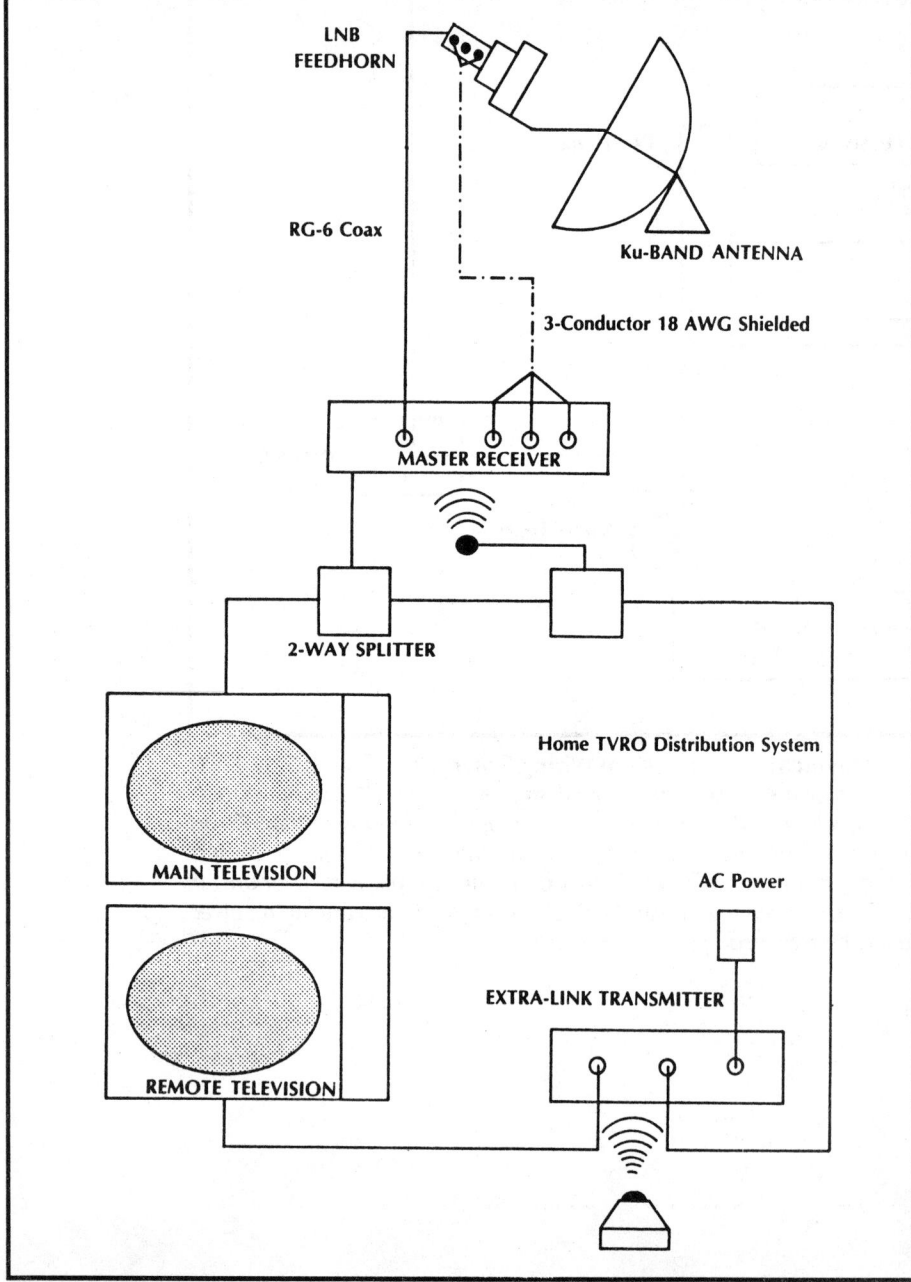

Figure 7-33. Single-Block Downconversion Receiver, Dual-Television Extra-Link System. *With the Extra-Link control system installed, satellite receiver channels, polarisation and audio format can be selected from remote TV locations anywhere in a home by sending pulses back through the distribution system. Its wide operational frequency range permits use of infrared transmitter frequency pulses from many of the hand-held remotes available on the market today.*

SYSTEM DESIGN

MULTIPLE RECEIVER/TV SYSTEMS

Dual Receiver, Single Polarity Headend

This system uses a splitter rated for an LNB output range to feed two receivers. Usually, one side of this splitter is dc blocked so that only the master receiver passes dc current to power the LNB. Either master or slave receiver can independently select any of the channels available on the downconverted block. Only the master receiver controls the polariser probe to select from either horizontal or vertical, or LHCP or RHCP polarisations. The slave unit is tuned to only those polarities selected by the master.

Similar systems can be designed using any number of slave receivers. All that is required is more splitters and, when necessary, a line amplifier rated for these high frequencies to boost signal strength. The master receiver must be always turned on unless an independent power supply and power insertion blocks are used. Otherwise, switching it off would interrupt power to the LNB.

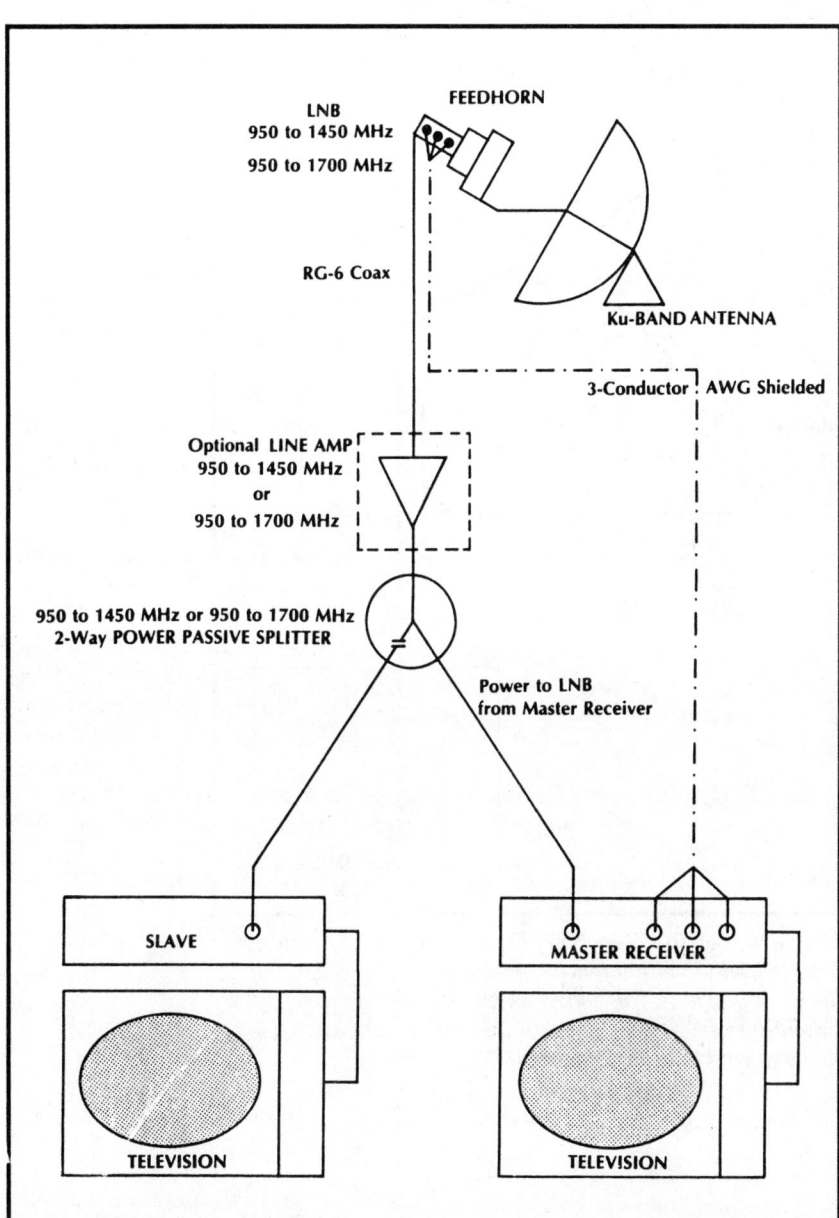

Figure 7-34. **Dual-Block Downconversion Receiver, Single Polarity, Master/Slave Configuration.** *The 950 to 1700 MHz (950 to 1450 in North America) output from the LNB is fed into an optional line amplifier before division by a two-way splitter. In this arrangement, the master receiver on the right controls power through the coax for the line amplifier and LNB as well as switches the polariser between horizontal and vertical polarities. The slave receiver must be tuned to the same polarisation as the master receiver.*

SYSTEM DESIGN

MULTIPLE RECEIVER/TV SYSTEMS

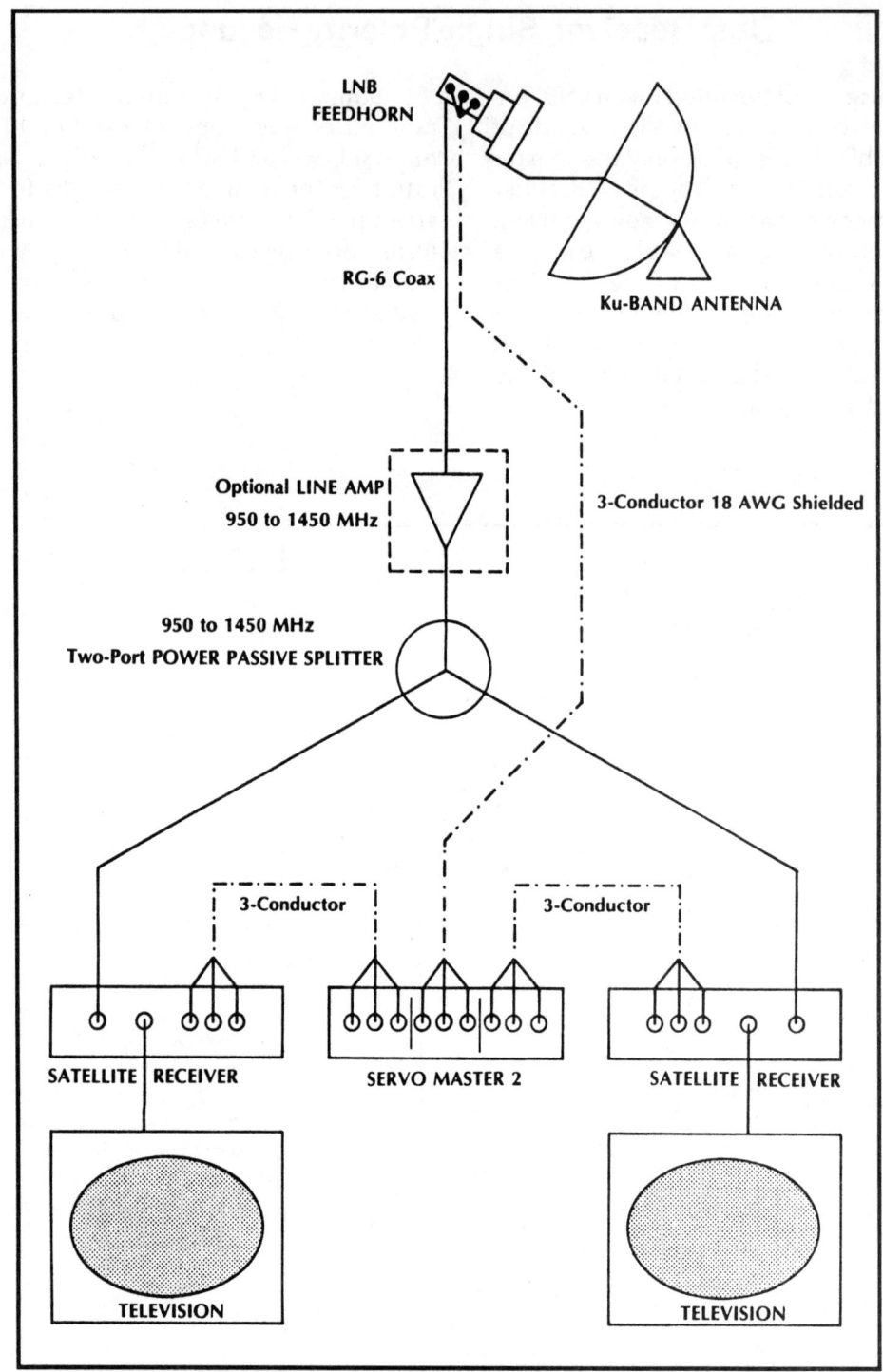

Figure 7-35. Dual-Block Downconversion Receiver, Dual-Polarity, Servo Master Configuration. *The Servo Master II allows either receiver to change polarity and cuts all power to the feed system after three seconds. This allows 24 channel viewing from either installed receiver. Some contention can arise because when either receiver changes channels it may alter the polarity to which the second receiver is tuned.*

MULTIPLE RECEIVER/TV SYSTEMS

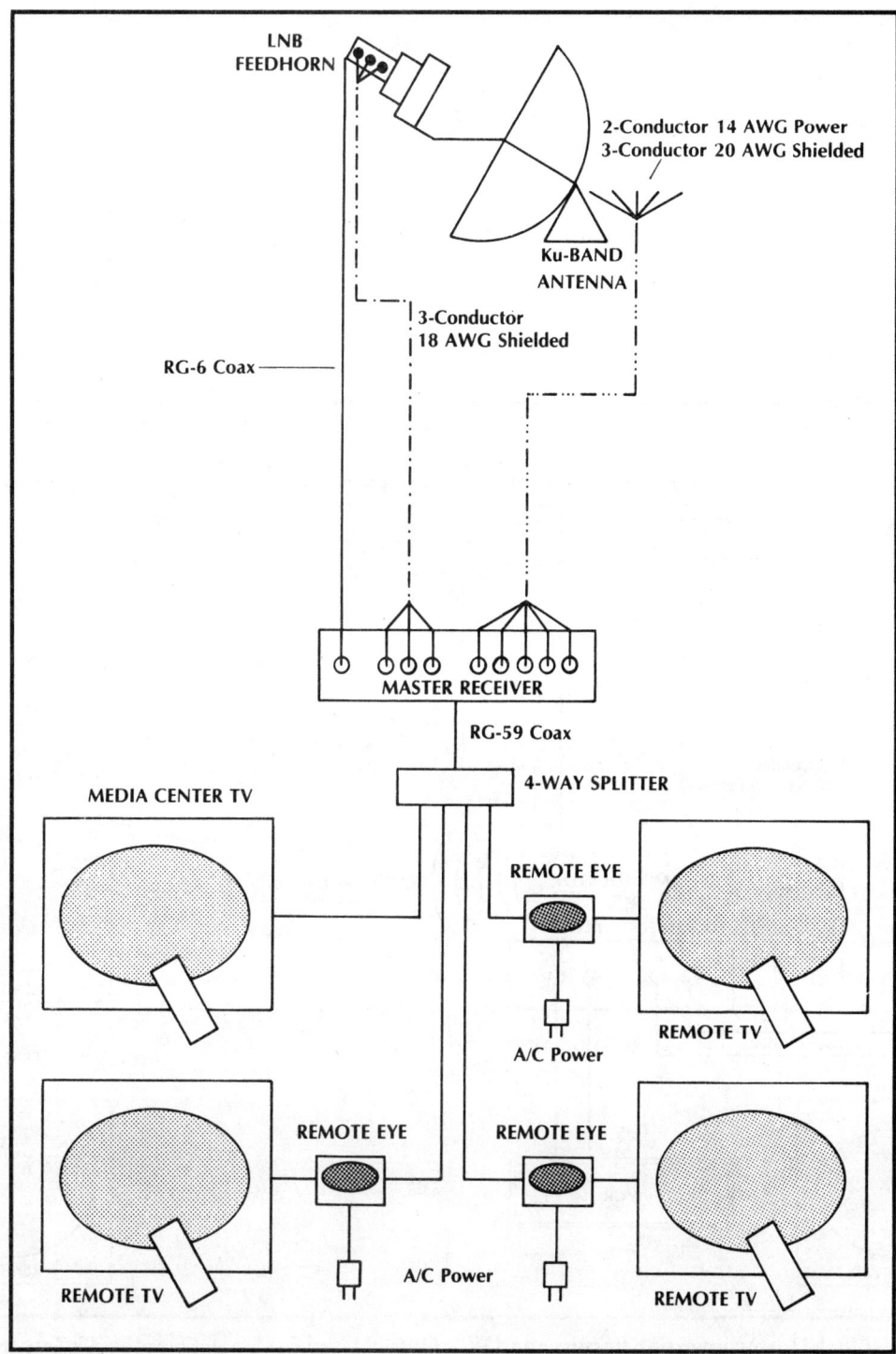

Figure 7-36. Sierra Remote-Eye Diagram. *The Chaparral Sierra or Cheyenne receivers can be controlled from the remote-eye located at any television throughout the video distribution system. All receiver and positioner functions including channel and satellite as well as parental lockout and C to Ku-band switching are displayed on every TV set with unique on-screen graphics.*

SYSTEM DESIGN

MULTIPLE RECEIVER/TV SYSTEMS

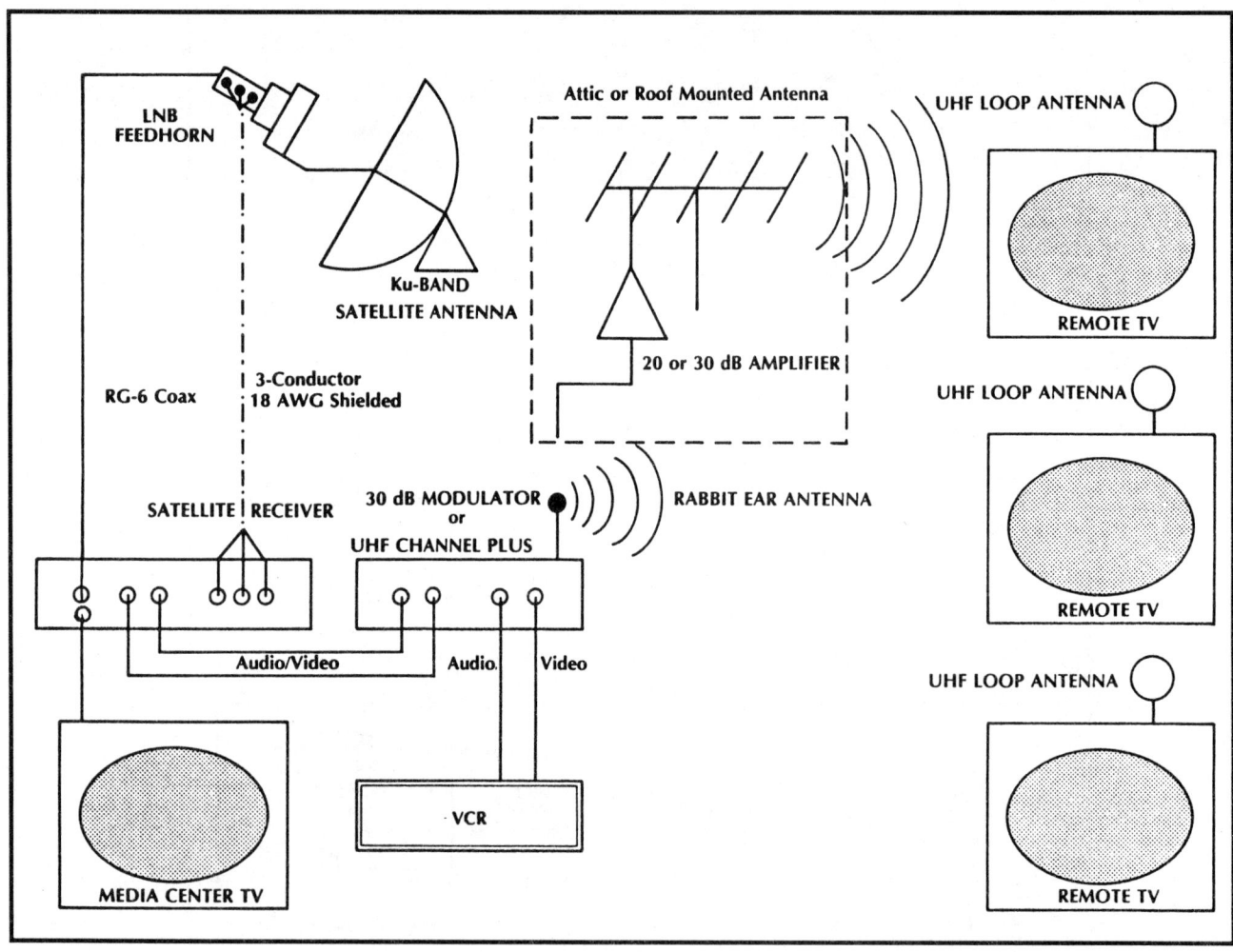

Figure 7-37. Single-Block Downconversion Receiver and VCR Output Used For Small Over-the-Air Transmitter.
The audio and video outputs from either a satellite receiver or VCR can be fed into a modulator or a Channel Plus multiplexer with its output connected to a set of rabbit ears or an external antenna located in the attic of a home. The output RF level should be 20 dBmv. If greater range is required, a higher power modulator or distribution amplifier can be installed with up to a 50 dBmv output. This will allow television sets up to 100 metres away to receive the broadcast with a simple rabbit ear antenna and eliminates the need to install coax. In some countries this "mini-transmission station" may conflict with local broadcasting ordinances. This system is most useful when all receiver and positioner functions can be controlled by a wireless UHF remote like a Houston Tracker VI or Maspro SR-3.

SYSTEM DESIGN

MULTIPLE RECEIVER/TV SYSTEMS

Multiple Receiver, Dual Polarity Headend

Two examples of multiple receiver, dual polarity headends are presented below. Both these systems incorporate special "all-in-one- container" components to accomplish all the required splitting, isolation and power blocking functions.

M/A COM Switching Matrix

The M/A COM switching matrix system uses dual LNBs, each tuned to one signal polarity from a dual orthomode feed, to drive any number of receivers. In this case, a 4-way receiver switch is diagrammed. A plus or minus 12 Vdc terminal on the rear panel of M/A COM receivers activates a network of splitters. As a result, all receivers can independently select any transponder irrespective of either polarity.

DX Switching Matrix

This system is designed like multi-home cable TV installations with taps to as many receivers as are necessary. This design is a first step taken when creating more extensive distribution SMATV systems used for apartments complexes, hotels, condominiums and other multi-dwelling situations. An optional line amplifier can be inserted for especially long cable runs. A separate 24 Vdc power supply feeds both LNBs.

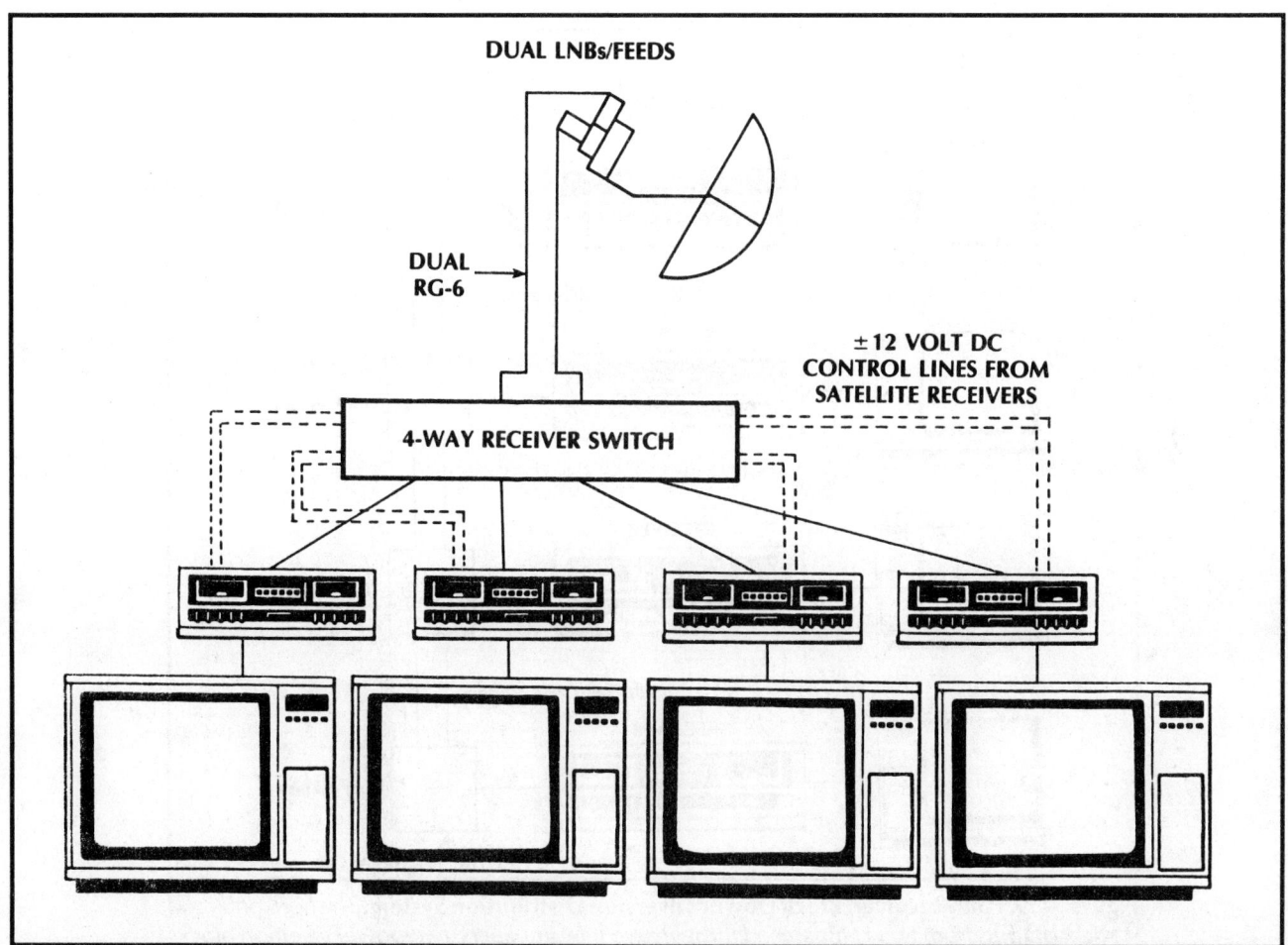

Figure 7-38. Two or More Block Downconversion Receivers with 24 Channel Independent Polarity Control. *The horizontal and vertical outputs from two LNBs are fed into a switching matrix composed of splitters and coaxial relays. Upon command from any of four receivers, the internal switch selects between horizontal or vertical polarity independently of all others. Each receiver must feed a single RG-6 and a two-conductor cable for switching voltage. These switches are not waterproof and should be installed inside a building or within a weatherproof box. Manufactures of these components include Pico, M/A COM, Gensat, Channel Master, and Uniden.*

MULTIPLE RECEIVER/TV SYSTEMS

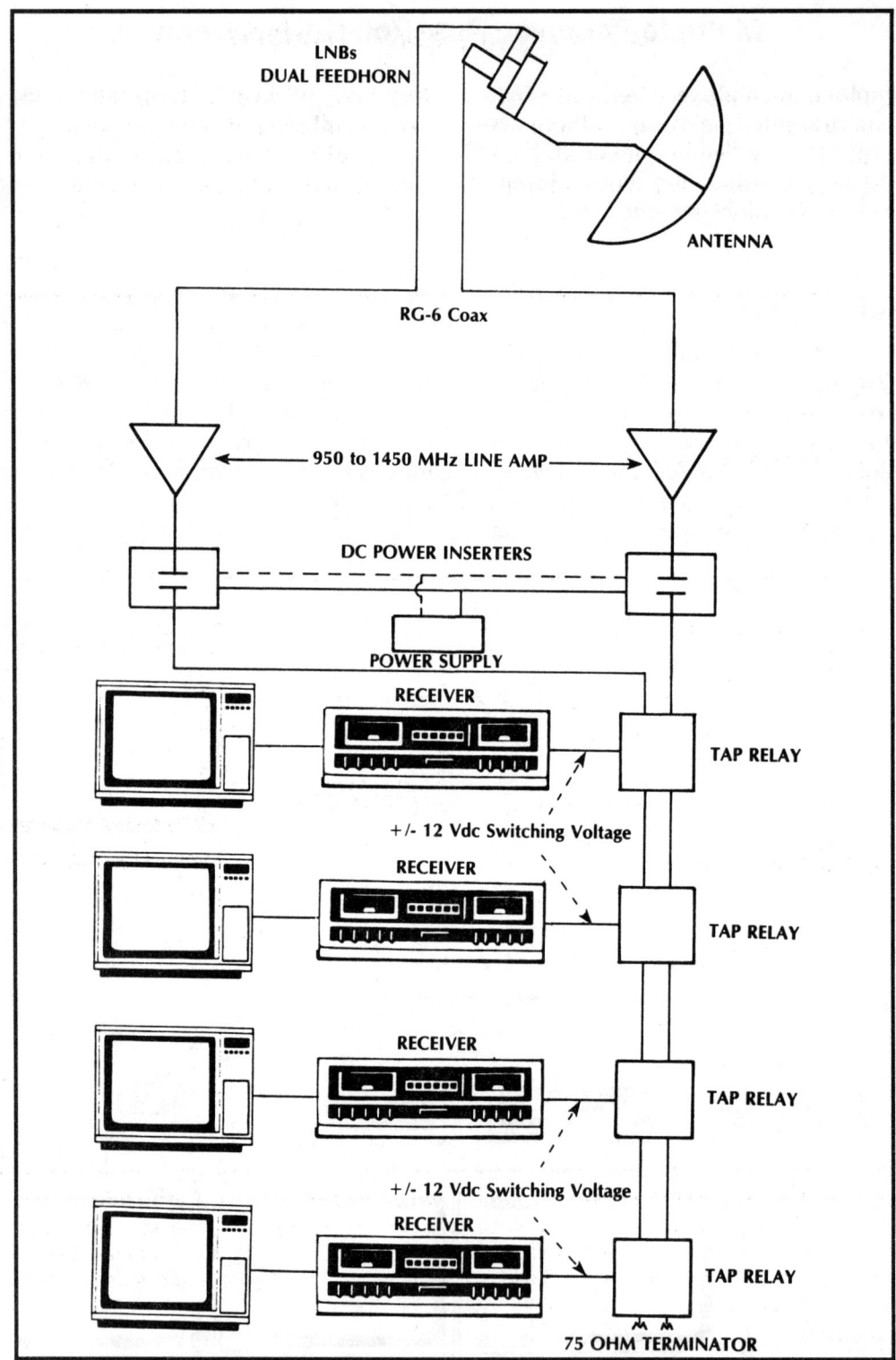

Figure 7-39. Four-Receiver, Block Downconversion Distribution System. *The horizontal and vertical outputs from two LNBs are fed through two line amplifiers, two power insertion blocks and through four tap relays. Care must be taken when designing block distribution systems because the signals at the upper end of the band will suffer high roll-off or attenuation. A spectrum analyser is a helpful tool when installing and troubleshooting these systems. The power supply in the centre constantly feeds amplifier and LNBs via power insertion blocks. If power were provided by a single receiver, switching it off would cut signals to all other receivers. Manufacturers of switch or taps include Gensat and DX Communications.*

MULTIPLE RECEIVER/TV SYSTEMS

Dual Frequency Band Headend

This system permits the user to view either C-band or Ku-band broadcasts over the same dual-band receiver and TV set. Since both the Ku-band and C-band LNB downconvert to the same 950 to 1450 (or 950 to 1700) MHz range, only one receiver is required. A power passing A/B switch activates either LNB. Companies such as DX and Chaparral manufacture such equipment.

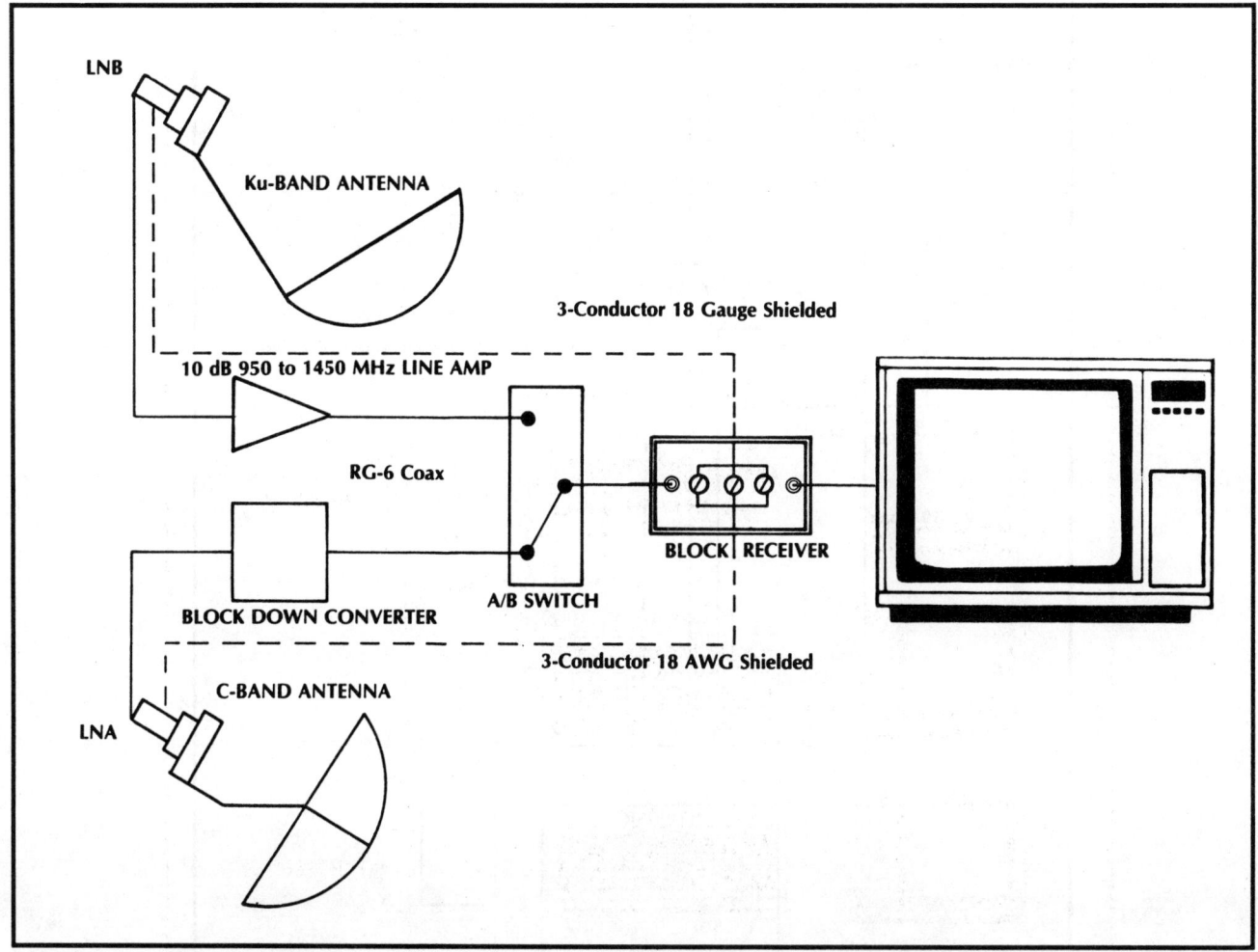

Figure 7-40. Single-Block Downconversion Receiver, Dual-Frequency Band System Using Two Antennas. *The output from both C and Ku-band LNBs are fed through an optional line amplifier to a manual A/B switch. This system requires that one RG-6 cable be connected from each antenna to the indoors A/B switch and receiver. A separate three-conductor, 18 gauge shielded cable is fed to each polariser thus allowing each antenna to have the capability to independently receive all channels transmitted on both polarities.*

SYSTEM DESIGN

MULTIPLE RECEIVER/TV SYSTEMS

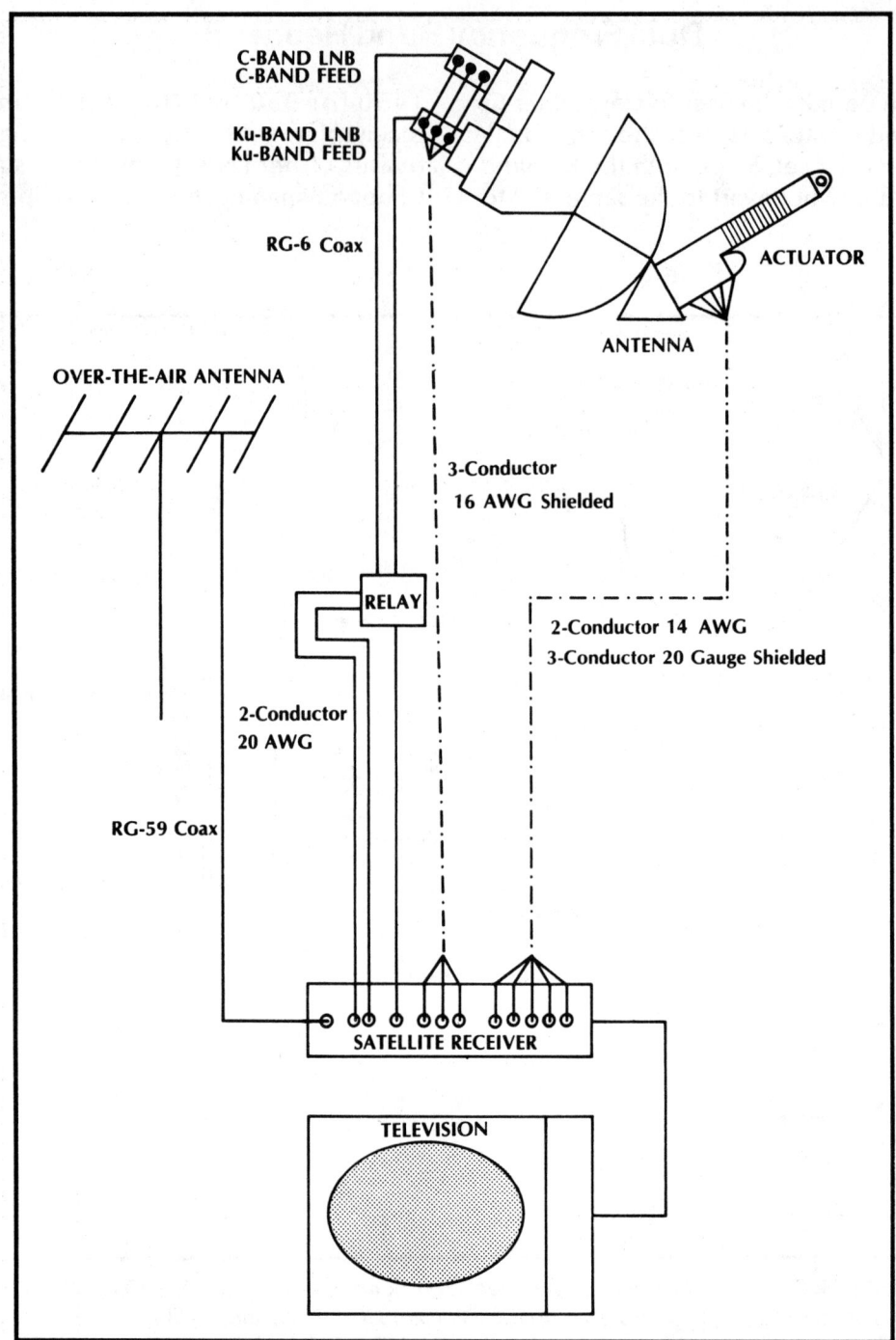

Figure 7-41. Single-Block Downconversion Receiver, Dual-Band, Single Antenna Configuration. C and Ku-band feedhorns can be installed either in a prime-prime or an offset arrangement with the Ku-band feedhorn centred. Both polariser motors are wired in parallel at the antenna. A heavier than normal 3-conductor, 18 gauge or thicker wire should be used for connection to the receiver. A capacitor may also have to be installed at the polariser motors to provide an additional power boost. A five-conductor cable consisting of a two-conductor, 14 gauge wire for the motor and a three-conductor, 20 gauge shielded wire for the actuator sensor is also used. Outputs from the C/Ku-band LNBs are fed into a coaxial relay so that switching between the two bands can be accomplished upon command from the satellite receiver. The over-the-air antenna is connected to either an A/B switch or the receiver VHF input. In this latter case, switching between satellite and over-the-air broadcasts will occur automatically when the receiver is turned off.

MULTIPLE RECEIVER/TV SYSTEMS

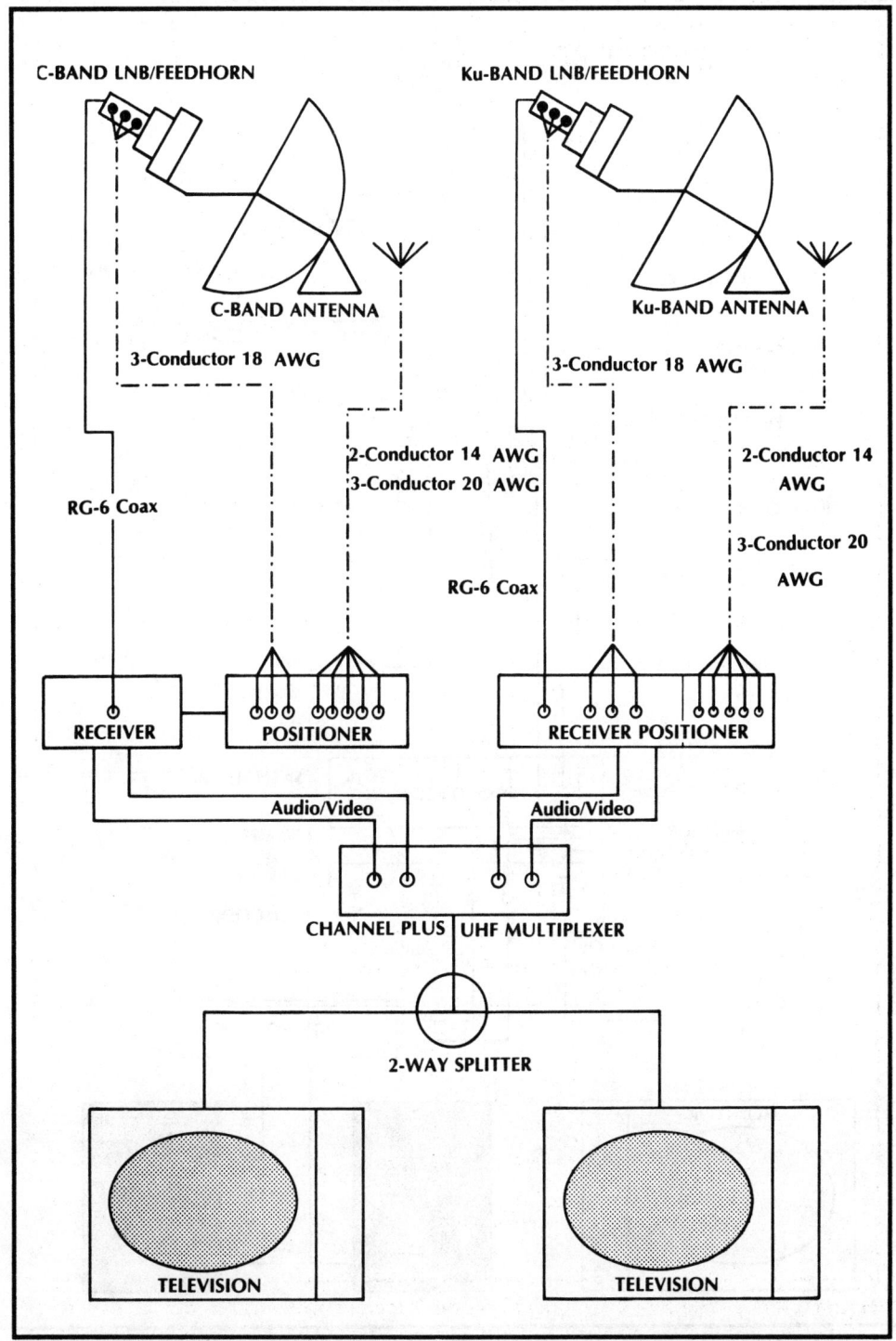

Figure 7-42. Dual-Block Downconversion Receiver, Dual-Antenna, Dual-Actuator C/Ku-Band System. *The audio and video outputs from both C/Ku-band receivers are fed into a Channel Plus UHF multiplexer and then to two television receivers. A receiver having a UHF remote control is useful in this system.*

SYSTEM DESIGN

MULTIPLE RECEIVER/TV SYSTEMS

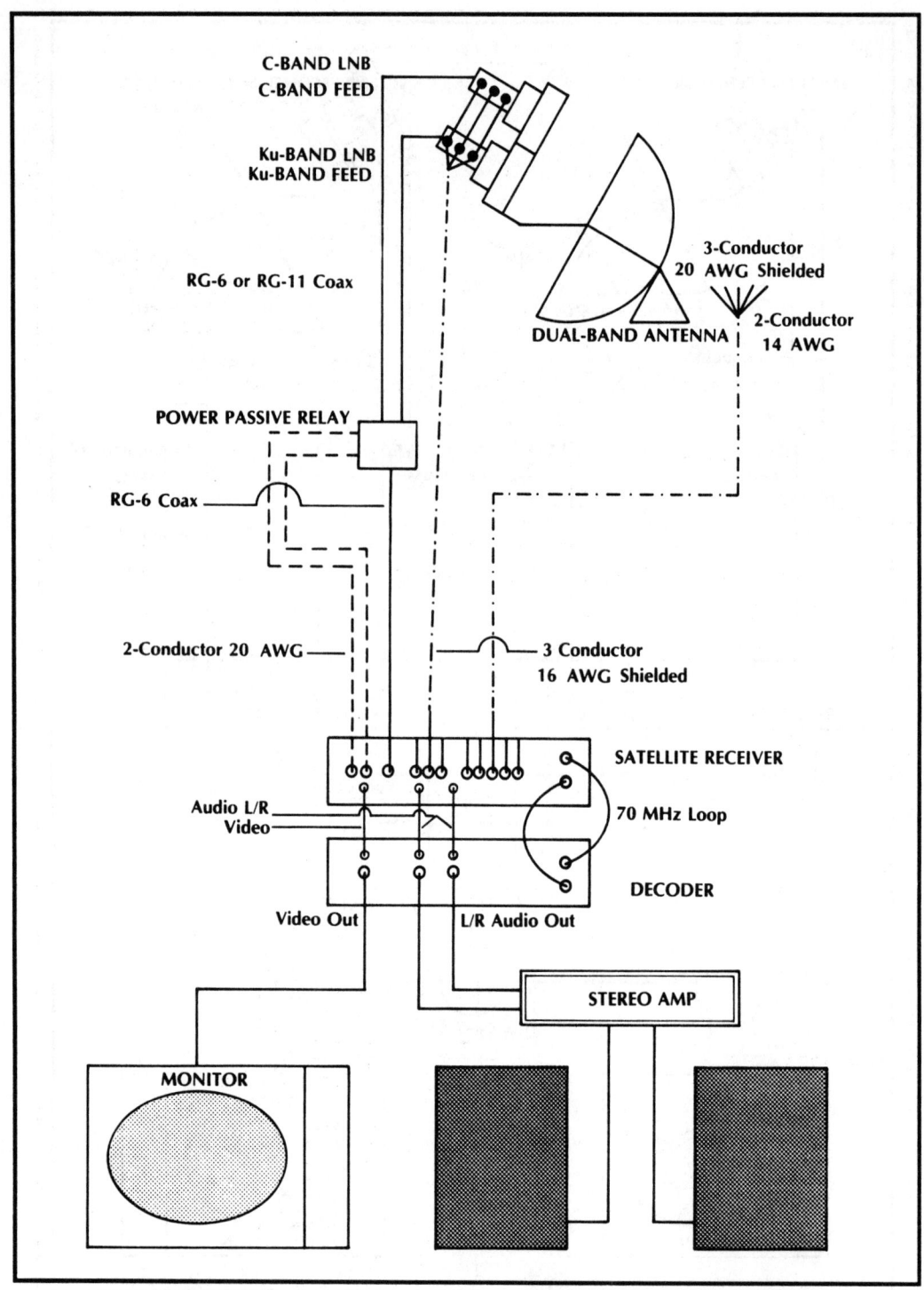

Figure 7-43. Single-Block Downconversion Receiver, C/Ku-band with Descrambler Interface.
The output from the C or Ku-band LNB is fed through an electronic A/B switch into the block input of a satellite receiver. The 70 MHz IF plus the audio/video outputs of the receiver are fed into the descrambler. In some satellite receiver designs the composite baseband output is used. The channel 3/4 decoder video output is fed to a television or a monitor and the audio is fed into a home stereo amplifier. With this arrangement a customer has the capability of decoding either C or Ku-band scrambled broadcasts.

MULTIPLE RECEIVER/TV SYSTEMS

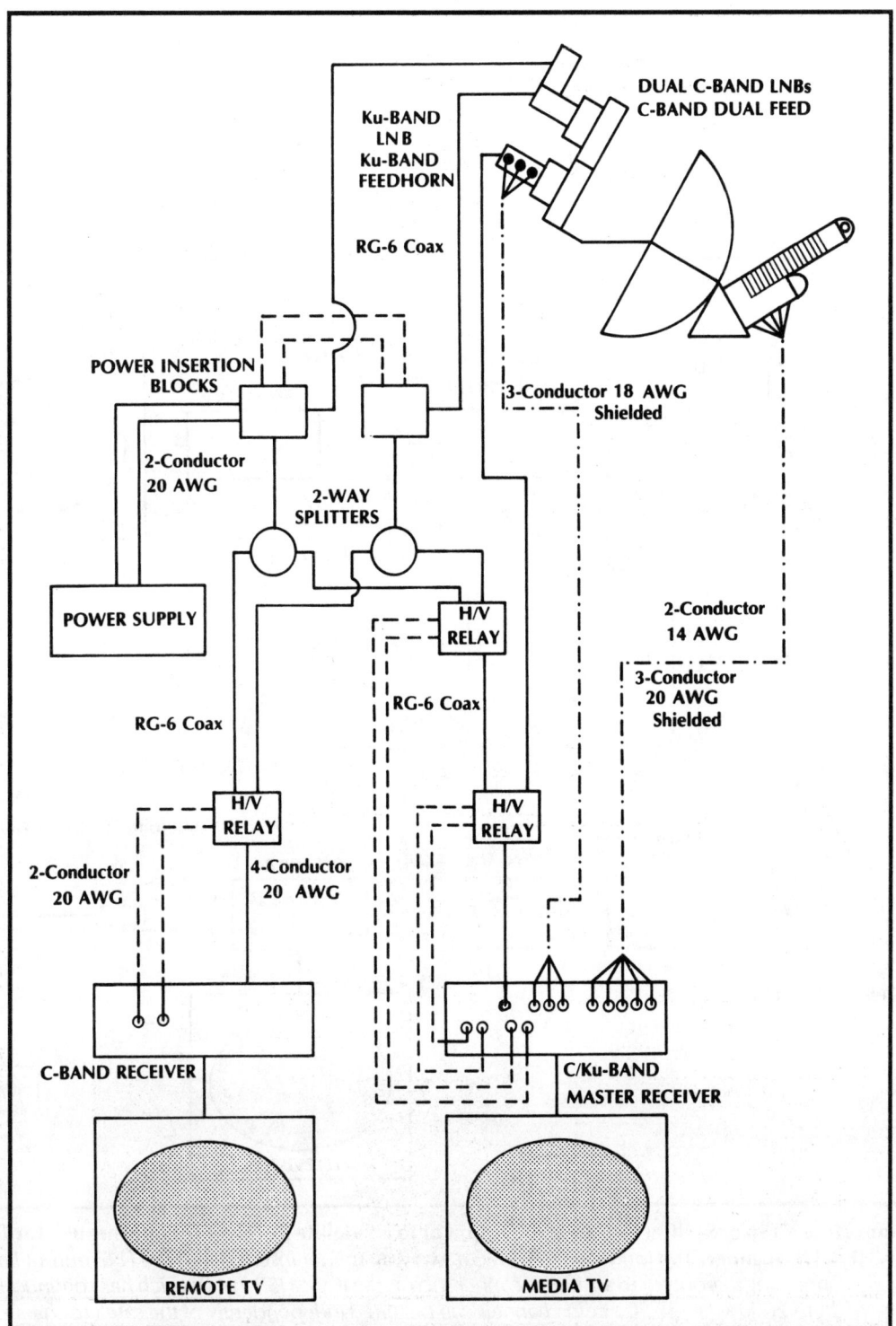

Figure 7-44. Dual-Block Downconversion Receiver, Single-Antenna C/Ku-Band Configuration. *In this wiring arrangement, the slave receiver is dedicated to one band. Both C-band polarities are relayed to both receivers. However, Ku-band signals are only relayed to the master receiver.*

MULTIPLE RECEIVER/TV SYSTEMS

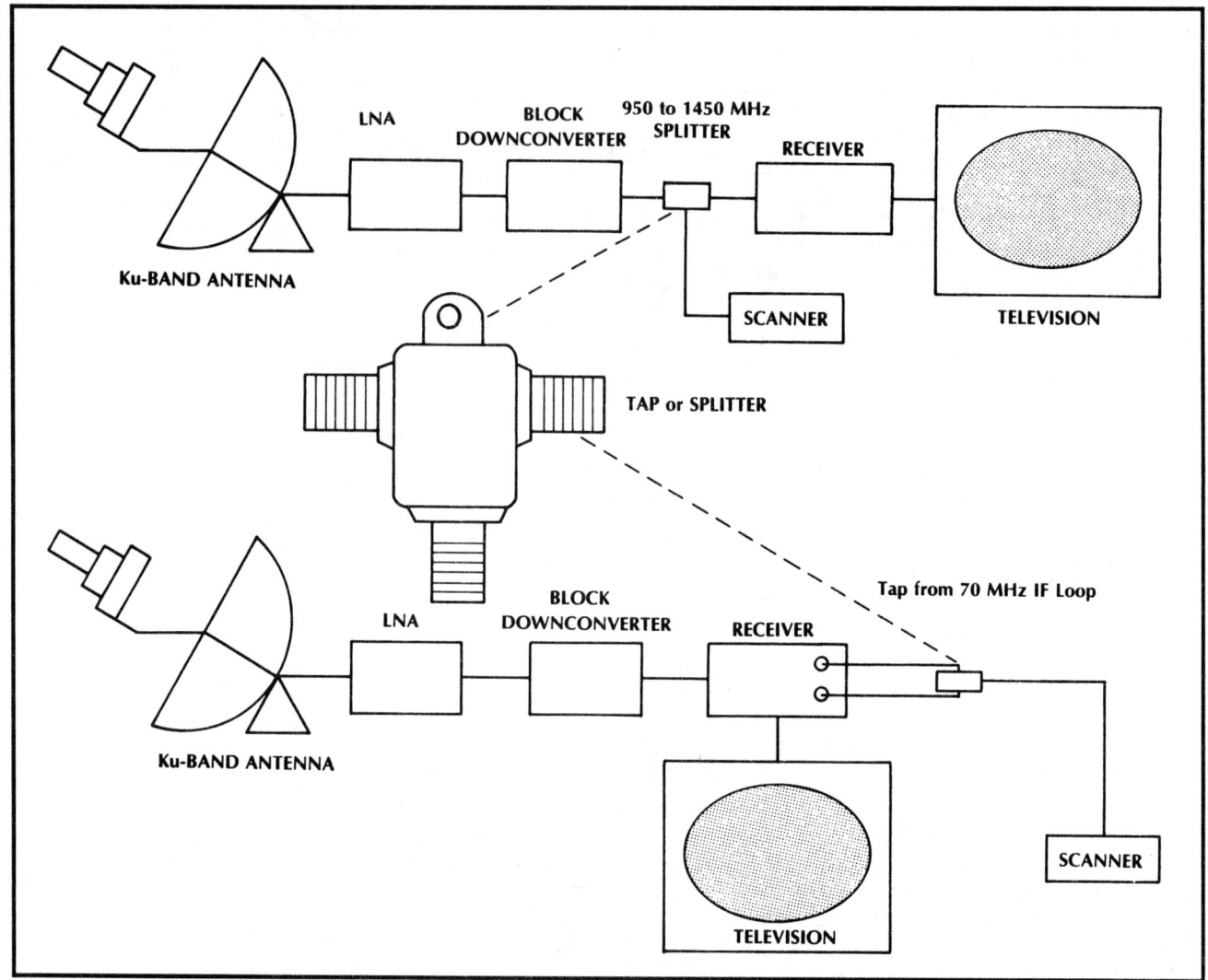

Figure 7-45. Connecting a Tap or Splitter into an LNB Output or to a Satellite Receiver IF Loopthrough for Tuning SCPC Services with a VHF/UHF Scanner. *The top drawing shows a two-way splitter inserted into the LNB output line before the satellite receiver. By connecting this output to an audio scanner such as the ICOM ISR-7000 which has continuous tuning over a frequency range of 25 to 2000 MHz, SCPC audio channels can be tuned independently of the satellite channel tuning. The bottom drawing shows the use of a tap inserted into the satellite receiver IF loop, which is typically 70 or 134 MHz. VHF/UHF scanner models such as the Regency model MS-5000 or the JIL SX-400 can tune over 500 MHz which exceeds the IF bandwidth of most satellite receivers. The only drawback when taping into the IF line is that the receiver must be tuned to the same transponder that the scanner is detecting. Tapping into the block IF output frees the satellite receiver to tune to any video channel independently of the scanner selection.*

MULTIPLE RECEIVER/TV SYSTEMS

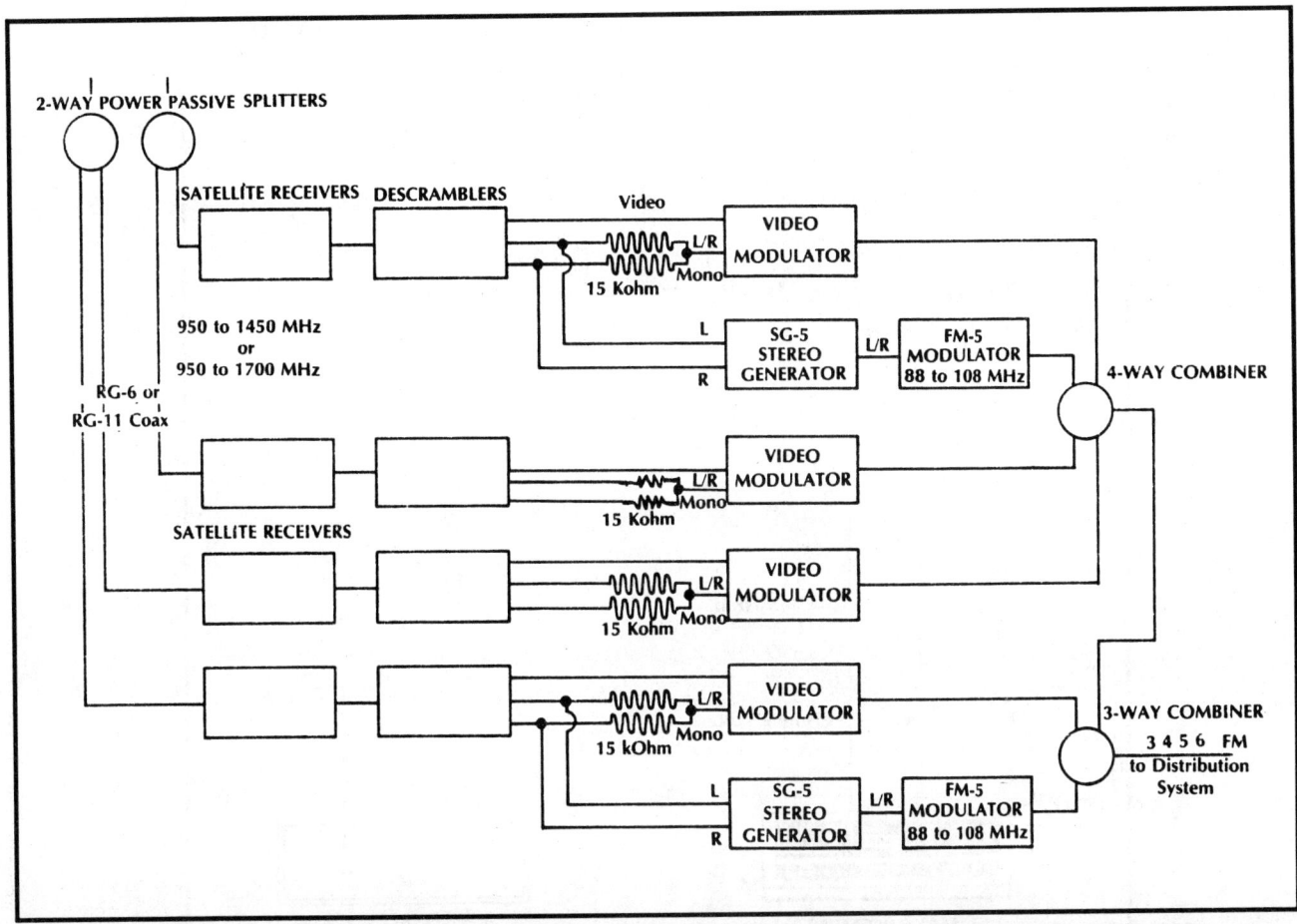

Figure 7-46. **FM Stereo Simulcast Schematic.** *Here the LNB block output is fed into satellite receivers. Their outputs can either be through the 70 MHz IF loop or the composite baseband output to a descrambler. The video output from a descrambler is fed into a video modulator. The left and right stereo audio outputs from a descrambler are fed into a stereo generator and video modulator. The SG-5 stereo generator multiplexes the left and right signals onto one output, which is fed into an FM modulator in the 88 to 108 MHz range. A single frequency different from any local FM broadcast must be selected. The video output from the modulator and the FM audio output from the FM modulator are mixed into a combining network and fed through the distribution system. A customer would tune his television to the appropriate channel, connect a stereo receiver through a tap to the FM antenna input and tune onto the appropriate FM audio channel for simulcast reception.*

MULTIPLE RECEIVER/TV SYSTEMS

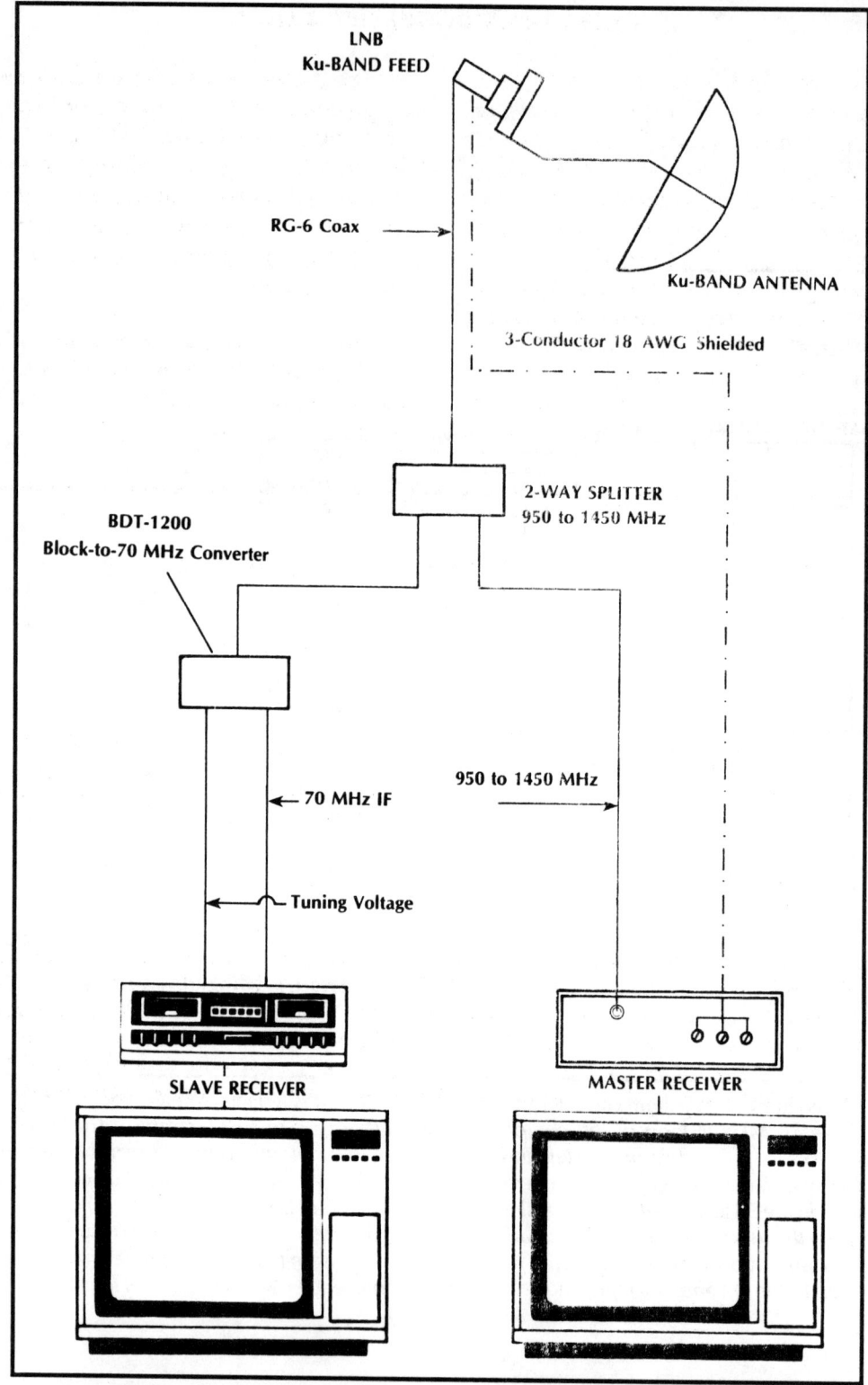

Figure 7-47. Converting a 70 MHz Receiver to Ku-band Reception Using the Gensat BDT 1200 Block Translator. *A standard 70 MHz satellite receiver can accept a 950 to 1450 Mhz LNB output if it is first fed into the Gensat BDT 1200. The receiver's IF output and a separate two-conductor cable from the 70 MHz receiver connects to the BDT 1200. The receiver can then be tuned to any satellite channel within this block downconversion band.*

MULTIPLE RECEIVER/TV SYSTEMS

4, 7 and 12 Channel Headends

A 4-channel headend might be located at a hotel in an area where no over-the-air television is available. In this case, four satellite TV channels are being fed into the distribution system. Following modulation, the channels are combined in a 4-way splitter before being added together on one output line. This system design protects adjacent channels from interfering with each other. Since all channels are not adjacent in frequency, lower cost modulators without bandpass filters can probably be used. In general, this is not advised when high quality results are expected.

The 7-channel headend combines four satellite channels with 3 over-the-air broadcasts onto one distribution network. In this case, care has to be taken in protecting channels 8 and 9 from interfering with each other. Signal levels must be properly balanced and a good quality bandpass filter, either as an extra component or built into the channel 9 modulator, is necessary.

The 11-channel headend occupies 11 of the 12 available VHF channels. An AC-5 combiner mixes all channels together and feeds them into the distribution network.

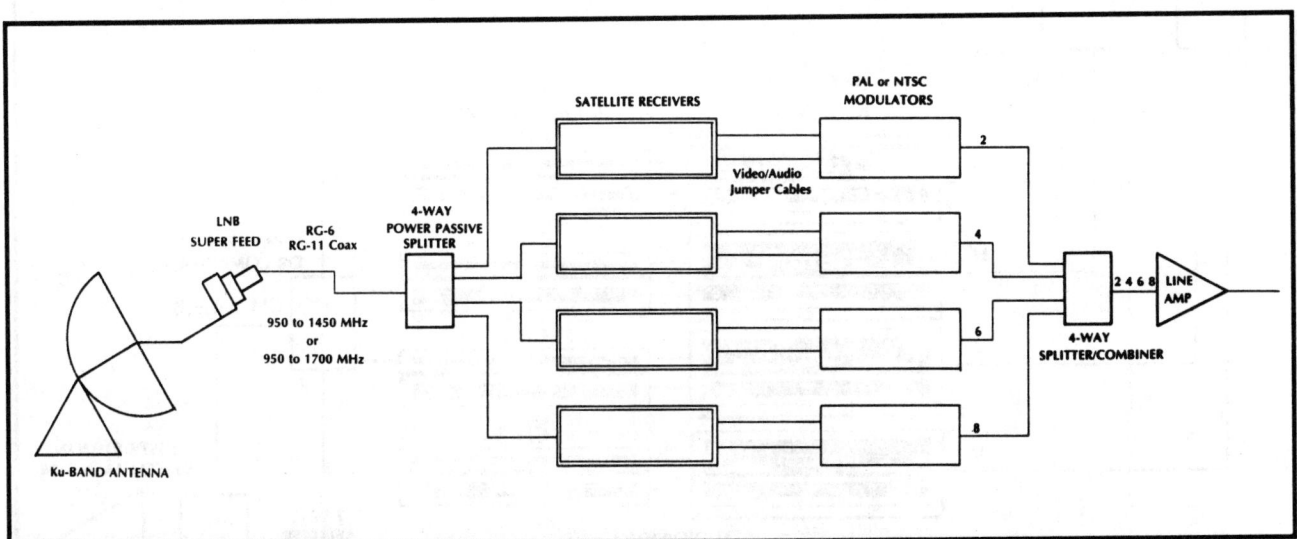

Figure 7-48. Four-Channel, Single-Polarity Headend. *The LNB block output is fed through RG-6 or RG-11 cable into a 4-way splitter located at the headend. The output from the splitter is fed into the satellite receivers' IF inputs. A single receiver will supply power through the power passive port on the splitter to the LNB. The audio and video receiver outputs are modulated onto channels not occupied by a local over-the-air broadcast. The modulated outputs are then added together using a 4-way splitter used in reverse as a combiner. The signal is coupled to a line amplifier where appropriate levels are transmitted through the distribution system.*

SYSTEM DESIGN

MULTIPLE RECEIVER/TV SYSTEMS

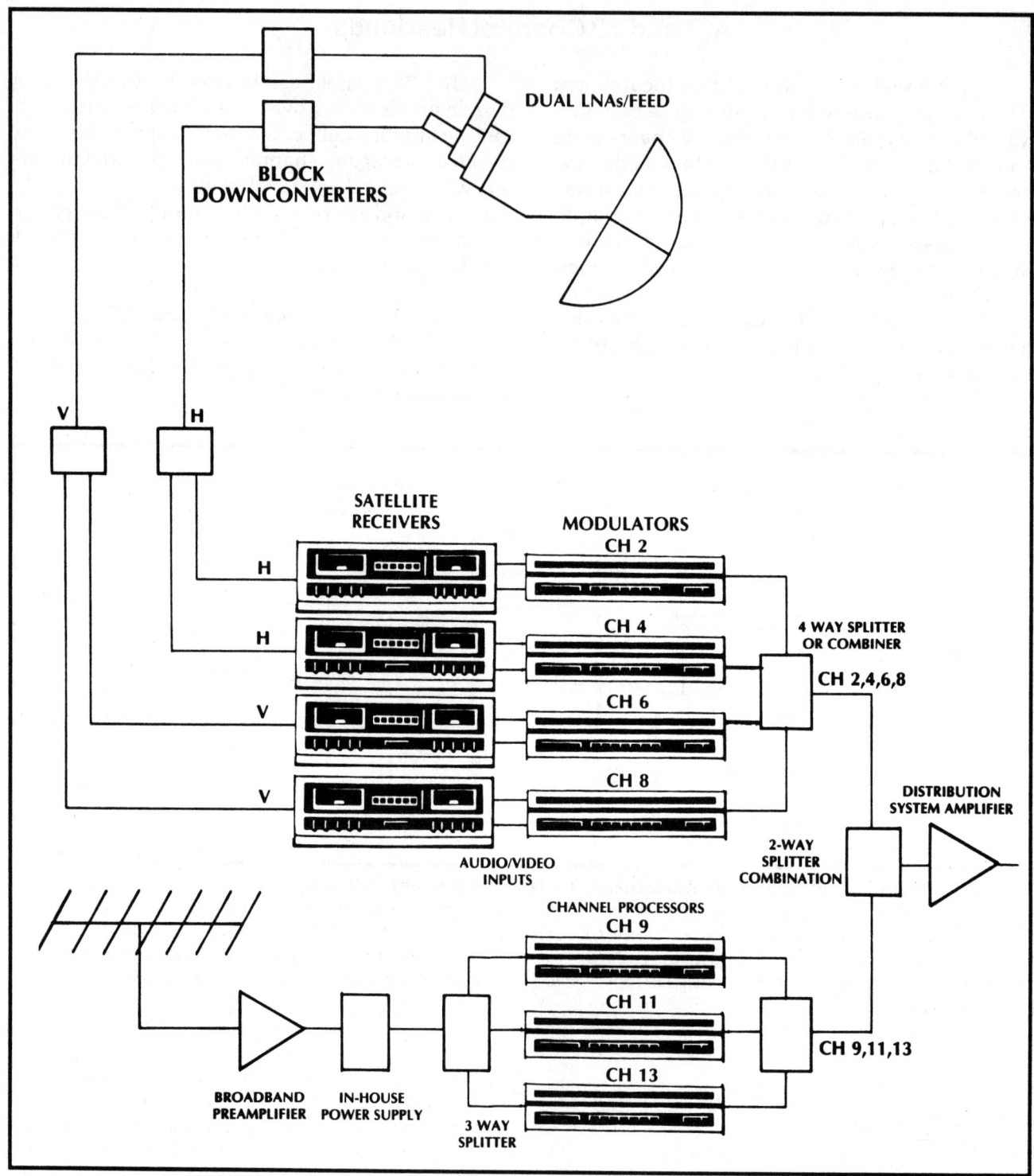

Figure 7-49. Seven-Channel Headend with Over-the-Air Television. *Four satellite receivers are mixed together like in Figure 7-69 except that two horizontal and two vertical inputs are fed into the satellite receivers from two LNBs and a dual feed. A local over-the-air broadcast is fed into the headend from a broadband antenna. Its output enters a pre-amplifier which is mounted at the antenna. The 10 to 20 dB of gain compensates for cable losses. A power supply feed via the coax to the pre-amp. The amplified signal passes via a 3-way splitter into a channel processor. This device reconstitutes the signal so that it has video of equivalent quality as the satellite channels. Outputs are then fed into a 3-way splitter and subsequently via a 2-way splitter to mix with the satellite modulated channels. The combined outputs of both satellite modulators and the over-the-air channels are then fed through a line amplifier to the distribution system.*

MULTIPLE RECEIVER/TV SYSTEMS

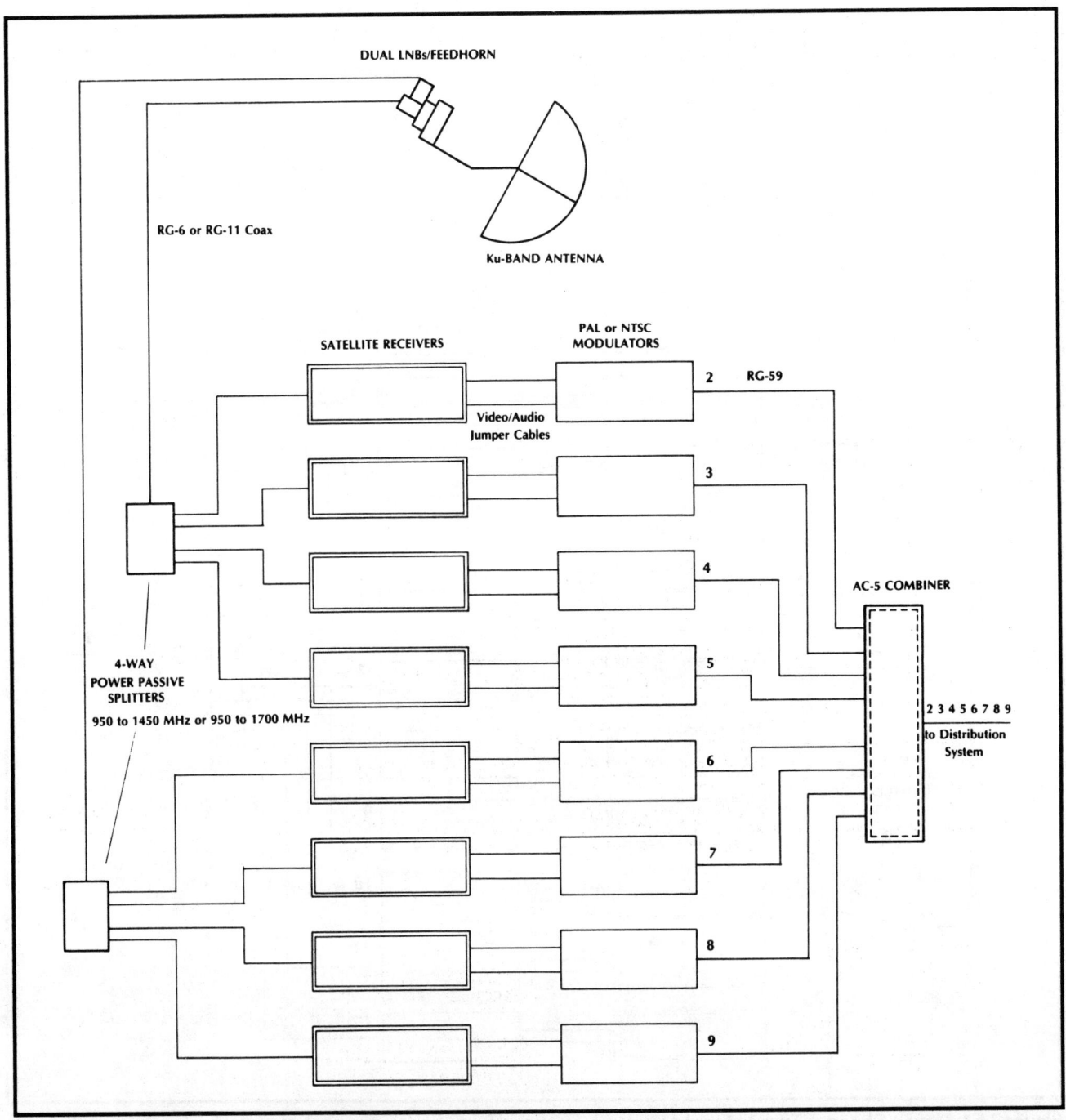

Figure 7-50. Eight-Channel Satellite Headend. *Outputs from the horizontal and vertical LNBs pass via coaxial cable to the headend. The signal is then split in two 4-way splitters which produces four vertical and four horizontal outputs for relay to the satellite receiver IF inputs. A single receiver in each bank supplies power through a power passive port on each one of the 4-way splitters. The audio and video outputs from the satellite receivers are then fed into the modulator inputs for appropriate channel output selection and subsequent mixing in a channel combiner. Most combiners have a 20 dB output which combined with the modulator gain outputs can provide as much as 50 dBmv to the distribution system. All of this equipment is rack mounted and installed in a well-ventilated room. Year-round active temperature control is recommended to maintain signal levels constant and to minimise frequency drift.*

MULTIPLE RECEIVER/TV SYSTEMS

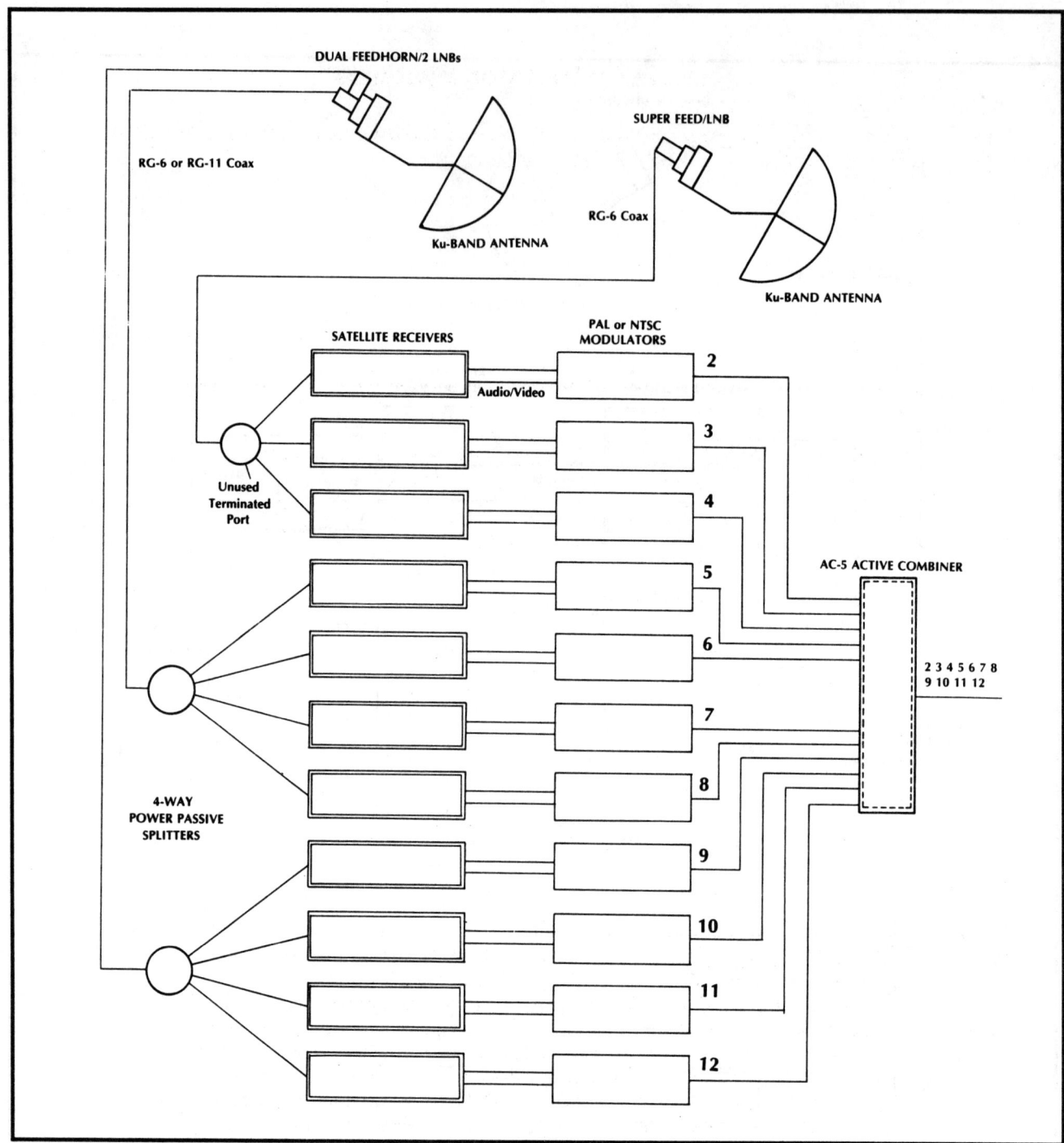

Figure 7-51. Eleven-Channel, Two-Antenna Headend. *The signal from the antenna on the left is mixed together in the same fashion as in Figure 7-71. The antenna on the right has a single polarity output selected by the feedhorn. The LNB output is fed via RG-6 or RG-11 cable into a 4-way, power-passive splitter with one port terminated. These three outputs pass into the satellite receiver IF inputs and subsequently to modulators. All these modulated outputs are then mixed via a channel combiner with the single coaxial cable output feeding into the distribution system. Signal levels on the modulator outputs should have approximately 5 dB of tilt between the low and the high channels. In this example, the outputs which are fed into an amplifier to compensate for any high frequency roll-off are:*

Channel Number	Modulator Output (dB)	Channel Number	Modulator Output (dB)
2	20	8	23
3	20.5	9	23.5
4	21	10	24
5	21.5	11	24.5
6	22	12	25
7	22.5		

MULTIPLE RECEIVER/TV SYSTEMS

Sample Distribution Systems

One simple distribution system using taps and splitters is shown below. This diagram clearly demonstrate how power levels can be traced to their final destination.

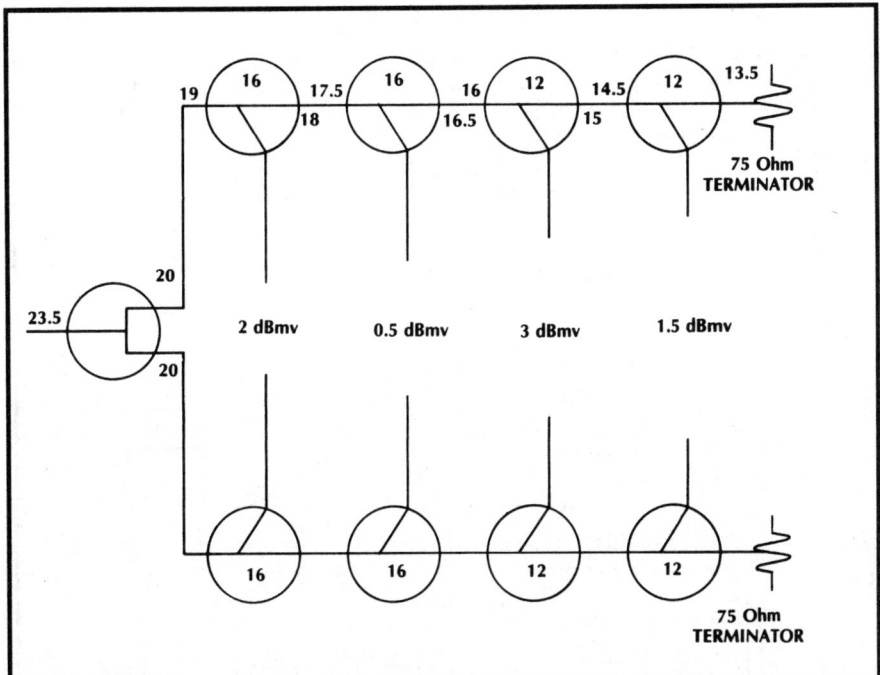

Figure 7-52. Eight-Drop Distribution System. *A broadcast distribution network can be compared to a water sprinkler system in which each outlet should have as much water pressure as any other. Each leg in a distribution network should have enough power to properly drive the audio and video equipment. In this example, the 23.5 dBmv signal would reach the 2-way splitter and feed 20 dBmv into each branch. A 1 dB cable loss is incurred on the way to the first tap resulting in a 19 dBmv input. Then a 16 dB tap which has an insertion loss of 1 dB lowers the signal level to 18 dBmv. Finally, 18 dBmv less the 16 dB tap value or 2 dBmv reaches the first television receiver. 17.5 dBmv arrives at the second tap after a 0.5 dB cable loss. Then 16.5 dBmv less 16 dB tap value equal to 0.5 dBmv enters the second TV. 16 dBmv is relayed to the input of the third tap. Inserting another 16 dB tap here would cause levels to fall below the designed 0 to 3 dBmv required television input, so therefore a 12 dB tap is used. Subsequently 15 dBmv less 12 dB or 3 dBmv is fed to the television receiver. 14.5 dBmv less a 1 dB tap insertion loss less the 12 dB tap value equal to 1.5 dBmv arrives at the final television set. A 75 ohm terminator is then used on the output.*

MULTIPLE RECEIVER/TV SYSTEMS

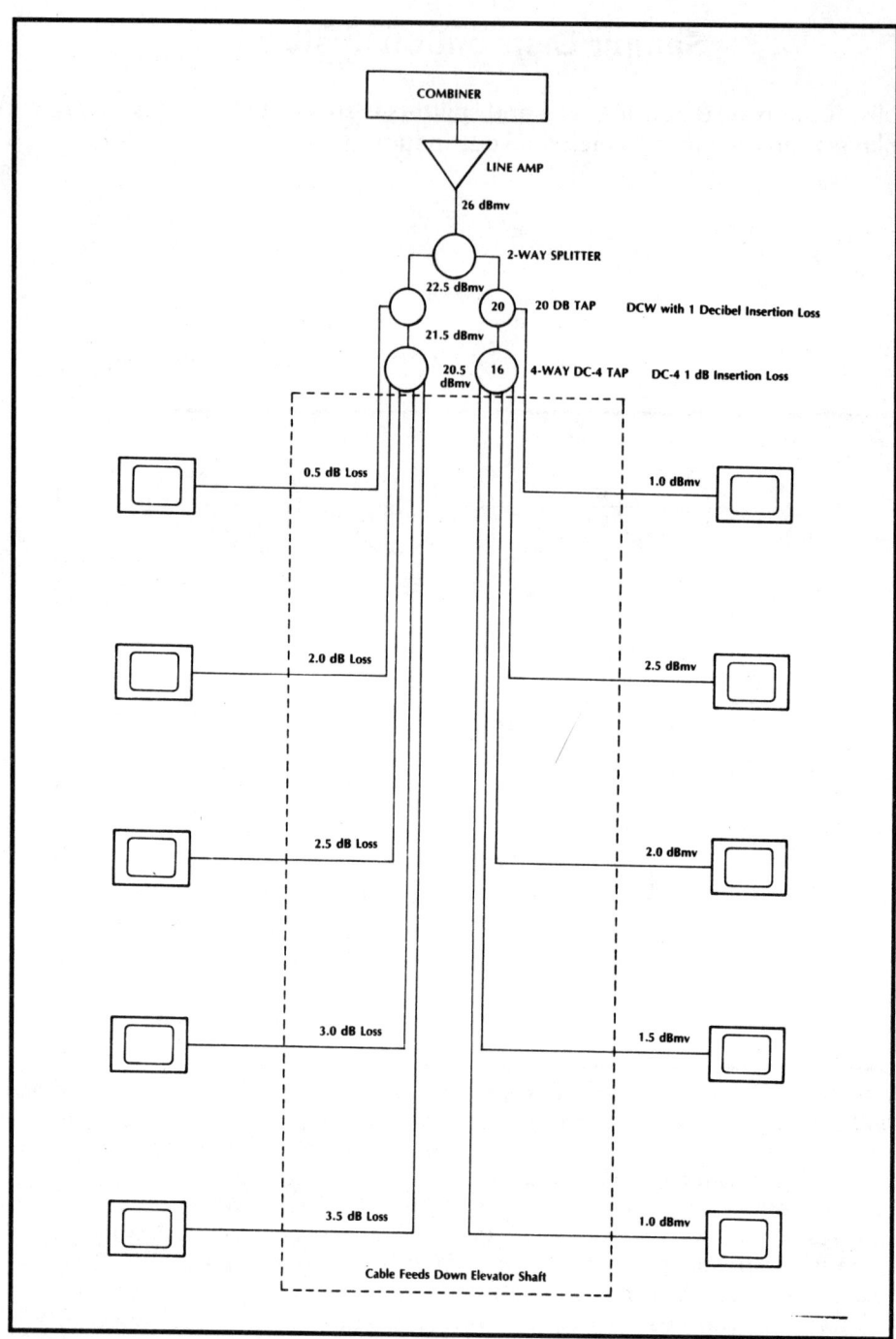

Figure 7-53. Ten-Drop, Homerun System. *The combined outputs from the modulators are fed into a line amplifier with its output set at 26 dBmv. The 2-way splitter drops the signal by 3.5 dB so that 22.5 dBmv is present on the output ports. These signals are fed into 20 dB taps each having 1 dB of insertion loss resulting in a 21.5 dBmv input to a 4-way, 16 dB tap, also having a 1 dB insertion loss. Cable losses to television receivers on the building first floor are 0.5 dB in each drop so that a 1.0 dBmv signal level arrives at the set. Second, third, fourth and fifth floor televisions receive 2.5, 2, 1.5 and 1 dBmv after 2, 2.5, 3 and 3.5 dB cable losses, respectively. All these signal levels are within the 0 to 3 dBmv input requirements for a TV set.*

MULTIPLE RECEIVER/TV SYSTEMS

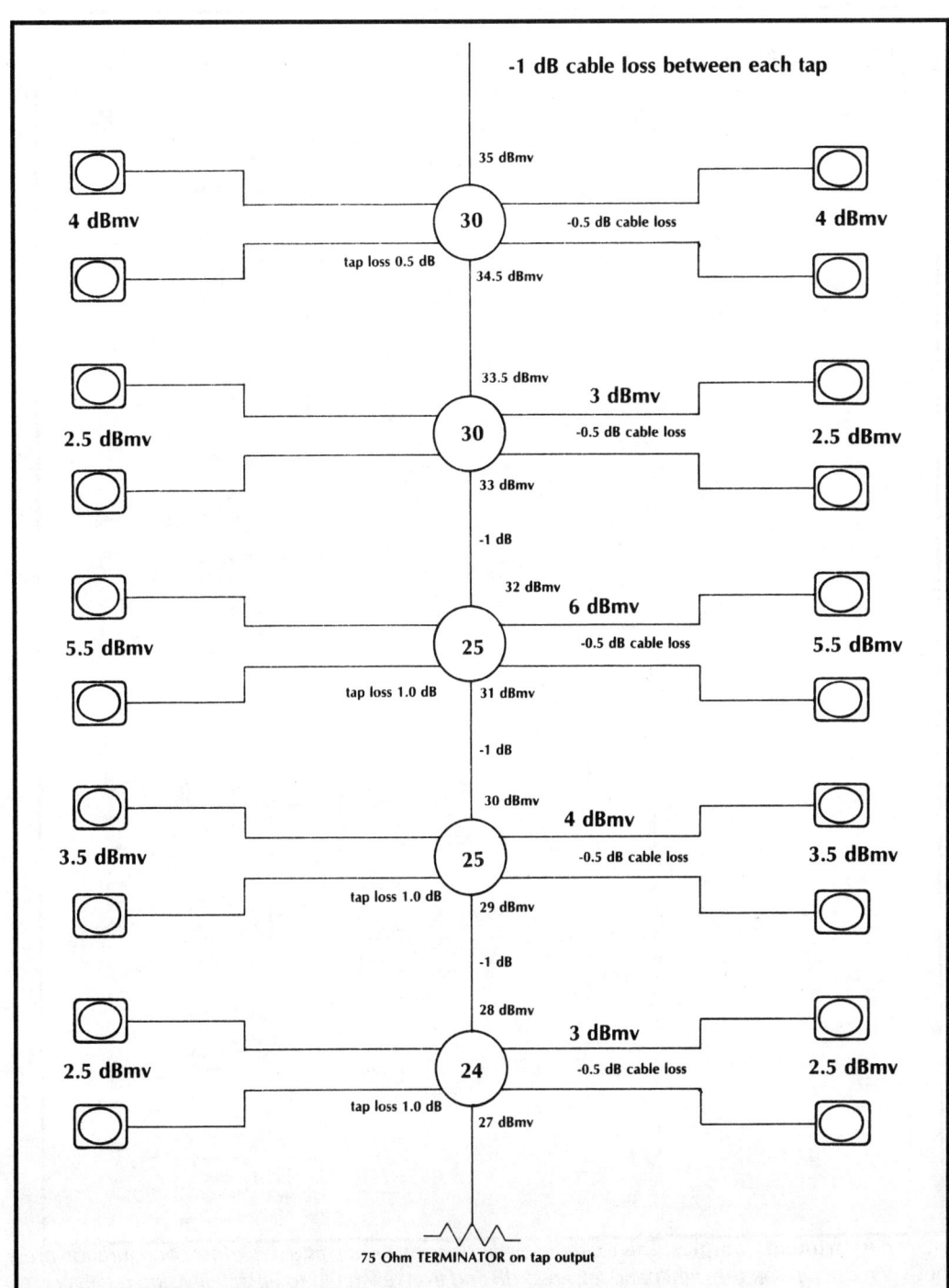

Figure 7-54. Twenty Drop-Loop Distribution System. *A 35 dBmv signal is received at the input of the first 4-way tap. The tap insertion loss is 0.5 dB giving an output of 34.5 dBmv. Each cable run from the tap to the TV set also has 0.5 dB loss. Therefore, 34.5 dBmv less 0.5 dB less the 30 dB tap value results in 4 dBmv reaching each television set. A 2-way splitter having a 3.5 dB insertion loss installed in each unit would yield a final signal level is 0.5 dBmv. There is 1 dB of cable attenuation loss between each tap. 33.5 dBmv enters the second 30 dB tap which has a 0.5 dB insertion loss. Therefore the tap output is 3 dBmv less the 0.5 dB cable loss so that 2.5 dBmv signal is available to these televisions. The input to the 25 dB tap is 32 dBmv and following an insertion loss of 1 dB the output is 6 dBmv less 0.5 dB for cable loss or 5.5 dBmv. 30 dBmv arrives at the second 25 dB tap which has a 1 dB insertion loss. Therefore, 4 dBmv less 0.5 dB cable losses or 3.5 dBmv enters these sets. A 28 dBmv signal arrives at the 24 dB tap. Following a 1 dB insertion loss and 0.5 dB cable loss the output is 2.5 dBmv. A 75 ohm terminator must be installed at the output of this final tap to avoid signal reflections and unwanted interference.*

MULTIPLE RECEIVER/TV SYSTEMS

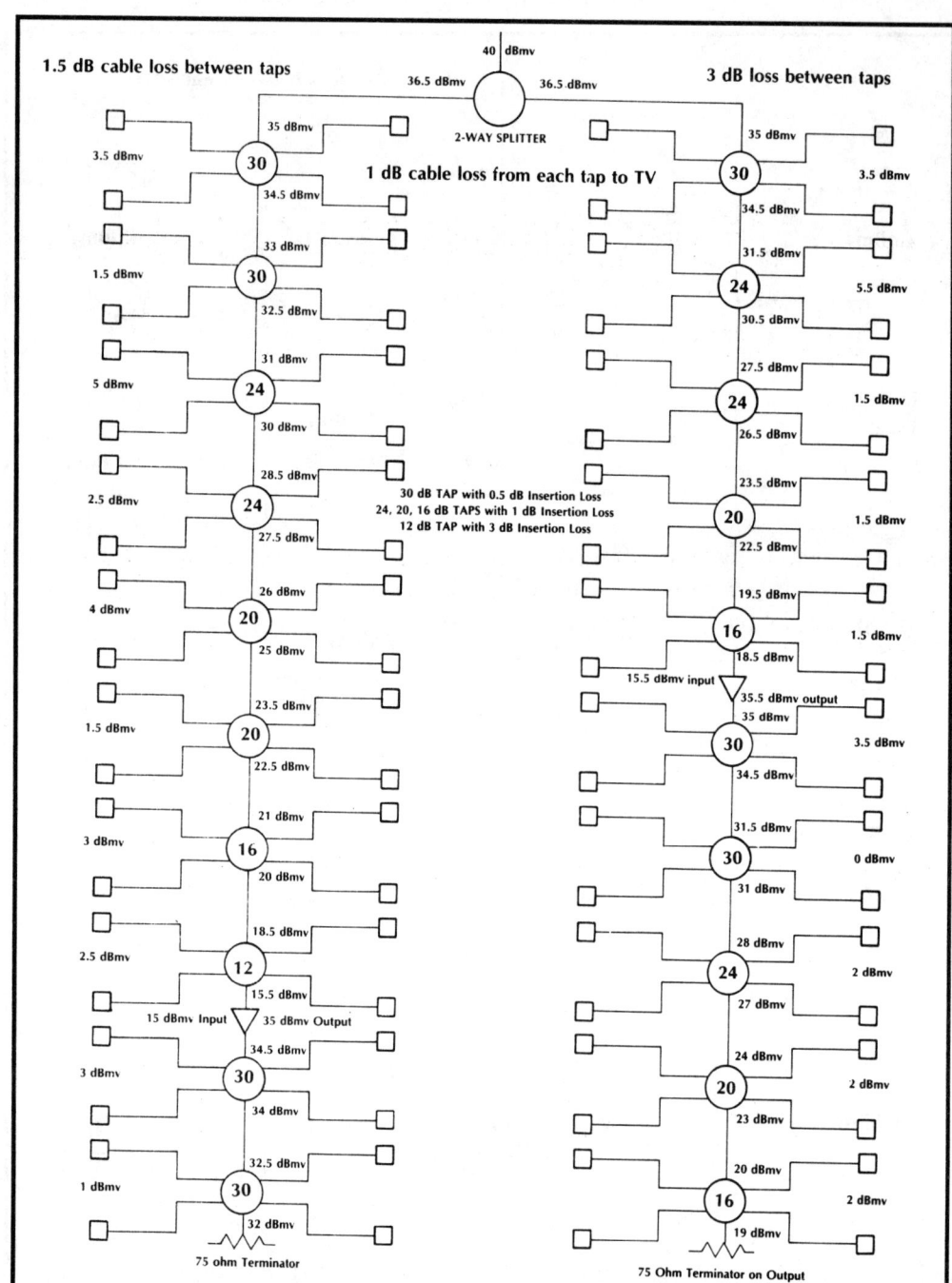

Figure 7-55. Eighty-Drop Apartment Complex. *This example expands upon the tap and cable loss configuration presented in Figure 7-54. Cable losses between the taps on the right-hand leg are 3 dB and from each tap to each television receiver are 1 dB. Insertion losses for the 30, 24, 20, 16 and 12 dB distribution taps are 0.5, 1.0, 1.0, 1.0 and 3.0 dB, respectively. The output power on the right hand leg following the fifth tap is, by subtraction, therefore 18.5 dBmv. An additional 3 dB is lost to cable attenuation. The net result is a 15.5 dBmv input to the line amplifier, whose gain must be chosen to be 20 dB in order to generate a 35.5 dBmv output for the remaining drops. Cable and insertion losses must be subtracted between the remaining five 30, 30, 20, 20 and 16 taps. A 75 ohm terminator is installed at the final output.*

The left hand leg has only half the cable attenuation of 1.5 dB between taps because the cable runs are shorter than those on the right hand leg. After the eighth 12 dB tap on the left hand side the signal level is 15.0 dBmv. A short jumper from the output of the tap to an amplifier results in another 0.5 dB loss. The amplifier output is 35 dBmv. Subsequently, signal levels decrease through each successive tap and cable run until the 75 ohm terminator installed on the output of the last 30 dB tap.

If additional apartment buildings were to be added to this system, the terminator could be removed and a further series of taps with drop cables running into television sets could be added until the signal level dropped to 15 dBmv. At this point, an additional line amplifier would be installed on this feeder line. For this example, it is clear that designing even the most complex system is simply a matter of addition or subtraction to account for all signal losses in cable runs and taps. Actual measurements are completed by a field strength meter, an essential tool when installing headend and distribution systems..

VIII. WORLDWIDE KU-BAND SATELLITE TELEVISION

Satellite technologies have evolved at a rapid pace during the past 20 years. While C-band links have become the backbone of television broadcasting and other forms of satellite communication, the growth of Ku-band systems has now surpassed the pioneer technology in most nations of the world. Even though C-band TVROs have enjoyed unexpected popularity during the past ten years in North America and the growth in the number of systems has been nothing short of spectacular, Ku-band technology has taken the spotlight. In Australia and European nations Ku-band systems are the dominant technology available and many other countries including the United States, Canada and Mexico are launching Ku-band spacecraft and broadcasting services.

The intent of this chapter is to provide a brief outline of activities under way in selected regions of the world (see Figure 8-1). This information is continually being revised and updated as new satellites are launched and as various countries re-formulate their communication policies. For more information on this topic, the reader should refer to *The World Satellite Almanac*.

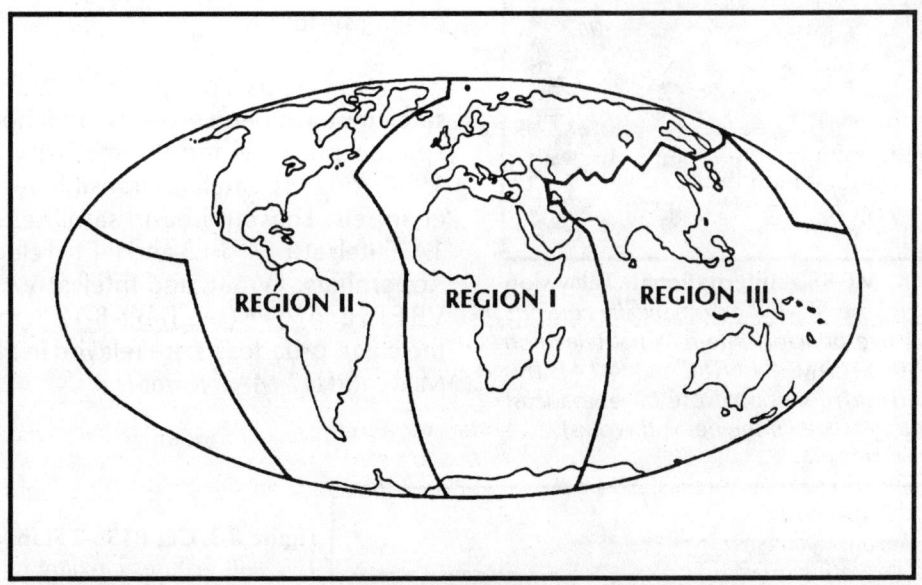

Figure 8-1. ITU Worldwide Zone Definitions. *The ITU (International Telecommunication Union) has subdivided the globe into three regions. Region 1: the USSR, Europe, the Middle East and Africa. Region 2: North and South America. Region 3: Asia, India, Australia and the South Pacific Islands.*

A. THE AMERICAS

Canada and the United States experimented with direct broadcast Ku-band technology in the mid-1970s. Experience gained with the Communication Satellite Technology (CTS) spacecraft proved the viability of using this frequency range. Today, numerous Ku-band audio, video and data services are being transmitted by spacecraft owned by Canada, Mexico and the United States as well as by regional Intelsat satellites. A complete listing of these satellites has been provided in Chapter I.

B. EUROPE

European nations have been at a worldwide C-band crossroads because they could observe broadcasts from a handful of American and Soviet spacecraft. For example, a 5 m antenna in Western Europe could receive NTSC broadcasts from four transponders onboard the North American satellite Satcom 2R. Colour SECAM transmissions from the Soviet C-band Gorizont and Raduga satellites and other very low power Intelsat broadcasts relayed to countries such as Nigeria, Morocco or Oman were also available. Given the wide range of programming sources and the use of NTSC, PAL and SECAM broadcast formats components such as multistandard televisions and standard converters were necessary (see Figures 8-2 and 8-3).

Reception of Ku-band broadcasts had been limited to those transmitted by European and regional Intelsat spacecraft because these footprints span more limited geographical areas. However, today a wide range of Ku-band satellite services are available via such pioneer satellites as Eutelsat and Astra. In fact, European are now among the leaders in developing an extensive regional Ku-band broadcasting system.

Numerous European Ku-band broadcast satellites have been recently launched. While three spacecraft provided the majority of services in 1988, today 10 satellites transmit over 60 television channels. These Ku-band satellites include Astra 1A, Eutelsat 1-F4, 1-F5 and II-F1, Telecom 1C, DFS1 Kopernikus, TV Sat, and Intelsat VA F11, VA F12, VB F15 and VI F4 (see Table 8-1). Scrambled and "in the clear" broadcasts are relayed in SECAM, PAL C-MAC and D2-MAC formats.

Figure 8-2. Merlin ME-888 International Television Standards Converter. *The ME-888 automatically converts between any of the five principal international television standards, NTSC, PAL, SECAM, 4.4 NTSC and PAL-M. This unit is manufactured for professional use in rebroadcast quality video. (Courtesy of Merlin Engineering Works)*

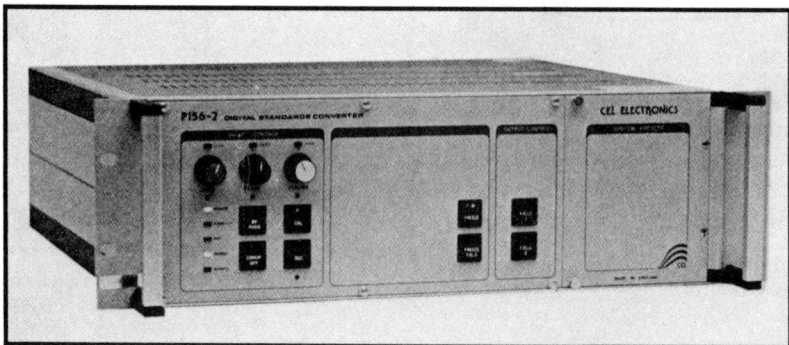

Figure 8-3. CEL P156-2 Standards Converter. *This unit features circuitry to correct for variations in picture geometry when converting between a 625 and a 525 line broadcast standard. It is intended for industrial, commercial and corporate applications. The P156-2 is available in two configurations: PAL/SECAM input with NTSC output; and NTSC input with PAL output. (Courtesy of CEL Electronics Limited)*

WORLDWIDE KU-BAND TELEVISION

TABLE 8-1. EUROPEAN KU-BAND SATELLITES

Location	Satellite Name	Owner
53°W	Intelsat VB F13	Intelsat
45°W	PanAmSat	PanAmSat
34.5°W	Intelsat V F4	Intelsat
31°W	Marcopolo 1&2	BSB Sky
27.5°W	Intelsat VI F4	Intelsat
24.5°W	Intelsat VI F2	Intelsat
21.5°W	Intelsat V F2	Inselsat
19.2°W	TV SAT	Germany
19°W	TDF-1&2	France
18.8°W	Olympus 1	ESA
18.5°W	Intelsat V F6	Intelsat
14°W	Statsionar 4	U.S.S.R.
11°W	Statsionar 11	U.S.S.R.
8°W	Telecom 1A	French PTT
5°W	Telecom 1C	French PTT
1°W	Intelsat VA F12	Intelsat
4°E	Eutelsat 1 F2	Intelsat
5°E	Tele-X	Nordic
7°E	Eutelsat I-F4	Eutelsat
8°E	Telecom F3	France
10°E	Eutelsat I-F5	Eutelsat
13°E	Eutelsat II-F1	Eutelsat
16°E	Eutelsat 1-F1	Eutelsat
19.2°E	Astra IA	Luxembourg
23.5°E	DFS-1 Kopernikus	Germany
28.5°E	E DFS-2Kopernikus	Germany
40°E	Statsionar 12	U.S.S.R.
53°E	Statsionar 5	US.S.R.
57°E	Intelsat V F7	Intelsat
60°E	Intelsat VA F11	Intelsat

Eutelsat, the European Telecommunications Satellite Organisation, was founded in 1977 by 19 member nations. The Eutelsat series of spacecraft were originally intended to relay primarily telephone signals. However, the rapid growth in demand for video services resulted in this and the successive Ku-band Eutelsat spacecraft being almost entirely dedicated to broadcasts of English, French, German and Italian language television programs.

Eutelsat 1-F4 and F5 transmit Atlantic, Eastern and Western European spot beams as well as a wide coverage Eurobeam so that broadcasts can be simultaneously accessed from anywhere within Western Europe as well as some locations along the northern coast of Africa. Twelve 72 MHz wide transponders operate in the 10.95 to 11.2 GHz and 11.45 to 11.7 GHz Fixed Satellite Service (FSS) bands. Transmission of vertically and horizontally polarised signals permits frequency re-use. These satellites also relay communications known as satellite multi-services (SMS) on two additional transponders operating in the 12.5 to 12.583 GHz range. SMS are point-to-point and point-to-multipoint high speed digital communications serving the European private and public sector.

Before the launch of the recent generation of satellites, the demand for video services in Europe had already gobbled up the entire planned capacity of the two orbiting and one future Eutelsat satellites. Excess business demand was handled by the Intelsat VA F11 and VA F12 spacecraft which operate in the 10.95 to 11.2 and 11.45 to 11.7 GHz FSS bands and transmits both vertically and horizontally polarised signals. Both east and west spot beam downlink antennas onboard the Intelsat V series of satellites are steerable by ground command. This capability allows a flexible response to demand.

Telecom 1A, the first in a series of three French domestic communication satellites, was launched by an Ariane rocket in 1984. It carried six 36 MHz wide Ku-band transponders which are capable of delivering 47 dBw signals to its French mainland boresight. Eutelsat had leased four circuits as part of its SMS high speed digital network. Telecom 1A operates in the 12.5 to 12.75 GHz band which had been allocated for digital business communications. However, the heavy demand for satellite distribution of cable TV signals led to use of this band for video broadcast transmissions. Telecom 1B and 1C were launched in May 1985 and September 1987.

Luxembourg's Astra 1A satellite, launched in December 1988, has been a major force in encouraging rapid growth in Ku-band broadcasting (see Figure 8-4). Antennas of less than 60 cm (24 inches) used in conjunction with 1.2 dB LNBs now are capable of delivering excellent quality pictures from the beam centre 50 dBw footprint. As component costs have fallen and installations have become easier, sales of home satellite systems have rapidly increased.

The European Broadcasting Union (EBU), a non-profit association of national Western European television and radio networks, has recommended that all European Ku-band direct broadcast services use the MAC transmission system developed by the Independent Broadcasting Authority (IBA) in Britain. In particular, the C-MAC and D2-MAC standards have been studied. D2-MAC is now relayed over some transponders on Astra 1A, TV-Sat and TDF 1. C-MAC is transmitted via Tele-X and Intelsat VA-F12. The motivation behind this decision is to institute one subcontinent-wide, high quality broadcast standard as well as to retain European control over the manufacture and distribution of TVRO hardware.

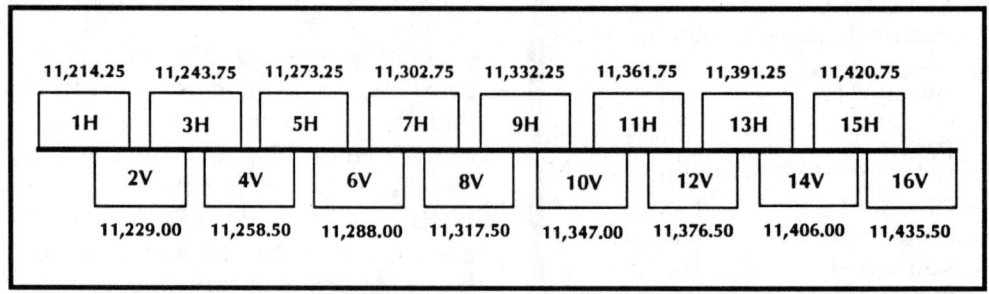

Figure 8-4. Astra Downlink Frequency Plan. *Luxembourg's Astra satellite carries 16 transponders each having a 26 MHz bandwidth. Co-polarised and cross-polarised channels are spaced 29.5 and 14.75 MHz apart, respectively.*

C. AUSTRALIA, NEW ZEALAND AND NEW GUINEA

In mid-1985 Australia launched the first in a series of three spacecraft designed to serve the entire continent. Previous to this, communication was provided by regional Intelsat satellites. Aussat Proprietary Limited, a company wholly owned by the Australian government, has responsibility for Aussat I, II and III which are stationed at 160, 156 and 164 degrees east, respectively. The third spacecraft, Aussat III, is an in-orbit spare. These satellites communicate in the 12.25 to 12.75 GHz band. Point-to-point non-broadcast services are transmitted from 12.25 to 12.5 GHz; the upper half of the band is shared among direct broadcast TV and radio services as well as other non-broadcast applications.

Each 15 channel satellite can receive signals on three beams designated as National A (NA); Papua New Guinea (PNG); and National B (NB). Horizontally polarised transponders 1 through 8 are connected to the NA beam while vertically polarised transponders 9 through 15 receive the NB beam. In addition, transponders 4,6 and 8 can be individually switched to receive on either the NA or PNG beam. On Aussat III these transponders can also be accessed by a Southwest Pacific beam.

Each satellite has seven transmit beams. The National A (NA), Papua New Guinea (PNG), Southeastern Australia (SE) and Western Australia (WA) transmit vertically polarised signals. The National B (NB), Northeastern Australia (NE) and Central Australia (CA) downlinked beams are horizontally polarised. Aussat III is also designed to transmit Southwest Pacific (SWP) beams on transponders 4,6 and 8. Ground controllers can switch the various footprints on command onto any of the transponders listed In Table 8-2.

The Aussat satellite communication network offers a wide range of services. These include expanded TV and radio feeds, multi-cultural educational television and radio programs, long-distance telephone, private voice and data communications, teleconferences and nationwide linking of major air traffic control and flight service centres.

Eight 30-watt transponders, four per satellite, have been assigned to direct Ku-band broadcasting to the rural Australian outback. The Homestead and Community Broadcasting Satellite Service (HACBSS) has the potential to reach 350,000 Australians living in areas which previously were beyond the reach of conventional services, as well as an additional one million who have had only marginal reception of conventional over-the-air radio and television. Programming will originate in each of the four spot beam areas. These are Sydney in the southeast, Darwin in central Australia, Brisbane in the northeast and Perth in western Australia.

Although the PAL format is presently being used in all conventional Australian broadcasts, the B-MAC transmission standard has been chosen and is being used for satellite broadcasts. This allows each transponder to deliver six high quality audio channels in addition to the video signal. One or more of these audio slots can also be used to transmit data. Satellite broadcasts will therefore be accompanied by stereo sound as well as by additional radio services and possibly a data channel capable of carrying teletext or other information. B-MAC was chosen instead of the conventional Australian PAL format because it allows a high level of picture resolution and colour fidelity, even under marginal operating conditions. The elimination of the audio subcarrier by using this sound-in-sync method makes it possible to improve the signal-to-noise ratio and to achieve the objective of obtaining high quality reception in clear weather with 1.2 to 1.5 m diameter antennas.

The Australian B-MAC transmission format is designed so that each transponder occupies a total bandwidth of 24 MHz. The PAL bandwidth could either be 22 or 40 MHz. An energy dispersal waveform is superimposed upon the uplinked signal so that receivers require clamping circuitry.

TABLE 8-2. AUSSAT TRANSPONDER AND BEAM ALLOCATIONS

Transponder Number	NA	SE	WA	PNG/SWP	NB	NE	CA
1	x	x					
2			x				
3	x	x					
4	x		x	x			
5		x					
6	x			x			
7	x	x	x				
8		x	x	x			
9					x	x	
10					x		x
11						x	
12					x		x
13					x		
14					x	x	x
15					x	x	x

KEY TO TABLE

NA	National A
SE	Papua New Guinea
WA	Western Australia
PNG/SWP	Papua New Guinea/S.W.Pacific
NB	National B
NE	Northeastern Australia
CA	Central Australia

D. JAPAN

The Japanese government's Ministry of Post and Telecommunications launched the experimental BSE direct broadcast satellite in 1978. It had two 100-watt transponders which operated only for half of their expected brief lifetime. Positive results from this experiment paved the way for an aggressive direct broadcast program. In early 1984, Japan's BS-2A was lifted into orbit at 110°E. It carried two broadcast transponders and one spare, each rated at 100 watts. The performance was disappointing because two of the three transponders failed shortly after launch while one provided less than predicted power. Presently NHK is using this one transponder for testing their new high definition TV system. The spare BS-2B is a replacement for the troubled satellite.

Japan has now geared up to manufacture its own 100 watt traveling wave tube amplifiers (TWTAs), the component which probably failed on the earlier satellites. BS-3, launched in 1989, has three operational and two backup transponders all rated at 100 watts.

E. THE SOVIET UNION

The first experimental transmissions from Soviet Luch (also known as Loutch) Ku-band satellite began in 1986 as a result of a program initiated in the mid-1970s. The development of Soviet Ku-band transponders was based on extensive experience with their C-band Statsionar system which includes the Raduga, Gorizont, Moliyna and Ekran series of spacecraft. Presently, the Luch I Ku-band hardware is carried onboard four Gorizont satellites including the Gorizont 7 vehicle located at 14°W. Experiments on the effects of factors such as rainfall on reception are presently under way.

The proposed orbital slots for the Luch and Luch P series which are outlined in Table 8-3 correspond to those of Statsionar 4, 5, 6 and 7 and Statsionar 8, 9, 3 and 10, respectively. Of course, both C and Ku-band satellites can be co-located with no concern for interference. Future Luch transponders may be carried onboard C-band vehicles or be on satellites dedicated solely to Ku-band broadcasting.

The Luch satellites have 10 right hand circularly polarised transponders downlinking in the fixed satellite service bands, 10.95 to 11.2 GHz and 11.45 to 11.7 GHz. The transponder bandwidth is 34 MHz and channels are expected to be spaced on 50 MHz centre frequencies. Therefore, for example, channel 1, 2, ...,10 will be at downlink frequencies of 10975, 11025, ..., 11675 MHz. It appears that the downlinked beam on the Luch 1 spacecraft spans a wide area and is not designed to be a targeted spot or zone beam. Initial measurements indicated that an EIRP of 41.5 dBw is obtainable in Great Britain and relatively large antennas must be used for reception. A second spot beam aimed at Havana, Cuba should make transmissions from this satellite accessible to TVROs in southern Florida and some of the nearby islands.

TABLE 8-3. SOVIET LUCH KU-BAND ORBITAL ASSIGNMENTS

Satellite	Orbital Position
Luch 1	14°W
Luch	253°E
Luch	390°E
Luch 4	140°E
Luch P1	25°W
Luch P2	45°E
Luch P3	85°E
Luch P4	170°W

WORLDWIDE KU-BAND TELEVISION

F. INTELSAT KU-BAND SATELLITES

Numerous Intelsat satellites also provide worldwide C- and Ku-band coverage with video, data and audio transmissions. Figure 8-5 and 8-6 show the beam steering ability on Intelsat V and VA satellites and the footprint of VA F11, respectively.

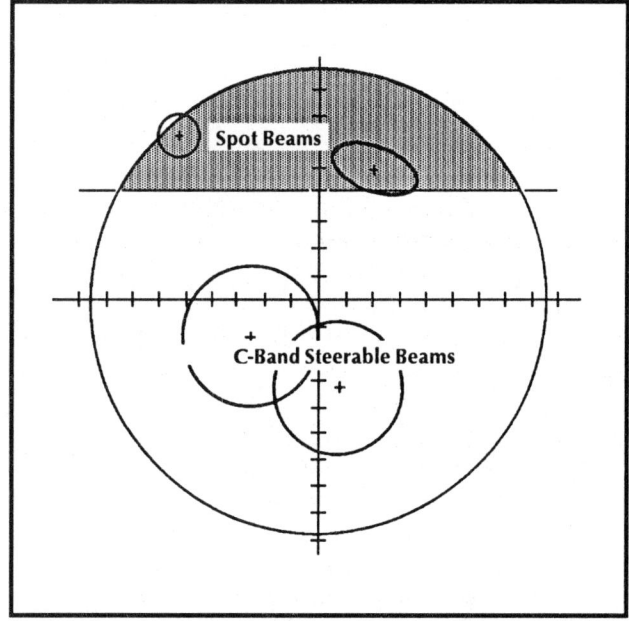

Figure 8-5. Intelsat V/VA Satellites - Beam Steering Ability. *This diagram shows the somewhat limited beam steering capabilities on the Intelsat V and VA series of satellites. Spot beams can be relayed only to the shaded areas. The circles show the EIRP contours of the east and west Intelsat VA F11 spot beams. For Intelsat spacecraft located between 24.5 to 27.5°W longitude, eastern North America and Western Europe can be illuminated by the western and eastern spot beams, respectively.*

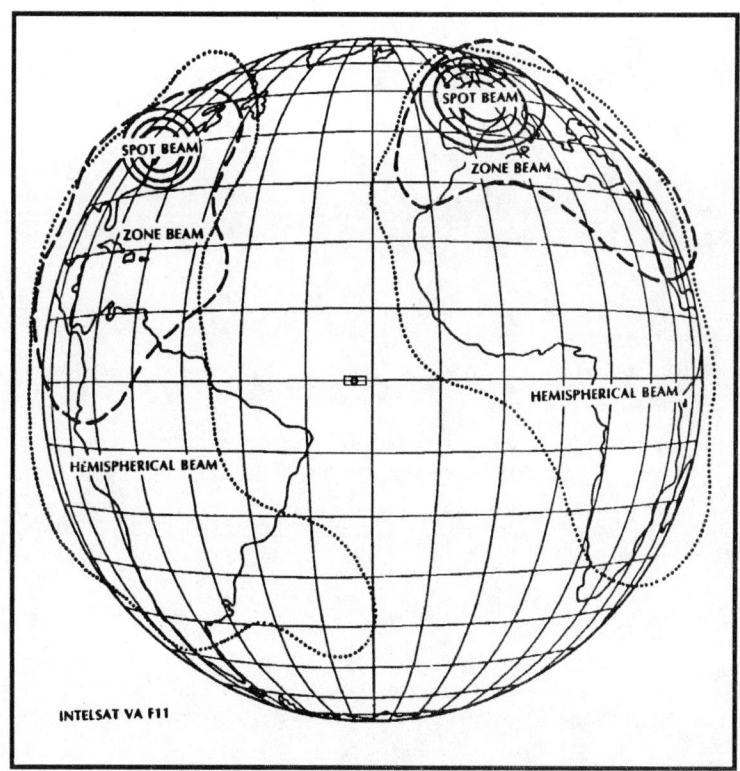

Figure 8-6. Intelsat VA F11 Footprint. *This illustrates the Ku-band spot beams and the associated zone and hemispherical C-band beams produced by the spacecraft.*

INTELSAT

WORLDWIDE KU-BAND TELEVISION

IX. TROUBLESHOOTING AND REPAIR

A. INTRODUCTION

It is inevitable that every home satellite system will eventually have operational problems of one sort or another. When this might occur depends to a certain extent upon whether equipment was properly selected and installed. Poor weatherproofing is the most common installation error. Although it is clear that no man-made device will last forever, all problems can be solved by thoughtful troubleshooting. The purpose of this chapter is to provide a consistent method to understand and rapidly solve any difficulties that may arise. Finding out what is wrong, the diagnostic procedure, can in itself be an interesting and rewarding exercise if a correct approach is followed.

A competent technician should expect breakdowns and thus factor service and repair calls into the sales price. It is important to realise that the cost for a service call including travel time, wear and tear on the service vehicle, fuel plus a normal hourly rate can be substantial. Clearly it is wise to have a clear and effective approach to troubleshooting and repair.

Troubleshooting can be relatively easy when only one component of a system goes bad. This can often happen a short time after an installation has been completed. Then diagnosis and repair can generally be quite simple, especially if the service call is made by the same technician who installed the system. The installation documentation should facilitate locating the fault. Unfortunately, if another person installed the system, diagnosis may be more difficult. Some outrageous mistakes have been discovered where servicing home satellite systems which were poorly installed.

However, most problems often arise over a longer period of time when several unrelated, minor faults that result in poor quality pictures simultaneously develop. For example, connectors can corrode, concrete may settle and throw a pole slightly off from a vertical orientation, a centre conductor in a coax may slightly contract and not make perfect contact, the antenna may have sagged slightly or shifted from its north/south alignment, or the feed may have moved noticeably over time. Eventually a customer may telephone to complain about poor reception. As a result, a check of antenna tracking, feed alignment and all connections would be required.

Troubleshooting during an installation can be much more difficult, especially if proper procedures have not been followed from the start. These "proper procedures" are very important in order to avoid future headaches. To illustrate, assume that the components had not been tested before installation and the LNB was faulty and that a cable which had already been buried also had an internal short. The problem with the LNB would have been easily located if it had been tested before installation. But once the installation is completed, trying to separate the two problems without going back to square zero could be difficult.

It is clear that as the number of installed systems continues to grow and as equipment in the field ages, the troubleshooter's role can only become more important and more profitable. In fact, during the brief but dramatic turn-down in the home satellite television business in the United States during 1986, many dealers survived by offering a professional repair service. The hallmark of a successful troubleshooter has been and will be an approach that combines technical knowledge with a systematic and methodical process for diagnosing, isolating and repairing TVROs.

B. A SYSTEMATIC APPROACH TO TROUBLESHOOTING

Diagnosing an operational problem in a satellite TV system should follow a logical path to a rapid solution. The strategy is to take the simplest route having the highest probability of solving the problem before proceeding with more complicated steps. However, even these more complicated steps can be approached with a similar perspective of tackling the most quickly solved problems.

The first step in a systematic approach to troubleshooting and repair is an in-depth interview, preferably over the telephone in order to save time and expense. Of course, this interview can be successful only if the underlying theory is clearly understood. This first step is aided by a technique known as the subsystems approach that can be quite useful in isolating problems in any satellite system. If necessary, the next step would be a visit to the installation site for a visual inspection. If a simple inspection does not isolate the problem, a series of technical troubleshooting procedures would be followed to eliminate the fault.

None of these procedures is magical or complicated. All that is required is knowledge of how a satellite TV system works and a clear, level-headed approach. No problem is too difficult to solve and, with experience, solutions come more quickly.

The Subsystems Approach

Every satellite system can be broken down into three subsystems: the mechanical, consisting of the mount, antenna, feed support and feedhorn, the electromechanical consisting of the actuator and the polariser, and the RF consisting of the LNB, satellite receiver, modulator and TV. Each can be diagnosed somewhat independently even though symptoms may be similar in nature. The only reason for defining and isolating these three subsystems is to simplify troubleshooting and repair procedures.

TABLE 9-1. SATELLITE TV SUBSYSTEMS

Mechanical	Electromechanical	RF
Mount	Actuator	LNB
Antenna	Polariser	Modulator/Receiver
Feed Support		Television Set
Feedhorn		

The Diagnostic Interview Identifying the Problem

A telephone conversation between the customer and an experienced dealer can often solve problems and eliminate the need for most service calls. This is especially true during the initial few weeks or months of ownership when customers are not familiar with many aspects of their systems. Troubles can stem from a television set that is incorrectly tuned to the modulated channel or from a customer's inexperience with a hand-held remote, stereo processor or operation of the parental lockout feature on a receiver.

In particular, a new owner generally has expectations about how the pictures should appear and would certainly notice any problems or irregularities. Complaints about poor picture quality are most typical.

The interview can also identify symptoms of a problem that may require a service call. Such information can prove extremely useful during the initial diagnosis as well as in assembling the proper equipment to isolate and repair the system. For example, if the customer had been out tilling his garden when the satellite system failed, certainly bring some extra cable and connectors. There is nothing worse than driving out to a site and then having to return to the shop for one small forgotten component such as an extra fuse or voltage regulator. If adequate documentation about an installation was on

TROUBLESHOOTING & REPAIR

file, questions can be guided by knowledge of this particular system. The documentation should have included information about antenna size and make, the location of cable runs, the type of electronics installed and the other factors outlined in Chapter V.

In summary, the interview is an important first step before anything else is done. Some sample questions are:

- When did the trouble first start?
- Was it a gradual change or a sudden occurrence?
- Describe the problems.
- Are these difficulties constant or do they occur at regular or irregular periods during the day or night?
- Was the equipment recently moved or has any component been replaced?
- Which channels and how many channels are coming in clearly?
- Can you receive all the satellites?
- Does the antenna move easily from satellite to satellite and does it make any unusual noises in doing so?
- Have you possibly cut or disconnected any cables by mistake?
- Does picture quality vary from channel to channel or from satellite to satellite?
- Is the audio or video most affected?
- Has there been any recent construction in front of or near the antenna?
- Did the problem first occur after a storm and did it persist when the weather cleared up?
- Do your problems occur more during rainstorms than on clear days or more during hot than during cold weather?
- Is it possible that a fuse has been blown in the receiver? Do the lights come on? Could a circuit breaker in the house providing power to the receiver have been tripped?
- Have there been any children playing around the antenna?
- Did you recently dig a garden or do any other excavation work?
- When the wind blows can you hear any part of the antenna rattling?
- Can your television set receive over-the-air channels?
- Did these problems begin to appear in spring (see Figure 9-1)?

A competent troubleshooter is like a doctor. The patient often knows what is wrong but simply does not have the vocabulary to express it. A well conducted interview will gather useful and important information. With a little experience, a complete list of questions can be created so that such an interview will proceed smoothly.

Many problems can be quickly solved by this technique. For example, a customer may complain that pictures are perfect except during heavy rain or snow storms. This might be caused by rain fading because the system was under-designed. Or a wet connector or leaky gasket between the feedhorn and LNB may be at fault. Also, for example, if every transponder went bad after a severe lightening storm, the cause could likely be an LNB with a blown stage or simply a blown fuse or circuit breaker.

Then there are occasional satellite problems or sun outages which happen infrequently enough that people forget that they are even possible. Sun outages occur twice a year during the spring and fall equinoxes when the sun lines up directly behind the targeted satellite and its RF radiation completely overpowers the broadcast for twenty to thirty minutes. Some technicians are simply not aware of these occurrences. This underlines the need to have first hand acquaintance with satellite TV by owning an operational system for comparison purposes.

C. VISUAL INSPECTION

Once the customer interview and a thoughtful evaluation of the assembled facts are completed, a visual inspection may be required. This is especially important if the system was installed by another technician.

The visual inspection can be guided by symptoms of the problem. For example, if the customer reports that he can receive television from only one satellite, the first step would be to check the actuator and its control box to see if the antenna is stuck in one position. If all channels are received with

TROUBLESHOOTING & REPAIR

Figure 9-1. Antenna with Obstructed View. *This antenna was installed in the middle of dense foliage. Although it may not have been completely screened during the winter when the surrounding trees had no leaves, chances are excellent that few satellites in the arc can be clearly received in the summer.*

crystal clarity on some satellites but poorly on others, the first visual check would be to determine if the reflector is warped, if the feed is moving off the focal point as satellites are being tracked or if the north/south alignment has been disturbed.

It is important to realise that many symptoms can arise from the same problem. It is easier to initially conduct a visual inspection when a not-so-obvious symptom is encountered than to jump into a more technical troubleshooting approach or to immediately start exchanging components. It is important not to create additional problems. But symptoms will usually give an indication of where to start the visual inspection. For example, if picture quality were poor, it would certainly make more sense to first try moving the antenna east and west with the actuator or a hand crank than to start measuring voltages. A second sensible step would be to check for loose or corroded connectors and a third one would be to check that the feed is properly centred and the antenna is correctly aligned. Perhaps the mount has rotated in a strong wind because the bolts on the collar were not tightly secured or pinned. Sight with a compass to be sure that the polar mount is still properly oriented in the north/south plane.

The questions below show the thought process underlying a simple visual inspection. Many more questions that can be asked will become evident as additional technical details are discussed below.

RF Subsystem

In general, the visual inspection should begin at the system control centre, the satellite receiver/television set. Turn on the TV and what do you see?

1. Does the television work properly on over-the-air channels?

2. Is the picture completely snowy or covered with "bug-fight" sparklies? If there is a weak picture in the background perhaps the satellite is targeted onto a side lobe.

3. Is the receiver turned on and are all the indicator lights working? Maybe a rear panel or internal fuse has blown or a 120 or 220 Vdc circuit breaker in the house has been tripped.

4. Is the television tuned to the correct channel that matches the modulator output?

5. Have any cables behind the receiver been disturbed? Perhaps the coax was improperly crimped and the cable is simply disconnected.

6. Are any other wires loose or disconnected? If this is a new installation make sure that all the wires are well attached to the proper place before powering on.

7. Is there a high signal strength meter indication but no picture? This would suggest that the modulator may be bad or that the TV set is not tuned to the right channel.

8. Is the receiver shutting off because it is overheating? Most receivers have ventilation slots that allow cooling by natural air flow. If heat is not carried away, the voltage regulator that generates substantial amounts of heat might be internally compensating and shutting down.

9. Are all the cable runs intact? Check all the cable routes from indoors to the antenna. Perhaps some over-enthusiastic gardener has severed the cable with a weed-eater, shovel or lawnmower. Maybe a dog or rodent has chewed through the cable.

10. Are all the connections at the antenna secure? Make sure that all connectors are tight and that no water has leaked in. Are drip loops installed in the correct places? Maybe a connector on the LNB has corroded. When a cable is used to carry power, water entry can cause corrosion similar to the build-up that can be seen on a car battery terminal. Cut the cable back and install a new connector past the point where the corrosion ends. Make sure that coaxial sealant has been used. Clean any suspected connectors with a solvent like carbon tetrachloride, dry them thoroughly and re-seal. Also, if a defective pigtail has a slightly short centre conductor, it may perform well until cold weather arrives. The cold could cause this lead to shrink and break contact. This phenomena is appropriately known as centre connector suck-out. If the centre conductor is too long, it may split the shroud on the centre pin when inserted into an LNB and cause it to short out against the inner connector wall.

11. If an LNB/feed cover has not been used, check to see if any water may be collecting in one of the waveguides. If a polariser has been used with the right angle waveguide fitting, water may also have collected and a cure might be to drill a small 1 or 2 millimetre drain hole on its underside. Maybe the gaskets between the various flanges have been omitted to further worsen the problem.

12. Does the system work intermittently? Sometimes a bad LNB will function only when it is cooled down to the point where its gain increases. A test for this malfunction is to use a hair drier to warm it up on a cold day or to use a can of freeze spray on a warm day to bring it back to life.

Electromechanical Subsystem

1. Does the antenna respond to the actuator? Maybe a voltage transient from a nearby lightning hit fried the Hall effect, reed switch or optical feedback sensor. If shielded cable was not used for the sensor leads, the actuator controller may be miscounting and aiming the antenna incorrectly. Possibly water has entered the arm and has frozen or rusted the internal assembly so that it does not move.

2. Make sure that the motor is pointed up so that the drain holes are pointed down. If there are no drain holes, drill some. Maybe the gasket between the motor cover and housing has been omitted.

3. Is the actuator tube bent or is there wear or scoring on the inner tube? Perhaps the original installer did not use ball joints on the actuator arm supports in order to minimise lateral force on the tube. Maybe he did not set both mechanical and electrical limits to prevent bending.

4. Is the polarity adjustment having any effect on picture quality? If it is a mechanical device, is the probe moving when channels are changed or when the skew adjustment is altered? Check to see if obstructions caused by moisture collection, birds or even wasps are interfering with signal reception in the throat of the feedhorn.

5. Is the skew adjustment having little or no effect? Maybe the probe has been bent or mechanically distorted by freezing and thawing water. Make sure that the waterproofing was ensured by correctly installed gaskets.

6. Is the probe motor bad? Have a spare hand-held controller because the 555 timer in the actuator controller or receiver may be at fault, not the motor. The 555 timer chip is a basic component of most actuator controllers.

7. Is the motor servohunting? This means that the motor is driving the probe back and forth unsuccessfully looking for the maximum signal. This can cause hum bars visible as wide horizontal regions of disturbances rolling across the screen. Have a 10 Vdc, 1000 microfarad electrolytic capacitor on hand and install it on the polariser motor at the antenna. Also, check that the wire feeding the motor is of sufficient gauge and thus thick enough to avoid this problem (see below for more details). When using two servo motors in a parallel dual-band configuration, a capacitor should be used to provide an extra current when simultaneously switching both probes.

Mechanical Subsystem

1. Does the antenna look as if it is mounted correctly and that all the angles are still set properly? If the reflector is facing north, the problem is obvious. The collar was not properly tightened, drilled and bolted.

2. Has the concrete shifted or settled under the pole or pad so that the pole is no longer level or plumb?

3. When using fiberglass antennas, perhaps the surface of the antenna has delaminated or is blistering resulting in poor quality reception.

4. Has the owner painted the antenna with a roughly applied metallic paint that has affected its reflective surface? Or maybe he has applied a highly reflective paint that is concentrating sunlight enough to overheat the components at the focus.

5. Shake the antenna from the outer rim for maximum leverage in order to check for loose nuts and bolts. Maybe the installer did not use lock washers or lock-tight and the wind had gradually loosened up the assembly.

6. Are any nuts or bolts missing?

7. Is there excessive wear at any attachment points? Scoring on the support pipe caused by such movement may be a telltale sign. If this is the case, set a bolt through the outer to the inner pipe when all the readjustments are completed. This will maintain a true north/south tracking axis.

8. Is the antenna warped? When sighting along one lip does the other side line up perfectly parallel? Do the two or three string test, if necessary, to check for warping. Maybe the mount support was not designed properly and the reflective surface has twisted.

9. Is the feed system secured and centred? This can be checked with a focal finder or with a tape measure. A loose feed could possibly move in the wind or be jarred out of place as the antenna rotates across the arc. Guys wires are recommended.

10. Is the feed opening free of bugs or wasp nests? A quick visual inspection will eliminate this possibility.

11. Does the actuator jack have complete freedom of movement across the whole belt of satellites?

D. TECHNICAL TROUBLESHOOTING

If the initial visual inspection does not provide a solution to a problem, then more technical approaches can be followed. There are always a number of ways to arrive at the same conclusion. It is important to understand that the intuitive approach is a valid one, not to be diminished in value. Troubleshooting certainly has its share of personal touch.

The tools used in technical troubleshooting are test instruments. A voltmeter, spectrum analyser, portable signal simulator or portable receiver can greatly simplify and speed up the diagnostic process. Switching components is included in this section as one of the technical troubleshooting methods because it involves more than casual observations of TVRO problems.

Test Instruments

Voltmeters and Multimeters

A voltmeter is by far the simplest, lowest cost and potentially most effective test instrument (see Figure 9-2). Most electrical breakdowns occur in power supplies that provide energy to operate the LNB, actuator and all other signal processing circuitry. A competent technician can isolate power supply and other voltage related problems in minutes with the proper use of a voltmeter.

Voltmeters, like most test instruments, have variable sensitivities. The maximum range for each setting is typically 200 millivolts, 2 volts, 20 volts, 200 volts and 1000 volts dc. Testing should always begin in a range which can manage the maximum voltage expected so the indicator will not be driven off the upper end of its scale.

Most voltmeters are designed to read resistance and ac voltage as well as dc voltage. These ca-

TROUBLESHOOTING & REPAIR

Figure 9-2. A Multimeter. *One of the most frequently used pieces of test equipment either in the field or on a test bench is a multimeter. It has the capability of reading dc and ac voltages or currents as well as taking resistance measurements for checking cable continuity.*

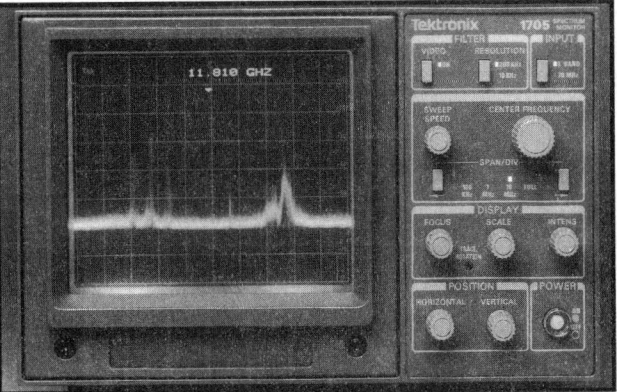

Figure 9-3. Tektronix 1705 Spectrum Analyser. *The 1705 features front panel controls to select between monitoring of L-band (900 to 1450 MHz) television downlink signals, of the 70 MHz band (45 to 100 MHz) television uplink exciter or of other IF signals such as those passing in a receiver's 70 MHz loop. The input frequency is displayed on-screen in both L-band and 70 MHz modes and the L-band read-out may be re-configured to display a satellite frequency in the range of 0.9 to 20.0 GHz. An operator of this instrument can identify the spectral pattern of a particular satellite and transponder to determine which video, audio and SCPC carriers are present. (Courtesy of Tektronix)*

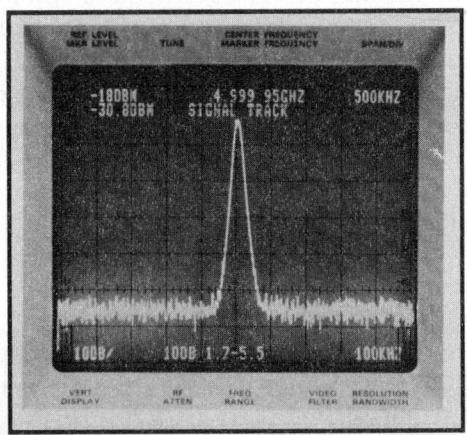

Figure 9-4. Spectrum Analyser Frequency Display. *The analyser screen shows the amplitude and frequency of a 4.99 GHz signal. (Courtesy of Tektronix)*

pabilities are extremely useful in testing for shorted wires, in checking line voltages or in troubleshooting other areas.

Spectrum Analysers

Spectrum analysers are powerful diagnostic tools designed so that incoming signals in a range of frequency bands can be observed and measured (see Figure 9-3). This instrument essentially plots a graph of signal power versus frequency (see Figure 9-4). Frequency increases from left to right on a CRT screen while signal amplitude is presented as a series of peaks and valleys. Incoming microwaves located anywhere within the entire band of satellite communication frequencies can be observed at once or fine details of the frequency and amplitude structure of any single carrier can be studied. The spectrum analyser is to an RF designer what an oscilloscope is to an ordinary engineer. A spectrum analyser displays the signals plotted versus signal frequency whereas an oscilloscope displays signals plotted against time.

TECHNICAL TROUBLESHOOTING

TROUBLESHOOTING & REPAIR

Portable spectrum analysers have long been used for cable television applications (see Figure 9-5). Many analyser manufacturers have marketed add-on modules designed to allow coverage of the satellite television IF block frequency ranges. Such analysers can be used to observe signals at any point in a satellite system. When the signal from an LNB, or preferably an LNB and feedhorn assembly, is fed into this instrument, microwaves of any chosen polarity can be detected and measured. By scanning a feedhorn or an antenna across the sky and observing the variation in detected power, the source can also be pinpointed. This capability makes a spectrum analyser invaluable in aiming an antenna and in peaking a feed. An LNB typically generates a signal having powers ranging from -60 to -80 dBm while the highest level signal at the modulator output is on the order of -40 dBm. One common portable, relatively low cost instrument, the Avcom PSA-35, has a sensitivity spanning the range of zero to -90 dBm and therefore can respond to RF signals produced at any point in a typical satellite reception system (see Figure 9-6).

Technicians at uplink facilities use on-line analysers to monitor the power, frequency, stability and composition of signals relayed to communication spacecraft. Powers are generally equalised across all transponders on-board a broadcast satellite. This is seen as a series of equal peaks spread across the entire C or Ku-band frequency range. A spectrum analyser monitoring the output of an LNB should display the same, equal distribution of transponder power.

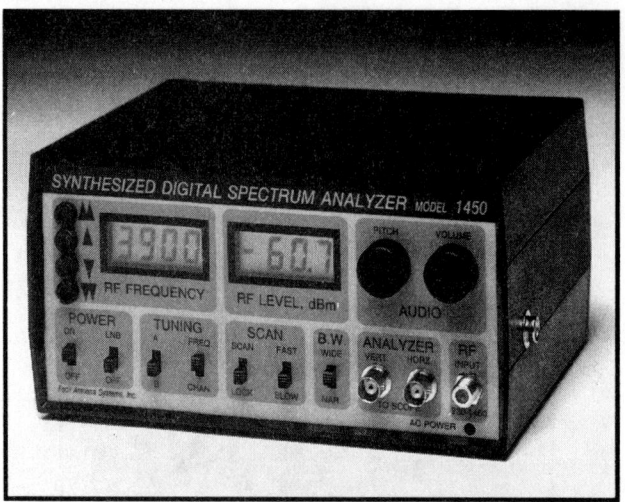

Figure 9-5. Focii Digital Spectrum Analyser. *This battery-powered unit accepts input signals in the range from 950 to 1450 MHz and has both digital, audio and oscilloscope outputs. Centre frequency is adjusted in steps of 1 or 20 MHz. (Courtesy of Focii Manufacturing Company)*

The ability of an analyser to create a visual display of signals over a wide range of frequencies makes it a valuable troubleshooting instrument which can pinpoint system or component problems (see Table 9-2). This tool can also be directly connected into the output of an LNB, the output of the cable leading from the LNB to the satellite receiver, the output of the satellite receiver or any other point in the RF system. Signals can therefore be traced enroute from the antenna to the television set in order to identify any faulty components or cable breaks.

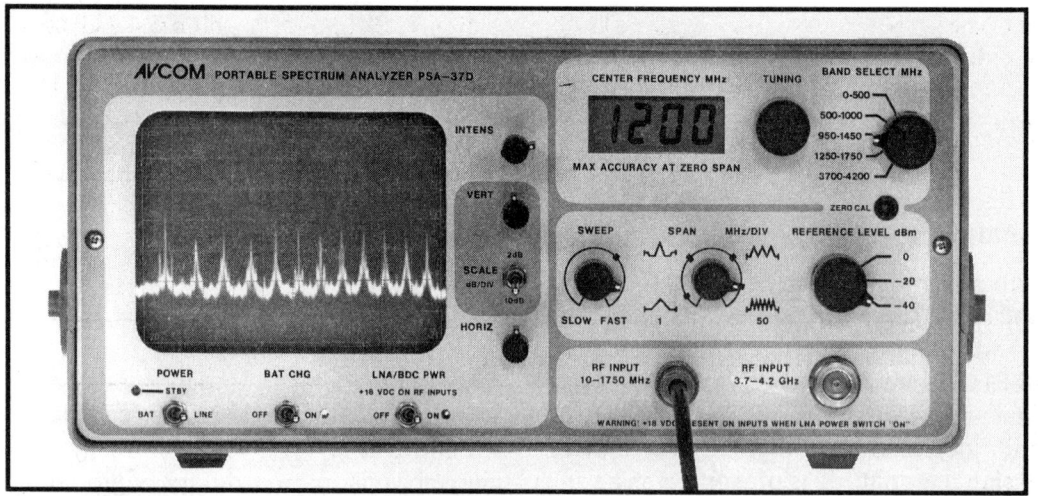

Figure 9-6. Avcom PSA-37D Spectrum Analyser. *This battery-operated analyser accepts input signals from 10 to 1750 MHz and from 3.7 to 4.2 GHz in five bands. It has both an LCD digital and CRT output. A built-in supply powers LNAs and LNBs. (Courtesy of Avcom of Virginia)*

TECHNICAL TROUBLESHOOTING

TABLE 9-2. SPECTRUM ANALYZER ANALYSIS REPORT

Customer Name_____Date_____

Address_____Telephone_____

_____

City_____State_____ZIP_____

System Description

Antenna: Manufacturer_____Diameter_____Model_____

Polarization/Feed: Manufacturer_____Model_____

LNB: Manufacturer_____Max. Noise Temperature_____Gain_____

Coaxial Cable: Length from LNB to Receiver_____ Type_____

Receiver(Main): Manufacturer_____Model_____

        Serial Number_____Location_____

Receiver(Slave): Manufacturer_____Model_____

        Serial Number_____Location_____

System Performance Statistical Data

Satellite-Targeted_____Transponder#_____

Output Levels: LNB_____dBm @ 3.7 GHz _____dBm @ 3.9 GHz _____dBm @ 4.2 GHz

        Coax_____dBm @____MHz _____dBm @____MHz _____dBm @____MHz

Terrestrial Interference: Frequency _____MHz Level _____dBm Bandwidth _____MHz

Notes

_____

_____

TROUBLESHOOTING & REPAIR

Two common symptoms observed on an analyser are tilt and wipe-out. If transponder power appears to decrease across the satellite band on the instrument's display, the feedhorn or LNB may be faulty. If the problem is a bad component, it might be identified by replacing it with a healthy one. Tilt can also be a result of attenuation in coaxial cables that have losses which increase with frequency. An analyser can be used to locate the source of high frequency roll-off in block downconversion distribution systems. Once the source has been isolated a solution might be to use a line amplifier with tilt compensation or to somehow shorten the cable runs. Another common problem, transponder wipe-out, which can also be traced with an analyser, appears as a sharp reduction in power of one or a group of transponders. One source of this problem may be an impedance mismatch that causes signals in a rather narrow frequency band to be reflected back to the antenna. Something as simple as a missing terminator on an unused splitter port or a faulty connector could be the cause.

The potential uses of an analyser are limited only by the creativity of the technician. Some unusual tasks can be created. For example, this instrument can even be used to test a hand-held remote which emits RF signals. As buttons are pushed on the remote, spikes should appear on the analyser's screen. Probably one of the most important uses of an analyser is to locate faint Ku-band satellites on narrow beamwidth antennas during alignment. An analyser can also allow visualisation of weaker signals that would not normally be seen with a satellite receiver and monitor. Thus an installer can detect those satellites that may not yet be transmitting television signals by observing their telemetry beacons.

Portable Signal Simulators

A portable signal simulator can be used to quickly diagnose faults in a satellite system. Such a device transmits a signal which mimics the television colour bar test pattern received from a satellite (see Figure 9-7 and 9-8).

Two instruments are used as examples here. Please note that neither of these are commercially available at this time. The Newton Electronics GBS 1600 (GBS stands for ground-based-satellite) was a good example of a simulator that was designed for testing C-band TVROs. The one designed by Satellite Test Equipment, the ET-15 Simulator, generates Ku-band as well as C-band test signals. These simulators also generate the same block of frequencies as the LNB output as well as some typical IFs and an AM modulated signal. These can therefore be used as diagnostic tools at any point in a satellite receiving system. For example, the ET-15 transmits colour bar signals having IFs of 70, 130 and 612 MHz. This unit also provides both normal and inverted video and has an output to test continuity of polarisers. Both units are battery powered thus eliminating the need for a bulky extension cord trailing behind during the troubleshooting procedure. The battery in the GBS-1600 lasts about two hours

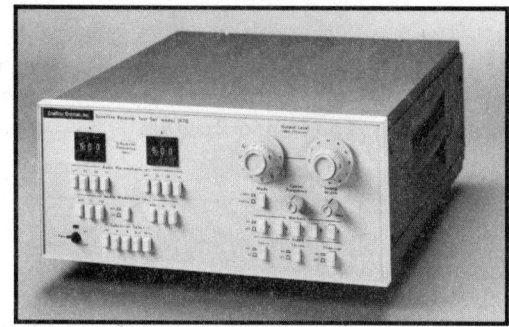

Figure 9-7. Comtest 1470 Satellite Receiver Test Set. *The Wavetek Model 1470 is a precision programmable test set designed to simulate a satellite television transmission for receiver bench testing. This complete test set operates as an IF sweep generator to enable troubleshooting and to identify receiver problems. It also operate a video IF source to verify receiver operation (Courtesy of Comtest Systems Inc.)*

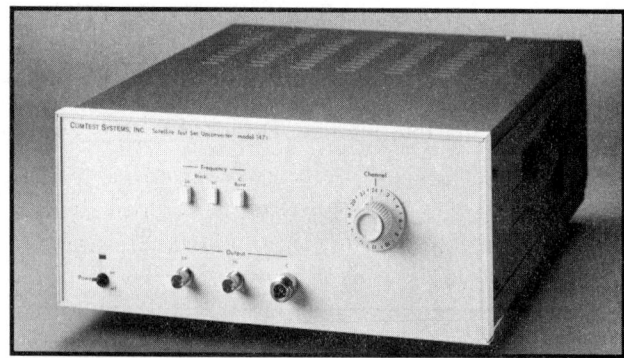

Figure 9-8. Comtest Model 1471 Satellite Test Converter. *The Model 1471 is designed to operate as an upconverter for the Comtest Model 1470 Satellite Receiver Test Set. It converts the 1470's 70 MHz test signals to the 3.7 to 4.2 GHz range for testing LNAs, LNBs and LNC. It also provides block IF outputs in the 440 to 940 and 950 to 1450 MHz ranges for testing of block downconversion receivers on a repair bench. (Courtesy of Comtest Systems Inc.)*

TECHNICAL TROUBLESHOOTING

TROUBLESHOOTING & REPAIR

before recharging is required. The ET-15 has the added advantage of being capable of generating test patterns in NTSC, PAL or SECAM formats.

If the output of the ET-15 is aimed into the antenna and the TVRO is functioning properly, clear colour bars should be seen on the television screen, assuming that the TV channel has been correctly tuned to match the modulator output frequency. The satellite receiver must be set to 1190 MHz in the 950 to 1450 block downconversion range as this instrument broadcasts on 1190 MHz. This test shows whether or not a satellite is targeted properly and can conclusively prove that the RF system is functioning. If colour bars are seen, the RF system is functioning well and it is probable that the antenna is not tracking properly or has its vision obstructed.

If colour bars are not seen, the LNB can be bypassed by connecting the 950 to 1450 MHz F-connector output from the simulator directly into the coaxial cable leading to the receiver. If colour bars are observed, the LNB is defective or is not receiving power. The signal simulator can also be used to ensure that the necessary 15 Vdc power is being received by the LNB. A light on its front panel would illuminate if there were adequate voltage at the LNB.

If the test pattern is still not detected on the TV, the simulator can be connected directly into the satellite receiver. This test is used to determine if the cable between the receiver and LNB is faulty. If this is not the problem, then the channel 3 (or 4) output of the ET-15 can be connected directly into the television receiver. This may show that the channel 3 (or 4) input or the set itself is faulty or the TV channel has not been properly fine tuned. If the television is functioning properly, then there is a problem with the satellite receiver.

If all these components are healthy, the problem may be something as simple as a tree screening the antenna and blocking the satellite signal, a non-functioning polariser or even a noisy, partially corroded connector that operates with the high levels of power generated by the test unit but may not be capable of detecting very weak satellite signals.

Another much simpler diagnostic tool, also a source of microwave energy and in a sense a portable signal simulator, is a portable fluorescent light (see Figure 9-9). If it is waved in front of a functioning LNB some change in noise level should be seen

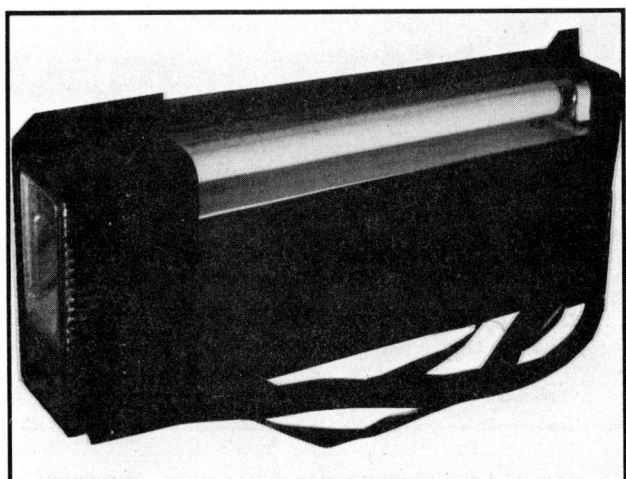

Figure 9-9. Portable Fluorescent Light. *This portable fluorescent light can be used as a simple tool to diagnose the condition of an LNB. In addition to generating light, it emits microwaves covering a broad frequency band. When waved in front of a healthy LNB, the noise level on the television screen should change and the signal level meter should indicate full scale reading.*

on the television and the signal strength meter should respond. If this does not occur, the LNB may not be receiving adequate power or it may be defective. Similarly, the human hand is a source of 300°K noise. When placed in front of a feedhorn, a sensitive meter should indicate an increase in detected power, assuming that the LNB is operating. Of course, this action will blank out any picture which might have been present if a satellite was targeted.

Portable Receivers

Portable test signal simulators are valuable, but will not help in aligning an antenna onto the satellite arc. This job is best accomplished with a portable receiver or other sensitive signal detection instruments described above and in Chapter V (see Figure 9-10). While most installers simply connect the customer's receiver or an extra unit from the shop, using a self-contained portable satellite receiver avoids the difficulty of hooking up 3 or 4 components as well as lugging a TV monitor or receiver out to the site each time a test is conducted.

TROUBLESHOOTING & REPAIR

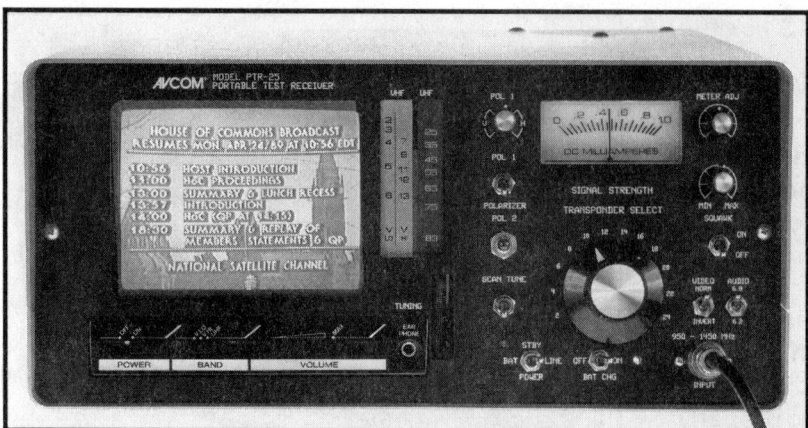

Figure 9-10. Avcom PTR-25 Portable Test Receiver. *Avcom's COM-2 and COM-3R satellite receiver circuits as well as PSD controls are included in this battery-operated unit. Outputs include connections to drive TV monitors, video recorders and audio amplifiers, an audible tone and a signal strength meter. It accepts an input frequency of 950 to 1450 MHz at a −70 to −25dBm level (Courtesy of Avcom of Virginia)*

Most portable receivers today are battery operated and are equipped with either a colour or a black and white monitor all packaged in a relatively lightweight container. The only extra equipment required to receive a satellite broadcast is an antenna and LNB/feed. This portable tool eliminates the risk of forgetting some necessary component back at the shop or of dropping a customer's prized set in the mud during testing.

Testing with a portable receiver proceeds similarly but in an opposite sense to that used with a signal simulator. The LNB output is first connected to the portable receiver 950 to 1450 MHz or 950 to 1700 MHz input. If an adequate picture is seen on the small TV monitor, the LNB is fine and any problem would lie downstream towards the customer's TV set. Next, the output of the coaxial cable can be connected to the portable test unit. In this fashion, by working from the antenna towards the television, the faulty component can ultimately be identified.

Testing with both portable signal simulators and receivers can also follow along the lines of a process called "half-split" troubleshooting. Instead of checking each component sequentially, the diagnosis begins somewhere in the middle of the system. This might save time if the input components seemed to be working fine but nothing was happening at the output, the television set.

Switching Components

One very simple method to troubleshoot a satellite system is to switch out components, one by one. Thus, if the LNB is suspected of being faulty, substitute a healthy one and see what happens. Remember, in this and all other similar situations when any component is switched, always be sure that the power is turned off and the receiver is unplugged to avoid causing further damage. Some brands of receivers continue to send voltages to the system even when the power switch is off.

This diagnostic method is direct but can be somewhat time-consuming. Also, if switching components is attempted during a new installation and more than one component happens to be bad, it may prove to be a frustrating endeavour.

Interchanging or shotgunning components is used mainly to diagnose problems in the RF system. It is easy to check if the actuator or polariser is not functioning properly by visual inspection and then to find the problem by voltage checks and other troubleshooting methods.

Power Supplies

Most electronic malfunctions can be traced to satellite receiver and actuator power supplies (see Figure 9-11). Unless adequate measures have been taken to suppress voltage spikes, it is common to have a fuse blow or a voltage regulator fail. Some brands of equipment have both external, screw-in and internal fuses so both must be checked. If nothing happens when a receiver or actuator is connected, the unit should be unplugged and the fuses examined. To be absolutely certain that fuses are blown, if this is not obvious by a visual inspection, check their continuity with an ohmmeter. When replacing fuses it makes good sense to install spike suppressers so this problem does not reoccur.

If fuses are repeatedly blown, a very inexpensive current limiter can be constructed with a 40 or 50 watt light bulb and a 1 ohm resistor connector in series with the power line into the receiver. ac voltage is measured across the resistor. If there is a direct short in the receiver, the bulb will glow brightly as it limits current to the receiver and protects the fuses. Make sure that this test setup is well insulated or severe shocks could occur.

Integrated circuit (IC) voltage regulators are often not adequately protected and can malfunction. Excess voltages can damage the receiver, actuator or LNB. In unusual cases, the antenna could even be damaged by being driven into the ground by a runaway actuator. However, if a regulator does fail, identification of the faulty part is easy with the voltage tracing methods outlined below. Replacement is just as simple.

The 7800 and LM340 series of positive regulators are capable of supplying up to 1 amp of positive current and are often mounted directly onto the receiver chassis. The 7900 and LM320 series are the complimentary negative components but must be isolated from the chassis and therefore mounted by use of a mica insulator and insulated screw. The output voltage is identified by the last two digits of the 7800 and 7900 regulators. For example, a 7818 regulator provides a positive 18 Vdc. Two other adjustable regulators, the LM317T and LM723, are also often used in satellite receivers.

Voltage Checking

Tracing voltages with a voltmeter is an effective method to rapidly identify a faulty component. When examining an RF problem, all voltage checking should begin at the receiver. It is the brains of a satellite system and it (1) powers the LNB, polariser and modulator, (2) selects channels by sending a correct voltage to an internal VTO and (3) communicates polarity, channel selection and audio format information to the user.

The satellite receiver relays from 15 to 24 Vdc to the LNB and typically 5 to 6 Vdc to the polariser. If this power does not reach its destination for one reason or another, the system will not function.

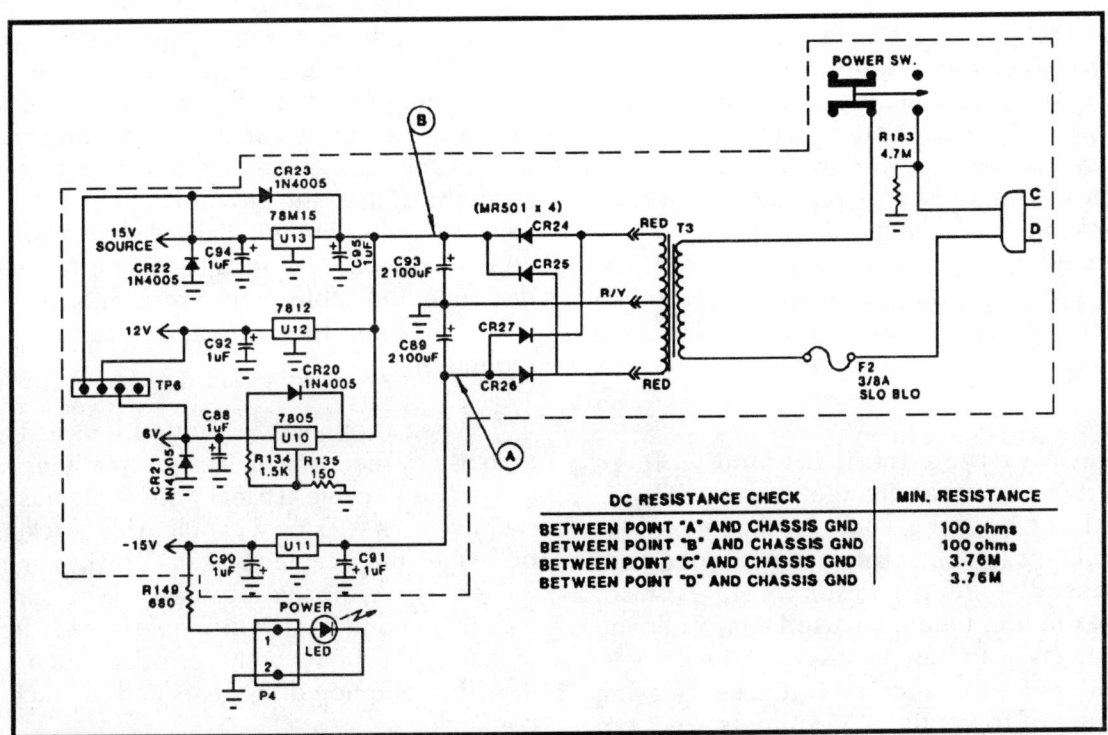

Figure 9-11. Receiver Power Supply - Block Diagram. ac power enters on points C and D from the wall plug. Notice that it has no ground connection. Input power flows through F2, a 3/8 amp slow-blow fuse to the primary side of transformer T3. Voltage is then stepped down to approximately 24 Vac and is fed into the rectifier diodes, CR24, 25, 26 and 27. These have a 24 Vdc unfiltered output. C93 and C89 filter out the remaining ac voltage ripples. Test point A yields a negative dc voltage; test point B yields a positive voltage. The filtered power is then connected to voltage regulators U13, U12, U11 and U10 which have outputs of 15, 12, -15 and 5 Vdc, respectively. (Courtesy of R.L. Drake Company)

TROUBLESHOOTING & REPAIR

Voltage checking should begin at the receiver output. The voltmeter is set on an appropriate range, from 20 to 200 Vdc. An output of 36 Vdc is common on the actuator circuit so the 200 Vdc scale would be used here. If the range is set too low, the indicator will be pushed off scale. In general, the voltmeter red lead is connected to power and the black lead to ground. Voltage is measured between the central conductor and ground at one port of a two-way, power passing splitter. The input coax is connected to the second port. It is important not to leave the splitter in the line because 3.5 dB will be lost by its insertion. An open port is also an invitation to ingress interference and causes signal losses because of reflection at this discontinuity. If no voltage or a low voltage is measured at the receiver IF input, turn the power off, unplug the receiver and disconnect the cable connected to the LNB. If a second reading obtained after powering back on indicates the correct voltage, there is a short in the system upstream towards the antenna. If the voltmeter still reads zero or near zero, 99% of the time an internal or external fuse or a voltage regulator is blown in the receiver.

If the disconnected receiver was delivering full power, but when connected to the cable the voltage was unduly low, there is probably a short or a bad component further upstream towards the antenna. Unplug the receiver and disconnect the coax at the LNB input. Then re-plug the receiver and turn it back on. If full voltage is measured across the disconnected cable, the fault lies in the LNB. If zero voltage is measured, then there is certainly a shorted connector or coax. It is possible that the cable could be crushed somewhere along its length.

Always carry extra fuses and voltage regulators. For pocket change, standard 7812, 7815, 7912 or 7915 regulators for the receiver or 7805 regulators for the polariser control circuit can be purchased at electronic supply stores. Also, have an extra 555 timer chip used to send timing pulses to many brands of polarisers. These can easily be soldered into a receiver in the field, saving many hours of travel time. 555 timers are 8-pin chips which look like a small centipede. They control the position of the polariser probe.

Whenever leads are disconnected or connected it is important that the satellite receiver be unplugged in order to avoid inadvertently destroying a good component during the troubleshooting procedure. Just because the receiver is turned off does not mean that power is not being sent to the various components.

It is also wise to have an extra receiver, actuator and LNB, as well as appropriate connectors on hand so a bad one can be temporarily replaced. Most customers will appreciate the loaned equipment during the time these faulty ones are being repaired.

Similar techniques can be used to ensure that the proper voltages are reaching the polariser as well as the actuator motor. The path can be traced from the controller or, if voltage is present at the peripheral equipment, it can be followed in reverse from the antenna to either the receiver or actuator control box. When checking voltages on the actuator motor two people are required, one to activate the motor via the control box and one to take the reading at the motor.

The voltmeter can also be used to check both continuity and shorting of cables. If a coax or other wire is suspected of having a break, follow this procedure. Set the volt/ohm meter on the ohms x 10 scale. Then use alligator clips to join the central conductor and ground together at one end. Make sure that the other end is disconnected from power and then attach the meter at this end. If the meter reads zero ohms, then the circuit is continuous; otherwise the cable is broken. If the alligator clips are not attached and a resistance reading is taken on the disconnected cable, a zero resistance indicates a short between the central conductor and the ground shield.

When a connector is shorted, look first to see if there is any mechanical cause. Sometimes water is soaked into a cable at a leaky connector just as wax is drawn up a wick on a candle. This "wicking" can pull water into a cable as far as 12 inches or more. If this is the case, the sign is usually a discolouration of the centre conductor. If the cable is made from aluminium, it can be severely corroded. Cut back until the discolouration disappears and reattach a new connector.

TECHNICAL TROUBLESHOOTING

TROUBLESHOOTING & REPAIR

Grounding Problems and Hum Bars

Occasionally problems arise that cannot be traced to a single faulty component but are related to how all the components and wiring fit together. Diagnosing and servicing such difficulties can be somewhat more time consuming but also more rewarding than simply replacing a bad LNB or finding a shorted cable.

Ground loops are a good example of this type of system problem. A ground loop results from having two grounding points which offer slightly different resistances to ground within a TVRO. In most installations all the electrical components are grounded to a common point through the wall outlet. However, if the cable run between the antenna and the indoor equipment is long enough, the resistance between the ground potential at the earth near the antenna and the coax's ground resistance can be substantially different. When a ground loop occurs, it may be necessary to "float" the indoors components from the wall outlet and use only a single common ground at the antenna.

Ground loops can create a series of problems including "hum bars" on the screen, difficulties with polarity control when a pin diode controller is used, actuator positioning errors or a low-level hum in the audio. Hum bars appear on a television screen as messy horizontal regions of varying width which may be stationary or which can move up or down on the screen. In effect, the difference in ground potential causes the picture to lose some of the necessary synchronising ability.

Ground loops can be broken into two categories: those that appear long after the installation is completed and are usually caused by bad connections; and those that begin during installation that can, on occasion, be tricky to cure. However, one of two methods can usually clear up the difficulty. The first cure is to systematically eliminate any bad connections that might exist in the TVRO wiring or, perhaps, in the home's ground connection wiring. The second is to tie all of the interconnected satellite equipment together via a heavy gauge wire with a common ground at the antenna by "floating" the indoor equipment. Or the antenna and LNB can be tied directly into the electrical service ground at the point where it enters the home with a heavy gauge (e.g. #6 gauge) or a flat welding cable.

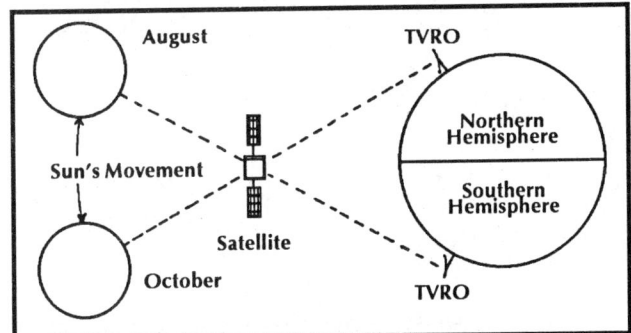

Figure 9-12. Solar Outages. *The timing of solar outages is directly related to the site latitude. They occur when the sun lines up directly along the antenna boresight. Solar outages always take place within 3 1/2 weeks of the equinoxes on March 21 and September 21, when the sun crosses the equator during its annual north/south journey. Solar outage in February, March and April begins at the northernmost latitudes and moves southward until satellite systems on the equator have outages at equinox time. Then earth stations in the southern hemisphere begin having outages until those at southernmost latitudes experience them 3 1/2 weeks after the equinox. In August, September and October, the entire pattern is reversed because the sun is moving in the opposite direction. Outage season lasts for a little over a week in any given location. During this period, there will be an outage daily on each satellite. It appears first as small amounts of video noise, rapidly becomes annoying interference and, on the days at the center of the outage season, peaks out as a total loss of incoming signal. Antennas should be moved away from this satellite during outages in order to avoid heat build-up on the feedhorn and LNB. Spun aluminum and shiny surfaced fiberglass antennas can raise the temperature at the focal point sufficiently to cause severe damage and possibly destroy components.*

Inadequate system voltages can have the same effect and cause hum bars by not providing enough power to generate the necessary signal to adequately power the receiver. This could happen if undersized wire is used on a long run to the LNB so that the cumulative cable resistance causes unacceptably large voltage drops. Note that the same symptom may be seen if the feedhorn probe servo hunts. This is also caused by inadequate voltages in undersized cables spanning long distances.

Solar Outages

When the sun lines up directly on to the axis of an antenna microwave noise overpowers any satellite transmission. Solar outages occur at brief periods in early spring and fall (see Figure 9-12).

TROUBLESHOOTING & REPAIR

E. UNDERSTANDING COMPONENT FAILURES

One approach to troubleshooting is to examine the operation of all components in order to understand how each may fail. This background knowledge is a useful aid in following the methods described earlier.

Satellite Receiver

The function and operation of a satellite receiver must be clearly understood by a competent troubleshooter. It is crucial to be versed in the alignment procedures for each unit installed to avoid many potential problems. Technicians should also be familiar with all front panel controls such as skew adjustment, subcarrier selection and video scanning.

A visual inspection of all feedback lights, fuses and connections and measurement of voltages will often reveal a simple problem. If this is not the case, more technical approaches can be used to determine any fault. One simple alternative is to swap an identical working unit and see if the trouble disappears.

Some more common problems encountered are listed below:

1. A high signal strength meter reading but no picture. Assuming that the TV is tuned to the correct channel and all other components are functioning properly, this fault can be caused by a bad modulator. Note that many receivers use VCR modulators and replacements that are available in electronic part stores. Be sure that a crystal controlled, "linear" modulator is purchased. Non-linear brands are intended for use in computers.

2. A zero signal strength indication with no picture. This may be caused by a shorted IF/voltage cable, lack of powering voltages or a problem somewhere else in the system such as poor antenna alignment.

3. The receiver shuts off when it is hot. A thermal overload is often caused by a voltage regulator overheating. A can of cold spray or a hair dryer may help isolate the defective component. The problem may be solved by simply clearing enough room to allow air to freely flow around the receiver vents. Occasionally a defective component may have to be repaired in the shop.

4. A herringbone pattern in the picture. This can be caused by a bad modulator or unwanted feedback from the IF line into the modulator. A strong, nearby FM station or transmitter may also cause the same problem if signals leak into the system through poorly shielded cables or components.

5. A lack of audio. This may be a problem in either the receiver or modulator. Check all connections and make sure the right audio subcarrier is being selected. If a faint sound can be heard when listening with a pair of headphones at the receiver's audio output port, the modulator is probably bad.

6. A buzz in the audio. This can be caused by a faulty modulator, incorrect audio tuning or an unduly high video level in the modulator. The appearance of saturated colours on the screen is also associated with an excessive video level.

7. Thick black bars across the TV screen. This is often a symptom of a faulty power supply caused by an irregular voltage or a ground loop. A 4700 microfarad, 25 volt capacitor applied across the dc rail may quickly solve this problem, but the underlying cause should also be found and corrected.

Aside from swapping out a voltage regulator or visually inspecting internal components, repair work on a satellite receiver of any complexity should be done in the shop or in a portable van equipped as a shop. There is nothing worse than scattering parts from a receiver all over a customer's home and then not having the components necessary to fix it when the problem is discovered.

Stereo Processor

Troubleshooting either a stand-alone stereo processor or one built into a receiver is usually limited to educating the user about its proper operation. Many may believe that since the system has a stereo processor, all audio should now be received in stereo. Most stereo processors have too many unfamiliar buttons for the novice user. This can easily lead to incorrect operation. Note that a satellite

TROUBLESHOOTING & REPAIR

TV guide should list the correct audio formats for use with stereo processors.

A first step is to tune to a known channel that employs a stereo format that the processor can demodulate. Most can manage the discrete format. Select matrix stereo position on the processor controls and tune both subcarrier channels to the correct position. If stereo is not heard, the problem may lie with the customer's stereo system. This can be ascertained by tuning their stereo receiver to a known source such as FM radio, a record or a tape.

In those cases where the customer's stereo is functioning properly but no stereo can be obtained, first make sure the leads are securely connected. A voltage reading can be taken to determine if a signal is actually reaching both ends of the audio cables. Make absolutely sure that the stereo processor is connected to the baseband output of the receiver, not the audio output which is used to drive an external modulator. If these tests are negative, the best course of action is to swap it for a working unit and take it back to the shop for necessary repairs.

Cable and Connectors

Coaxial cable and connectors can easily be overlooked as potential sources of trouble. But incorrect installation procedures and improper selection of cables can often cause substantial deterioration in overall performance.

A poorly grounded cable or a connector which leaks moisture can introduce unwanted noise or can severely attenuate signals. A cable that is bent too sharply can result in signal losses because its impedance changes at the bend and causes signals to be partially reflected. Improperly installed direct burial cables and leaky connectors can corrode in surprisingly short times.

When installing cables and connectors, it is imperative to securely crimp the casing onto the shield wires with the proper tool, a cable crimper. All locations where water may enter must be sealed and protected as well as possible. Coaxial sealant can be used to prevent water ingress at all outside, exposed locations.

Troubleshooting cables and connectors can be as simple as a visual inspection. If necessary, a continuity test will make sure that signals can pass without serious or complete attenuation. If wire or cable having too fine a gauge has been used for long runs, insufficient voltages reaching the LNB or polariser may cause faulty operation.

LNB

A faulty LNB can cause problems ranging from excessively sparklie pictures to complete whiteout. Some common symptoms and their causes are:

1. Lack of voltage to the LNB. This can result from a blown fuse or other fault in the receiver, a corroded connector or a shorted or broken coax. A simple method to check that the full 15 to 24 Vdc is reaching the LNB is to disconnect the input line and measure voltage between ground and its central conductor. Be careful not to short the leads together.

2. An open circuit at the LNB output. One or more of the LNB stages may be blown. There will be no signal strength reading at the receiver and the TV picture will be blank white.

3. Excessively sparklie pictures. One or more LNB stages may be faulty, resulting in noise temperatures in excess of rated values. Switching in a good LNB or using a signal simulator is a quick troubleshooting procedure if this is suspected.

4. Presence of hum bars. Hum bars, horizontal lines of varying thickness moving up or down across the screen, can be caused by insufficient voltage at the LNB. Most LNBs require a minimum of at least 15 volts dc and 150 milliamps of current to function correctly.

5. A blank TV screen. This may be caused by a faulty LNB. To determine if the LNB is powered and operating, connect and disconnect the LNB cable while observing the television screen. If the noise level on the picture does not change, this amplifier is probably at fault or is not receiving power. Another check for a faulty LNB is to wave a fluorescent light in front of the feedhorn. If the noise level does not change, the LNB is probably faulty or is not receiving power.

6. Picture deteriorates in hot or cold weather. This problem can be isolated by using a can of freeze spray or a hair dryer to change the external case temperature of the LNB.

COMPONENT FAILURES

TROUBLESHOOTING & REPAIR

LNBs are finely tuned pieces of equipment. A faulty unit should be returned to the manufacturer for repair. Although they are carefully encased to eliminate water ingress, if a leak does develop, the moisture can destroy sensitive circuits by oxidizing copper tracks on circuit boards. The early signs of water damage can be an overly noisy signal and the resulting sparklies.

A faulty LNB can show up as a blown fuse in the receiver. However, if fuses keep on blowing even after the LNB is disconnected, the cable or connectors are probably at fault.

Feedhorn

Feedhorns are precisely machined waveguides designed to efficiently capture microwaves. Any obstructions in their throat can detune the waveguide and seriously impair performance. Some commonly encountered symptoms of problems are:

1. Excessively sparklie pictures. This may be caused by wasps, water, ice or other obstacles lodged in the waveguide throat. If water is not removed it could freeze and crack the housing. When using the 90° waveguide elbow with the polariser, a cure is to drill a very small hole (less than 2 millimetres) in the lower corner of the right angle guide to allow water drainage.

Such performance impairments may also be caused by a bent or mechanically distorted probe. But never readjust the probe position even if it appears to be bent. These are finely tuned devices.

2. Signals of only one polarity are received. This probably indicates a bad servo or dc probe motor. Burnout can be caused by improperly installing servo motors without allowing for full 180° freedom of motion at the correct position. This can also be caused by using a wire gauge that is too small for the cable length used or by not using shielded wires.

3. Oscillation or servo hunting of the probe. A random oscillation in a servo motor polariser can appear in the form of black bars or white lines on a TV screen. In severe forms, it may show up as a jumping between two adjacent channels of opposite polarity as the probe hunts intermittently around its axis. This can be easily differentiated from channel drifting because the same two channels are always seen. An easy method to prove the problem is caused by servo hunting is to disconnect the polariser connections. If the problem disappears, it's cause is evident.

Most often this occurs when cable runs exceed 25 metres and an undersized wire or parallel servo motors are installed. The solution to this problem is to install a 1000 microfarad, 10 Vdc electrolytic capacitor between the red +5 volt wire and ground which is black on a servo motor. The capacitor should be connected so the plus on the capacitor is connected to the +5 volt and the minus side to ground. To prevent oscillation, the suggested minimum wire sizes mentioned in Chapter V should be used.

Antenna Actuators

A well designed and manufactured antenna actuator is usually quite reliable when correctly installed. However, there can be large quality differences between different brands, and proper equipment selection will save troubleshooting and repair time.

The actuator consists of a mechanical assembly and its electronic control circuitry. The mechanical component serves to hold the antenna in place and to provide movement across the polar arc while the controller keeps track of the antenna position. Some common symptoms and underlying causes for mechanical failures are as follows:

1. Inaccurate pointing of the antenna. If there is too much play in the movement of either the inner or outer tubes (or gears or a chain drive in the case of a horizon-to-horizon assembly) then some wearing or loosening of parts has occurred. If the antenna has more than one centimetre of play in either direction at its rim, the mount needs attention. The bolts supporting the pivot points or jack attachment should have double nuts, lock washers and lock-tight, and be tightly set.

2. Wear or scoring on the jack tube. This may result when the internal O-rings are damaged by lateral stress being applied when the arm is not installed using ball joints. Waterproofing should be adequate so that water does not leak into the casing and motor assembly.

3. The jack tube is binding or even bending as satellites are tracked. This may occur if both ends of

COMPONENT FAILURES

the jack arm have not been properly aligned and mounted on self-adjusting ball joints. A single ball joint on one end is not enough. The jack should track through its entire range of possible motion without binding on either the antenna or mount structure. Therefore, if the electrical east/west limits fail, damage to the arm will not occur. The motor may also have failed possibly due to water damage. It makes sense to have a spare motor or brushes for placement around the armature.

4. Water entering the motor housing or the tube. Water can collect inside the arm or housing if they have not been properly sealed by gaskets or a rubber boot. Even so, condensation can cause water to collect, freeze, seize the motion and even damage the housing. Drain holes must be in their correct position to prevent this chain of events from occurring. A neoprene rubber accordion sleeve with a hole at either end to allow condensed water to escape is an excellent method to prevent water ingress.

5. A grid of dots appears on the TV screen. Some control boxes cause interference on channels 2 or 3. The internal 3 to 6 MHz clock crystal that runs the microprocessor makes a grid of evenly spaced dots appear of the screen. This effect sometimes occurs when the actuator is set on top of the TV set and beats directly into the tuner. Or if 300-ohm flat, twin lead cable is used to feed over-the-air broadcasts into the TV, it may also pick up the signal from the clock crystal. These dots can also appear on satellite TV channels, but can be easily eliminated by using a 10 dB line amplifier on the receiver modulator output to overpower the interference.

The most common types of feedback circuits used for controlling actuator position are Hall effect sensors, reed relay switches, 10- turn potentiometers and photo optical read-outs. Most are coupled to a central microprocessor. Each has its own set of potential problems and solutions.

Feedback potentiometers receive a constant voltage and divide it into a smaller portion depending upon the antenna position. This returned voltage is calibrated and fed onto a read-outs on the face of the controller.

If the potentiometer, also familiarly known as a pot, breaks or becomes disengaged from the drive shaft in the motor, an antenna can be driven too far and the jack tube can be bent. A preset electronic and mechanical limit is an important feature for such control systems.

A method to troubleshoot this type of actuator is to use a spare 5 K-ohm potentiometer. It can be temporarily installed in place of the permanent one by simply disconnecting it and reconnecting the two or three wires using alligator clips. If read-out numbers change on the control box when the temporary pot is adjusted, then the problem is in the motor; if not, the problem is in the control box.

To troubleshoot a unit with a reed switch or Hall effect transistor in the motor, attach an alligator clip between the pulse and ground connector at the control box after disconnecting the pulse input line. As the drive arm is moved in either east or west direction, make and break the connection on the pulse and ground wires to simulate drive pulses. If the controller works properly and does not display an error, then the problem is in the sensor or the cable connection between the motor and control. Troubleshoot for cable continuity.

One method of troubleshooting a Hall effect sensor at either the controller or motor is by reading voltage between the pulse and ground points. A drop of about 0.75 Vdc should be measured when the east or west button is pressed. If not, then the Hall effect transistor is bad. If the sensor itself is not faulty, check that the magnets on the motor have not moved or fallen out of place. A spare Hall sensor and reed switch are useful additions to a troubleshooter's spare parts kit. Simulated pulses can also be produced by hooking the polariser pulse and ground outputs from a receiver or a hand-held polariser controller to the sensor and ground inputs on the positioner. This will cause the actuator motor to move and will indicate a bad sensor. Pulses will continue until this connection is broken. Be careful not to drive the antenna into the ground.

It is important to always use shielded sensor cable on Hall effect or reed sensors. These devices will give spurious readings if they pick up outside noise. Even the inductive action of a starting or stopping motor can be detected as a "spike" in the sensor. This added to the control count would cause a cumulative error in measuring antenna position. After the problem has been corrected, the actuator control must be resynchronised because each satellite will probably be missed by the same amount.

TROUBLESHOOTING & REPAIR

Make sure that the magnets surrounding these two types of sensors have not been dislodged. If they have, remember that magnets have a positive and negative end and should be reinserted all facing in the correct direction.

The ground lug on the controller power cord must never be cut to try to fit the three prong plug into a two prong outlet. The microprocessor requires a reliable ground to accurately set its counting circuitry. Always use some type of surge protection on the power line into an actuator to protect the microprocessor. Note that if a problem does develop in the internal circuitry of an actuator, the best bet is to return it to the manufacturer for repair.

Antenna and Mount

Servicing an antenna and mount is a relatively easy matter of conducting an educated visual inspection. Reflectors should be checked for warping and surface imperfections. Mounts must be examined for stability, alignment and secure connections. The feedhorn/LNB support structure can be checked for centring and stability. Methods to perform all these checks have been outlined earlier in Chapter V.

Television

A television is a complex piece of electronic equipment. However, troubleshooting a TV is easy when over-the-air channels of adequate signal strength are present. If the set works well with conventional reception, then about the only problem encountered is when the television is not tuned onto the modulated channel.

There are major differences in quality between different brands of televisions sets. It would be wise for a well-equipped troubleshooter/installer to carry a small, portable, excellent quality television or monitor to demonstrate if there ever is a question about the performance of a satellite system. After all, most judgments are made on the basis of picture quality. Also, if a monitor is used, a faulty modulator internal to a receiver can easily be diagnosed since it would be bypassed.

F. SPARE PARTS KIT

If a site visit is necessary, it is essential to have the proper tools on hand. Driving 60 kilometres to discover that a 50 cent voltage regulator is missing can be expensive.

All the light tools of the trade should be taken on a service call. This usually excludes concrete mixers and other large items. In addition, a complete, healthy spare system, possibly including a small test antenna, should be part and parcel of the troubleshooting kit. This kit is composed of:

- Feedhorn with polariser
- Extra servo or dc motor for polariser
- Modulator
- Inexpensive receiver
- Ample supply of applicable fuses
- Appropriate voltage regulators
- Extra reed switches or Hall sensors
- Spare brushes, motor or a complete actuator
- 5 K-ohm potentiometer
- Fluorescent light
- 1000 microfarad electrolytic capacitor
- Scotch locks
- Electrical tape, solder
- Splitters
- Assorted connectors
- 10 dB line amplifier
- Coaxial sealant and shrink tubing

G. A SYMPTOM/CAUSE MAP

Table 9-2 has been developed to aid a troubleshooter in making a rapid diagnosis of any potential system problem. This tool should be used as a complement and not a replacement to the detailed explanations which have already been presented earlier in this chapter.

TROUBLESHOOTING & REPAIR

Components and Symptoms	Problems	Symmetrical Dots on T.V. Screen	Interference on Channels 2, 3, and 4	Every Other Channel Good	Motor Boat Sound in Audio	Buzz in Audio and Bright Colors on Video	One Polarity Only Received	Picture Comes and Goes	No Picture but High Signal Strength Reading	Actuator Jack Stuck	Good Video but No Audio	Good Audio but No Video	Receiver Blows Fuses	Lines in Picture	Horizontal Hum Bars and Line Rolling	Receiving Only Some Satellites	Satellites in Arc Centre Snowy	Poor Video with Sparklies	Channel Drifting	Jack Moves but Controller Not Counting	Herringbone in Video	Flutter in Video	All Satellites Snowy	Video Good Only During Daytime	Video Bad at Night	Video Bad on Hot Day
Dish																		X								
Mount								X								X	X									
Feed Support																	X									
Jack, Actuator or Motor Sensor								X		X										X						
LNA or LNB													X	X	X		X	X					X	X	X	X
Downconverter													X	X	X			X							X	X
Feedhorn				X			X						X	X												
RG-6 or RG-59 Coax													X		X										X	X
Cable Problems and Corrosion			X															X					X			
Satellite Receiver					X	X		X		X	X	X	X		X			X	X		X	X				
Modulator					X			X		X	X										X					
Polarisation Controller							X																			
Actuator Controller		X						X		X				X						X						
MATV Signal Distribution			X											X	X											
Terrestrial Interference			X	X				X	X								X									
Customer TV Set								X			X	X														
Grounding														X	X											
Water In Actuator									X																	
Water In Cable																			X				X			
Water In Feedhorn																	X						X			
Incorrect Voltages															X											
Dish Support																									X	
Polar Arc																X	X								X	
Scrambled Signal											X	X														

TROUBLESHOOTING & REPAIR

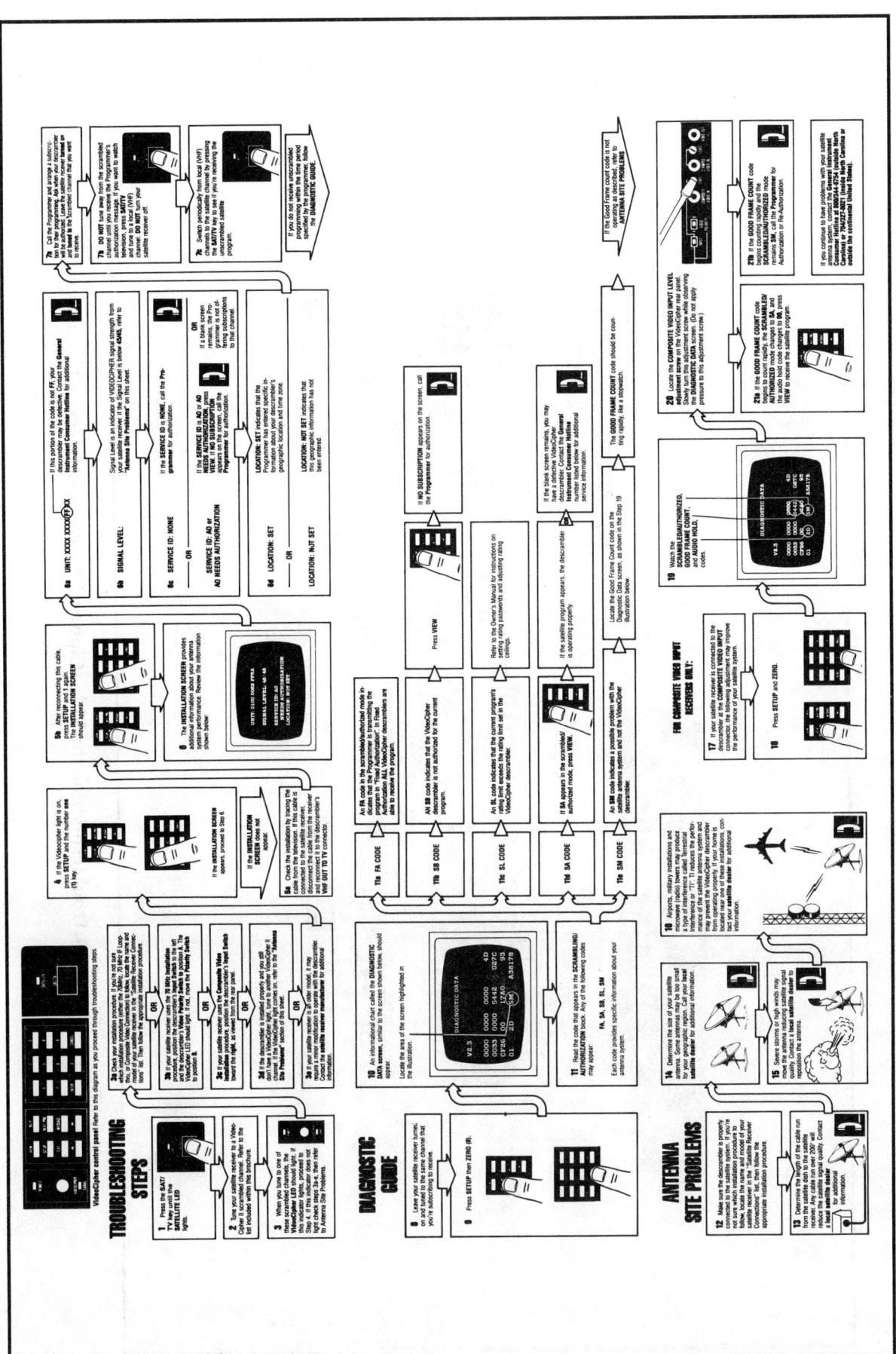

TABLE 9-3. Troubleshooting the VideoCipher II. *(Courtesy of General Instruments)*

TROUBLESHOOTING & REPAIR

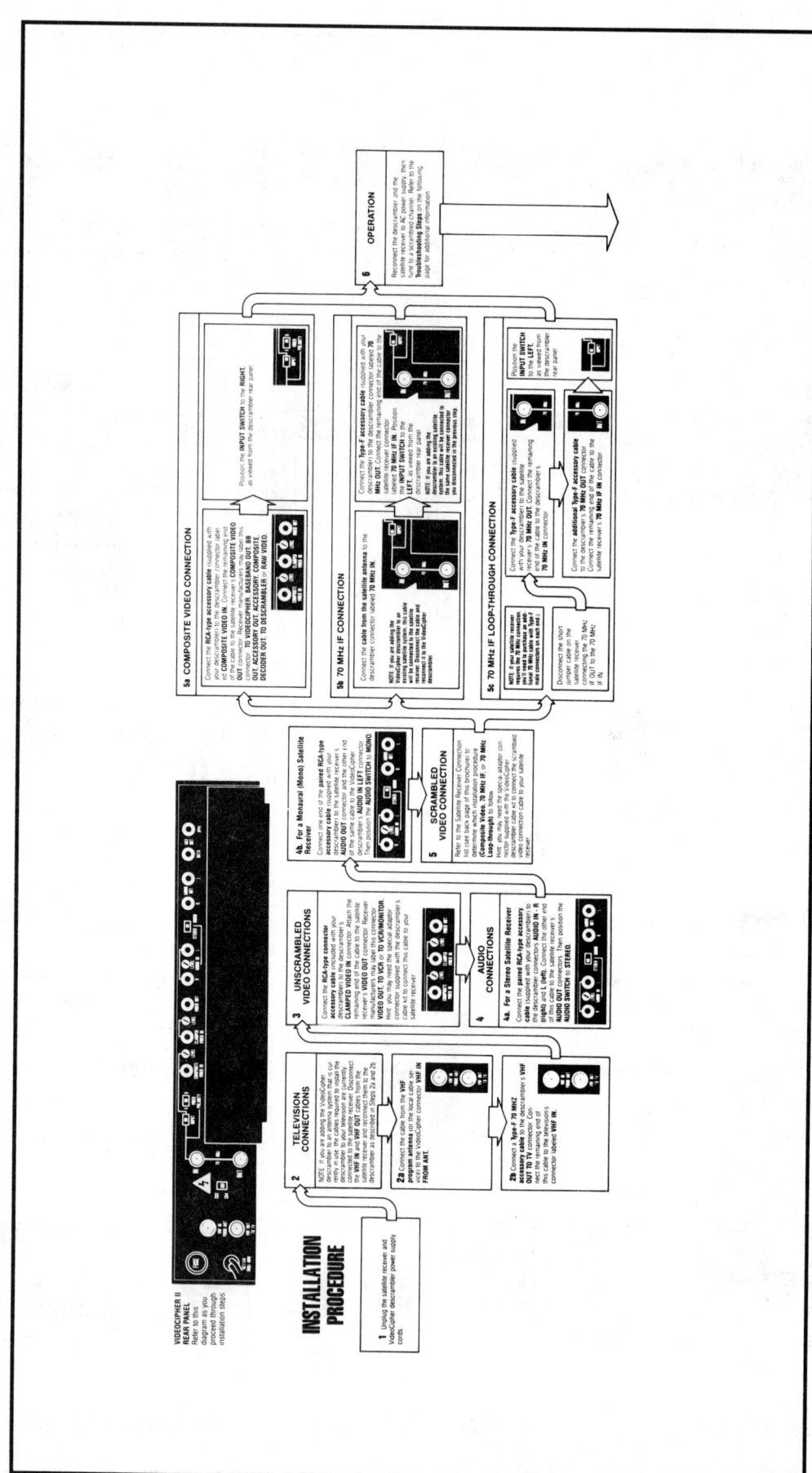

TABLE 9-4. Troubleshooting the VideoCipher II. *(Courtesy of General Instruments)*

APPENDIX A.
THE DECIBEL NOTATION

Decibels (dB) are used to express the relative values of two signals. The logarithmic scale is used to compress large differences in numbers to a more manageable range. Decibels are defined by the following equation:

Decibel difference = 10 log (signal A/signal B)

For example, if signal A is 1000 watts and signal B is 10 watts, then signal A is 20 dB stronger than signal B because:

Decibel difference in power = 10 log (1000/10)
= 10 x 2
= 20 dB

Therefore, if an amplifier received a signal of 10 watts and increased its strength by a factor of 100 to 1000 watts, it would have a gain of 20 dB. Similarly, if a 10 watt signal was increased by a factor of 1,000,000 to 10 million watts, the gain would be 60 dB.

Decibels are also expressed relative to a reference value such as watts, milliwatts or millivolts. The abbreviations dBw, dBm and dBmv mean the relative increase in power relative to one watt, to one milliwatt and to one millivolt, respectively. For example, 60 dBw means a power of 1 million watts.

The definition of decibels relative to one millivolt (or ampere) differs from that used to express power differences. A signal is expressed in dBmv by

20 log (signal in millivolts / 1 millivolt).

Therefore, for example, 20 dBmv is equal to a signal of 10 millivolts. The difference in the definition arises because power levels are proportional to the square of the voltage (or current).

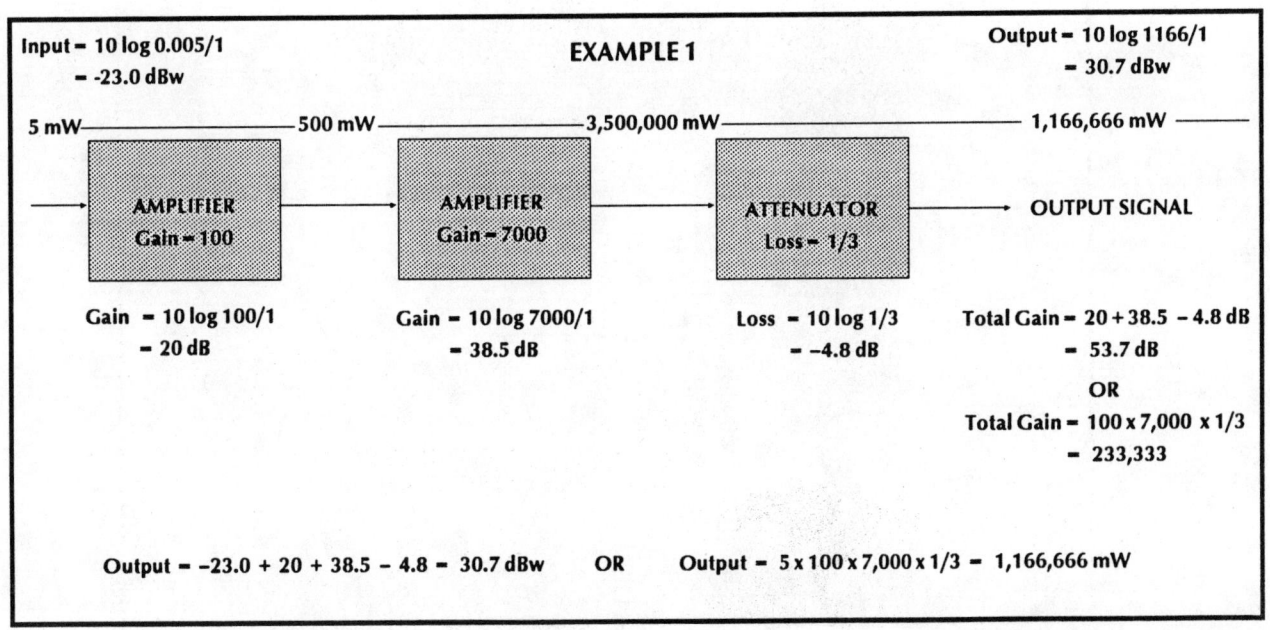

APPENDIX A

DECIBEL NOTATION

APPENDIX B.
SATELLITE TV EQUATIONS

The performance of a satellite TV system can be theoretically calculated by using some general equations. Some of these are presented below.

Link Equations

The link equations are used to calculate the ratio of carrier-to-noise (C/R) power reaching the input of a satellite receiver. The link equation is:

$$C/R = EIRP - \text{Path Loss} + G/T_{sys} - 10\,kTB$$

$$C/R = EIRP - \text{Path Loss} + G/T_{sys} - 10\log B + 228.6$$

EIRP is the effective isotropic radiated power directed by a downlink antenna to a location below. It is expressed in dBw, decibels relative to one watt.

Path Loss and Slant Range

Path loss measures how much signal is lost on the journey from the communication satellite to the receiving antenna. Losses are mainly due to the "spreading out" of the signal on its long journey. The amount of spreading out is determined by the distance the signal travels from a satellite to a receiving antenna. Path loss in decibels is given by:

$$\text{Path Loss} = 20 \log 4\pi S f$$

where S is the slant range and f is the signal frequency in Hertz. Slant range in kilometres is given by:

$$S = [R^2 + (R+h)^2 - 2R(R+h)\cos\Phi\cos\Delta]^{1/2}$$

where R equals 6,367 km, the radius of the earth, h equals 35,803 km, the distance of a satellite above the centre of the earth, Φ is the site latitude and Δ is the absolute difference between the site and satellite longitude. Substituting R and h in this equation gives:

$$S = 1000\,[1018.32 - 53.69\cos\Phi\cos\Delta]^{1/2}$$

If this value of S is substituted into the above equation for path loss we find:

$$\text{Path Loss} = 185.05 + 10\log[1 - 0.295\cos\Phi\cos\Delta] + 20\log f$$

where f is expressed in GHz. At 12 GHz the path loss equals 205.11 dB at an earth station located on the equator directly below a satellite. This equation also shows that signals from a Ku-band satellite 10 degrees in longitude away from an earth station at 40 degrees latitude would suffer a 205.54 dB free space path loss.

Atmospheric absorption causes additional path losses. Absorption increases with slant range because the satellite signal must pass through a greater thickness of atmosphere. As described in Chapter III, variations between rainy or overcast days and clear days are substantial for Ku-band broadcasts. An absorption loss of 0.5 dB is assumed in most calculations for reception on average clear days.

G/T_{sys} – System Figure of Merit

G/T_{sys}, the ratio of antenna gain to system noise temperature, is the figure of merit of an antenna/feed/LNB system. It is expressed in decibels per degree Kelvin (dB/K) as:

$$G - 10\log T_{sys}$$

The system noise temperature primarily depends upon both the antenna and LNB noise temperatures. However, components further

APPENDIX B

downstream towards the receiver also contribute small amounts of noise. This term is given by:

$$T_{sys} = T_{ant/feed} + T_{LNB}/G_{feed} + \frac{T_{rec/coax}}{G_{LNB} + G_{feed}}$$

where G refers to gain. The gain of a typical feed is in the vicinity of 0.99 while LNB gain is usually about 50 dB, equal to a factor of 100,000. This equation clearly shows why the noise contributed to T_{sys} by the satellite receiver and coaxial cable are negligible. The LNB amplifies both the signal and noise so much that any later contributions are of minor importance.

The second of last term in the link equations above adjusts for the system bandwidth and the final is a constant, called Boltzman's constant.

Antenna Gain

Antenna gain relative to an "isotropic" antenna, one that radiates equally in all directions, is given by:

$$G = E(\pi D/\lambda)^2$$

where E is the antenna efficiency, D is the diameter and λ is the wavelength of the incoming radiation. The wavelength in centimetres can be simply calculated by dividing 30 by the frequency expressed in GHz. The wavelength of 12 GHz microwaves is 2.50 centimetres or just under one inch.

For example, a 2 metre with an efficiency of 55% operating at 12 GHz has a gain given by:

$$G = 0.55 \times (3.14 \times 200 \text{ cm} / 2.5 \text{ cm})^2 = 34{,}706$$

Expressed in decibels this gain or signal concentration of a factor of 34706 relative to an isotropic antenna is given by:

$$G = 10 \log 34706 = 45.4 \text{ dBi}$$

Loss of Gain with Reflector Surface Irregularities

The decrease of gain relative to a perfect antenna having no surface irregularities is given by:

$$\text{Loss of Gain} = e^{-8.80 \text{RMS}/\lambda}$$

where RMS is the root mean square deviation from a perfect geometrical shape and λ is the wavelength of the incoming signal. The RMS is a measure of the "tightness" of the surface or its average tolerance.

For example, a Ku-band antenna operating at 12 GHz, where wavelength equals 2.5 centimetres, having a RMS tolerance of 0.15 centimetre has a decrease in gain relative to a perfect antenna given by:

$$\text{Loss of Gain} = e^{-8.80 \times 0.15/2.50}$$

$$= e^{-0.53}$$

$$= 0.59$$

or a 41 percent decrease in gain. This equates to a 2.3 decibel loss of gain given by:

$$\text{Loss in Gain (dB)} = 10 \log 0.59$$

$$= -2.3 \text{ dB}$$

Antenna Beamwidth

An approximate but very useful formula for 3 dB antenna beamwidth is:

$$\text{Beamwidth} = 70 \lambda / D$$

where λ is the wavelength of the microwave radiation and D is antenna diameter. For example, a 2 metre antenna has a 3 dB beamwidth given by:

$$\text{Beamwidth} = 70 \times 2.5 / 200 = 0.88°$$

Similarly, a 1 metre antenna would have a calculated beamwidth given by:

$$\text{Beamwidth} = 70 \times 2.5 / 100 = 1.75°$$

Noise Temperature and Figure

The noise any system generates is proportional to its ambient temperature and the bandwidth of the signal it processes. The larger either of these two quantities, the greater the contributed noise.

$$\text{Noise} = kTB$$

where k is Boltzman's constant, T is the ambient temperature and B is the system bandwidth.

A quantity called noise factor is defined by the ratio of the noise at the output on an electronic component to the noise at its input. This quantity measures, in essence, the amount of noise internally generated in any device. In a perfect device whose electronic circuits added no extra noise to a signal, the noise factor would be one.

SATELLITE TV EQUATIONS

APPENDIX B

Noise Factor = (Ideal Noise + Internal Noise) / Ideal Noise

$$= (kBT_{Ideal} + kBT_{Eq}) / kBT_{Ideal}$$

$$= (T_{Ideal} + T_{Eq}) / T_{Ideal}$$

$$= 1 + T_{Eq} / T_{Ideal}$$

$$= 1 + T_{Eq} / 290$$

T_{Eq}, termed the equivalent noise temperature. The reference noise temperature, T_{Ideal}, is usually taken to be 290°K, equal to an average room temperature of about 63 °F.

Noise figure is the decibel equivalent of noise factor as is given by:

Noise Figure = 10 log Noise Factor

For example, if the noise figure is 1.9 dB, the equivalent noise temperature is:

$$1.9 = 10 \log (1 + T_{Eq} / 290)$$

turning this equation inside out:

$$1 + T_{Eq} / 290 = 10^{0.19}$$

$$= 1.55$$

$$T_{Eq} / 290 = 0.55$$

$$\therefore T_{Eq} = 159°K$$

The Effect of Bandwidth on System Noise Power

The noise power in any communication system is given by:

System noise power = $kT_{sys}B$

where T_{Sys} is the system noise temperature in degrees Kelvin mainly determined by antenna and LNB noise, k is Boltzman's constant equal to 1.381×10^{-23} and B is the communication bandwidth. The change in noise power between two systems can be computed as follows:

Change in noise power = (kT_1B_1/kT_2B_2)

$$= T_1B_1/T_2B_2$$

Therefore, if the noise temperature remains constant the change in noise power is simply the ratio of bandwidths. If the bandwidth were cut from 36 to 18 MHz as would be the case in half transponder formats, the noise power would be reduced by 50 percent or 3 decibels. The resulting doubling of the signal-to-noise ratio sometimes makes the difference between a watchable picture and a sparklie, faint ghost of a picture. But reducing the bandwidth will also result in a "softening" of the video as well as smearing and streaking of pictures having fast changes.

Declination Angle

The declination angle for a polar mount can easily be found from the tables and figures in the early chapters. It can also be calculated from:

$$\text{Declination} = \text{Tan}^{-1} \frac{3964 \sin L}{22300 + 3964(1 - \cos L)}$$

where L is the site latitude. The two numbers in this equation are the radius of the earth and the distance from the surface of the earth to the arc of satellites. For example, the declination angle at 40 degrees latitude is given by:

$$\text{Declination} = \text{Tan}^{-1} \frac{3964 \sin 40}{22300 + 3964(1 - \cos 40)}$$

$$= \text{Tan}^{-1} 0.11$$

$$= 6.26°$$

Azimuth and Elevation Angles

Antenna pointing angles can be calculated in degrees from true north from the following equations:

Azimuth Angle = $\tan^{-1}[\tan\Delta/\sin\Phi]$

Elevation Angle = $\tan^{-1}[(\cos Y - 0.15116)/\sin Y]$

$Y = \cos^{-1}[\cos\Phi\cos\Delta]$

where Δ is the absolute value of the difference between satellite and TVRO site longitudes and Φ is the site latitude.

SATELLITE TV EQUATIONS

APPENDIX B

Voltage Standing Wave Ratio

The voltage standing wave ratio, VSWR, is a measure of the amount of input signal reflected back and lost. A perfect device would have no reflective losses and have a VSWR of 1:1. Table B-1 shows how reflected signal power and transmission losses vary with VSWR:

TABLE B-1.

VSWR	Percentage of Reflected Signal	Transmission Loss(dB)
1.0:1	0	0
1.1:1	0.2	0.01
1.2:1	0.9	0.03
1.3:1	1.6	0.07
1.5:1	4.0	0.18
2.0:1	11.0	0.50

Antenna Parabolic Geometry

The basic equation for a parabolic reflector is given by:

$$y = x^2/4f$$

where f is the focal distance. Another useful formula gives the focal distance f in terms of the antenna diameter and depth:

$$f = \text{diameter}^2/16 \times \text{depth}$$

Wind Loading

Winds can have very dramatic effects on antennas and their supporting structures. The following tables, kindly calculated and provided by Dick Zlotky, president of Earthbound, Inc., (913-273-1345) indicate the strength and direction of the forces that can be expected in a real-world environment.

Design wind loads on parabolic reflectors were calculated from wind tunnel coefficient data complied by the Jet Propulsion Laboratory at the California Institute of Technology as presented in report number JPL 78-16. Wind tunnels tests were performed on several models of parabolic reflectors, both solid and porous, at azimuth angles ranging from 0°, wind directly along the antenna axis, to 180°, as well as at elevation angles from zero to 90° at each azimuth angle. Axial, normal, side force, moment and torsional coefficients were determined at each position of the reflector.

All coefficients compiled in JPL 78-16 act at the apex of the reflector as shown in Figure B-1. These are all positive in sign. Since all the force coefficients are perpendicular and parallel to the longitudinal axis of the reflector, it is necessary to resolve these forces and moments in planes perpendicular and parallel to the ground. These must be then transferred to the centreline of the pole in order to use them for the pole and foundation structural design. The resolution of these forces and moments is shown in Figure B-2. It is assumed that the centreline of the pole is at a distance of 15.2 cm (6 inches) from the apex of the reflector. The forces and moments presented in Tables B-2, B-3 nd B-4 are centred about point "A" in Figure B-2. This point is located at the intersection of the pole centreline and a horizontal plane through the apex of the reflector at a height H above ground level.

An antenna having an f/D of 0.313 was used in the tests. Antennas having surface porosities of 0% (solid) and 25% were tested. The wind loads shown in the tables below are calculated from these two different surfaces. Note that in order to achieve the crucial pointing accuracy in Ku-band reception, the support poles and foundations must have sufficient rigidity so that deflections in torsion or twisting as well as bending are minimised.

JPL 78-16 states that the amount of reduction of wind loads offered by a porous panel such as a perforated reflector is primarily affected by its t/d ratio, where t is the thickness of the panel and d is the diameter of its holes. As t/d approaches 3, the efficiency nears zero and the drag characteristics approach that of a solid reflector. The t/d for models used in the JPL wind tunnel tests approached 100% efficiency in reducing drag. In other words, the t/d values were low and therefore representative of the performance of today's perforated and mesh antennas.

The information in the tables below is presented in the metric system. The conversions to the English system use the following:

1 cm = 0.394 inches
1 kg = 2.205 pounds
1 kg-cm = 0.869 inch-pounds
1 km/hr = 0.622 mph

SATELLITE TV EQUATIONS

APPENDIX B

TABLE B-2. SOLID ANTENNA

FORCES & MOMENTS	ANTENNA DIAMETER (m)					
	1.0	1.2	1.8	2.4	3.0	3.7
$F_{X'}$ (kg)	154	222	499	886	1384	2107
$F_{Y'}$ (kg)	-5	-7	-16	-28	-44	-66
$F_{Z'}$ (kg)	-55	-80	-180	-320	-500	-761
Torsion (kg-cm)	-450	-756	-2432	-5589	-10759	-19946
$M_{Y'Y'}$ Moment (kg-cm)	9	263	2257	6952	15478	31766
$M_{X'X'}$ Moment (kg-cm)	-138	-238	-807	-1898	-3721	-6977

Assumptions: Wind Velocity = 160.0 k/hr

TABLE B-3. 25% POROUS ANTENNA

FORCES & MOMENTS	ANTENNA DIAMETER (m)					
	1.0	1.2	1.8	2.4	3.0	3.7
$F_{X'}$ (kg)	74	110	248	439	686	1044
$F_{Y'}$ (kg)	15	21	48	84	132	201
$F_{Z'}$ (kg)	-24	-34	-78	-137	-215	-327
Torsion (kg-cm)	616	1006	3026	6728	12647	23024
$M_{Y'Y'}$ Moment (kg-cm)	207	467	2160	5779	12135	23927
$M_{X'X'}$ Moment (kg-cm)	147	255	866	2036	3991	7469

Assumptions: Wind Velocity = 160.0 km/hr

TABLE B-4. MAXIMUM TORSION

MAXIMUM TORSION (kg-m)	ANTENNA DIAMETER (m)					
	1.0	1.2	1.8	2.4	3.0	3.7
Solid Antenna	1647	2735	8776	16646	38533	71289
25% Porous Antenna	1467	2450	7754	17716	33895	62629

Assumptions: Wind Velocity = 160.0 km/hr
Azimuth = 120°
Elevation = 20°

APPENDIX B

Additional Calculations

The forces and moments given in Tables B-2 through B-4 are for 160.9 km/hr winds exerted at a height H above ground level. Forces and moments for other wind velocities are easily calculated by multiplying the 160.9 km/hr velocity by the square of the new velocity divided by 10,000, namely

$$(v/100)^2$$

For example, to find $F_{X'}$ exerted by a 112.6 km/hr wind on a 1.8 m solid antenna, multiply the 499 kg found in Table B-2 by $(112.6/100)^2$. The result is a force of 245 kg at the lower velocity.

To find the forces at the point where a pole attaches to the foundation, point B in Figure B-2, it is necessary to transfer the forces and moments from point A to point B. All components of the forces have the same values at both points A and B. For example, from Table B-2 the following values are found for a 1.8 m solid antenna at H of 121.9 cm above ground level:

$F_{X'}$ = 499 kg
$F_{Y'}$ = −16 kg
$F_{Z'}$ = −180 k
Torsion = 02432 kg-cm
$M_{Y'Y'}$ = 2257 kg-cm
$M_{X'X'}$ = −807 kg-cm

The loads at ground level (point B) are found by

$M_{Y'Y'} = F_{X'} \times H + M_{Y'Y'}$ = 499 × 121.9 + 2257
       = 63085 kg-cm

$M_{X'X'} = F_{Y'} \times H + M_{X'X'}$ = −16 × 121.9 − 807
       = 2,757 kg-cm

The resultant overturn moment at point B is then given by the resultant of these two moments

$= [(63085)^2 + (-2757)^2]^{1/2}$
$= 63,145$ kg-cm

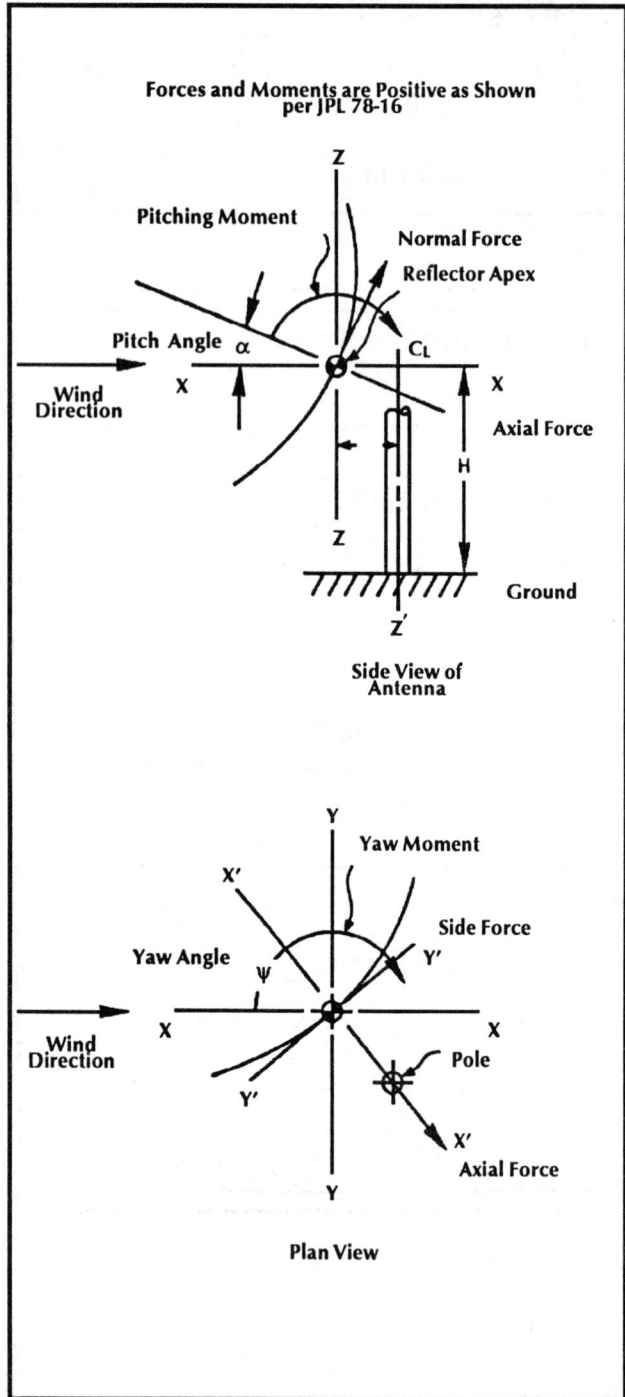

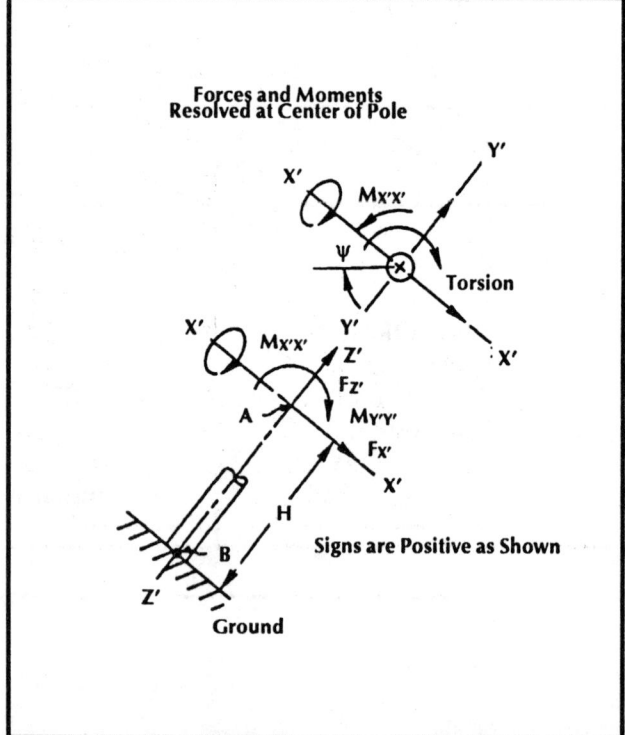

SATELLITE TV EQUATIONS

APPENDIX C. SATELLITE FOOTPRINTS AND AIMING CHARTS

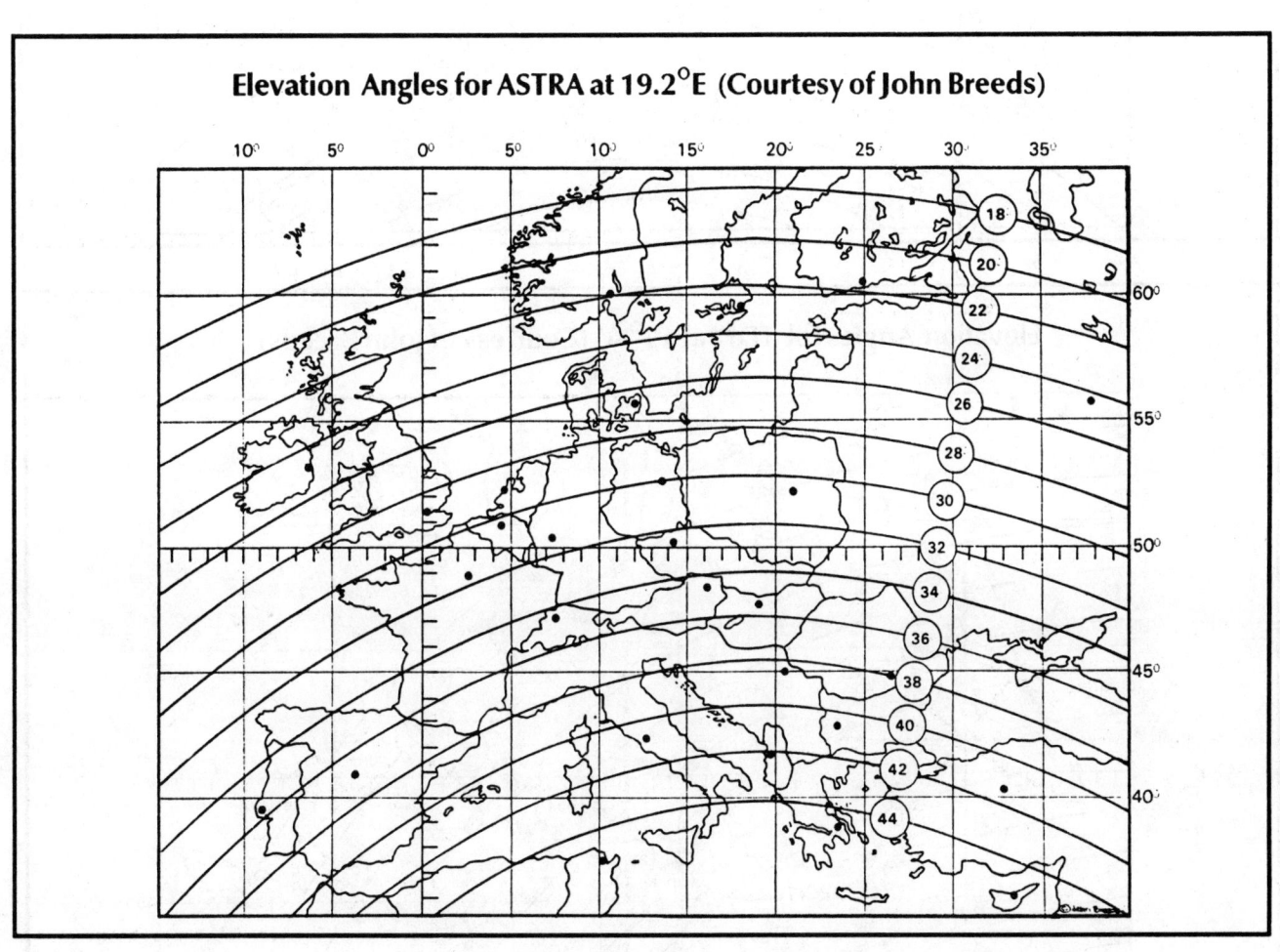

APPENDIX C

Elevation Angles for Eutelsat at 13°E (Courtesy of John Breeds)

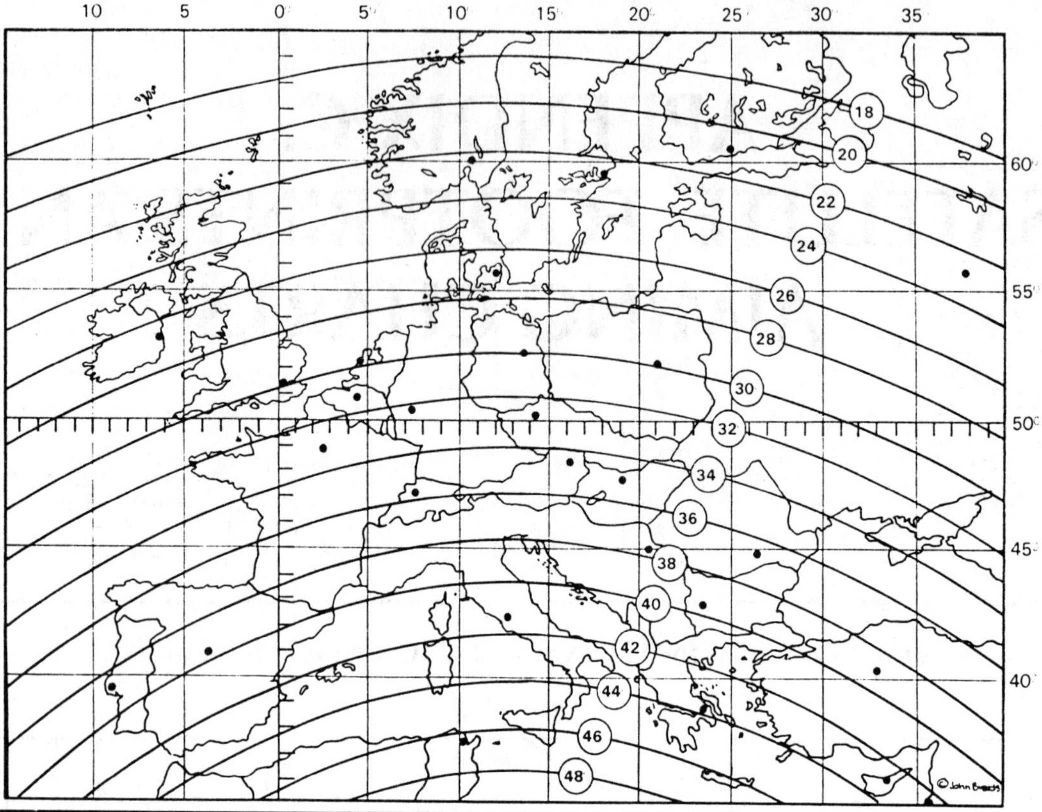

Elevation Angles for TDF1 at 19°W (Courtesy of John Breeds)

SATELLITE FOOTPRINTS

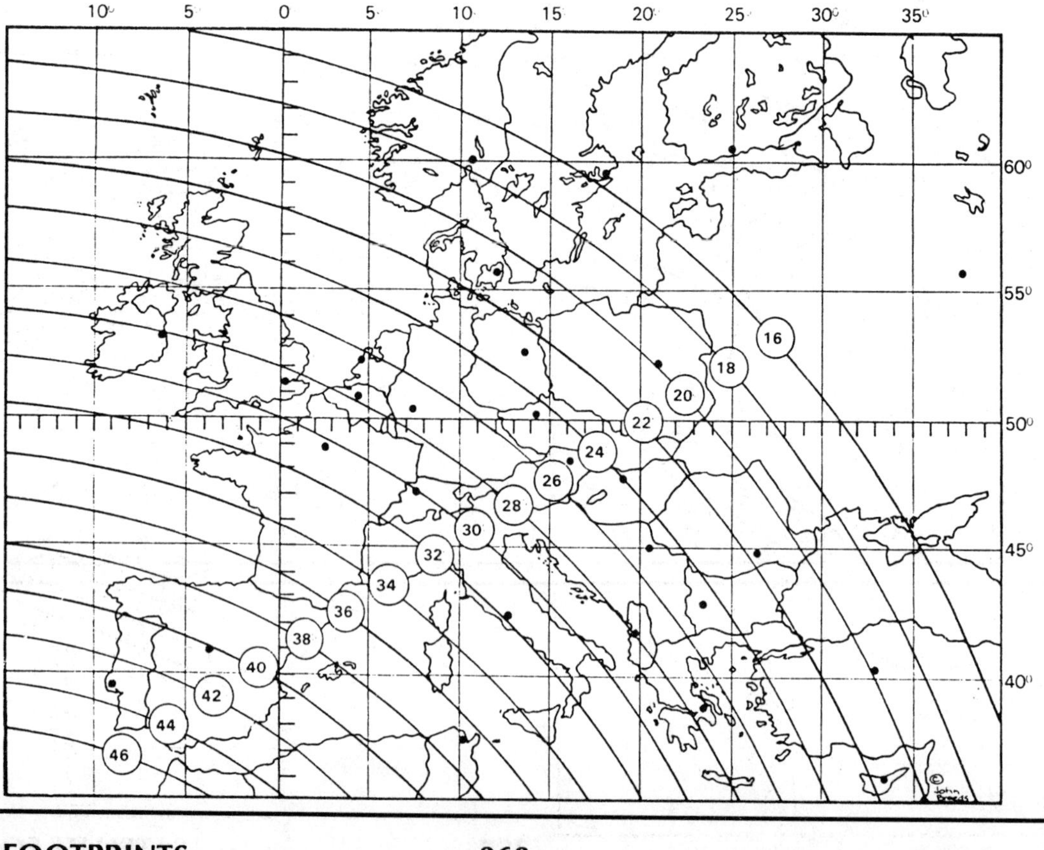

APPENDIX C

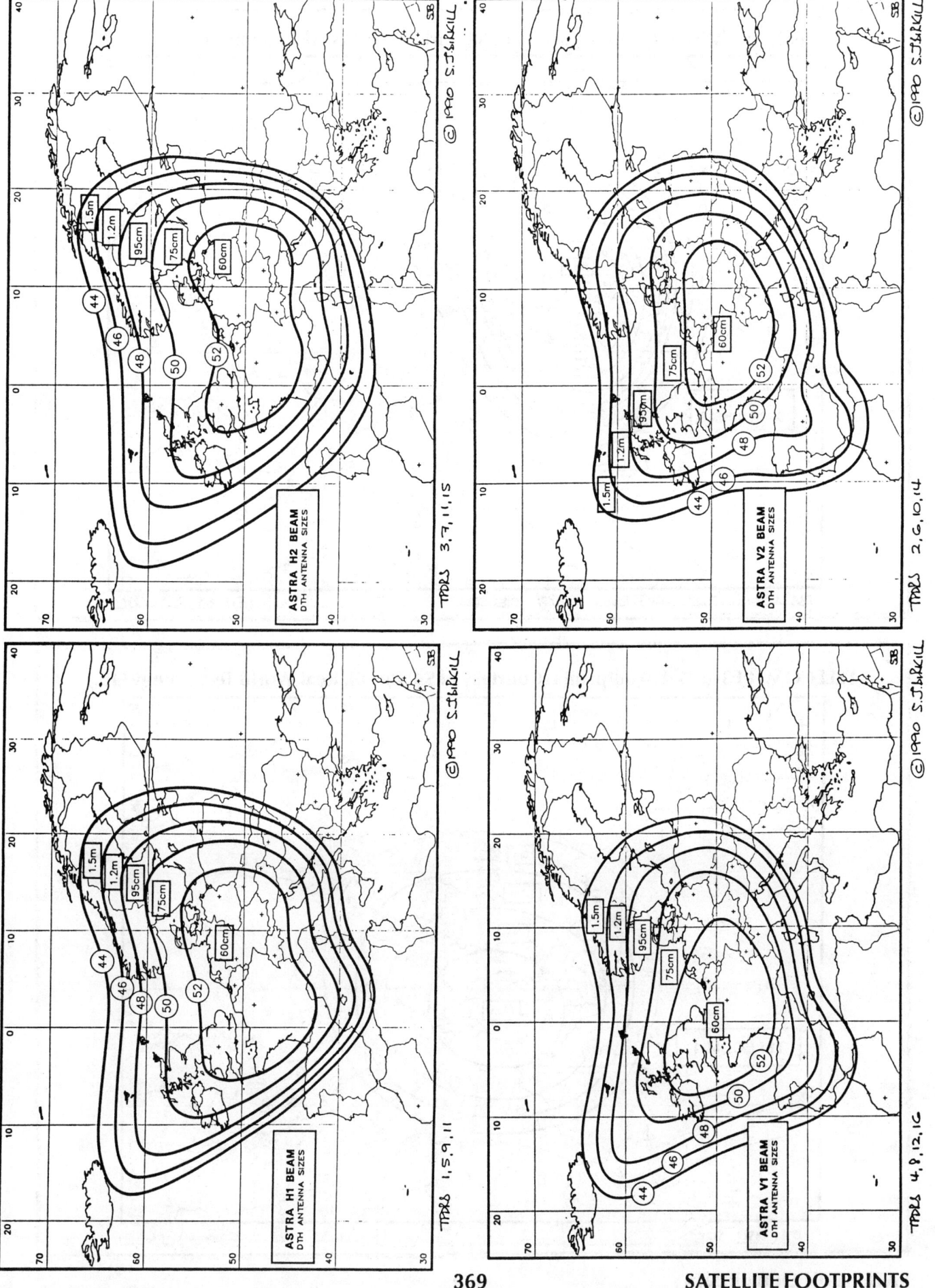

SATELLITE FOOTPRINTS

APPENDIX C

BSB (31°W) Footprint (Courtesy of S.J. Birkill, Real World Technology Ltd.)

DBS ANTENNA SIZES, HALF-POWER 5-TPDR OPERATION © 1990 S.J. BIRKILL

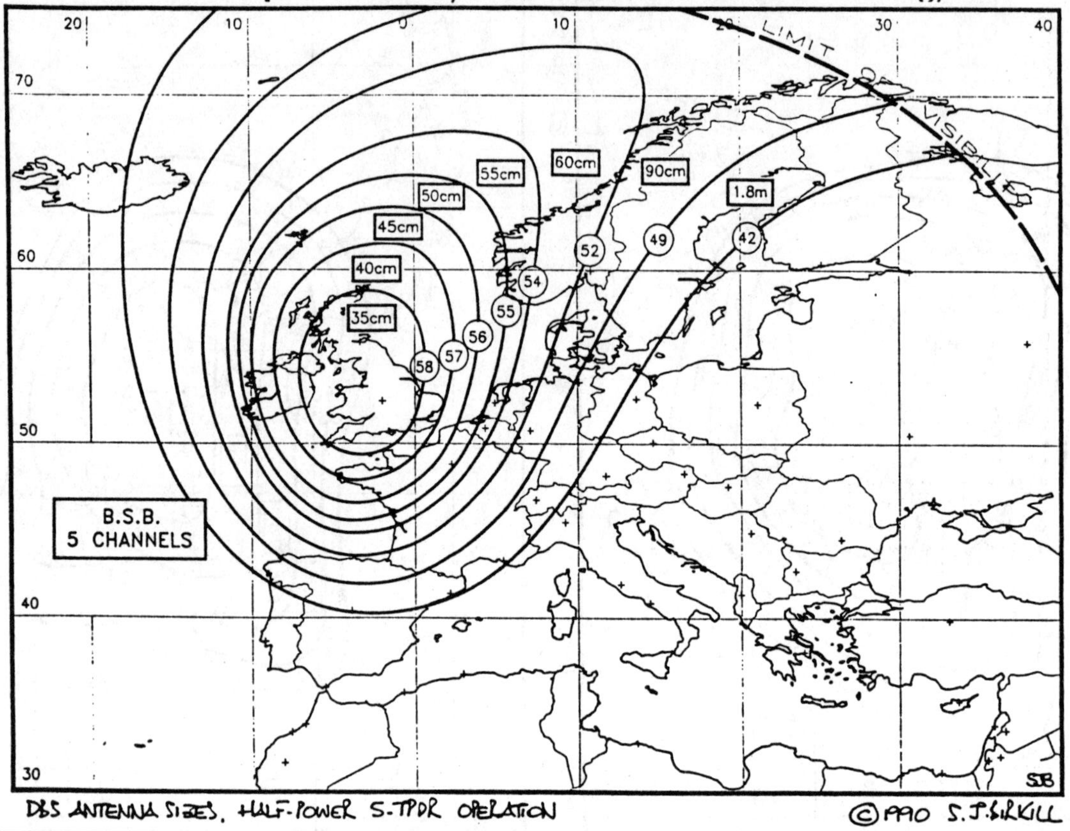

TELECOM-1 F3 (3.5°E) Footprints (Courtesy of S.J. Birkill, Real World Technology Ltd.)

DTH DISH SIZES © 1990 S.J. BIRKILL

SATELLITE FOOTPRINTS

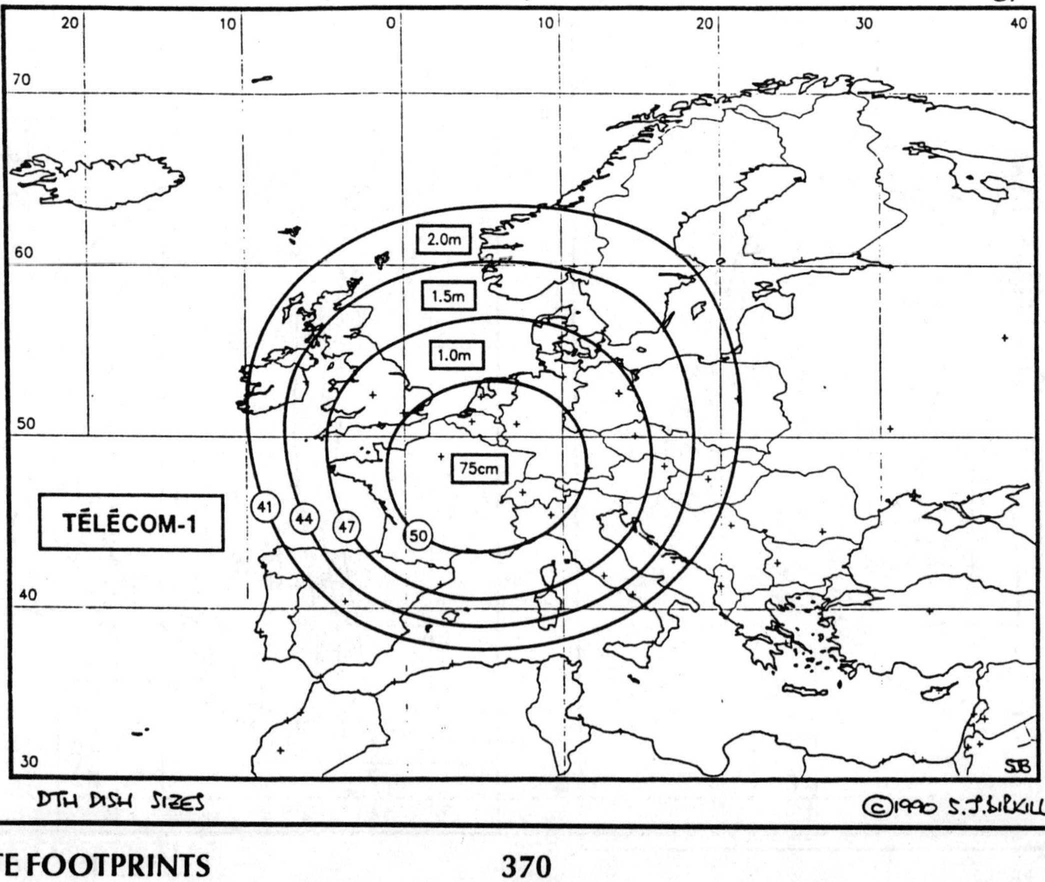

APPENDIX C

EUTELSAT II (13°E) HG BEAM (Courtesy of S.J. Birkill)

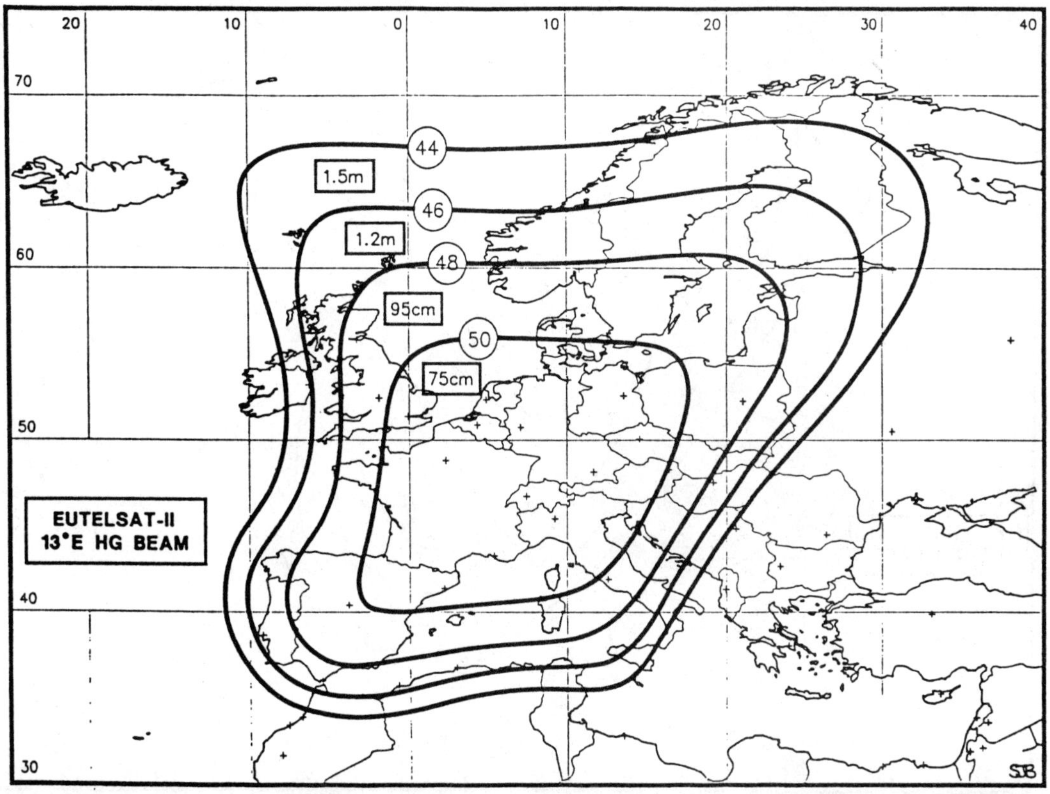

DTH DISH SIZES, TYPICAL FULL TPDR.   © 1990 S.J.Birkill

TDF-1 (19°W) Footprint (Courtesy of S.J. Birkill)

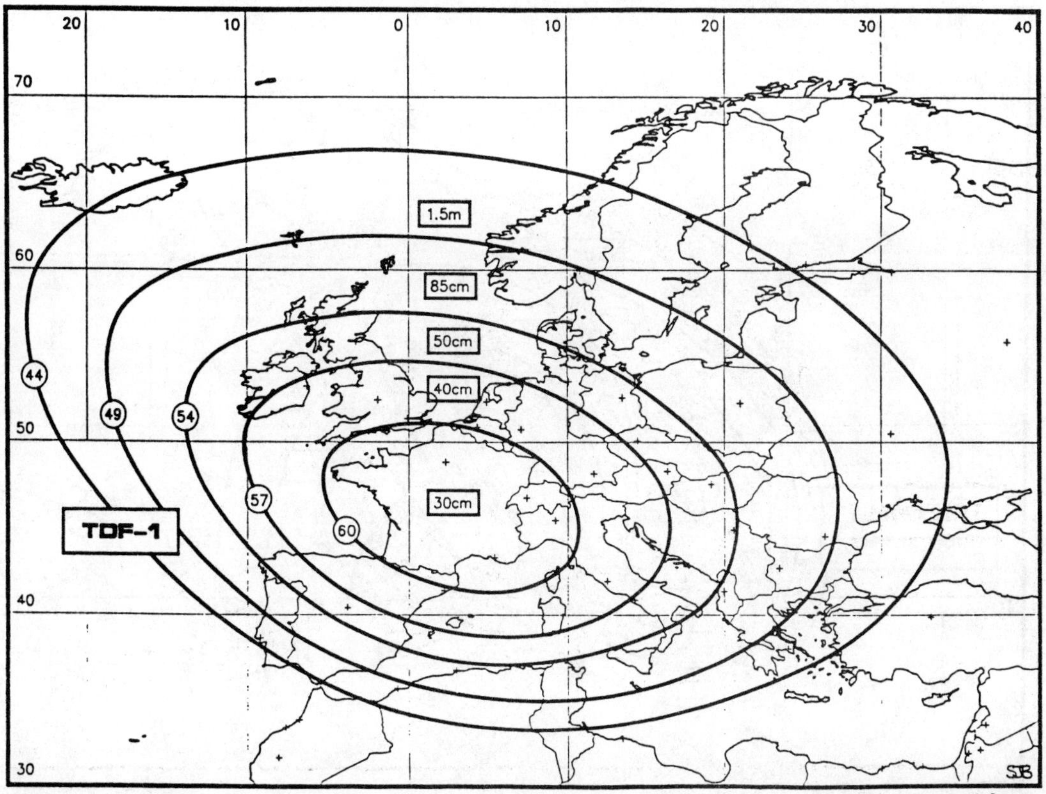

DBS DISH SIZES   © 1990 S.J.Birkill

371   SATELLITE FOOTPRINTS

APPENDIX C

DFS/KOPERNIKUS (23.5°E) Footprint (Courtesy of S.J. Birkill)

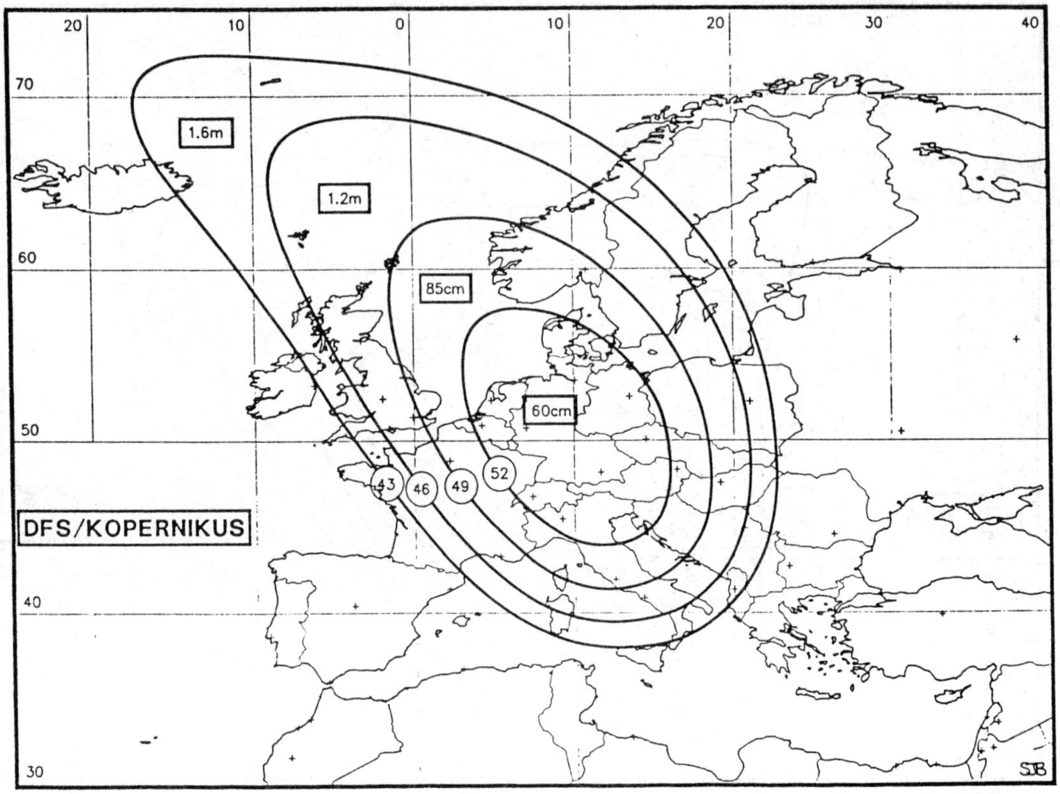

DTH DISH SIZE ZONES, FULL TRANSPONDER. © 1990 S.J.BIRKILL

OLYMPUS (19°W) Footprint (Courtesy of S.J. Birkill)

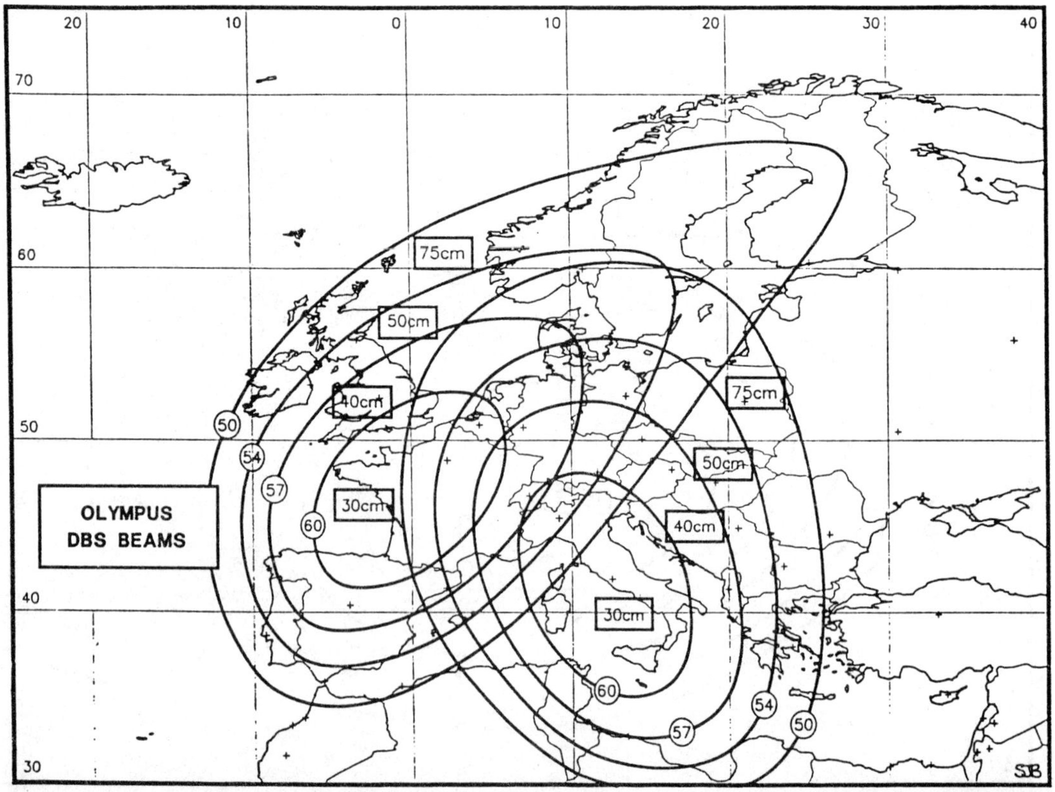

DBS DISH SIZES, NOMINAL BORESIGHT. © 1990 S.J.BIRKILL

SATELLITE FOOTPRINTS

APPENDIX C

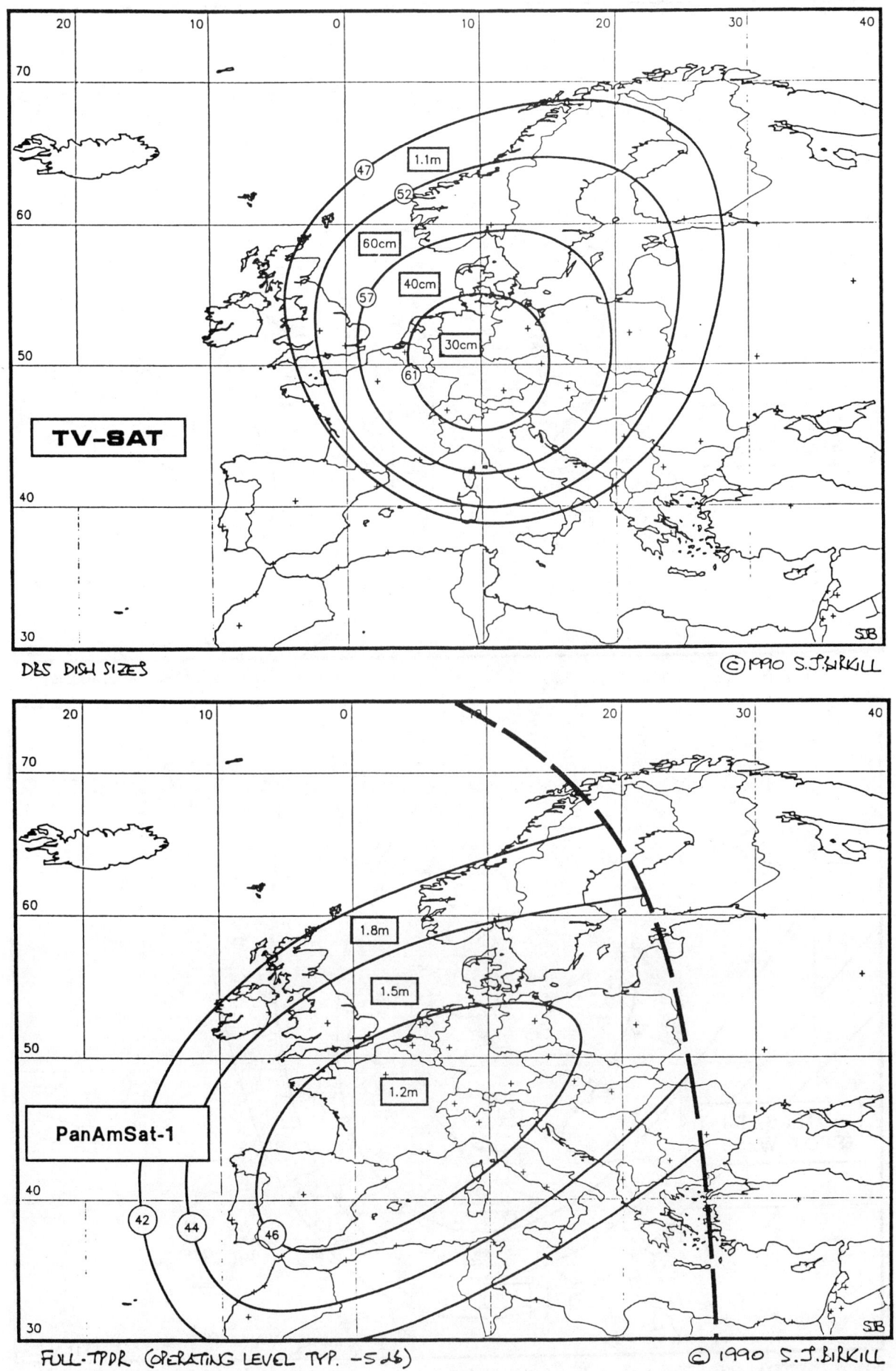

SATELLITE FOOTPRINTS

APPENDIX C

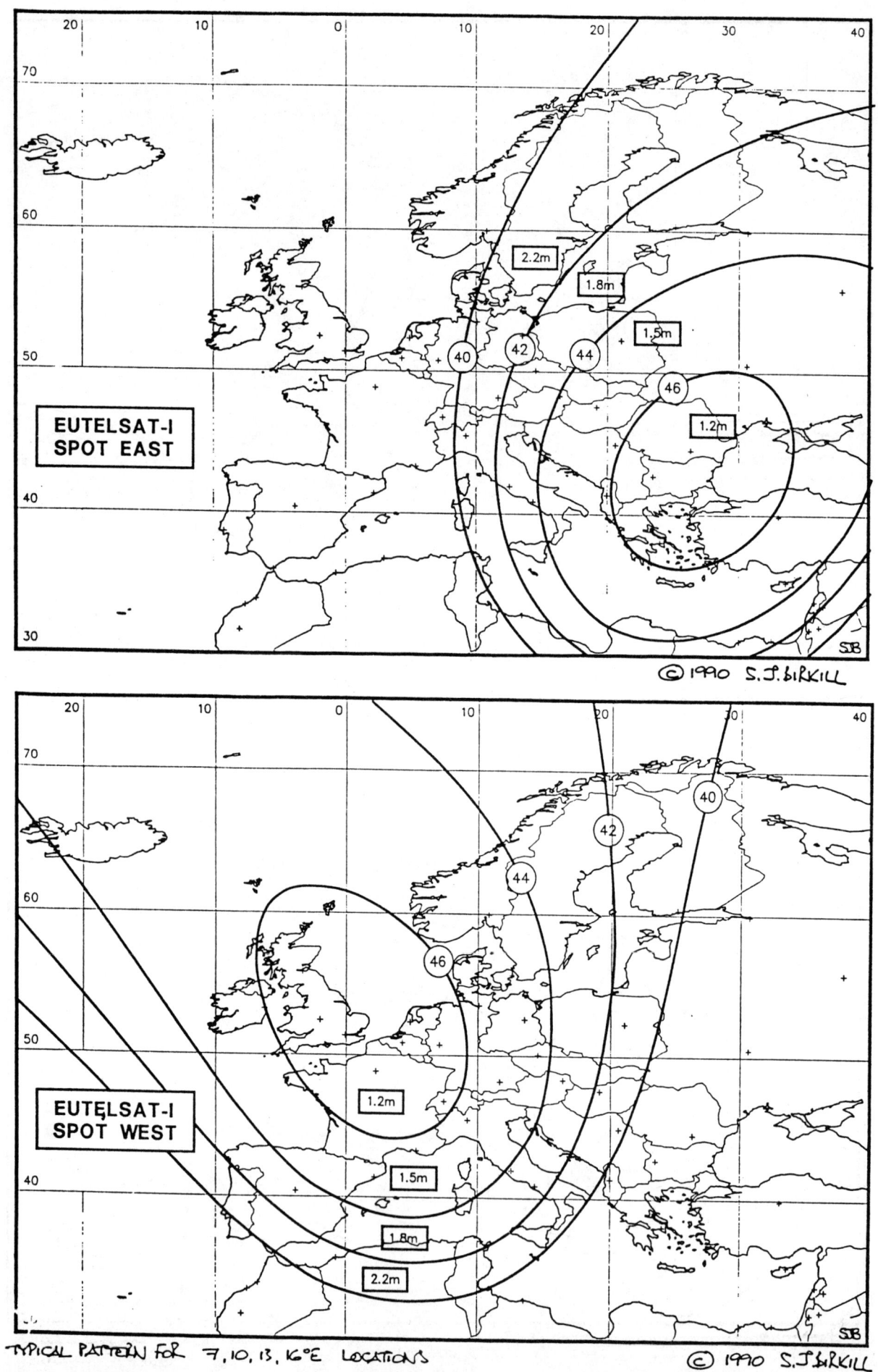

TYPICAL PATTERN FOR 7, 10, 13, 16°E LOCATIONS

SATELLITE FOOTPRINTS

APPENDIX C

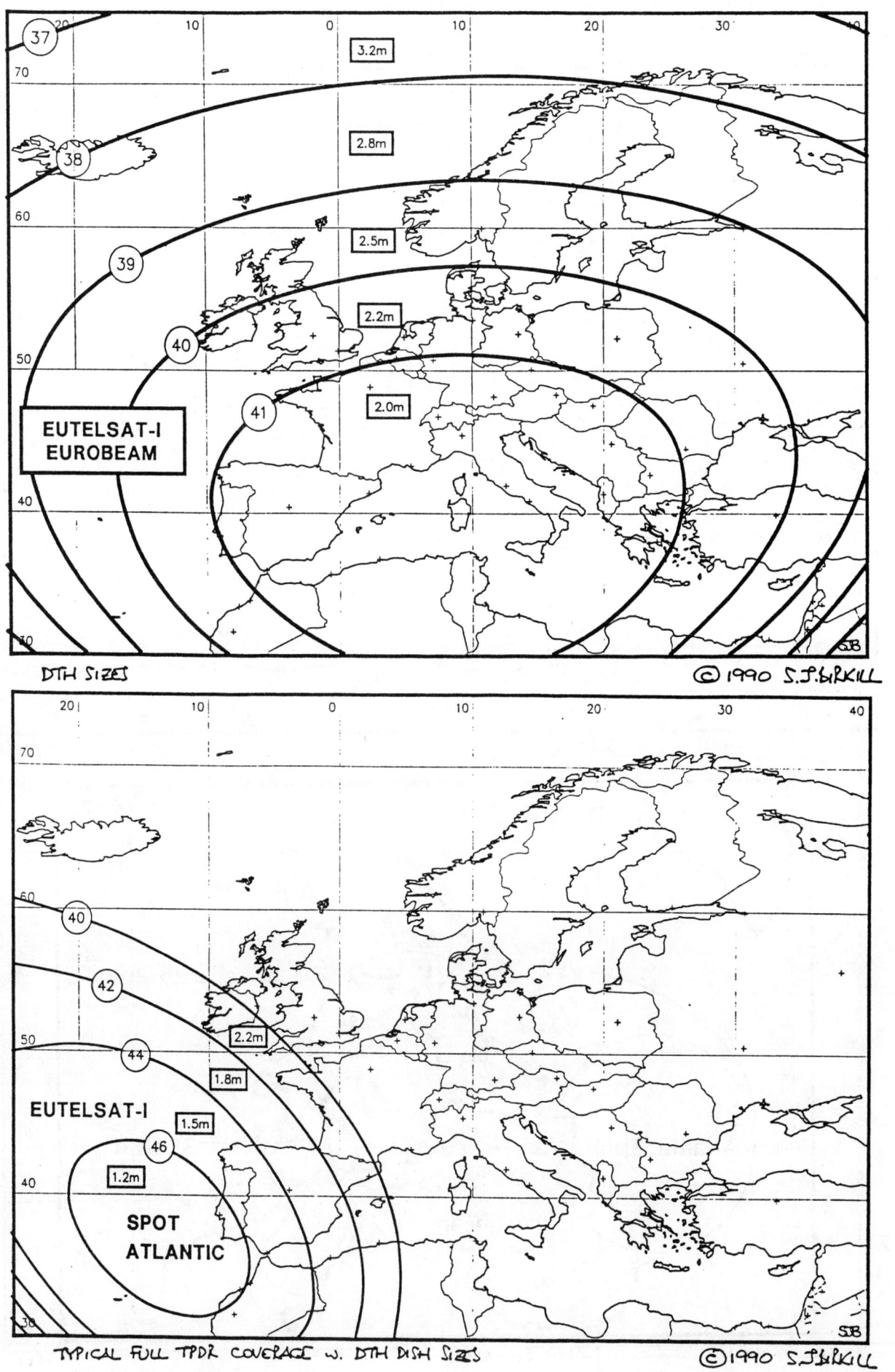

SATELLITE FOOTPRINTS

APPENDIX C

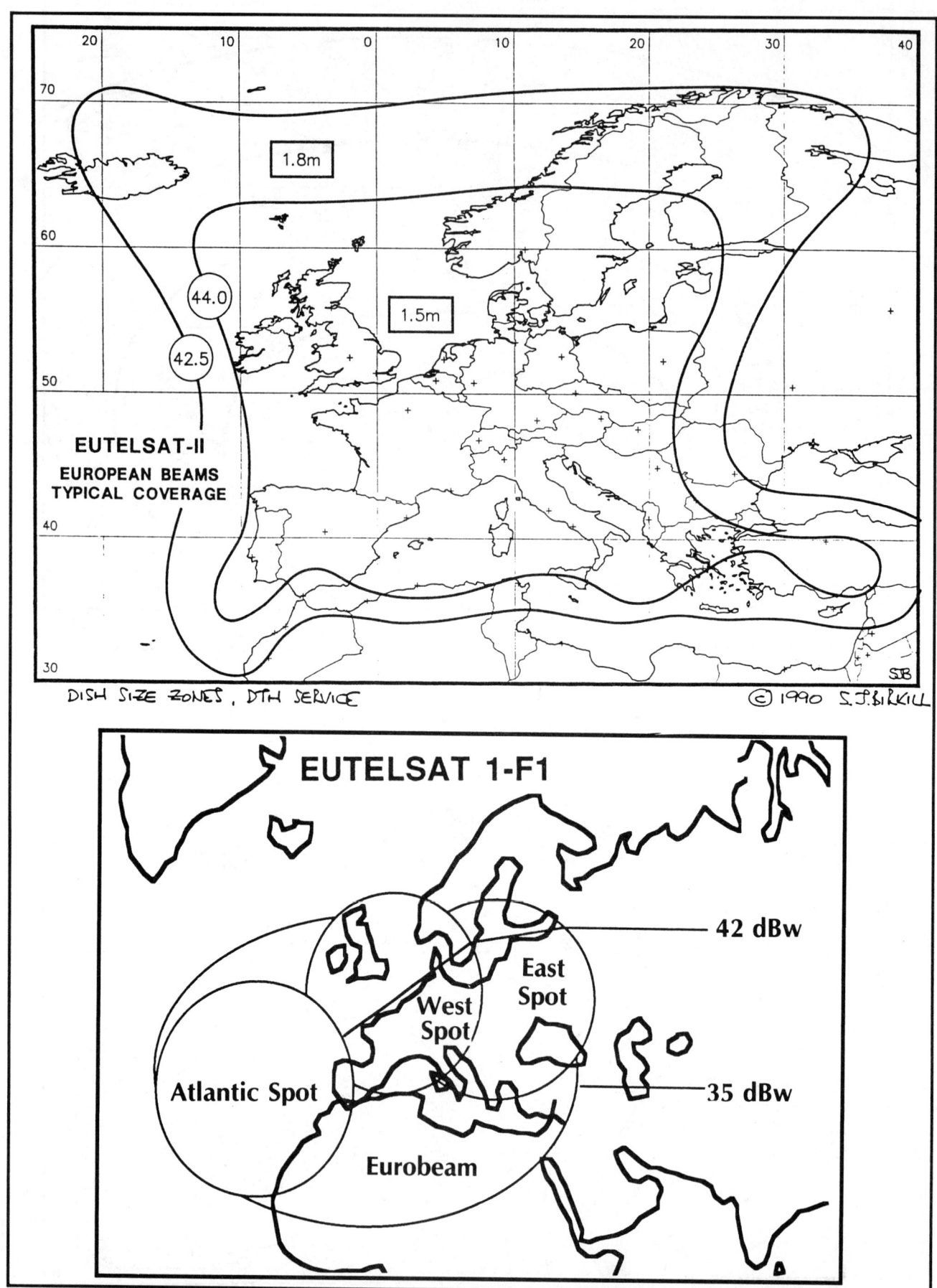

SATELLITE FOOTPRINTS

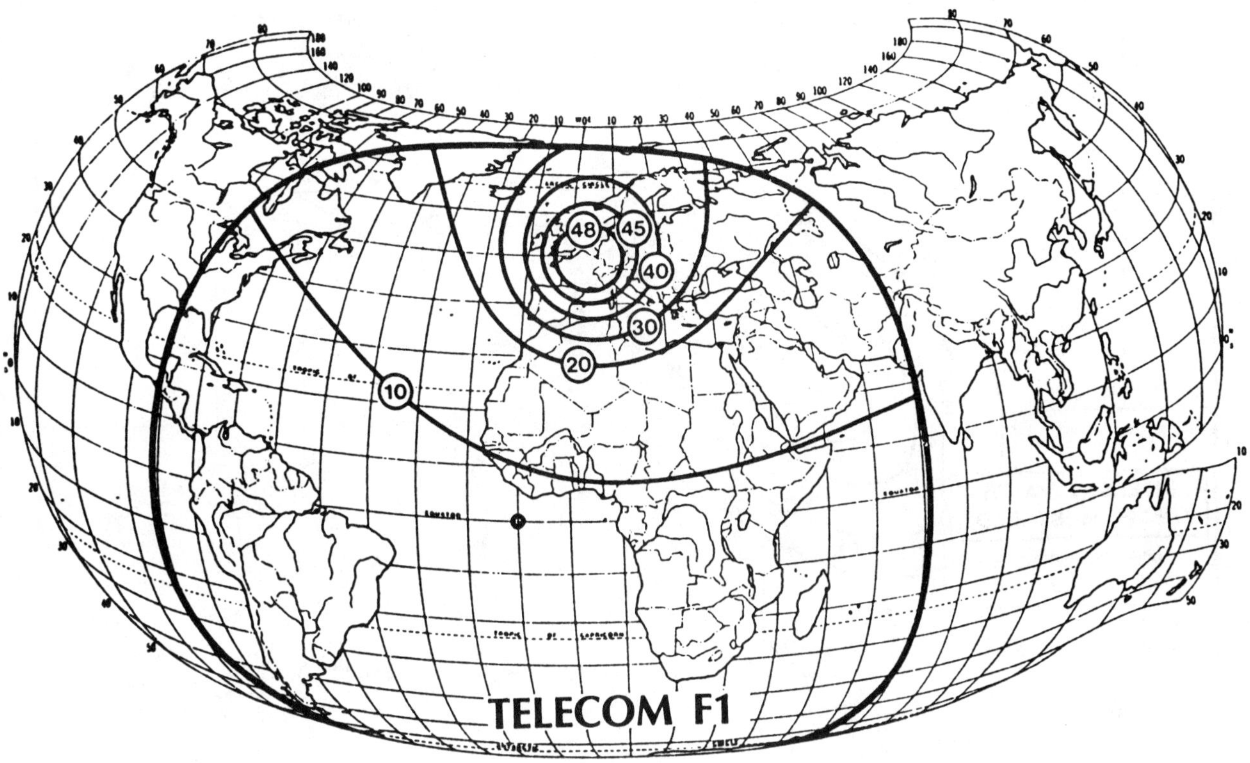

APPENDIX C

TELECOM F1

INTELSAT VA F12 (60°E) West Spot Beam (Courtesy of S.J. Birkill, Real World Technology Ltd.)

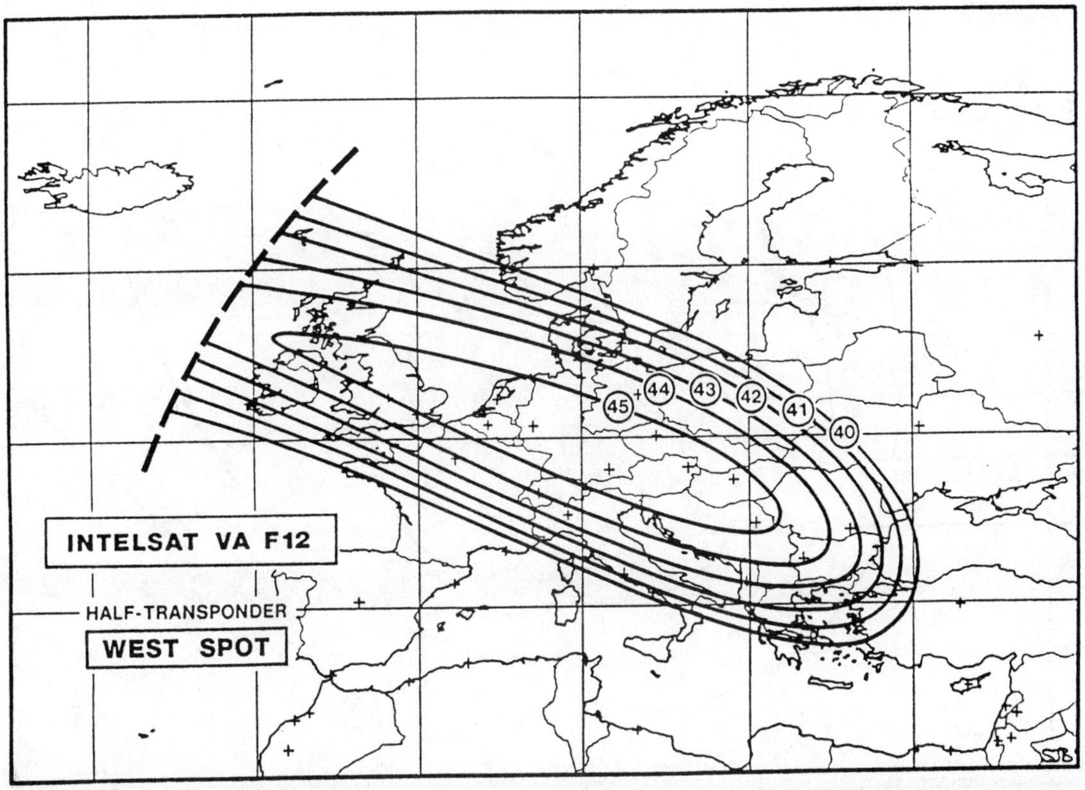

INTELSAT VA F12
HALF-TRANSPONDER
WEST SPOT

377 SATELLITE FOOTPRINTS

APPENDIX C

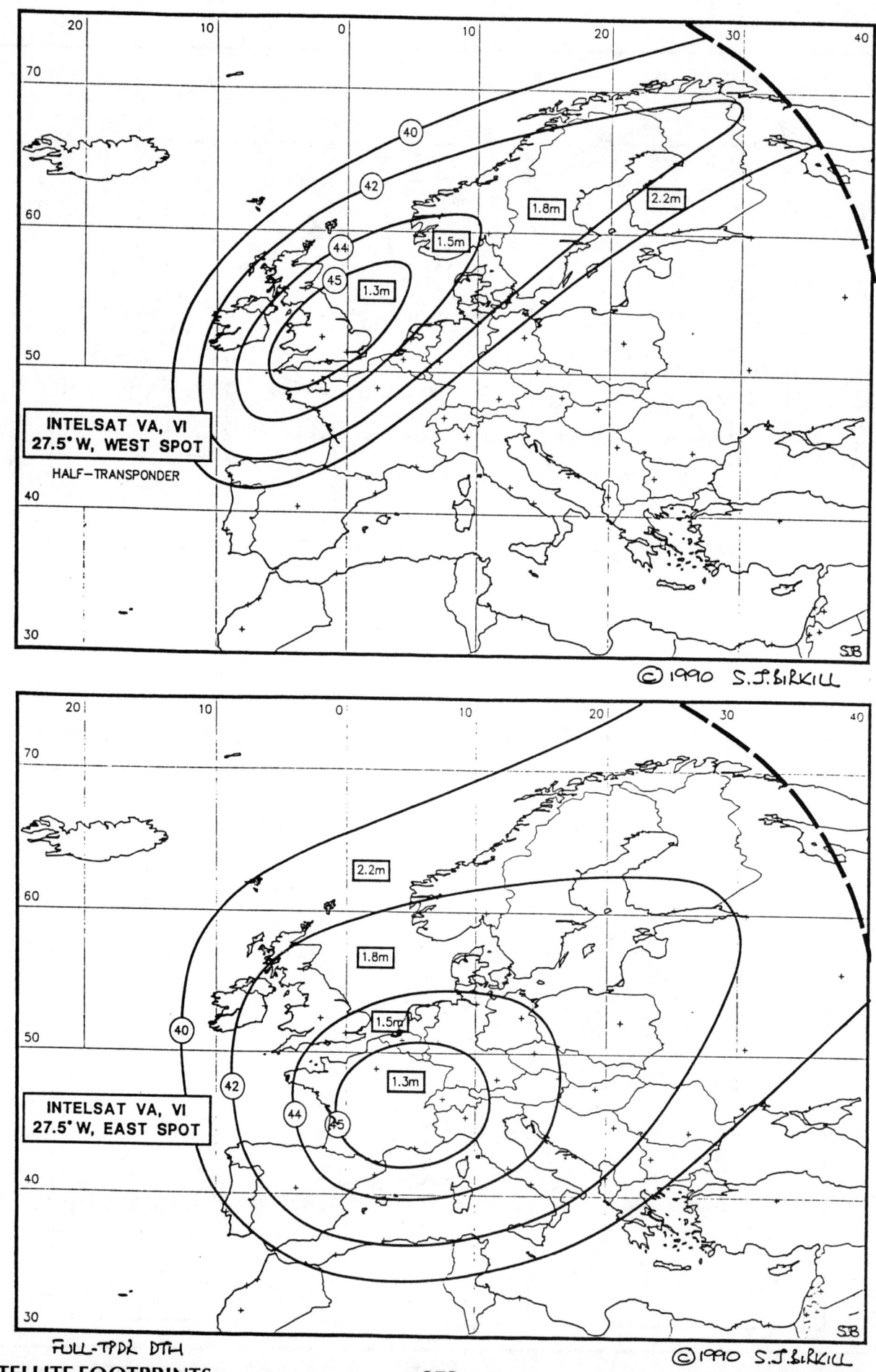

SATELLITE FOOTPRINTS

APPENDIX C

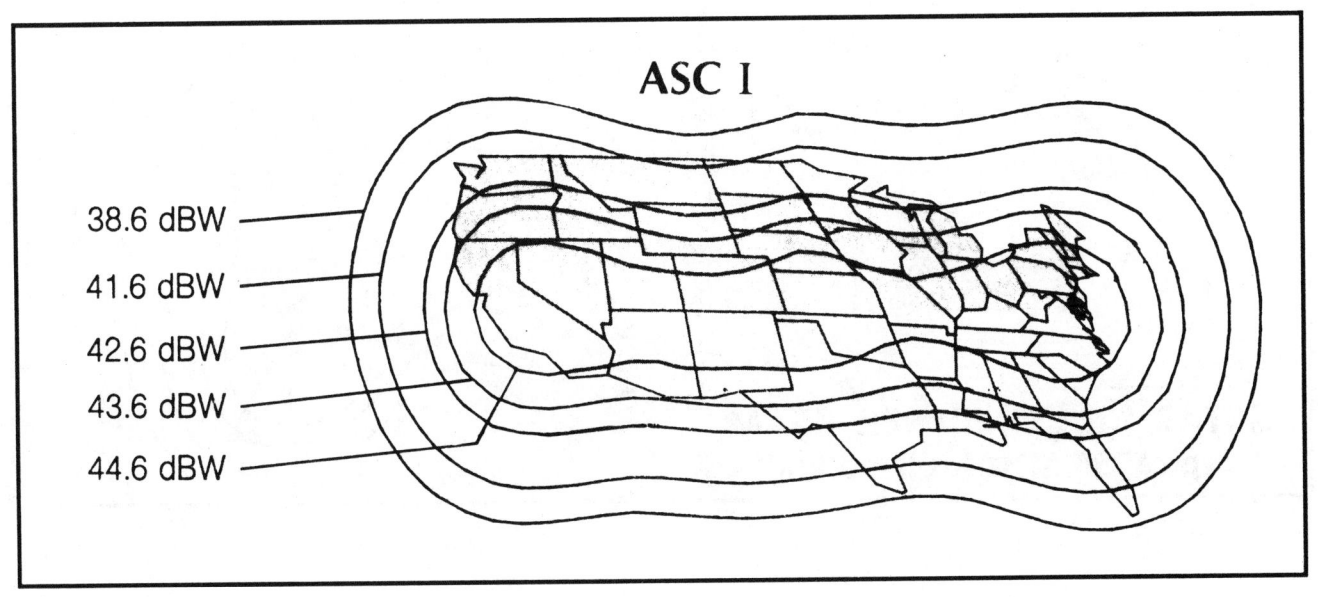

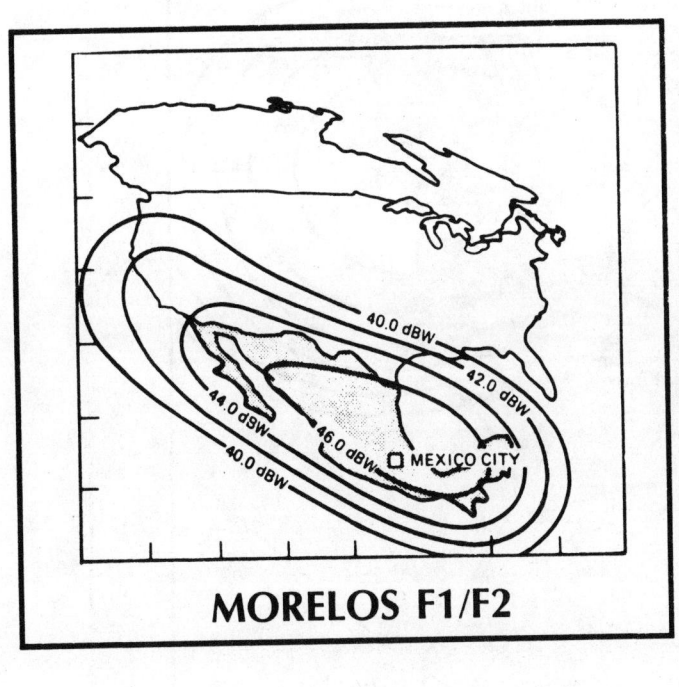

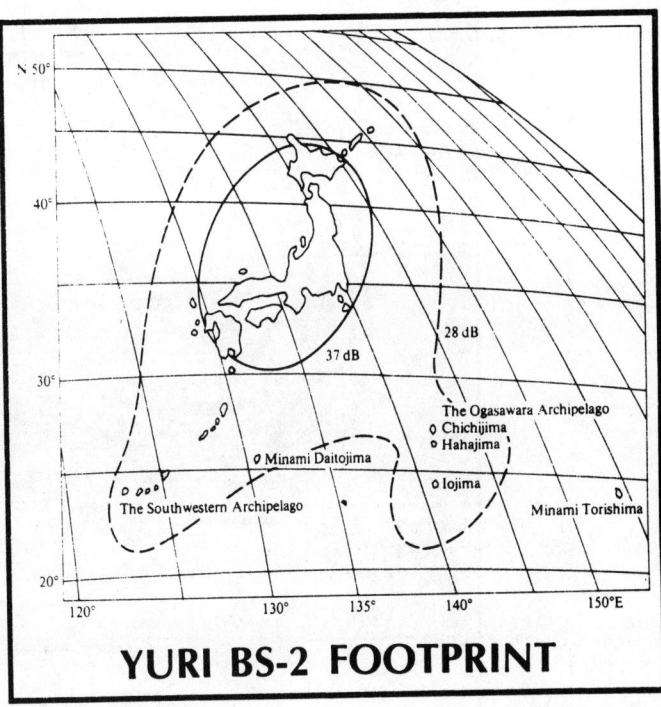

379 SATELLITE FOOTPRINTS

APPENDIX C

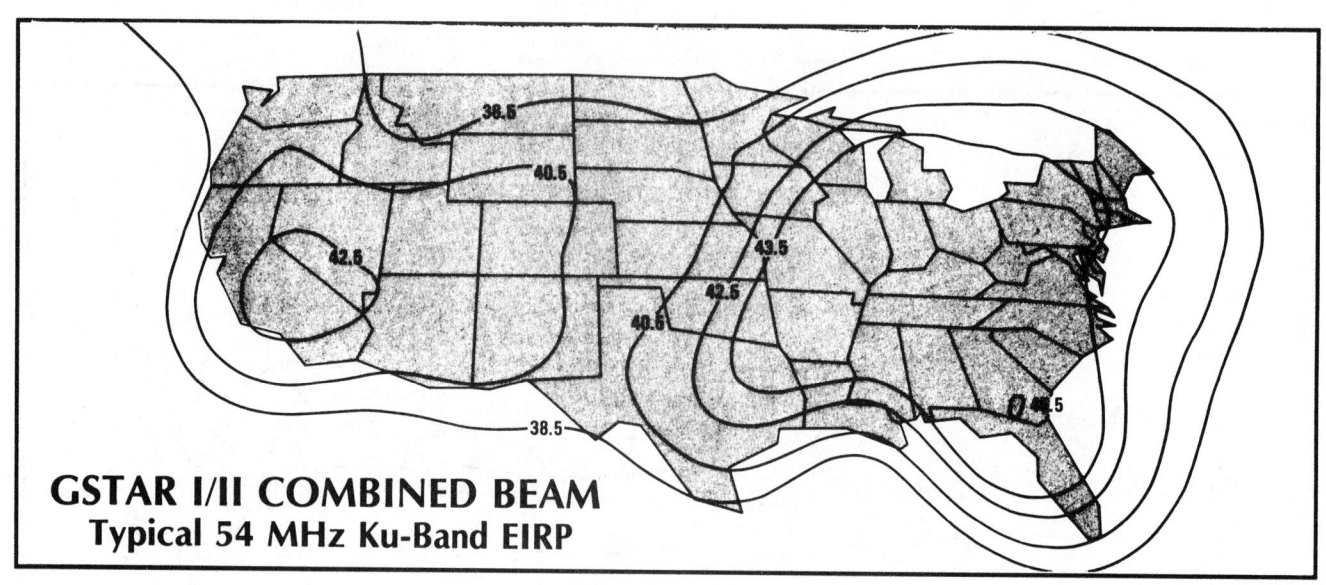

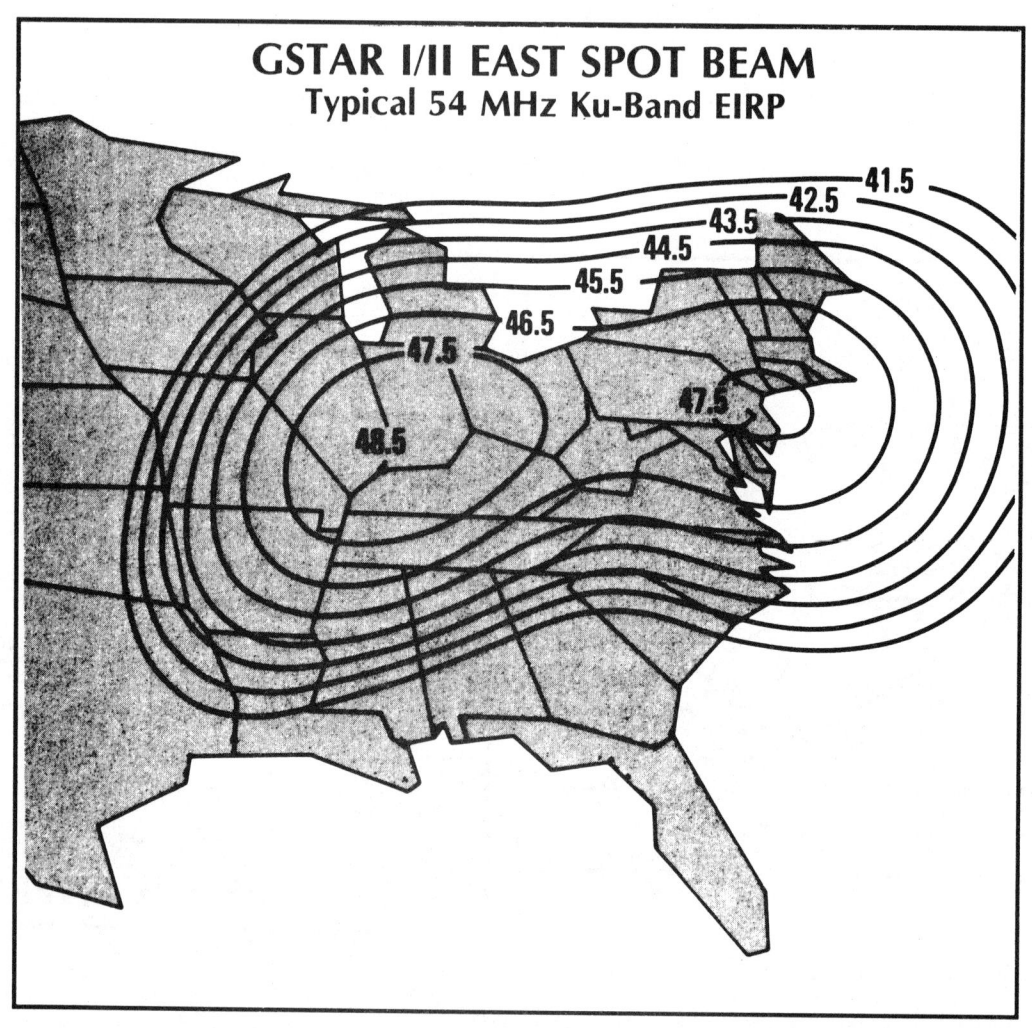

SATELLITE FOOTPRINTS

GSTAR I/II EAST REGIONAL BEAM
Typical 54 MHz Ku-Band EIRP

GSTAR I/II WEST REGIONAL BEAM
Typical 54 MHz Ku-Band EIRP

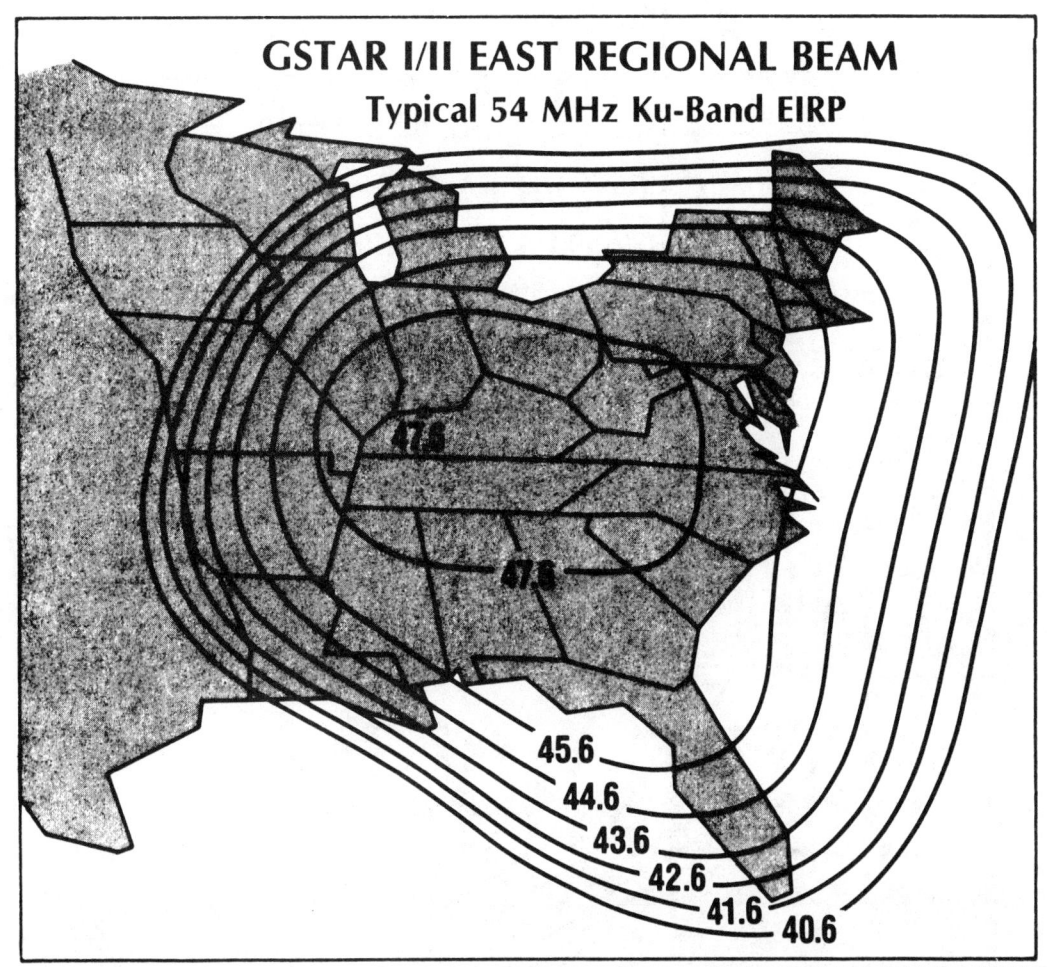

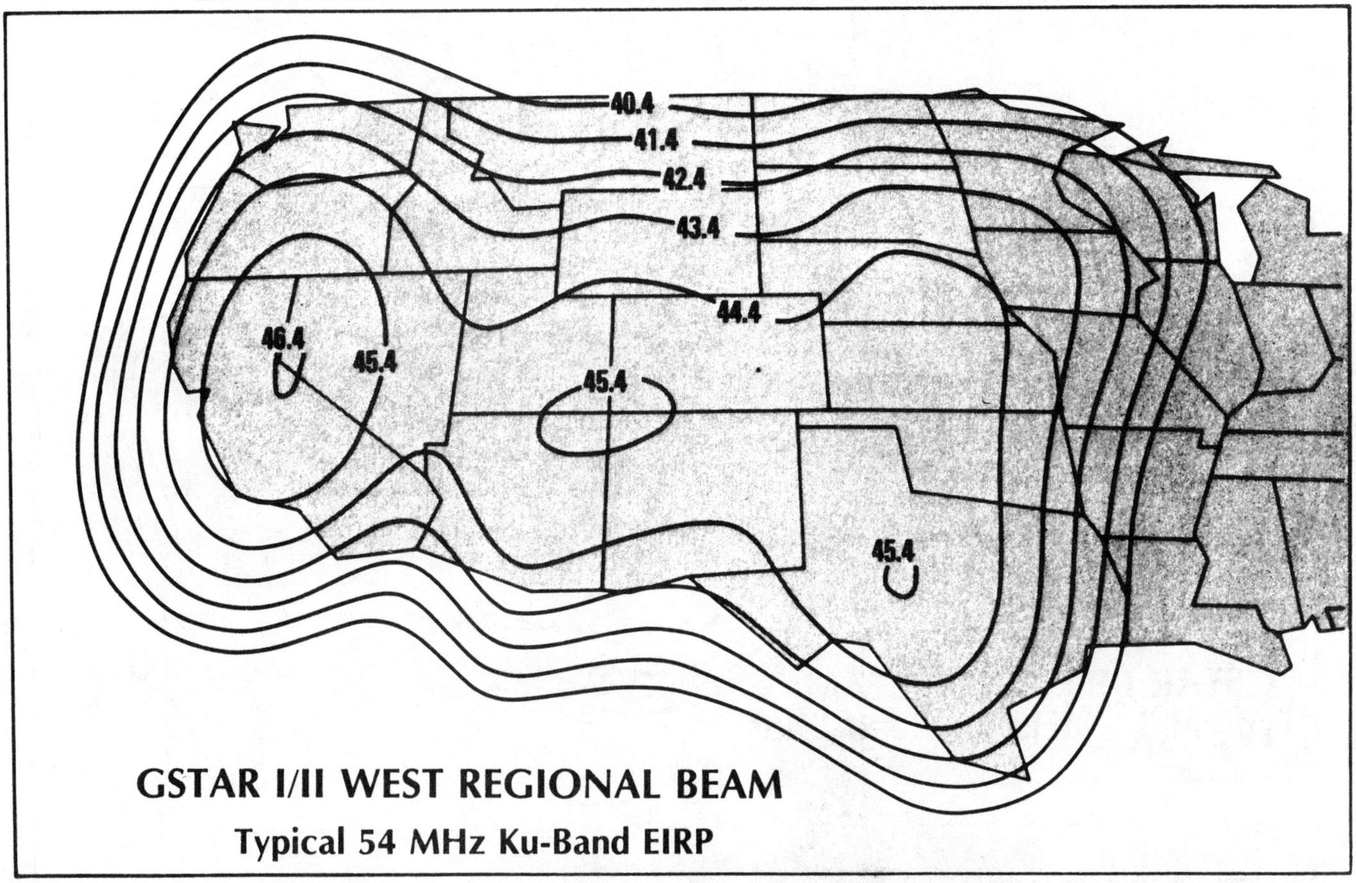

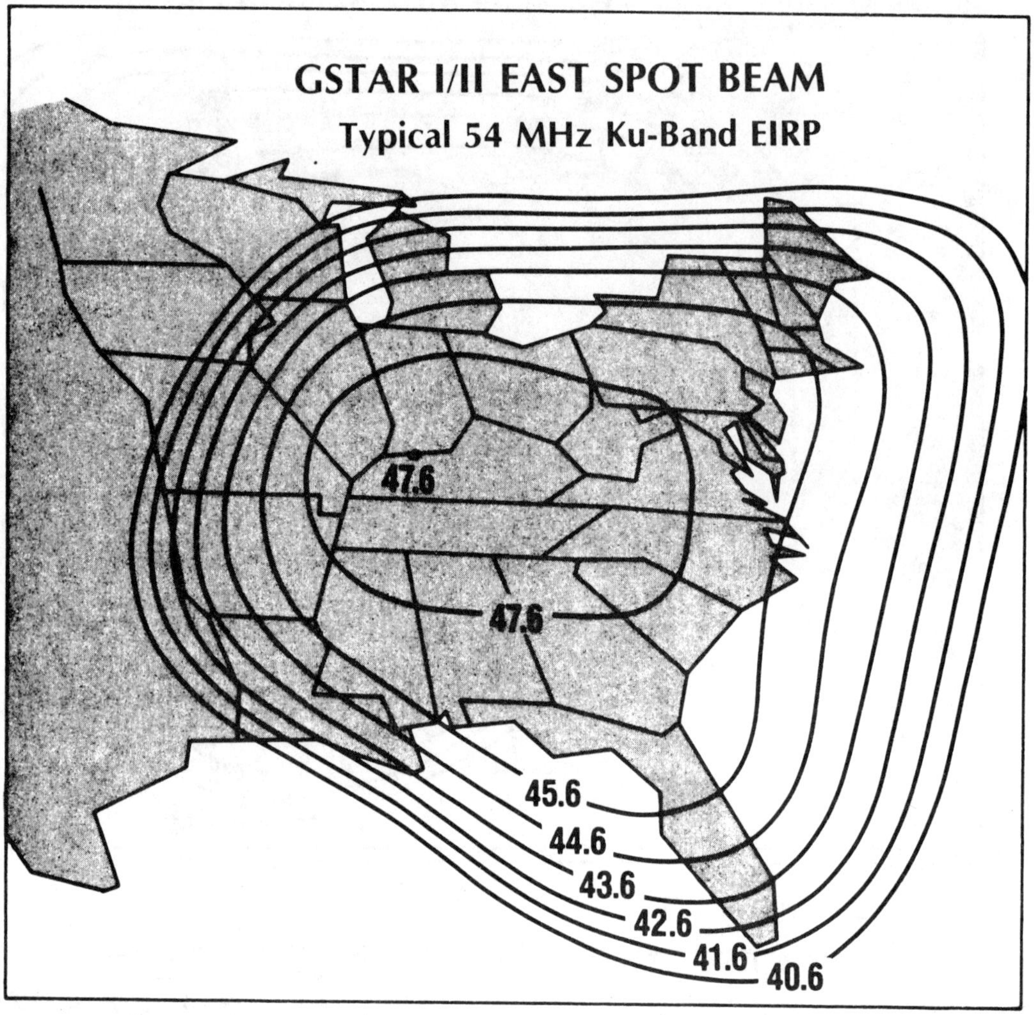

APPENDIX C

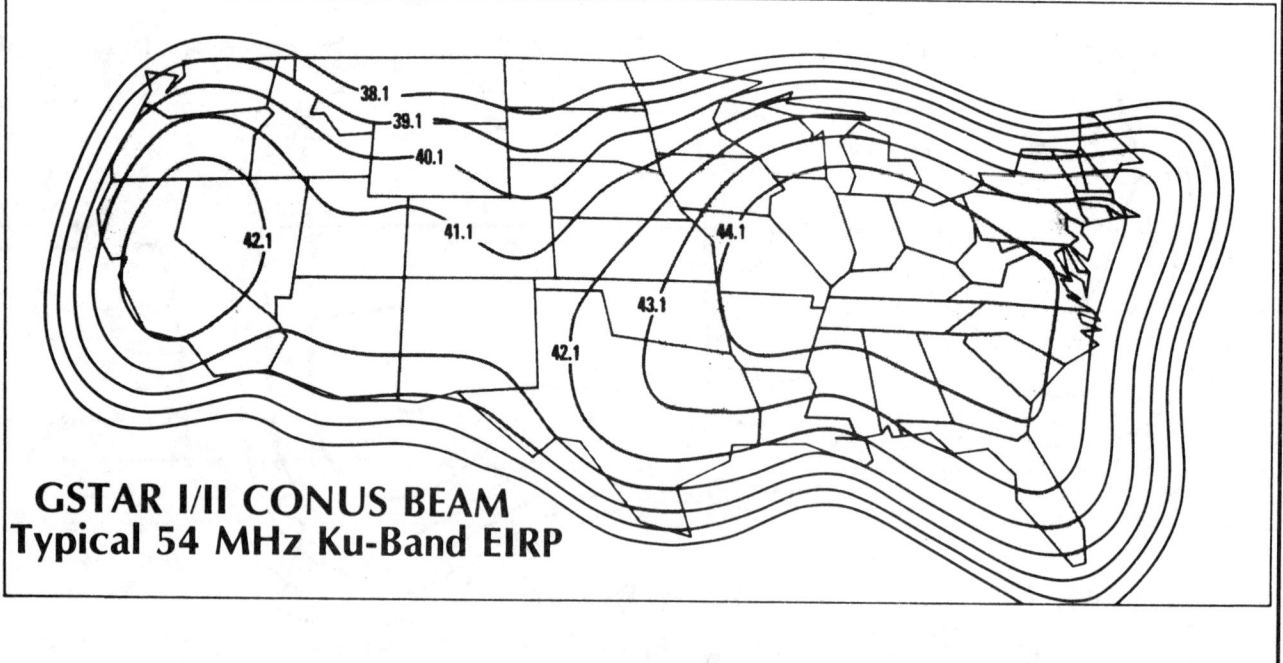

SATELLITE FOOTPRINTS

APPENDIX C

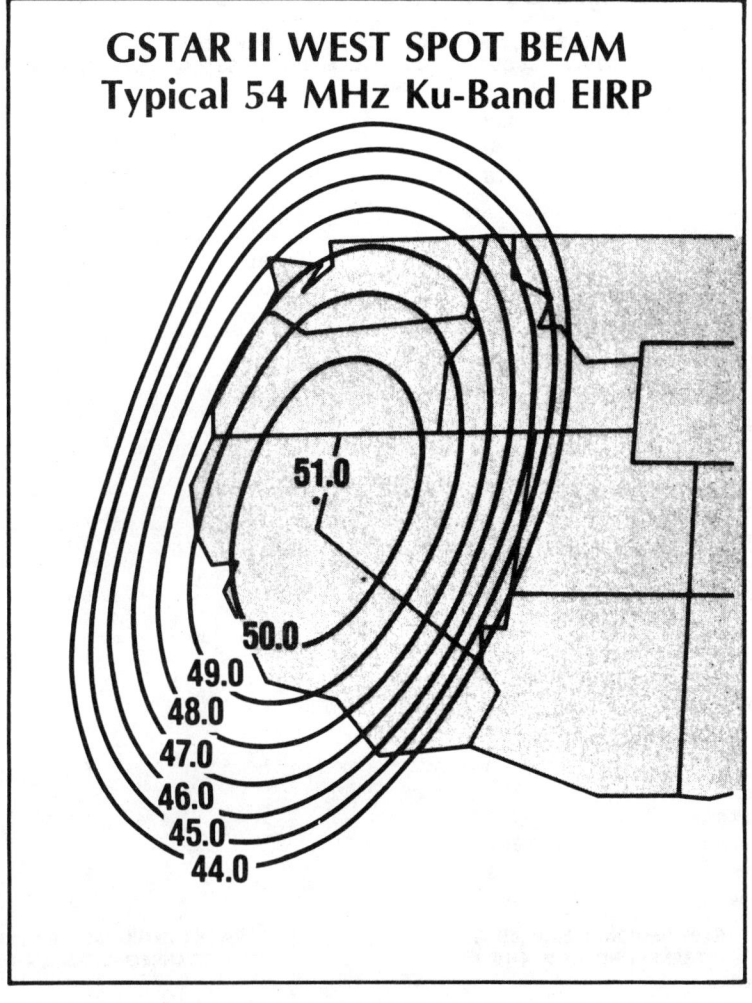

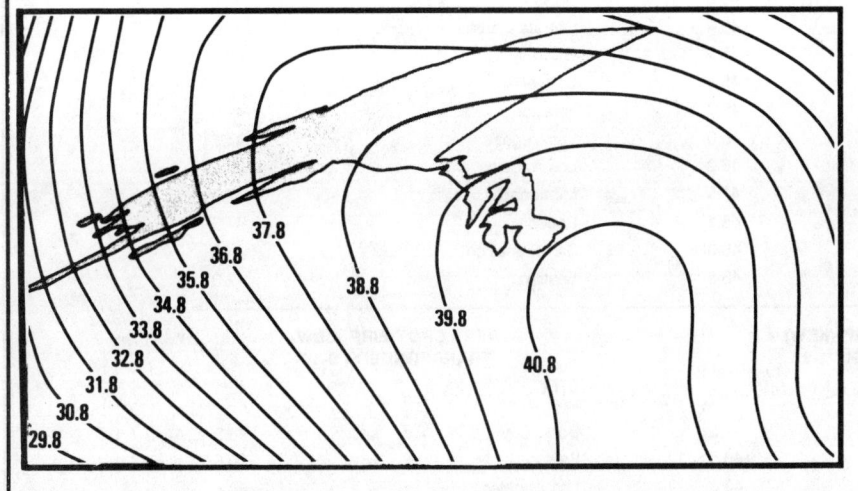
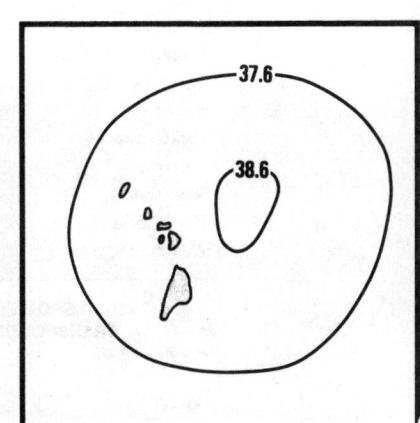

GSTAR I/II COMBINED BEAM
EIRP Performance in Alaska & Hawaii

SATELLITE FOOTPRINTS

Expected GSTAR Performance

CITY	50-STATE EIRP (dBW) TRANSPONDERS 1 AND 3	CONUS EIRP (dBW) TRANSPONDERS 2,4-16	G/T(dB/K) TRANSPONDERS 1-16	SFD(dBW/m²)*
Atlanta	45.0	44.2	2.5	-95.6
Boston	43.3	43.1	2.3	-95.4
Buffalo	44.0	42.9	1.9	-95.0
Chicago	44.8	43.7	3.6	-96.7
Corpus Christi	40.4	40.5	-1.2	-91.9
Dallas	43.6	42.6	1.3	-94.4
Denver	41.6	40.8	0.4	-93.5
Houston	42.2	41.7	1.1	-94.2
Las Vegas	43.2	42.0	1.0	-94.1
Los Angeles	43.1	42.0	0.4	-93.5
Melbourne	43.9	43.0	1.8	-94.9
Miami	41.1	40.6	-0.5	-92.6
Minneapolis	40.8	41.4	0.1	-93.2
Nashville	44.9	44.4	2.3	-95.4
New York	43.7	43.8	2.8	-95.9
Phoenix	41.8	41.2	0.7	-93.8
Portland, ME	43.4	42.5	2.0	-95.1
San Francisco	42.6	42.0	0.4	-93.5
Seattle	41.2	39.8	.0	-93.1
St. Louis	45.1	44.6	3.3	-96.4
Washington, DC	43.8	44.1	2.9	-96.0
Anchorage	38.7		-4.4	-88.7
Fairbanks	38.0		-5.7	-87.4
Honolulu	38.3		-3.5	-86.9
Nome	38.0		-5.0	-88.1

EAST REGIONAL EIRP (dBW) TRANSPONDERS 2, 4-16		WEST REGIONAL EIRP (dBW) TRANSPONDERS 2, 4-16	
CITY		CITY	
Atlanta	46.6	Boise City	42.6
Boston	45.4	Corpus Christi	42.2
Buffalo	45.3	Dallas	43.4
Chicago	45.4	Denver	43.8
Columbia, SC	46.7	Houston	41.5
Detroit	46.0	Las Vegas	44.5
Jacksonville, FL	46.2	Los Angeles	44.6
Nashville	47.5	Minneapolis	42.9
New York	46.1	Phoenix	44.3
Philadelphia	46.3	San Francisco	44.6
Washington, DC	46.3	Seattle	42.3

EAST SPOT EIRP (dBW) TRANSPONDERS 1-8		WEST SPOT EIRP (dBW) TRANSPONDERS 9-16	
CITY		CITY	
Boston	47.2	Boise City	48.5
Buffalo	46.5	Fresno	50.5
Chicago	46.8	Los Angeles	48.4
Cleveland	47.2	Reno	51.0
Detroit	46.9	Sacramento	50.6
Hartford	47.5	San Diego	45.9
Nashville	47.4	San Francisco	49.9
New York	47.7	Seattle	48.5
Philadelphia	47.7	Portland, OR	49.1
Washington, DC	47.3		

APPENDIX C

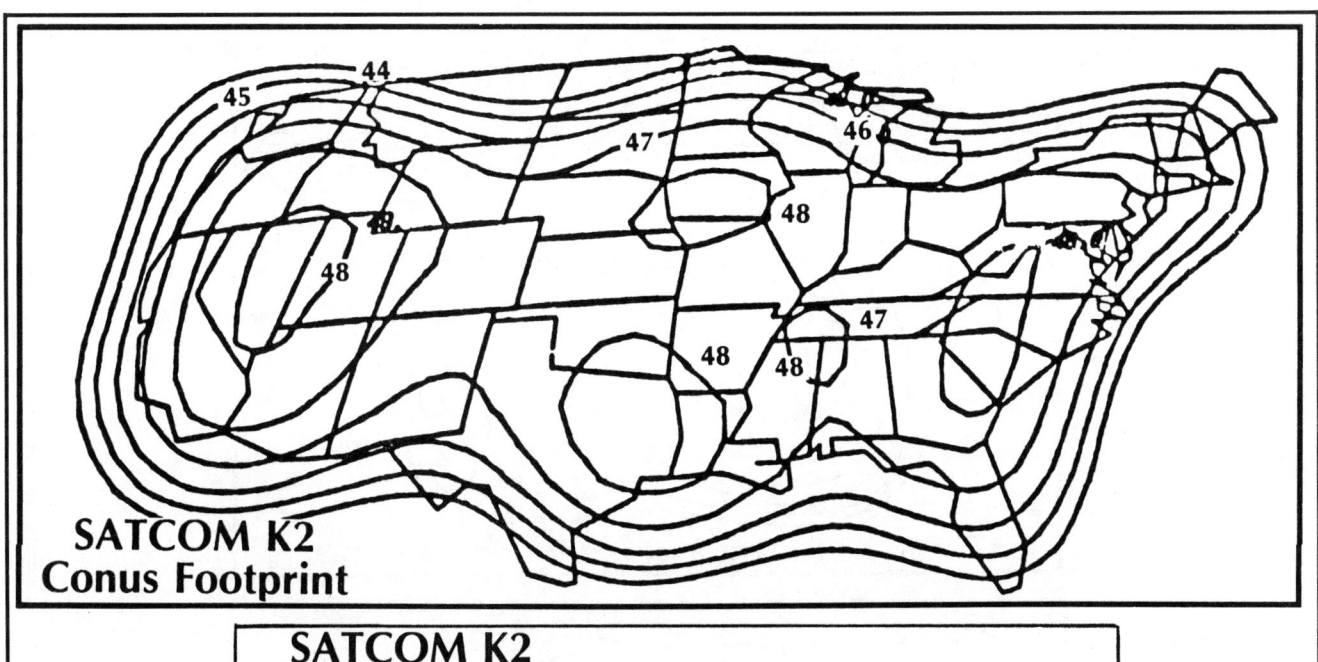

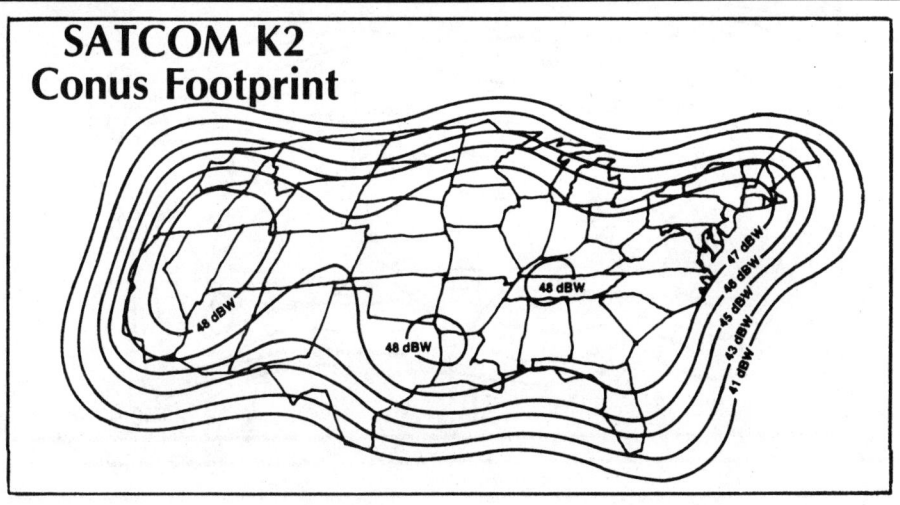

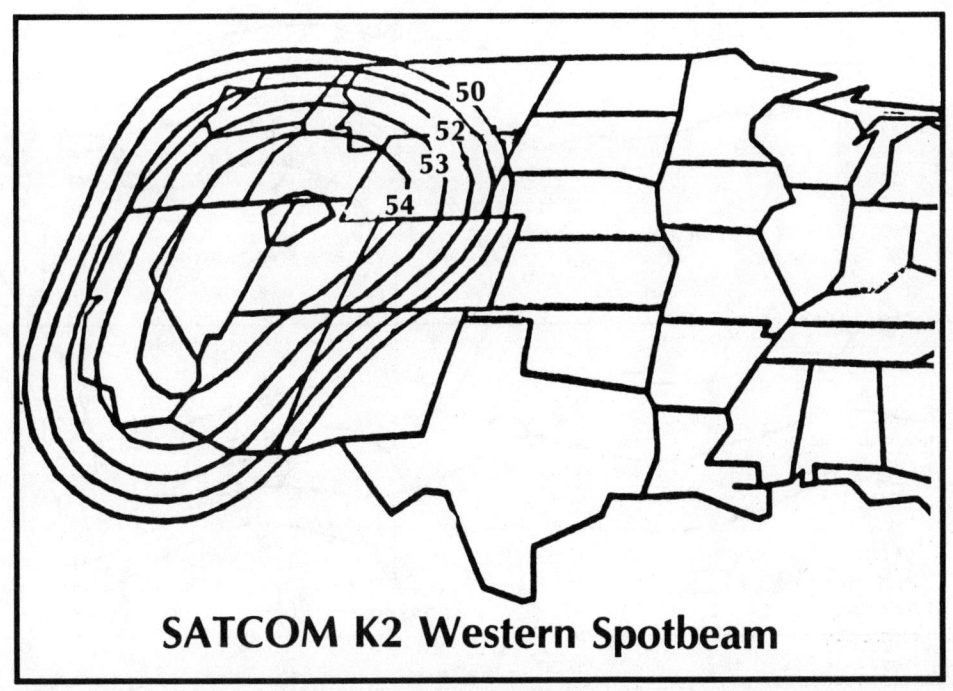

SATELLITE FOOTPRINTS

APPENDIX C

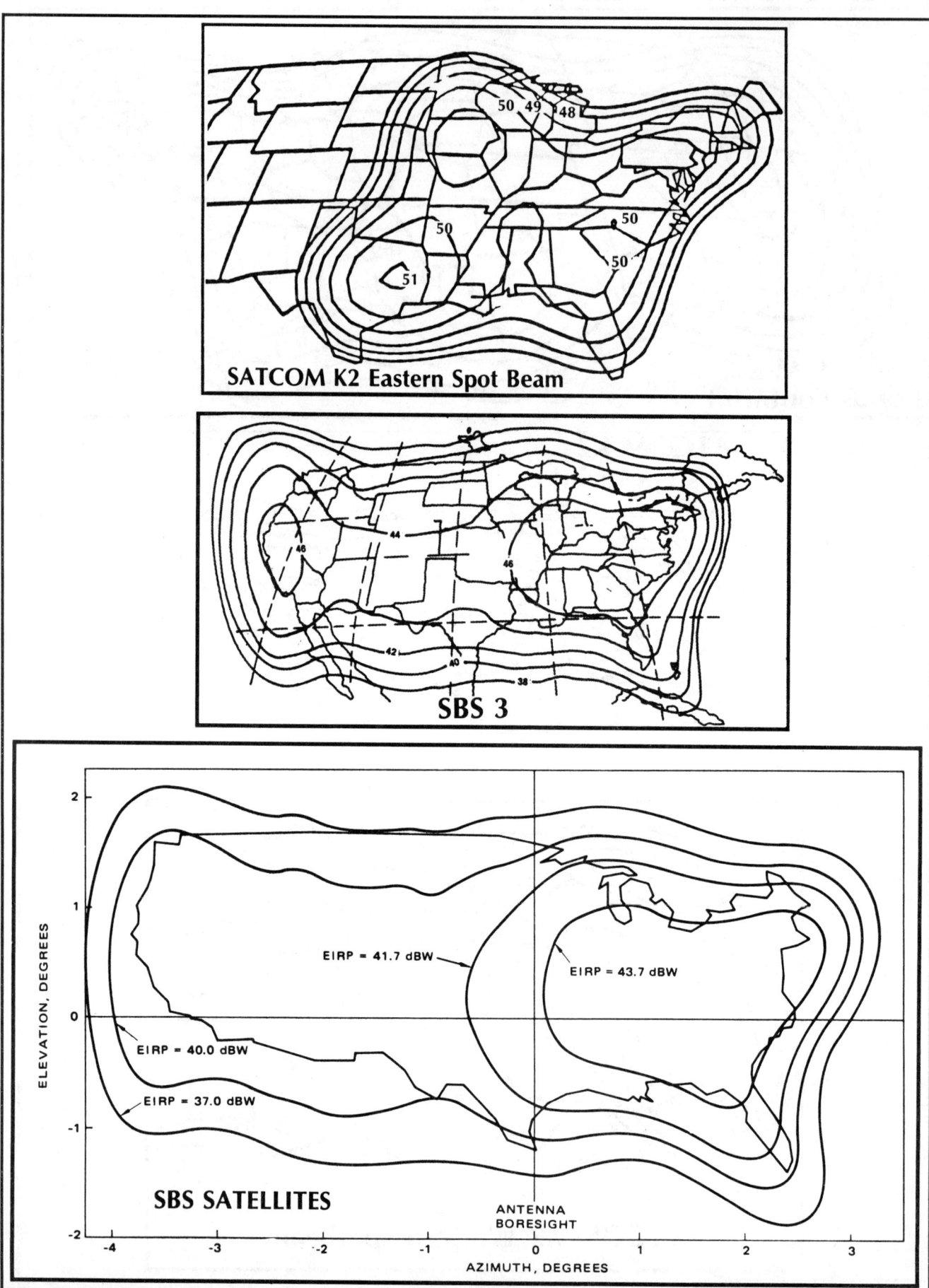

SATELLITE FOOTPRINTS

APPENDIX C

SPACENET 1
Expected 72 MHz Ku-Band EIRP

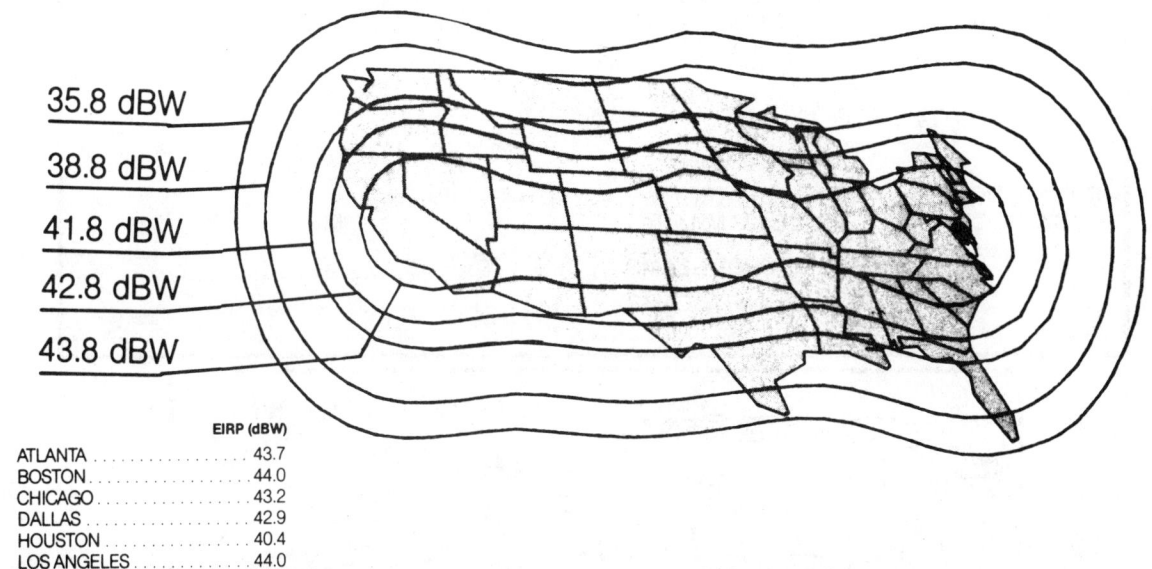

35.8 dBW
38.8 dBW
41.8 dBW
42.8 dBW
43.8 dBW

	EIRP (dBW)
ATLANTA	43.7
BOSTON	44.0
CHICAGO	43.2
DALLAS	42.9
HOUSTON	40.4
LOS ANGELES	44.0
NEW YORK	44.3
ORLANDO	40.3
SAN FRANCISCO	43.8
SEATTLE	40.5

SPACENET II
Expected 72 MHz Ku-Band EIRP

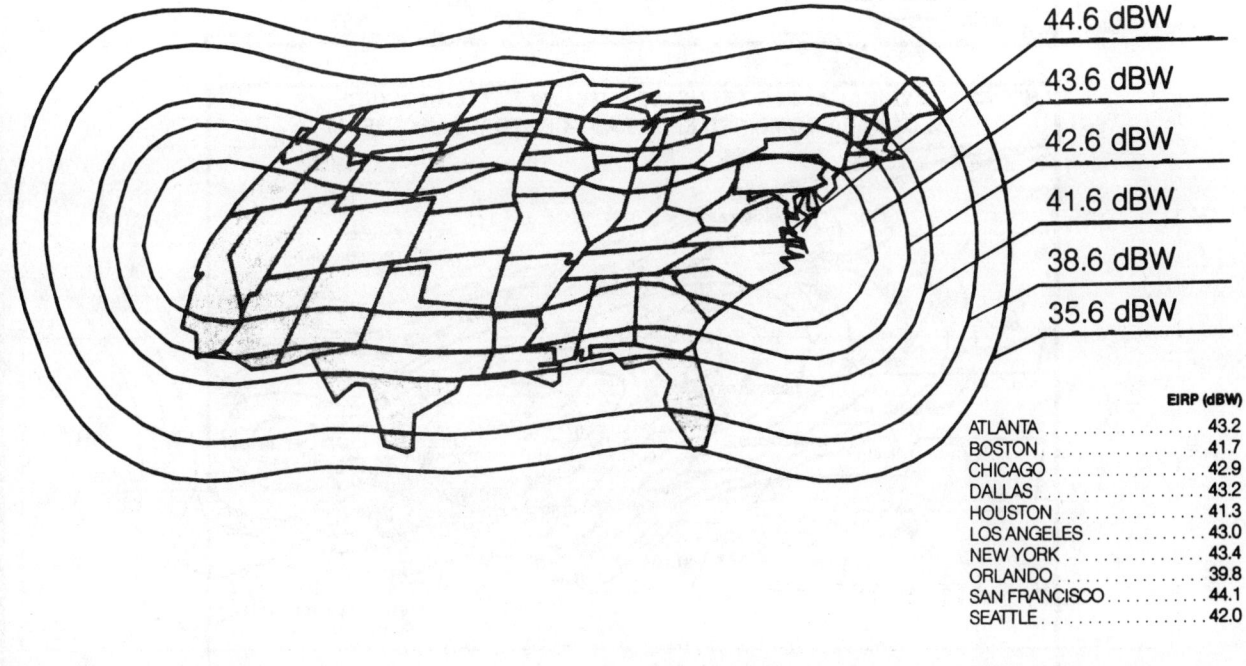

44.6 dBW
43.6 dBW
42.6 dBW
41.6 dBW
38.6 dBW
35.6 dBW

	EIRP (dBW)
ATLANTA	43.2
BOSTON	41.7
CHICAGO	42.9
DALLAS	43.2
HOUSTON	41.3
LOS ANGELES	43.0
NEW YORK	43.4
ORLANDO	39.8
SAN FRANCISCO	44.1
SEATTLE	42.0

SATELLITE FOOTPRINTS

APPENDIX C

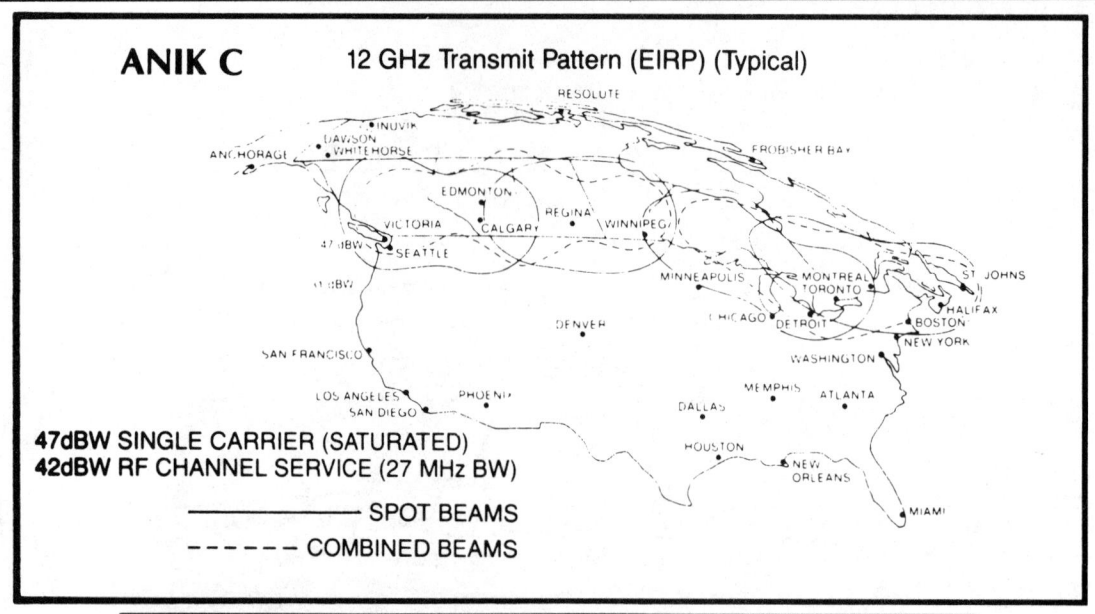

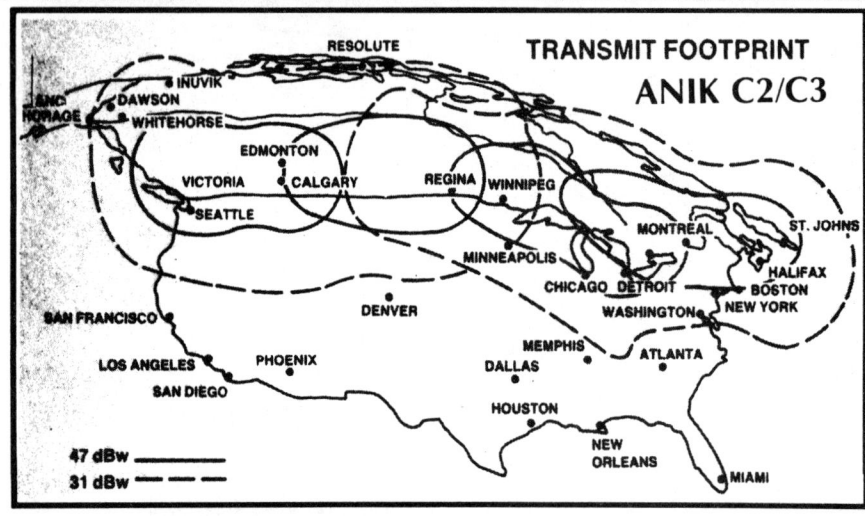

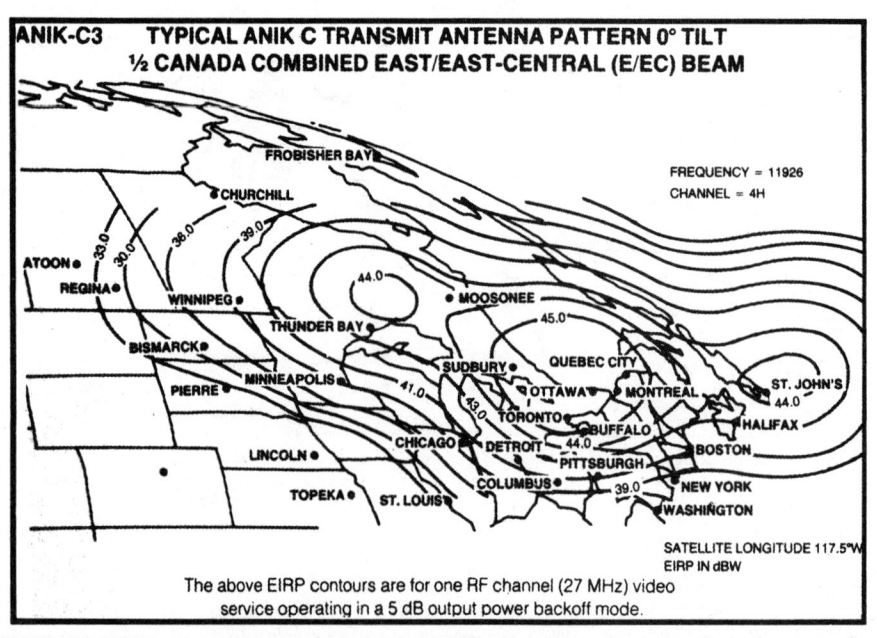

SATELLITE FOOTPRINTS

APPENDIX C

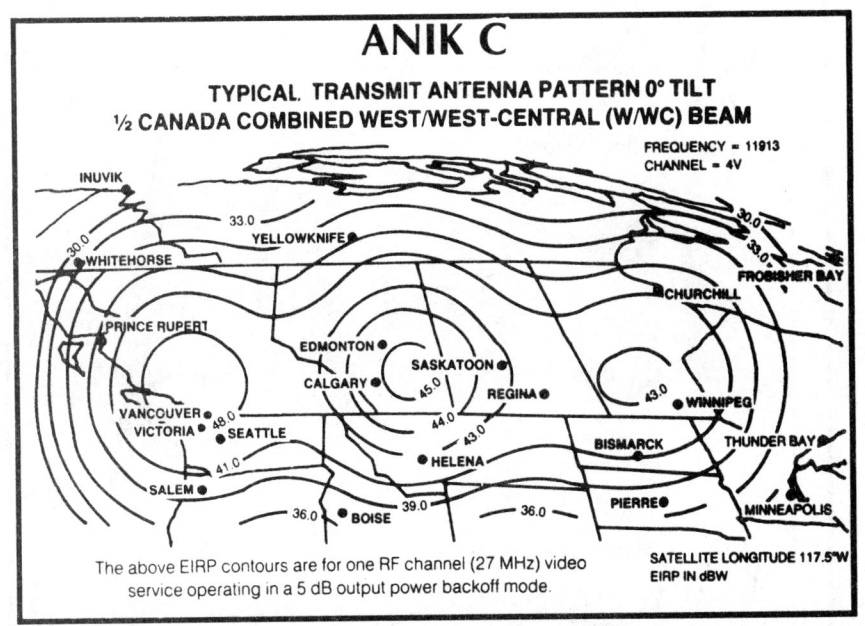

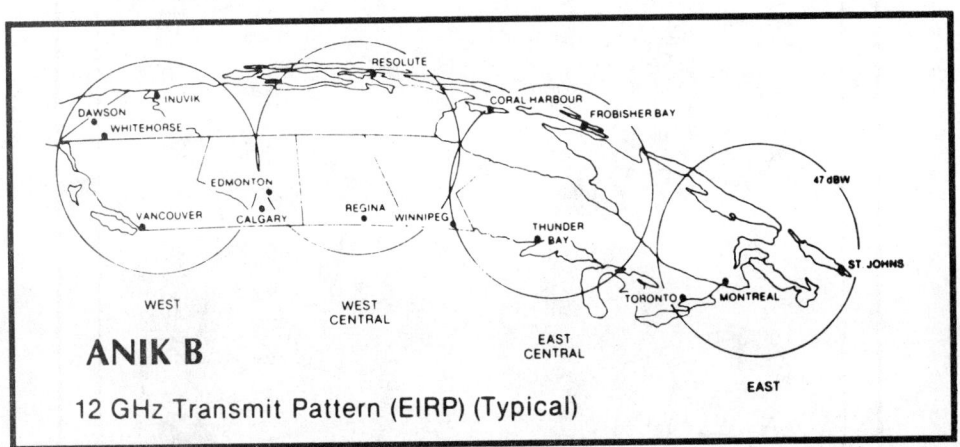

APPENDIX C

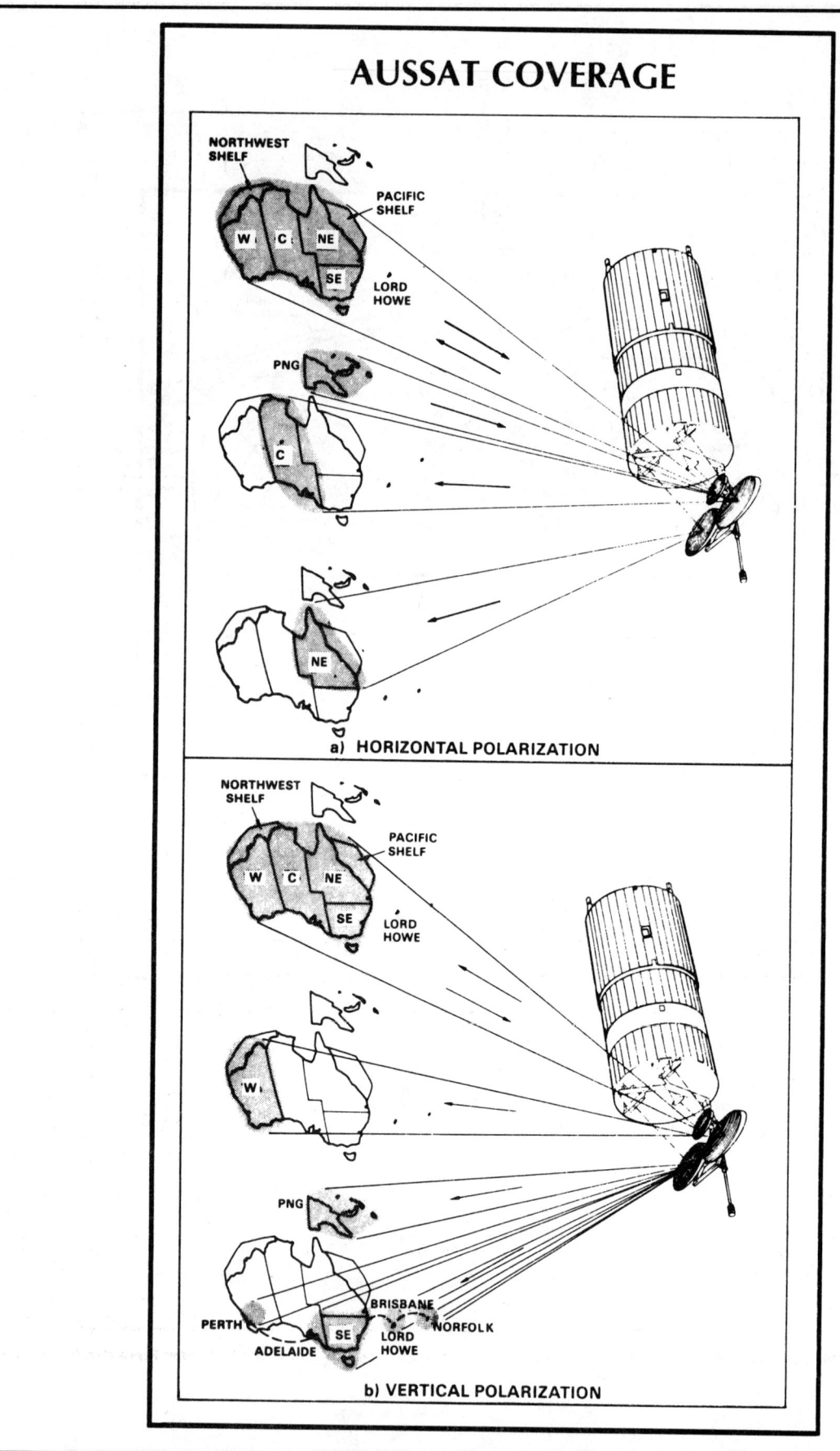

AUSSAT COVERAGE

a) HORIZONTAL POLARIZATION

b) VERTICAL POLARIZATION

SATELLITE FOOTPRINTS

APPENDIX C

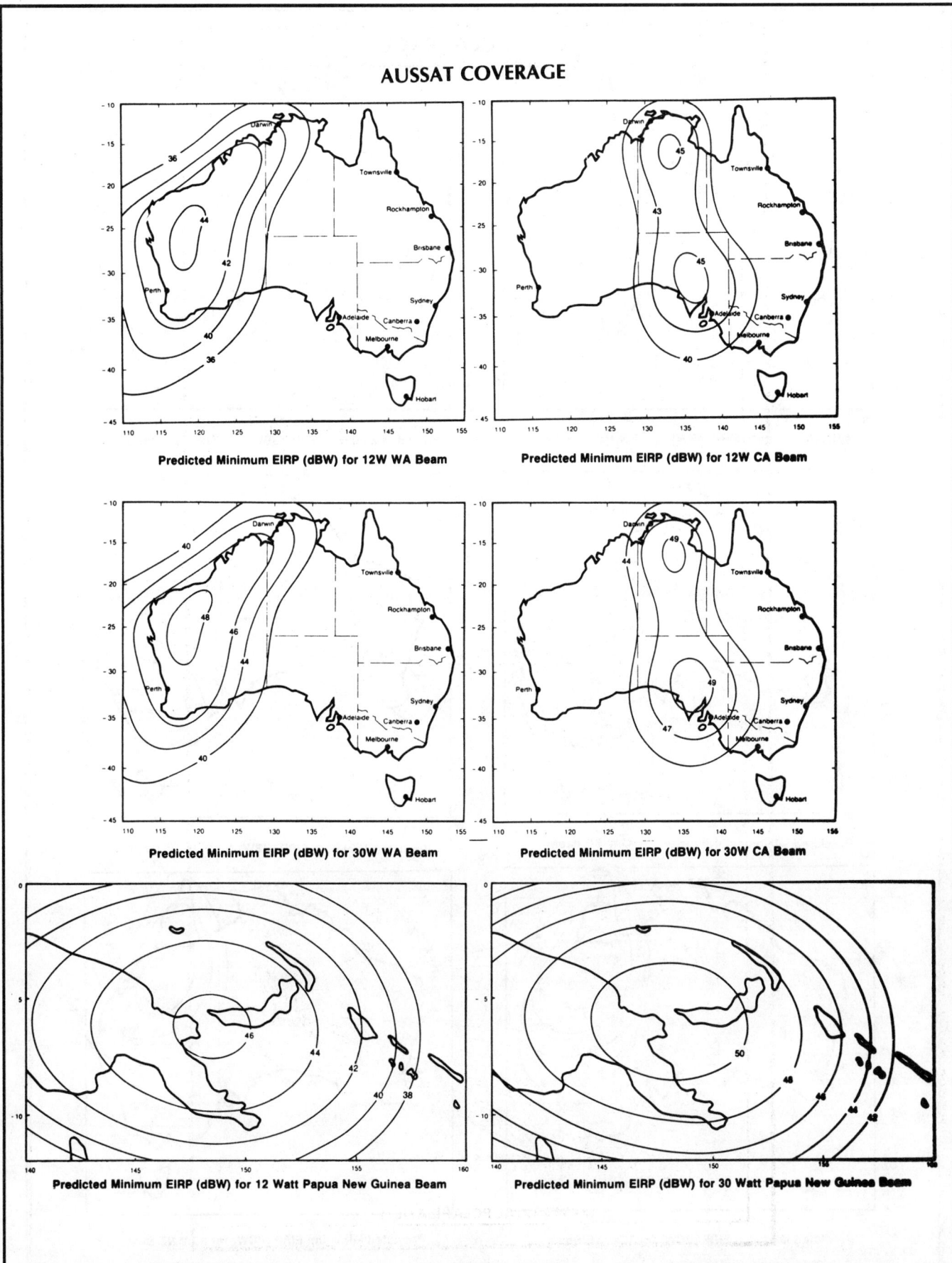

SATELLITE FOOTPRINTS

APPENDIX C

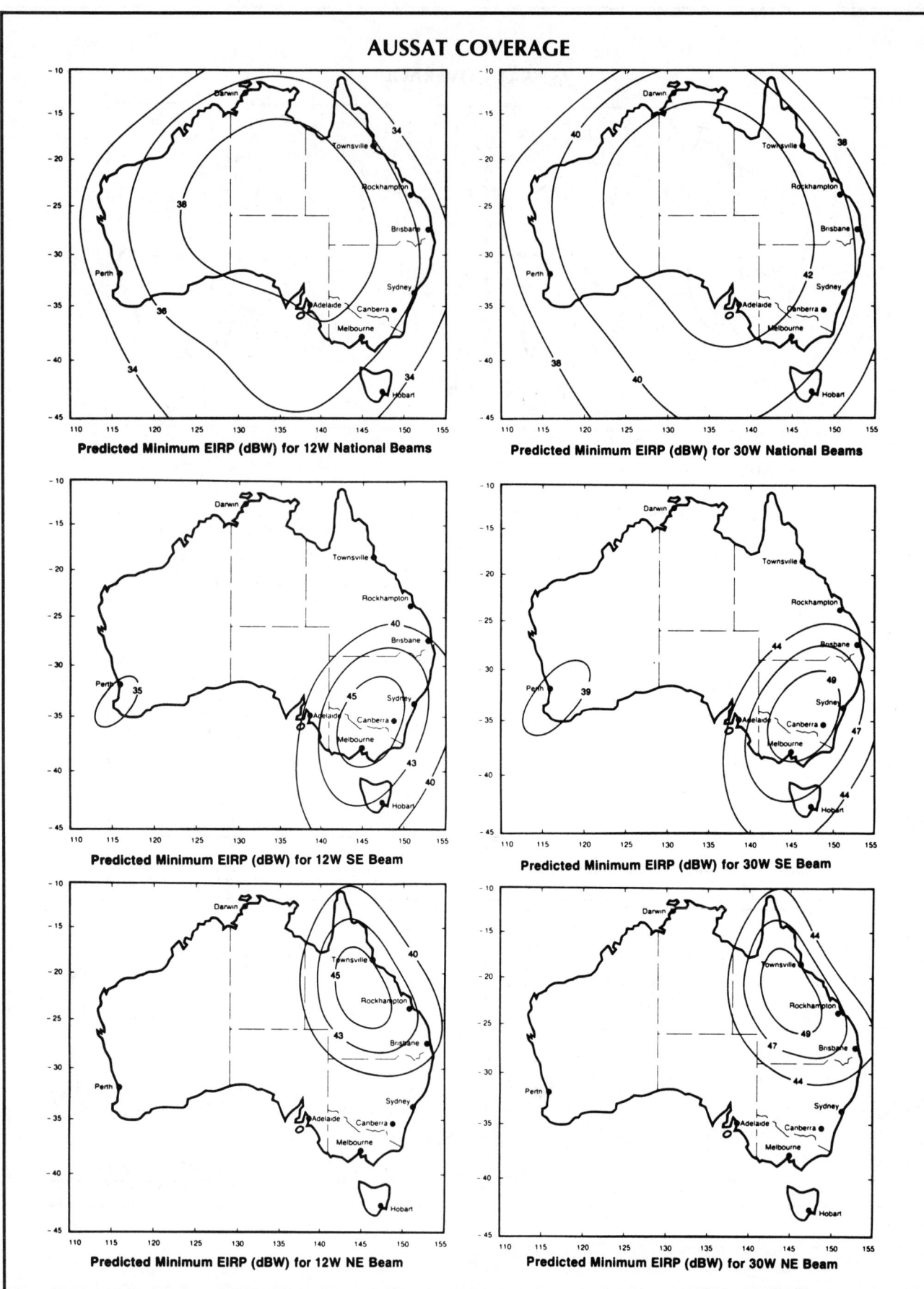

SATELLITE FOOTPRINTS

APPENDIX D. GLOSSARY OF COMMON BROADCAST AND SATELLITE TELEVISION TERMS

A/B Switch

A switch that selects one of two inputs (A or B) for routing to a common output while providing adequate isolating between the two signals.

AFC (Automatic Frequency Control)

A circuit which locks an electronic component onto a chosen frequency.

AGC (Automatic Gain Control)

A circuit that uses feedback to maintain the output of an electronic component at a constant level.

Absolute Zero

The coldest possible temperature at which all molecular motion ceases. It is expressed in degrees Kelvin as measured from absolute zero. Zero degrees Kelvin equals minus 273.16 °C or minus 459.69 °F.

Adjacent Channel

An adjacent channel is immediately next to another channel in frequency. For example, NTSC channels 5 and 6 as well as 8 and 9 are adjacent. However, channels 4 and 5 or channels 6 and 7 are separated by signals used by non-TV media.

Agile Receiver

A satellite receiver which can be tuned to any desired channel.

Alignment

The process of fine tuning a dish or an electronic circuit to maximize its sensitivity and signal receiving capability.

Ambient Temperature

The existing dry bulb temperature.

Amplifier

A device used to increase the power of a signal.

Analog

A system in which signals vary continuously in contrast to a digital system in which signals vary in discrete steps

Analog-to-Digital Converter

A circuit that converts analog signals to an equivalent digital form. The varying analog signal is sampled at a series of points in time. The voltage at each of these points is then represented by a series of numbers, the digital value of the sample. The higher this sampling frequency, the finer are the gradations and the more accurately is the signal represented.

Antenna

A device that collects and focuses electromagnetic energy, i.e., contributes an energy gain. Satellite dishes, broadband antenna and cut-to-channel antennas are some types of antennas encountered in private cable systems. In the case of satellite antennas, gain is proportional to the surface area of the microwave dish.

Antenna Efficiency

The percentage of incoming satellite signal actually captured by an antenna.

Antenna Illumination

Describes how a feedhorn "sees" the surface of a dish as well as the surrounding terrain.

Aperture

The collection area of a parabolic antenna.

Aspect Ratio

The ratio of television screen width to height. The standard aspect ratio is 4 to 3.

Attenuation

The decrease in signal power that occurs in a device or when a signal travels to reach a destination point (path loss).

Attenuator

A passive device which reduces the power of a signal. Attenuators are rated according to the amount of signal attenuation.

APPENDIX D

Audio Subcarrier

The carrier wave that transmits audio information within a video broadcast signal. Satellite transmissions can relay more than a single audio subcarrier in the frequency range between 5 and 8.5 MHz.

Automatic Brightness Control

A television circuit used to automatically adjust picture tube brightness in response to changes in background or ambient light.

Automatic Fine Tuning

A circuit that automatically maintains the correct tuner oscillator frequency and compensates for drift and for moderate amounts of inaccurate tuning. Similar to AFC.

Automatic Frequency Control (AFC)

A circuit that locks onto a chosen frequency and will not drift away from that frequency.

Automatic Gain Control (AGC)

A circuit that locks the gain onto a fixed value and thus compensates for varying input signal levels keeping the output constant.

Azimuth-Elevation (Az-El) Mount

An antenna mount which tracks satellites by moving in two directions: the azimuth in the horizontal plane; and elevation up from the horizon.

Azimuth

A compass bearing expressed in degrees of rotation clockwise from true north. It is one of the two coordinates (azimuth and elevation) used to align a satellite antenna.

Back Match

The matching of the resistive values of the input and output of electronic devices to reduce signal reflection and ghosting. Also known as impedance matching.

Back Porch

That portion of the horizontal blanking pulse that follows the trailing edge of the horizontal sync pulse.

Band

A range of frequencies.

Band Separator

A device that splits a group of specified frequencies into two or more bands. Common types include UHF/VHF, Hi/Lo-band and FM separators. This device is essentially a set of filters.

Bandpass Filter

A circuit or device that allows only a specified range of frequencies to pass from input to output.

Bandwidth

The frequency range allocated to any communication circuit.

Baseband

The raw audio and video signals prior to modulation and broadcasting. Most satellite headend equipment utilises baseband inputs. More exactly, the composite unclamped, non-de-emphasised and unfiltered receiver output. This signal contains the complete set of FM modulated audio and data subcarriers.

Beamwidth

A measure used to describe the width of vision of an antenna. Beamwidth is measured as degrees between the 3 dB half power points.

Bird

Jargon or nickname for communication satellites.

Blanking Pulse Level

The reference level for video signals. The blanking pulses must be aligned at the input to the picture tube.

Blanking Signal

Pulses used to extinguish the scan illumination during horizontal and vertical retrace periods.

Block Downconversion

The process of lowering the entire band of frequencies in one step to some intermediate range to be processed inside a satellite receiver. Multiple block downconversion receivers are capable of independently selecting channels because each can process the entire block of signals.

BNC Connector

A weatherproof twist lock coax connector standard on commercial video equipment and used on some brands of satellite receivers.

Boresight

The direction along the principle axis of either a transmitting or a receiving antenna.

Broadband

A device that processes a signal(s) spanning a relatively broad range of input frequencies.

Buttonhook Feed

A rod shaped like a question mark supporting the feedhorn and LNA. A buttonhook feed for use with commercial grade antennas is often a hollow waveguide that directs signals from a feedhorn to an LNA behind the antenna.

CATV

An abbreviation for Community Antenna Television - another name for cable TV.

CCD

Charge coupled device. In this device charge is stored on a capacitor which are etched onto a chip. A number of samples can be simultaneously stored. Used in MAC transmissions for temporarily storing video signals.

C-Band

The 3.7 to 4.2 GHz band of frequencies at which some broadcast satellites operate.

Carrier

A pure-frequency signal that is modulated to carry information. In the process of modulation it is spread out over a wider band. The carrier frequency is the center frequency on any television channel.

GLOSSARY

Carrier-to-Noise Ratio (C/R)

The ratio of the received carrier power to the noise power in a given bandwidth, expressed in decibels. The C/R is an indicator of how well an earth receiving station will perform in a particular location, and is calculated from satellite power levels, antenna gain and the combined antenna and LNA noise temperature.

Cassegrain Feed System

An antenna feed design that includes a primary reflector, the dish, and a secondary reflector which redirects microwaves via a waveguide to a low noise amplifier.

Channel

A segment of bandwidth used for one complete communication link.

Chrominance

The hue and saturation of a color. The chrominance signal is modulated onto a 4.43 MHz carrier in the PAL television system and a 3.58 MHz carrier in the NTSC television system.

Chrominance Signal

The color component of the composite baseband video signal assembled from the I and Q portions. Phase angle of the signal represents hue and amplitude represents color saturation.

Circular Polarity

Electromagnetic waves whose electric field uniformly rotates along the signal path. Broadcasts used by Intelsat and other international satellites use circular, not horizontally or vertically polarized waves as are common in North American and European transmissions. Circularly polarised waves are used for satellite telephony because Faraday rotation does not alter their behaviour.

Clamp Circuit

A circuit that removes the dispersion waveform from the downlink signal.

Clarke Belt

The circular orbital belt at 22,247 miles above the equator, named after the writer Arthur C. Clarke, in which satellites travel at the same speed as the earth's rotation. Also called the geostationary orbit.

Color Bars

A test pattern of specifically colored vertical bars used as a reference to test the performance of a color television.

Coaxial Cable

A cable for transmitting high frequency electrical signals with low loss. It is composed of an internal conducting wire surrounded by an insulating dielectric which is further protected by a metal shield. The impedance of coax is a product of the radius of the central conductor, the radius of the shield and the dielectric constant of the insulation. In an SMATV system, coax impedance is 75 ohms.

Color Sync Burst

A "burst" of 8 to 11 cycles in the 4.43361875 MHz (PAL) or 3.579545 MHz(NTSC) color subcarrier frequency. This waveform is located on the back porch of each horizontal blanking pulse during color transmissions. It serves to synchronize the color subcarrier's oscillator with that of the transmitter in order to recreate the raw color signals.

Composite Baseband Signal

The complete audio and video signal without a carrier wave. Satellite signals have audio baseband information ranging in frequency from zero to 3400 Hertz. NTSC video baseband is from zero to 4.2 MHz. PAL video baseband ranges from 0 to 5.5 MHz.

Composite Video Signal

The complete video signal consisting of the chrominance and luminance information as well as all sync and blanking pulses.

Companding

A form of noise reduction using compression at the transmitting end and expansion at the receiver. A compressor is an amplifier that increases its gain for lower power signals. The effect is to boost these components into a form having a smaller dynamic range. A compressed signal has a higher average level, and therefore, less apparent loudness than an uncompressed signal, even though the peaks are no higher in level. An expander reverses the effect of the compressor to restore the original signal.

CONE

An abbreviation for the European continent.

Contrast

The ratio between the dark and light areas of a television picture.

Conus

An abbreviation for the continental United States.

Cross Modulation

A form of interference caused by the modulation of one carrier affecting that of another signal. It can be caused by overloading an amplifier as well as by signal imbalances at the headend.

Cross Polarisation

Term to describe signals of the opposite polarity to another being transmitted and received. Cross-polarisation discrimination refers to the ability of a feed to detect one polarity and reject the signals having the opposite sense of polarity.

Crosstalk

Interference between adjacent channels often caused by cross modulation. Leakage can occur between two wires, PCB tracks or parallel cables.

dc Power Block

A device which stops the flow of dc power but permits passage of higher frequency ac signals.

Decibel (dB)

The logarithmic ratio of power levels used to indicate gains or losses of signals. Decibels relative to one watt, milliwatt and millivolt are abbreviated as dBw, dBm and dBmV, respectively. Zero dBmV is used as the standard reference for all SMATV calculations.

Declination Offset Angle

The adjustment angle of a polar mount between the polar axis and the plane of a satellite antenna used to aim at the geosynchronous arc. Declination increases from zero with latitude away from the equator.

APPENDIX D

Decoder
A circuit that restores a signal to its original form after it has been scrambled.

De-emphasis
A reduction of the higher frequency portions of an FM signal used to neutralize the effects of pre-emphasis. When combined with the correct level of pre-emphasis, it reduces overall noise levels and therefore increases the signal-to-noise ratio.

Demodulator
A device which extracts the baseband signal from the transmitted carrier wave.

Detent Tuning
Tuning into a satellite channel by selecting a preset resistance.

Digital
Describes a system or device in which information is transferred by electrical "on-off," "high-low," or "1/0" pulses instead of continuously varying signals or states as in an analog message.

Digital-to-Analog Converter
A circuit that converts digital signals into their equivalent analog form.

Direct Broadcast Satellite (DBS)
A term commonly used to describe Ku-band broadcasts via satellite directly to individual end-users. The DBS band ranges from 11.7 to 12.2 GHz.

Dish
Jargon for a parabolic microwave antenna.

Distribution System
A communication system consisting of coax but occasionally of line-of-sight microwave links that carries signals from the headend to end-users.

Domsat
Abbreviation for domestic communication satellite.

Downconverter
A circuit that lowers the high frequency signal to a lower, intermediate range. There are three distinct types of downconversion used in satellite receivers: single downconversion; dual downconversion; and block downconversion.

Downlink Antenna
The antenna on-board a satellite which relays signals back to earth.

Drifting
An instability in a preset voltage, frequency or other electronic circuit parameter.

Dual-Band Feedhorn
A feedhorn which can simultaneously receive two different bands, typically the C and Ku-bands.

Earth Station
A complete satellite receiving or transmitting station including the antenna, electronics and all associated equipment necessary to receive or transmit satellite signals. Also known as a ground station.

Effective Isotropic Radiated Power (EIRP)
A measure of the signal strength that a satellite transmits towards the earth below. The EIRP is highest at the center of the beam and decreases at angles away from the boresight.

Elevation Angle
The vertical angle measured from the horizon up to a targeted satellite.

Energy Dispersal
The modulation of an uplink carrier with a triangular waveform. This technique disperses the carrier energy over a wider bandwidth than otherwise would be the case in order to limit the maximum energy compared to that transmitted by an unclamped carrier. This triangular waveform is removed by a clamp circuit in a satellite receiver.

Equalizing Pulses
A series of six pulses occurring before and after the serrated vertical sync pulse to ensure proper interlacing. The equalizing pulses are inserted at twice the horizontal scanning frequency.

F-connector
A standard RF connector used to link coax cables with electronic devices.

FCC
The Federal Communications Commission, the regulatory board which sets standards for communications within the United States.

f/D Ratio
The ratio of an antenna's focal length to diameter. It describes antenna "depth."

Feedhorn
A device that collects microwave signals reflected from the surface of an antenna. It is mounted at the focus of all prime focus parabolic antennas.

Field
One half of a complete TV picture or frame, composed of 262.5 scanning lines. There are 60 fields per second for black/white TV and 59.94 fields per second for color TV in NTSC transmission. In the PAL broadcast system there are 50 fields per second.

Filter
A device used to reject all but a specified range of frequencies. A bandpass filter allows only those signals within a given band to be communicated. A rejection filter, the mirror image of a bandpass filter, eliminates those signals within a specified band but passes all other frequencies.

Focal Length
The distance from the reflective surface of a parabola to the point at which incoming satellite signals are focused, the focal point.

GLOSSARY

APPENDIX D

Footprint
The geographic area towards which a satellite downlink antenna directs its signal. The measure of strength of this footprint is the EIRP.

Forward Error Correction
FEC is a technique for improving the accuracy of data transmission. Excess bits are included in the out-going data stream so that error correction algorithms can be applied upon reception.

Frame
One complete TV picture, composed of two fields and a total of 525 and 625 scanning lines in NTSC and PAL systems, respectively.

Frequency
The number of vibrations per second of an electrical or electromagnetic signal expressed in cycles per second or Hertz.

Front Porch
The portion of the horizontal blanking pulse that precedes the horizontal sync pulse.

Gain
The amount of amplification of input to output power often expressed as a multiplicative factor or in decibels.

Gain-to-Noise Temperature Ratio (G/T)
The figure of merit of an antenna and LNA. The higher the G/T, the better the reception capabilities of an earth station.

Geostationary Orbit
See Clarke Belt.

GigaHertz (GHz)
1000 MHz or one billion cycles per second.

Global Beam
A footprint pattern used by communication satellites targeting nearly 40% of the earth's surface below. Many Intelsat satellites use global beams.

Ground Noise
Unwanted microwave signals generated from the warm ground and detected by a dish.

Hall Effect Sensor
A semiconductor device in which an output voltage is generated in response to the intensity of a magnetic field applied to a wire. In an actuator, the varying magnetic field is produced by the rotation of a permanent magnet past a thin wire. The pulses generated serve to count the number of rotations of the motor.

Hardline
A low-loss coaxial cable that has a continuous hard metal shield instead of a conductive braid around the outer perimeter. This type of cable was used in the pioneer days of satellite television.

Headend
The portion of an SMATV system where all desired signals are received and processed for subsequent distribution.

Heliax
A thick low-loss cable used at high frequencies; also known as hardline.

Hertz
An abbreviation for the frequency measurement of one cycle per second. Named after Heinrich Hertz, the German scientist who first described the properties of radio waves.

High Definition Television (HDTV)
An innovative television format having approximately twice the number of scan lines in order to improve picture resolution and viewing quality.

High Power Amplifier (HPA)
An amplifier used to amplify the uplink signal.

Horizontal Blanking Pulse
The pulse that occurs between each horizontal scan line and extinguishes the beam illumination during the retrace period.

Horizontal Sync Pulse
A 5.08 microsecond (4.7 microsecond in the PAL system) rectangular pulse riding on top of each horizontal blanking pulse. It synchronizes the horizontal scanning at the television set with that of the television camera.

Hum Bars
A form of interference seen as horizontal bars or black regions passing across the field of a television screen.

I Signal
One of the two color video signals which modulate the color subcarrier. It represents those colors ranging from reddish orange to cyan.

Impulse Pay-Per-View
Impulse pay-per-view (IPPV) is a feature of a decoder that allows an authorized subscriber to purchase a one-time scrambled program at will. IPPV shows are selected by a button on the decoder or its remote control unit.

Inclinometer
An instrument used to measure the angle of elevation to a satellite from the surface of the earth.

Interference
An undesired signal intercepted by a TVRO that causes video and/or audio distortion.

Insertion Loss
The amount of signal energy lost when a device is inserted into a communication line. Also known as "feed through" loss.

Interlaced Scanning
A scanning technique to minimize picture flicker while conserving channel bandwidth. Even and odd numbered lines are scanned in separate fields both of which when combined paint one frame or complete picture.

Intermediate Frequency (IF)
A middle range frequency generated after downconversion in any electronic circuitry including a satellite receiver. The majority of all signal amplification, processing and filtering in a receiver occur in the IF range.

INTELSAT
The International Telecommunication Satellite Consortium, a body of 154 countries working towards a common goal of improved worldwide satellite communications.

GLOSSARY

APPENDIX D

Isolator
A device that allows signals to pass unobstructed in one direction but which attenuates their strength in the reverse direction.

Isolation Loss
The amount of signal energy lost between two ports of a device. An example is the loss between the feed through port and the tap/drop of a top-off device.

Kelvin Degrees (°K)
The temperature above absolute zero, the temperature at which all molecular motion stops, graduated in units the same size as degrees Celsius (°C). Absolute zero equals -273 °C or -459 °F.

Kilohertz (kHz)
One thousand cycles per second.

Ku-Band
The microwave frequency band between approximately 11 and 13 GHz used in satellite broadcasting.

Latitude
The measurement of a position on the surface of the earth north or south of the equator measured in degrees of angle.

Line Amplifier
An amplifier in a transmission line that boosts the strength of a signal.

Line Splitter
An active or passive device that divides a signal into two or more signals containing all the original information. A passive splitter feeds an attenuated version of the input signal to the output ports. An active splitter amplifies the input signal to overcome the splitter loss.

Local Oscillator
A device used to supply a stable single frequency to an up-converter or a downconverter. The local oscillator signal is mixed with the carrier wave to change its frequency.

Longitude
The distance east or west of the prime meridian, $0°$, as measured in degrees.

Low Noise Amplifier (LNA)
A device that receives and amplifies the weak satellite signal reflected by an antenna via a feedhorn. C-band LNAs typically have their noise characteristics quoted as noise temperatures rated in degrees Kelvin. K-band LNA noise characteristics are usually expressed as a noise figure in decibels.

Low Noise Block Downconverter (LNB)
A low noise microwave amplifier and converter which downconverts a block or range of frequencies at once to an intermediate frequency range, typically 950 to 1450 MHz or 950 to 1750 MHz.

Low Noise Converter (LNC)
An LNA and a conventional downconverter housed in one weatherproof box. This device converts one channel at a time. Channel selection is controlled by the satellite receiver. The typical IF for LNCs is 70 MHz.

Magnetic Variation
The difference between true north and the north indication of a compass.

Master Antenna TV (MATV)
Broadcast receiving stations that use one or more high-quality centrally located UHF and/or VHF antennas which relay their signals to many televisions in a local apartment/condo or group-housing complex.

Match
The condition that exists when 100 percent of available power is transmitted from one device to another without any losses due to reflections.

Matching Transformer
A device used to match impedance between devices. A matching transformer is used, for example, when connecting a 75 ohm coax to a television 300 ohm input terminal.

MegaHertz (MHz)
One millions cycles per second.

Microprocessor
The central processing unit of a computer or control system, either on a single integrated (IC) circuit chip or on several ICs.

Microwave
The frequency range from approximately 1 to 30 GHz and above.

Mixer
A device used to combine signals together.

Modulation
A process in which a message is added or encoded onto a carrier wave. Among other methods, this can be accomplished by frequency or amplitude modulation, known as AM or FM, respectively.

Monochrome
A black and white television picture.

Mount
The structure that supports an earth station antenna. Polar and az-el mounts are the most common variety.

Multiple Analog Component (MAC) Transmissions
An innovative television transmission method which separates the data, chrominance and luminance components and compresses them for sequential relay over one television scan line. There are a number of systems in used and under development including A-MAC, B-MAC, C-MAC, D-MAC, D2-MAC, E-MAC and F-MAC.

Multiplexing
The simultaneous transmission of two or more signals over a single communication channel. The interleaving of the luminance and chrominance signals is one form of multiplexing, known as frequency multiplexing. MAC transmissions make use of time division multiplexing.

N-Connector
A low-loss coaxial cable connector used at the elevated C-band microwave frequencies.

GLOSSARY

NTSC
The National Television Standards Committee which created the standard for North American TV broadcasts.

NTSC Color Bar Pattern
The standard test pattern of six adjacent color bars including the three primary colors plus their three complementary shades.

Negative Picture Phase
Positioning the composite video signal so that the maximum level of the sync pulses is at 100% amplitude. The brightest picture signals are in the opposite negative direction.

Negative Picture Transmission
Transmission system used in North America and other countries in which a decrease in illumination of the original scene causes an increase in percentage of modulation of the picture carrier. When demodulated, signals with a higher modulation percentage have more positive voltages.

Noise
An unwanted signal which interferes with reception of the desired information. Noise is often expressed in degrees Kelvin or in decibels.

Noise Figure
The ratio of the actual noise power generated at the input of an amplifier to that which would be generated in an ideal resistor. The lower the noise figure, the better the performance.

Noise Temperature
A measure of the amount of thermal noise present in a system or a device. The lower the noise temperature, the better the performance.

Odd Field
The half frame of a television scan which is composed of the odd numbered lines.

Offset Feed
A feed which is offset from the center of a reflector for use in satellite receiving systems. This configuration does not block the antenna aperture.

Orthomode Coupler
A waveguide, generally a three-port device, that allows simultaneous reception of vertically and horizontally polarised signals. The input port is typically a circular waveguide. The two output ports are rectangular waveguides.

PAL
Phase Alternate Line. The European color TV format which evolved from the American NTSC standard.

Pad
A concrete base upon which a supporting pole and antenna can be mounted.

Path Loss
The attenuation that a signal undergoes in traveling over a path between two points. Path loss varies inversely as the square of the distance traveled.

Parabola
The geometric shape that has the property of reflecting all signals parallel to its axis to one point, the focal point.

Pay-Per-View
Pay-per-view is a method of purchasing programming on a per-program basis.

Persistence of Vision
The physiological phenomena whereby a human eye retains perception of an image for a short time after the image is no longer visible.

Phase
A measure of the relative position of a signal relative to a reference expressed in degrees.

Phase Distortion
A distortion of the phase component of a signal. This occurs when the phase shift of an amplifier is not proportional to frequency over the design bandwidth.

Picture Detail
The number of picture elements resolved on a television picture screen. More "crisp" pictures result as the number of picture elements is increased.

Polar Mount
An antenna mount that permits all satellites in the geosynchronous arc to be scanned with movement of only one axis.

Polarisation
A characteristic of the electromagnetic wave. Four senses of polarisation are used in satellite transmissions: horizontal; vertical; right-hand circular; and left-hand circular.

Positive Picture Phase
Positioning of the composite video signal so that the maximum point of the sync pulses is at zero voltage. The brightest illumination is caused by the most positive voltages.

Preamplifier
The first amplification stage. In an SMATV system, it is the amplifier mounted adjacent to an antenna to increase a weak signal prior to its processing at the headend.

Pre-emphasis
Increases in the higher frequency components of an FM signal before transmission. Used in conjunction with the proper amount of de-emphasis at the receiver, it results in combating the higher noise detected in FM transmissions.

PSD
An abbreviation for polarity selection device.

Primary Colors
Red, green and blue.

Prime Focus Antenna
A parabolic dish having the feed/LNA assembly at the focal point directly in the front of the antenna.

Q Signal
One of two color video signal components used to modulate the color subcarrier. It represents the color range from yellowish to green to magenta.

APPENDIX D

Radio Frequency

The approximately 10 kHz to 100 GHz electromagnetic band of frequencies used for man-made communication.

Raster

The random pattern of illumination seen on a television screen when no video signal is present.

Reed Switch

A mechanical switch which uses two thin slivers of metal in a glass tube to make and break electrical contact and thus to count pulses which are sent to the antenna actuator controller. The position of the slivers of metal is governed by a magnetic field applied by a bar or other type of magnet.

Reference Signal

A highly stable signal used as a standard against which other variable signals may be compared and adjusted.

Return Loss

A ratio of the amount of reflected signal to the total available signal entering a device expressed in decibels.

Retrace

The blanked-out line traced by the scanning beam of a picture tube as it travels from the end of any horizontal line to the beginning of either the next horizontal line or field.

SAW (Surface Acoustic Wave) Filter

A solid state filter that yields a sharp transition between regions of transmitted and attenuated frequencies.

Satellite Receiver

The indoors electronic component of an earth station which downconverts, processes and prepares satellite signals for viewing or listening.

Scanning

The organized process of moving the electron beam in a television picture tube so an entire scene is drawn as a sequential series of horizontal lines connected by horizontal and vertical retraces.

Scrambling

A method of altering the identity of a video or audio signal in order to prevent its reception by persons not having authorized decoders.

Screening

A metal, concrete or natural material that screens out unwanted TI from entering an antenna or a metal shield that prevents the ingress of unwanted RF signals in an electronic circuit.

Serrated Vertical Pulse

The television vertical sync pulse which is subdivided into six serrations. These sub-pulses occur at twice the horizontal scanning frequency.

Servo Hunting

An oscillatory searching of the feedhorn probe when use of inadequate gauge control cables results in insufficient voltage at the feedhorn.

Side Lobe

A parameter used to describe an antenna's ability to detect off-axis signals. The larger the side lobes, the more noise and interference an antenna can detect.

Single Channel Per Carrier (SCPC)

A satellite transmission system that employs a separate carrier for each channel, as opposed to frequency division multiplexing that combines many channels on a single carrier.

Signal-to-Noise Ratio (SNR)

The ratio of signal power to noise power in a specified bandwidth, usually expressed in decibels.

Skew

A term used to describe the adjustment necessary to fine tune the feedhorn polarity detector when scanning between satellites.

Slant Range

The distance that a signal travels from a satellite to a TVRO.

Snow

Video noise or sparklies caused by an insufficient signal-to-noise input ratio to a television set or monitor.

Solar Outage

The loss of reception that occurs when the sun is positioned directly behind a target satellite. When this occurs, solar noise drowns out the satellite signal and reception is lost.

Sparklies

Small black and/or white dashes in a television picture indicating an insufficient signal-to-noise ratio. Also known as "snow."

Spherical Antenna

An antenna system using a section of a spherical reflector to focus one or more satellite signals to one or a series of focal areas.

Splitter

A device that takes a signal and splits it into two or more identical but lower power signals.

Subcarrier

A signal that is transmitted within the bandwidth of a stronger signal. In satellite transmissions a 6.8 MHz audio subcarrier is often used to modulate the C-band carrier. In television, a 3.58 MHz subcarrier modulates the video carrier on each channel.

Surface Acoustic Wave

A sound or acoustic wave traveling on the surface of the optically polished surface of a piezoelectric material. This wave travels at the speed of sound but can pass frequencies as high as several gigahertz. See SAW Filter.

Synchronizing Pulses

Pulses imposed on the composite baseband video signal used to keep the television picture scanning in perfect step with the scanning at the television camera.

TVRO

A television receive-only earth station designed only to receive but not to transmit satellite communications.

Tap

A device that channels a specific amount of energy out of the main distribution system to a secondary outlet.

GLOSSARY

APPENDIX D

Television Receive-Only (TVRO)
A satellite system that can only receive but not transmit signals.

Terrestrial Interference (TI)
Interference of earth-based microwave communications with reception of satellite broadcasts.

Tilt
The uneven attenuation of a broadband signal as it travels through a coaxial cable. In general, attenuation increases as signal frequency increases.

Thermal Noise
Random, undesired electrical signals caused by molecular motion, known more familiarly as noise.

Trace
The movement of the electron beam from left to right on a television screen.

Threshold
A minimal signal to noise input required to allow a video receiver to deliver an acceptable picture.

Transponder
A microwave repeater, which receives, amplifies, downconverts and retransmits signals at a communication satellite.

Trap
An electronic device that attenuates a selected band of frequencies in a signal. Also known as a notch filter.

UHF
Ultrahigh frequencies ranging from 300 to 3,000 MHz. North American TV channels 14 through 83. European TV channels 21 to 69.

Upconverter
A device that increases the frequency of a transmitted signal.

Uplink
The earth station electronics and antenna which transmits information to a communication satellite.

VHF
Very high frequencies in the range from 54 MHz to 216 MHz, NTSC TV channels 2 through 13.

VSWR (Voltage Standing Wave Ratio)
The ratio between the minimum and maximum voltage on a transmission line. An ideal VSWR is 1.0. Ghosting can result as the VSWR increases. It is also a measure of the percentage of reflected power to the total power impinging upon a device.

Vertical Blanking Pulse
A pulse used during the vertical retrace period at the end of each scanning field to extinguish illumination from the electron beam.

Vertical Sync Pulse
A series of pulses which occur during the vertical blanking interval to synchronize the scanning process at the television with that created at the studio. See also Serrated Vertical Pulse.

VHF
Very high frequency range from 30 to 300 MHz

Video Signal
That portion of the transmitted television signal containing the picture information.

Voltage Tuned Oscillator (VTO)
An electronic circuit whose output oscillator frequency is adjusted by voltage. Used in downconverters and satellite receivers to select from among transponders.

Video Monitor
A television that accepts unmodulated baseband signals to reproduce a broadcast.

APPENDIX E. AMERICAN CHANNELS, WAVELENGTHS & BANDWIDTHS

Channel	Video Frequency (MHz)	Wavelength (Inches)
SUB CHANNELS		
01	15.75	749.7
02	21.75	542.9
03	27.75	425.5
04	33.75	349.9
05	39.75	297.1
VHF-LO CHANNELS		
2	55.25	213.7
3	61.25	192.8
4	67.25	175.6
5	77.25	152.9
6	83.25	141.8
FM	88.0-108.0	134.2-109.3
VHF-HI CHANNELS		
7	121.25	97.4
8	127.25	92.8
9	133.25	88.6
10	139.25	84.8
11	145.25	81.3
12	151.25	78.1
13	157.25	75.1
SUPERBAND CHANNELS		
J	217.25	54.5
K	223.25	52.9
L	229.25	51.5
M	235.25	50.2
N	241.25	49.0
O	247.25	47.8
P	253.25	46.6
Q	259.25	45.6
R	265.25	44.5
S	271.25	53.5
T	277.25	42.6
U	283.25	41.7
V	289.25	40.8
W	295.25	40.0
HYPERBAND CHANNELS		
AA	301.25	39.2
BB	307.25	38.4
CC	313.25	37.7
DD	319.25	37.0
EE	325.25	36.3
FF	331.25	35.7
GG	337.25	35.0
HH	343.25	34.4
II	349.25	33.8
KK	361.25	32.7
LL	367.25	32.2

Channel	Video Frequency (MHz)	Wavelength (Inches)
MM	373.25	31.6
NN	379.25	31.1
OO	385.25	30.7
PP	391.25	30.2
QQ	397.25	29.7
RR	403.25	29.3
SS	409.25	28.9
TT	415.25	28.4
UU	421.25	28.0
VV	427.25	27.6
WW	433.25	27.3
UHF CHANNELS		
14	471.25	25.1
15	477.25	24.7
16	483.25	24.4
17	489.25	24.1
18	495.25	23.8
19	501.25	23.6
20	507.25	23.3
21	513.25	23.0
22	519.25	22.7
23	525.25	22.5
24	531.25	22.2
25	537.25	22.0
26	543.25	21.7
27	549.25	21.5
28	555.25	21.3
29	561.25	21.0
30	567.25	20.8
31	573.25	20.6
32	579.25	20.4
33	585.25	20.2
34	591.25	20.0
35	597.25	19.8
36	603.25	19.6
37	609.25	19.4
38	615.25	19.2
39	621.25	19.0
40	627.25	18.8
41	633.25	18.7
42	639.25	18.5
43	645.25	18.3
44	651.25	18.1
45	657.25	18.0
46	663.25	17.8
47	669.25	17.6
48	675.25	17.5
49	681.25	17.3
50	687.25	17.2
51	693.25	17.0
52	699.25	16.9
53	705.25	16.7
54	711.25	16.6

APPENDIX F. EQUIPMENT MANUFACTURERS

MANUFACTURERS OF ENCODING EQUIPMENT

GENERAL INSTRUMENTS CORP
2200 ByBerry Road
Hatboro, PA 19040
(215) 674-4800

HAMLIN
13610 First Avenue South
Seattle, WA 98168
(206) 246-9330

M/A COM - LINKABIT
3033 Science Park Drive
San Diego, CA 9212
(619) 457-2340

OAK COMMUNICATIONS
Satellite Systems Division
P.O. Box 517
Crystal Lake, IL 60014
(815) 459-5000

OAK COMMUNICATIONS
Cable Division
16935 West Bernardo Drive
Rancho Bernardo, CA 92127
(619) 485-9880

SCIENTIFIC ATLANTA
4356 Communications Drive
Norcross, GA 30093
(404) 925-5778

ZENITH ELECTRONICS CORP
1000 Milwaukee Avenue
Glenview, IL 60025
(312) 391-8338

NORTH AMERICAN SATELLITE TV EQUIPMENT MANUFACTURERS

A&E TRAVEL-SAT
44389 Portofino Court
Palm Desert, CA 92260
Telephone: (619) 568-0666
FAX: (619) 341-6565
RV Systems

AJAK INDUSTRIES, INC.
112 South Robinson
Florence, CO 81226
Telephone: (719) 784-6301
FAX: (719) 784-6763
Horizon-to-horizon Mounts

ANDERSON MANUFACTURING
3125 North Yellowstone Highway
Idaho Falls, ID 85301
Telephone: (208) 523-6460
Antennas & Mounts

ANDREW CORPORATION
10500 West 153rd Street
Orlando Park, IL 60462
(312) 349-3300
Commercial Antennas

ASP/RCI
10679 Windmer
Lexena, KS 66215
Telephone: (913) 469-4125
TI Filters

ASTRA
792 East 93rd Street
Brooklyn, NY 11236
Telephone: (718) 346-1200
Mesh Antennas & Mounts

AVCOM OF VIRGINIA
500 Southlake Blvd.
Richmond, VA 23236
Telephone: (804) 794-2500
FAX: (804) 794-8284
Satellite Receivers & Test Equipment

BESTIN MANUFACTURING
2422 North Lee Avenue
South El Monte, CA 91733
Telephone: (818) 444-8170
Satellite Antenna Systems

BIRDVIEW SATELLITE SERVICE
1407 East Spruce
Olathe, KS 66061
Telephone: (913) 824-6240
Service and Repair of Equipment

CALIFORNIA AMPLIFIER
460 Calle San Pablo
Camarillo, CA 93010-8506
Telephone: (805) 987-9000
FAX: (805) 987-8359
Low Noise Amplifiers

CHANNEL MASTER/DIV. OF AVNET
P.O. Box 1416
Smithfield, NC 27577
Telephone: (919) 934-9711
Complete Satellite Systems

CHAPARRAL COMMUNICATIONS
2450 North 1st Street
San Jose, CA 95131
(408) 435-1530
Receivers, Feedhorns

APPENDIX G

COAST HITECH CORPORATION
6021-F North Figueroa Street
Los Angeles, CA 90042
Telephone: (800) 782-8344
　　　　　　(213) 255-5085
FAX: (213) 255-4708
Integrated LNB/Feedhorn

COMTECH ANTENNA CORP
3100 Communications Road
St. Cloud, FL 32769
(305) 892-6111
Antennas

CT SYSTEMS / WAVETEK RF
5808 Churchman Bypass
Indianapolis, IN 46203
Telphone: (317) 787-5721
　　　　　　(800) 245-6356
Test Equipment

DH SATELLITE, INC
P.O. BOX 239
PRAIRIE DU CHIEN, WI 53821-9990
Telephone: (800) 392-6884
Satellite Antennas

DX COMMUNICATIONS, INC.
10 Skyline Drive
Hawthorne, NY 10532
Telephone: (914) 347-4040
Commercial Receivers

DIAMOND PERFORATED METALS
28976 Hoplins Street
Hayward, CA 94545
Telephone: (415) 783-7444
Perforated Metals

E-Z TRENCH MANUFACTURING CO
Route 3, Box 78-B
Loris, SC 29569
Telephone: (803) 756-6444
Trenching Equipment

EARTHBOUND, INC.
3220 West Topeka
Topeka, KS 66611
Telephone: (913) 266-4944
Mounting Systems

ECHOSPERE CORP.
90 Inverness Circle East
Englewood, CO 80112
Telephone: (303) 799-8222
Complete Systems

FOCII MANUFACTURING CO
1324 South Kansas
Topeka, KS 66612
Telephone: (913) 234-6721
Test Equipment

FORD AEROSPACE
3939 Fabian Way
Palo Alto, CA 94303
Telephone: (415) 852-6980
Satellites

FORT WORTH TOWER COMPANY
1901 East Loop 820 South
Fort Worth, TX 76112
Telephone: (817) 457-3060
Antenna Support Towers

FUJITSU GENERAL
P.O. Box 2101
Chatsworth, CA 91313-2101
Telephone: (818) 341-5400
FAX: (818) 718-2938
Satellite Receivers

GARDINER COMMUNICATIONS
3605 Security Street
Garland, TX 75042
(214) 348-4747
Low Noise Amplifiers

GTE SATELLITE CORPORATION
170 Old Meadow Road
McLean, VA 22102
(703) 790-7700
Satellites

GENERAL INSTRUMENT CORP.
6262 Lust Boulevard
San Diego, CA 92121
Telephone: (619) 535-2545
Satellite Receivers and Decoders

GOURMET ENTERTAINING
3915 Carnavon Way
Los Angeles, CA 90027
Telephone: (213) 666-2728
Alignment Tools

HARRIS CORPORATION
Satellite Communications Division
P.O. Box 1700
Melbourne, FL 32901
(305) 724-3689
Commercial Systems

HERO COMMUNICATIONS
2290 West 8th Avenue
Hialeah, FL 33010
Telephone: (305) 887-3203
Satellite Antennas

HOUSTON TRACKER SYSTEMS.
90 Inverness Circle East
Englewood, CO 80112
Telephone: (303) 790-4445
Satellite Receivers

HUGHES AIRCRAFT COMPANY
P.O. Box 92919
Los Angeles, CA 90009
Satellites

INT'L ELECTRONIC WIRE & CABLE
520 Business Center Drive
Mt. Prospect, IL 60056
Telephone: (312) 299-0021
Cables

KAUL-TRONICS, INC.
1140 Sextonville Road
P.O. Box 637
Richland Center, WI 53581
Telephone: (608) 647-8902
Satellite Antennas

LEAMING INDUSTRIES
180 McCormick Avenue
Costa Mesa, CA 92626
Telephone: (714) 979-4511

MB SALES
P.O. Box 787
Wauconda, IL 60084
Telephone: (312) 526-5310

MICRODYNE CORP
P.O. Box 7213
Ocala, FL 32672
(904) 687-4633
Commercial Systems

MICROWAVE FILTER COMPANY
6743 Kinne Street
East Syracuse, NY 13057
Telephone: (315) 437-3953
TI Filters

MICROWAVE SYSTEMS ENG.
4221 East Raymond
Phoenix, AZ 85040
Telephone: (602) 437-9040
Commercial Systems

MIRALITE CORPORATION
4050 Chandler
Santa Ana, CA 92704
Telephone: (717) 641-7000
Satellite Antennas

MULIPLEX TECHNOLOGY
251 Imperial Highway
Fullerton, CA 92635
Telephone: (714) 680-5848
Video Equipment

NATIONAL ADL
255-G Easy Street
Simi Valley, CA 93065
Telephone: (805) 526-5249
Feedhorns

NEXUS ENGINEERING
4181 McConnell Drive
Burnaby, BC V5A 3J7 Canada
Telephone: (206) 644-2371
Commercial Electronics

NORSAT INTERNATIONAL, INC.
707 Johnson Way
Blaine, WA 98230
Telephone: (604) 597-6200
Satellite Receivers and LNBs

MANUFACTURERS

APPENDIX G

OAK COMMUNICATIONS, INC.
SATELLITE SYSTEMS
100 South Main Street
Crystal Lake, IL 60014
Telephone: (815) 459-5000

ODOM ANTENNAS
P.O. Box 1017
2502 DeWitt Henry Drive
Beebe, AR 72012
Telephone: (501) 882-6485
Satellite Antennas

ORBITRON
351 South Peterson Street
Spring Green, WI 53588
Telephone: (608) 588-2923
Satellite Antennas and Mounts

PTS CORPORATION
5233 South Highway 37
Bloomington, IN 47459
Telephone: (812) 824-9331
Service and Repair

PANAREX ELECTRONICS
13012 Saticoy Street, #4
North Hollywood, CA 91605
Telephone: (818) 764-0375
Ku-Band Satellite Systems

PANASONIC
One Panasonic Way 2A-2
Secaucus, NJ 07094
Telephone: (201) 348-7846
Satellite Receivers

PARACLIPSE, INC.
3711 Meadowview Drive
Redding, CA 96002
Telephone: (916) 365-9131
Satellite Antennas

PHANTOM ENGINEERING, INC.
16840 Joleen Way, Bldg. E-3
Morgan Hill, CA 95037
(408) 779-1616

PICO MACOM, INC.
12500 Foothill Boulevard
Lakeview Terrace, CA 91342
Telephone: (800) 421-6511
          (818) 897-0028
Electronic Components

PRECISION R.V. SATELLITES
5468 North Salinas Avenue
Fresno, CA 93722
Telephone: (209) 275-1780
RV Systems

PRO BRAND INTERNATIONAL, INC.
1900 West Oak Circle
Marietta, GA 30062
Telephone: (404) 423-7072
FAX: (404) 423-7075
Actuators

RCA AMERICAN COMMUNICATIONS
400 College Road, Suite E
Princeton, NJ 08540
(609) 734-4000
Satellites

RCA ASTRO-ELECTRONICS
P.O. Box 800, MS 54
Princeton, NJ 08540
(609) 426-2711
Satellites

R.L. DRAKE COMPANY
9111 Springboro Pike
Miamisburg, OH 45342
Telephone: (513) 866-2421
Satellite Receivers & LNBs

RMS ELECTRONICS, INC.
50 Antin Place,
Bronx, NY 10462
(212) 892-1000

ROHN
6718 West Plank Road
Peoria, IL 61656
Telephone: (309) 697-4400
Antenna Towers & Roof Mounts

SEA TEL
1035 Shary Court
Concord, CA 94518
Telephone: (415) 798-7979
Marine Satellite Systems

SATELLITE TECHNICAL SERVICES
1409 Washington Avenue
St. Louis, MO 63103
Telephone: (314) 567-0304
Satellite Receivers

SEAVEY ENGINEERING ASSOCIATES
155 Kind Sreet, P.O. Box 44
Cohasset, MA 02025
Telephone: (617) 383-9722
Feedhorns

SCIENTIFIC ATLANTA, INC.
P.O. Box 105027
Atlanta, GA 30348
Telephone: (404) 492-1111
Commercial Systems

STANDARD COMMUNICATIONS CORP
P.O. Box 92151
Los Angeles, CA 90009
Telephone: (800) 824-7766
Commercial Systems

SUPERIOR ANTENNA MANUFACTURING
Route 2, Box 465
Judsonia, AR 72081
Telephone: (501) 729-3103
Satellite Antennas

TEE-COMM
775 Main Street East
Milton, ON L9T 3Z3
Canada
Telephone: (416) 878-8181
Satellite Receivers

THOMPSON SAGINAW BALL SCREW CO
P.O. Box 9550
Saginaw, MI 48608
Telephone: (517) 776-5111
Actuators

TOSHIBA AMERICA CONSUMER PROD.
1010 Johnson Drive
Buffalo Grove, IL 60089-6900
Telephone: (312) 541-9400
FAX: (708) 541-1927
Satellite Receivers

TRAVEL-STAR
3100 West Segerstrom
Santa Ana, CA 92704
Telephone: (714) 540-6444

TSIGER PLANAR, INC.
2448 Waynoka Road
Colorado Springs, CO 80915
Telephone: (719) 591-7900
Satellite Antennas

UNIDEN AMERICA CORP.
4700 Amon Carter Blvd.
Ft. Worth, TX 76155
Telephone: (817) 858-3300
Satellite Receivers & LNBs

UNIMESH
Highway 367, Box 338
Ward, AR 72176
Telephone: (800) 843-6517
          (501) 843-6517
Satellite Antennas

VIDEO LINK
12950 Bradley Avenue
Sylmar, CA 91342
(818) 362-0353
Remote Controllers

VON WEISE GEAR COMPANY
500 Chesterfield Center
St. Louis, MO 63017
Telephone: (314) 532-3505
Actuators & Gearmotors

WINEGARD COMPANY
3000 Kirkwood Street
Burlington, IA 52601
Telephone: (319) 753-0121
Satellite Antennas

ZENITH ELECTRONICS CORP.
11000 West Seymour Avenue
Franklin Park, IL 60131
Telephone: (708) 671-2043
Satellite Receivers

WORLD SATELLITE TV AND SCRAMBLING METHODS
The Technician's Handbook – 3rd Revised Edition

by Frank Baylin, Richard Maddox, John McCormac

This thorough text is a must-buy for technicians, satellite professionals and do-it-yourselfers. The design, operation and repair of satellite antennas, feeds, LNBs & receivers/modulators are examined in detail. An in-depth study of scrambling methods and broadcast formats is the backdrop to a discussion of all current American and European satellite TV technologies including the VideoCipherII, Oak Orion, FilmNet, Sky Channel, EuroCypher, D2 MAC, BSB and Teleclub Payview III. Circuit and block diagrams of all components are presented and clearly explained throughout the handbook. This information is a prelude to the chapters on troubleshooting and setting up a test bench. This expert guidance on testing, servicing and tuning is complimented by a wealth of detailed illustrations.

PARTIAL CONTENTS: The Satellite System. Early Designs. **Antennas. Feeds** Polarization Formats. Polarity Selection & Control. Troubleshooting. **Low Noise Amplifiers.** Waveguides. Troubleshooting. Downconversion Methods. **Cables and Connectors.** Coax. Insulated Wire. Custom Cables. Sealing. **Positioning Systems**. Basic Motor Controllers. Feedback Circuits. End Point Limiters. Troubleshooting. **Power Supplies.** Actuator Controllers. IC Regulators. Troubleshooting. **IF Circuits.** Amplifiers. Bandpass Filters. Limiter Circuits. Troubleshooting. **Video Processing.** Demodulator Circuits. Video Processing. **Audio Processing.** Audio Subcarrier Specifications. Typical Audio Circuitry. Other Audio Detection Circuits. Stereo Broadcast Methods. Audio Companding. Dolby Noise Reduction. Troubleshooting & Aligning Audio Circuits. **RF Modulators.** Broadcast Formats. RF Modulator Circuits. Crystal Modulators. RF Interference. Troubleshooting. **Miscellaneous Receiver Circuits.** Indicator & LED Circuits. LED Read-out Displays. Remote Controls. **Complete Circuit Descriptions** European & American Receiver Circuits and Comparisons. **Television Operation. Broadcast Formats** NTSC, PAL, SECAM & MAC. Digital Audio. The Digit 2000 TV Receiver. Teletext. **Basic Scrambling Methods.** Video Techniques. Audio Techniques. Digital & Analogue Methods. DES Algorithm. Smart Cards. **Pioneer Scrambling Systems.** Telease/SAVE. Zenith SSAVI. **Case Studies.** RITC Discrete. OAK Orion. IRDETO. Sound-in-Sync. Lorentz PCM2. Filmnet. Telease SAVE. Payview III, VideoCrypt. VideoCipher II and II Plus. MAC Variants **Decoder Connections.** Loopthough and Chaining. **Future Developments.** European & Worldwide Standards. Security & Cost. Legalities. **Setting Up a Test Bench.** Equipment. Synthesized Tuned TV. **Troubleshooting.** Field Checking Microwave Components. Substitution. Troubleshooting Microprocessors. **Specialized Components.** Diodes. Transistors. ICs. Hybrid Components. SAWs Filters. **Appendices.** Active Component Guide. Equations. Glossary. Channel Allocations. Manufacturers.

356 pages / 8.5 x 11" / over 200 photos, diagrams, wiring schematics / 16 tables / appendices, index / 0-917893-15-8

SATELLITE & CABLE TV
Scrambling & Descrambling - 2nd edition

by Brent Gale and Frank Baylin

An important beginners aid in understanding encryption technology for satellite and cable television broadcasts. Fundamental issues and concepts are carefully explained in a comprehensive view of how televisions operate, a topic critical to understanding modern scrambling systems. Various methods used to relay stereo over satellite links and cable lines, including the over-the-air MTS stereo standard, are studied. The principles and techniques of encryption systems are examined in detail. A clear explanation of the innovative B-MAC broadcast system in relation to worldwide television standards is also provided. A wide selection of scrambling systems and equipment is explored, with particular attention to the more sophisticated cable television addressable systems and to the Oak Orion and M/A COM VideoCipher encoding/decoding devices.

PARTIAL CONTENTS: Introduction. Underlying Concepts. Principles of Television. Scrambling Techniques. Cable TV Scrambling Systems. Satellite TV Scrambling Systems. APPENDICES: Worldwide Broadcast Formats & Television Channel Structures. The Decibel Notation. Glossary. Reference Materials. Manufacturers/Distributors.

280 pages / 7 x 9-1/4" / 120 photos and illustrations / appendices, glossary / 0-917893-07-7

Congratulations to the authors of ... this much needed and excellent publication! ... the information contained in this new book is accurate, concise and timely ... the authors have created a work that artfully guides the reader through all facets of this industry. Steve Johnson, American Wireless Systems

WIRELESS CABLE and SMATV
(Replaces "Satellite, Off-Air & SMATV")

by Frank Baylin and Steve Berkoff

A comprehensive study of the new broadcast method, Wireless Cable, and the closely related field of satellite master antenna TV (SMATV) systems. This thorough manual clearly presents the concepts behind private cable systems as well as technical details of construction and operation. Private cable systems are installed in apartment complexes, hotels and motels, condominiums, hospitals, mobile home parks, and auditoriums as well as in many other multi-user environments.

The book explores the background and history of this rapidly evolving field and outlines the steps required to legally purchase and resell satellite entertainment for profit. Three chapters are devoted to the details of the site survey, planning and design phases of a private cable system. Off-air and satellite headends and all components from antennas to processing and mixing electronics are studied in detail. The chapter on distribution systems explores the components required to supply a high quality signal to every television set. Numerous examples are provided as illustrations of each stage of design. Following a treatment of the basics of bidding projects, the construction and installation tasks are detailed and the critical choice between in-house or subcontracted labor is discussed. Proven methods of locating and managing competent subcontractors are explained. Complex design issues such as inserting locally originated signals, two-way services and satellite audio reception are also studied. The chapter on systems operations presents methods to manage one or more systems as well as a logical approach to troubleshooting.

SECTION I. BASICS and SMATV SYSTEMS. Overview and History of Private Cable TV. A Brief History of Satellite Television. The Development of Community Antenna TV Systems. SMATV versus Franchised Cable TV Systems. The Development of SMATV and Wireless Cable TV. The Advantages of SMATV. **Economics, Sales Contracts and Regulations.** Economic Options for Installers, Operators and Owners. Sales Contracts. Regulations. **Programming.** Available Satellite Programming. The Mechanics of Purchasing Programming. Current Trends and Issues in the U.S. **The Site Survey and Planning.** The Site Survey Process. Background Technical Information. The Satellite Programming Survey. The TVRO Site Survey. Avoiding Terrestrial Interference. The Off-Air Site SUrvey. Off-Air Interference and Reception Problems. Headend Survey. Distribution System Survey. **Headend System Design.** The Basics of Headend Design. Off-Air Antennas. Off-Air Electronic Components. Solving Off-Air Signal Level Problems. Satellite TV Reception Components. Designing the Satellite Headend. A Strategy for Combating TI. Examples of Headend Designs. **Distribution System Design..** The Basic Components of Distribution Systems. Cable TV Design Issues Relating to SMATV Systems. Television Input SIgnal Requirements. Television Input Signal Requirements. Midband and Higher Frequency Distribution Systems. Sample System Designs and Calculations. 18 GHz Microwave Line-of-Sight Transmission Systems. Laser Links. **Project Bidding.** Bid Preparation. The Bid Document. Negotiations. **Construction and Installation.** The Permit Process. Subcontracting the Construction. In-House Construction. Local Origination and Additional SMATV Services. Stereo Audio Services. **System Operations.** The Customer Relations Office. Computer Systems. The Technical Support Group. The Overall Management Structure. Marketing The Programming. **Troubleshooting and Test Instruments.** Troubleshooting. A Catalog of System Problems. The Field Strength Meter. The Spectrum Analyzer. Point-by-Point Diagnostics. **SECTION II. WIRELESS CABLE /MMDS SYSTEMS. Wireless Cable System Overview.** Technical Overview. History. Frequency Allocations, Channel Availability and Regulations. **Wireless Cable Components.** The Transmitter Facility. Signal Processing and Transmission. Operations Facilities. Receive Sites. Single and Multi-Unit Receive Site Configurations. Viewer Quality Assessment and S/N Ratio. **Wireless Signal Coverage.** Wireless Coverage and Limitations. Coverage Predictions. Extended Reception Techniques. **Receive-Site Installation.** Installation Planning and Site Survey. Mast, Antenna and Downconverter Installation. Grounding and Weatherproofing. Cable and Converter Installation. Customer Education. Service Procedures and Safety. Safety. **System Planning and Operation.** The Business Plan. System Operations Planning. **A Legal Perspective on Wireless Cable in the United States.** An Historical Perspective. The Application Process for an MDS Operation. Recent Trends. The WIRELESS DEAL in the U.S. The Wireless Deal. The Wireless Cable Association International, Inc. **SECTION III. PRIVATE CABLE SECURITY SYSTEMS.** Private Cable Security. Wireless Cable Security Systems. SMATV Security Systems. **Appendix A. Additional Example of Distribution Systems. Appendix B. Equations and Technical Details.** Satellite and Cable TV Equations. Wireless Cable System Parameters. **Appendix C. Programmers. Appendix D. Satellite TV Guides and Publications. Appendix E. Off-Air and Cable TV Channels. Appendix F. Glossary. Appendix G. Manufacturers.** Manufacturers of SMATV Components. Manufacturers of Wireless Components.

386 pages / 8.5 x 11 / photos, illustrations, diagrams, tables / appendices, glossary / 0-917893-17-4

One of the best books on the subject that I've read to date...
John McCormac – well-known author, columnist and hacker

THE "HOW-TO" OF SATELLITE COMMUNICATIONS

by Dr. Joseph Pelton

This excellent book by a seasoned veteran of the industry is a brilliant excursion through the world of satellite communications. For any serious user of satellite services, it gives even a novice a knowledge of how to design, evaluate and purchase any type of data, audio or video services. It covers topics including:

- How geosynchronous satellites work
- The design, construction or purchase of satellite capacity
- The fundamentals of earth station operation
- Global and ISDN standards
- One-way and two-way spread spectrum VSATs and interactive technology
- How and why to buy satellite networks and services
- Voice and data networks
- Video services and digital television
- The future of satellite communications

250 pages / photos / 5.5 x 8.5" / illus / index / 0-936361-24-7

HIDDEN SIGNALS ON SATELLITE TV
Revised 3rd Edition

by Thomas P. Harrington

The third, updated edition is devoted entirely to the hidden, non-video services relayed by communication satellites, a topic of interest to satellite buffs worldwide. Most satellites and their transponders are known by their video transmissions and only enthusiasts are aware that satellites also transmit large amounts of non-video information including

- audio world news services
- news teletypewriter channels
- high speed, multiplex data systems
- long distance telephone circuits
- teletext and teletype services
- stock market and commodity news reports
- telephone and business data channels
- audio subcarrier and single channel per carrier (SCPC) audio services

While many of these services, such as SCPC, can be received with simple, readily-obtainable equipment, others require microprocessors to detect intelligent signals. Most of the well-known audio signals as well as the hi-tech services are outlined and explained in this book.

PARTIAL CONTENTS: 1. Understanding the Satellite Signals. 2. Audio Subcarriers on Satellite TV (SBC/FM). 3. Satellite Telephone Systems (SSB/FDM). 4. Single Channel Per Carrier (SCPC). 5. Satellite Networks. 6. Teletex Basics. 7. Other VBI Services. 8. Misc. Satellite Services. 9. Ku-band and Hidden Signals. Glossary. Appendix: Position charts for world satellites, C- and Ku-band.

234 pages / 8.5 x 11" / over 200 photos, illustrations, tables / appendix, glossary / 0-91661-04-0

Nuestro Libro de Mayor Venta en Inglés y en Español
Si Ud. está interesado en sistemas de satélites domésticos, ya sea para la venta o para su uso personal, le recomiendo que compre y lea este libro. Los autores han producido una obra que es fácil de leer para los aficionados sin prescindir la información técnica que necesitan los instaladores e ingenieros. – JLC VIDEO SPECIALIST NEWSLETTER

TELEVISION DOMESTICA VIA SATELITE
Manual de Instalación y de Localización de Averías

Una fuente de información muy valiosa tanto para los dueños de sistemas de televisión por satélite domésticos así como para los instaladores profesionales. Con más de 45.000 ejemplares publicados, este completo libro de referencia se ha transformado en el estándar de la industria. Examina en forma detallada los principios de la comunicación por satélite y la operación de las estaciones terrestres, explica cómo seleccionar y evaluar componentes de TV por satélite, expone los procedimientos de instalación incluyendo conexiones de múltiples receptores y televisores y antenas grandes, y delinea una estrategia detallada para localizar averías en cualquier sistema de TV por satélite. El manual es una referencia completa que explica en forma clara conceptos tales como los formatos transmisor-receptor medio y la polarización circular. Es una herramienta muy valiosa que presenta todos los detalles necesarios para instalar, hacer funcionar y mantener un sistema de satélite.

TABLA DE MATERIAS: Vistazo a la Teoría de Comunicación por Satélite. Funcionamiento de los Diferentes Componentes. Elección del Equipo de TV Vía Satélite. Cómo Instalar Correctamente un Sistema de Televisión vía Satélite. Sistemas de Distribución y de TV vía Satélite con Múltiples Receptores. Sistemas y Antenas Grandes. Localización de Averías. Profesionalismo. Apéndices.

341 páginas / 8-1/2 x 11" / 300 ilustraciones, fotografías y tablas / apéndices, glosario / 0-917893-03-4

TAMBIÈN EM PORTUGUESE

THE HOME SATELLITE TV INSTALLATION VIDEO TAPE

The perfect companion to our best-selling *Home Satellite TV Installation and Troubleshooting Manual*. Designed to make owning and operating your satellite TV system a pleasure, it demonstrates how to install a system as well as how to keep it tuned up for best reception. An excellent instructional tool, the video demonstrates professionally accepted satellite TV installation techniques, details how to perform a site survey, detect and avoid terrestrial interference, install the pole and antenna mount, assemble the various components, complete the necessary electrical connections and track the arc of satellites. Illustrated with NASA footage, this high-quality video follows the sequence of the *Manual* so viewers can benefit from the carefully coordinated text and visual aid.

43 minutes / VHS or Beta / 0-917893-04-2

CINTA MAGNETICA DE VIDEO
SOBRE LA INSTALACION DE LA TELEVISION VIA SATELITE

El complemento perfecto de nuestro libro de mayor venta. Está diseñado para hacer que poseer y operar un sistema de TV por satélite sea un placer. Demuestra cómo instalar un sistema y cómo mantenerlo a punto para conseguir la mejor recepción. Este video es una excelente ayuda didáctica, demuestra técnicas de instalación aceptadas por los profesionales, explica en detalle cómo efectuar una exploración de emplazamiento, cómo detectar y evitar la interferencia terrestre, cómo instalar el mástil y el soporte de la antena, cómo armar los diferentes componentes, cómo efectuar las conexiones eléctricas necesarias y cómo seguir el arco de los satélites. Este video de alta calidad incluye tomas de la NASA y sigue la misma secuencia del Manual para que los espectadores se beneficien de la coordinación entre las imágenes y el texto.

45 minutos / VHS o Beta / 02-0-917893-04-2

Our Best-Seller – Now in Arabic and Portuguese
The Home Satellite TV "Bible"

HOME SATELLITE TV INSTALLATION & TROUBLESHOOTING MANUAL – 3rd Revised Edition

This book is so informative and well written that anyone considering buying a system or who currently owns one would benefit from reading it. Every aspect of satellite TV is covered completely
– STV Magazine

The completely revised third edition is an invaluable sourcebook for owners of home satellite TV systems and professional installers alike, this comprehensive reference has become the industry standard with more than 53,000 copies in print. It thoroughly explores the fundamentals of satellite communications and earth station operation, explains how to select and evaluate satellite television components, details installation procedures including multiple receiver and multiple-television hookups and large antennas, and sets out a complete strategy for troubleshooting any satellite TV system. The *Manual* is a complete reference that clearly explains such subjects as half transponder formats and circular polarization. An excellent working tool, it presents all the details anyone needs to install, operate and maintain a satellite system.

PARTIAL CONTENTS: Satellite Communication Theory. System Components. Terrestrial Interference. Selecting New Satellite TV Equipment. Installing Satellite TV Systems. Upgrading a System. Multiple Receiver Systems. Troubleshooting. Large Antenna Systems. **APPENDICES:** Decibel Notation. Satellite Footprints. Satellite TV Equations. Basic Satellite-Finding Program. Glossary of Terms. Reference Materials.

324 pages / 85 x 11 inches / over 300 illustrations, photographs and tables / appendices, glossary / 0-917893-12-3

INSTALL, AIM and REPAIR YOUR SATELLITE TV SYSTEM
2nd Revised Edition

by Frank Baylin

This booklet, a shortened version of *The Home Satellite TV - Installation and Troubleshooting Manual*, explores how to install a satellite TV system, aim the dish at the arc of satellites as well as how to troubleshoot and repair the system if a problem arises. It also covers the periodic maintenance work that is required to keep a system tuned up and aligned onto the arc of satellites.

64 pages / 8.5 x 11" / 48 figures & photos, 3 tables / 0-917893-16-6

LEARN ALL THE FACTS ABOUT DBS

MINIATURE SATELLITE DISHES
The New Digital Television

This affordable book, written in the concise style that has become the trademark of Dr. Frank Baylin, explores all aspects of this exciting new field. Topics covered include:

- The background of DBS. How does it relate to large-dish TVRO systems
- How these satellites and receive systems function and installation methods
- The major players. Learn details of their systems including their satellites and future plans.
- A comparison of the competing programming packages.
- An exciting new strategy to upgrade large-dish TVROs to dual-band, DBS/TVRO operation.
- A study of the encryption systems. Are they secure? Will the pirates be held in check?
- Learn to connect DBS systems to home entertainment centers
- What the future holds in store.

128 pages / 6 x 11" / illustrations

The Future is Here Today!

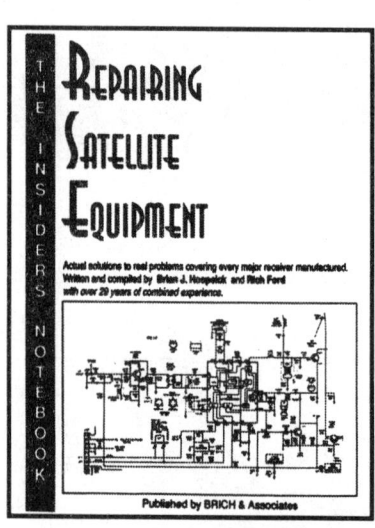

REPAIRING SATELLITE EQUIPMENT
"The Insider's Notebook" by Rich Ford & Brian Hoopsick

Learn the "how-to" of repairing nearly every brand of consumer home satellite TV receiver ever used in North America, including current models. This 130-page book in a 3-ring binder, is not theoretical but offers practical, real-world repair solutions. It is organized in a convenient format of PROBLEM in the left column, SOLUTION in the right. Actual repair procedures indicate components or components that need replacement.

Avoid shipping customers' receivers to repair centers and paying hundreds of dollars

REPAIR THAT OLD SATELLITE RECEIVER FOR PENNIES!

CHAPARRAL – CHANNEL MASTER – DRAKE – GENERAL INSTRUMENT – HOUSTON TRACKER – LUXOR – NORSAT – PANASONIC
STS TEE-COM TOSHIBA UNIDEN – EQUUS – KENWOOD – MTI – ECHOSPHERE – FUJITSU JANEIL – MA/COM – PICO
SCIENTIFIC ATLANTA – PROSAT – ZENITH

Annual Updates Availlable

THE SATELLITE BOOK – 2nd Edition
A Complete Guide to Satellite TV Theory and Practice

Available Only From Our UK Office

Edited by John Breeds

This book, written specifically for the European market, presents a complete overview of satellite television written by 22 authors and edited by John Breeds.

CONTENTS: Geostationary Satellites. Tools for the Trade. Working with Ladders. Wall Fixing Systems for Dishes. Cables for Satellite TV Installations. The SCART Connector. Carrier Level Meters. Equipment Installation: A Guideline. Customer Care. Microwave Basics. Ferrite Polar Selectors. Signal vs. Noise. Frequency Modulation. Satellite Antennas. Polyrod Lens Feed. Flat Plate Antenna. Link Budget Analysis. Introduction to SMATV. Intermediate Frequency Distribution Systems. British Telecom's Role. The Future for EUTELSAT. The ASTRA Satellites. The MAC Transmission Standard. The VideoCrypt Encryption System. Eurocrypt for MAC.

292 pages / 21.1 x 29.7 cm / 300 illustrations, photographs, and tables / glossary / index / 1-872567-02-9

WHAT'S on SATELLITE?
Video Activity Report on all Satellites

This accurate and detailed report, published three times a year by Design Publishers, lists the video services on active transponders for all the world's communication satellites. This information allows you to keep abreast of new programming, satellite launches, rapidly changing new services and satellite relocations. Includes:

- Location of each satellite
- Name, transponder number, polarity, bandwidth, center frequency and language of each service.
- Any associated audio subcarriers and/or SCPC (single channel per carrier) frequencies.
- Auxiliary radio services which are transmitted either by means of auxiliary audio subcarriers or independent SCPC carriers
- Estimated beam-center EIRP levels for each satellite TV service

One-Year Subscription – $95.00 U.S.

First Class Air Postage: USA – $8.00, Other Countries – $30.00

3 Issues – January, May, and September

FUTURE TALK
COMMUNICATIONS, TECHNOLOGY and SOCIETY in the 21st CENTURY

by Dr. Joseph N. Pelton

Future Talk, an insightful 222-page book by Dr. Joseph Pelton, explores his vision of the future of telecommunications. He presents a thoughtful study of topics such as the telepower revolution, the global electronic machine, living at super speed, the future of work, "smart energy," money goes electronic, life in the telelcity, electronic education, freedom, justice and telepower, redefining the world with telepower, telewar and the global brain.

222 pages / 6 x 9 inches / selected bibliography / published November 1992

1995 INTERNATIONAL SATELLITE DIRECTORY

The **COMPLETE REFERENCE** source of worldwide communication satellites, manufacturers, service providers, and users of satellite services. This brilliant new edition contains more than 25,000 entries, conveniently organized in 10 tab-indexed chapters.

CONTENTS: International Organizations. Government Regulators and Administrators. U.S. State Agencies. **Satellite Operators.** International Systems: Inmarsat, Intelsat & Intersputnik. Regional Satellite Systems: Arabsat, Astra, Eutelsat, Palapa, etc. Domestic Satellite Systems. Planned Satellite Systems: Alascom, Aussat, Brasilsat, Galaxy, GTE, Spacenet, etc. Planned Systems: ACTS, Cygnus, etc. **Manufacturers of Space Equipment. Manufacturers of Satellite Ground Equipment. Commercial and Domestic Systems. Users of Satellite Services.** Satellite Programmers. Transponder Usage. Location of Programmers by Interest. Broadcasters. Telephone Companies. Cable Companies. **Providers of Satellite Services.** Transmission Services.
Transponder Brokers. Teleports. Videoconferencing. VSATs. **Services for the Satellite Industry.** Distributors. Technical and Consulting Services
 Associations. Legal Services. Insurance Services. Research Centers.
Educational and Technical Centers. Publishers and Publications. Financial Institutions. **Geosynchronous Satellites.** List of Satellites. Satellite Database, EIRP Maps, Detailed Information on Satellite Spacecraft. **Index.**

Over 1250 pages / 8-3/8 x 10-7/8" / index / ISSN 1041-4541

THE AFFORDABLE STANDARD REFERENCE BOOK
1995 WORLD SATELLITE YEARLY
Footprints, Programming and Technical Use

This affordable book, written in the concise style that has become the trademark of Dr. Frank Baylin, provides the information required to easily determine the satellite programming available and the equipment necessary to receive these satellite signals from any point on our globe. It is organized into four easy-to-use sections that are separated by tabs.

SECTION I outlines required:
- METHODS to receive satellite audio and video signals
- to interpret satellite footprints
- to size antennas and select equipment
- to aim an antenna
- AS WELL AS the latest in audio and video compression
- an overview of scrambling methods and broadcast standards

SECTION II presents:
- a complete listing of footprint maps and other information about over 350 worldwide satellites – past, present and planned
- easy-to-read and accurate footprints of all active satellites

SECTION III lists:
- programming available on each active satellite
- world video standards and scrambling systems

SECTION IV lists:
- Addresses and telephone numbers of satellite manufacturers, service companies, programmers and major satellite system operators

780 pages / 8-1/2 x 11" / illustrations / glossary / index

NEW, POWERFUL, EASY-TO-USE & COMPREHENSIVE

SATELLITE TOOLBOX SOFTWARE – VERSION 1.2
Designed by Dr. Frank Baylin

PROFESSIONAL, EASY-to-USE, MENU-DRIVEN SOFTWARE
*Worldwide Satellite Programming
Antenna Aiming & Sizing
Receive-Site Analysis
Graphs, Tables & Reports*

This remarkable software does it all, and with surprising ease:

SATELLITE DATABASE: Lists of all the world's communication spacecraft and the programming they relay. Select any of the world's satellites to view all the video and audio programming available as well as transponder bandwidths, center frequencies or polarization formats. Or enter site latitude and longitude and print a list of those satellites viewable from your receive site.

AIMING SECTION: To activate simply enter your site coordinates. The software *automatically calculates magnetic variation*, compass heading, azimuth bearing as well as all the aiming angles for a polar or modified polar mount to any satellite within view or any group of satellites that you select. A datafile of major cities worldwide with latitudes and longitudes is provided. In addition, a similar database on every city, town and village in ITU regions I, II and III is available at a small additional cost.

ANTENNA SIZING: This component makes antenna sizing, choice of LNB noise temperature and other system parameters simple. Choose between NTSC, PAL or other broadcast formats to calculate G/T, S/N, C/N, etc... Print detailed analyis reports with your business name as the source. Save an unlimited number of files for each customer site anywhere on the globe.

ANALYSIS: This component allows you to easily calculate virtually every parameter relating to satellite reception including antenna gain, side lobe gain, focal distance as well as path loss, slant distance, G/T, C/N, video and audio S/N and much more. Create graphs or tables of how any one variable affects the outcome of any other. How would changing antenna size or LNB noise temperature effect audio signal-to-noise ratio? Answer a multidude of other pertinent questions.

SYSTEM REQUIREMENTS
BASIC SYSTEM
MSDOS 3.0 or higher version
A minimum of 640 kb of RAM
and
3 MB of harddrive space

ADDITIONAL DATABASES
ITU1 – 2.09 MB
ITU2 – 2.43 MB
ITU3 – 3.42 MB

MAIN PROGRAM........ $95

DATABASES:
ITU1, 2 or 3 .... $70
Any 2 databases ... $105
All 3 databases $120

YOU WILL BE ASTONISHED BY THE BEAUTY, POWER and SIMPLICITY of this SOFTWARE

The program runs without a hitch and the documentation is excellent. In light of the wide range of calculations and data, the program ... would be a bargain at twice Baylin Publication's price – *Satellite Retailer Magazine*

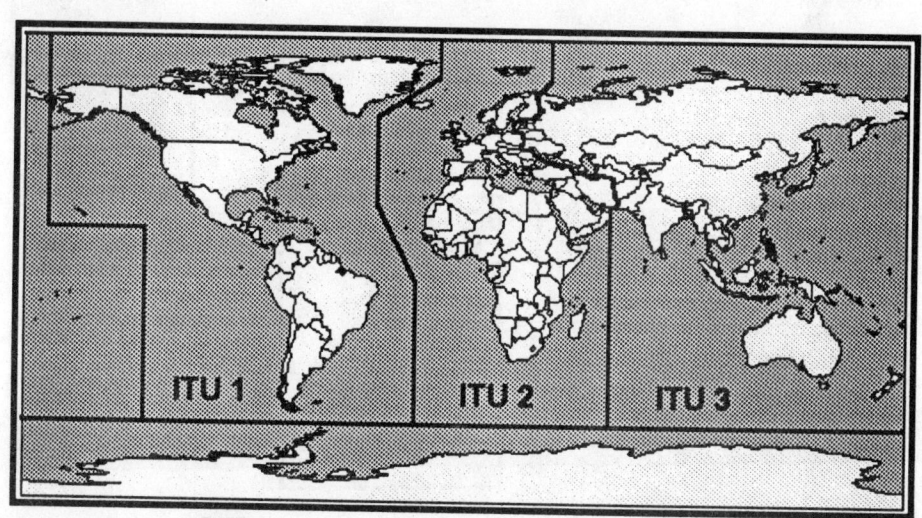

TOOLS for SATELLITE INSTALLATIONS

1. COAXIAL CABLE CUTTER & STRIPPER. This lightweight tool cuts and strips coaxial cable to prepare it for fittings. Adjust- able to fit RG-58, RG-59 and RG-6 coax. **$26.00**

5. DIGITAL COMPASS Works as easily as point-click-read. Has pistol sights, information display, memory and timer control, and bearing trigger. Memory holds 9-bearings **$167**

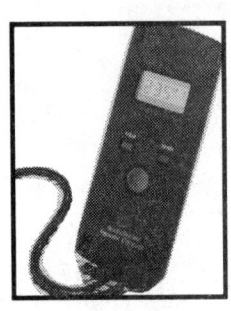

2. SILVA POLARIS COMPASS Lifetime guaranteed compass with magnetized Swedish steel needle, sapphire bearing. Dial reads in 2° increments. **$17**

6. ANGLE FINDER (LARGE) Level with magnetic base to measure angles from 0 to 90 degrees easily and accurately in any quadrant. **$23**

3. BIRD FINDER The perfect tool for sighting satellites. Built-in compass and inclino- meter on sturdy tripod. Aluminum construction – 36 oz. **$174**

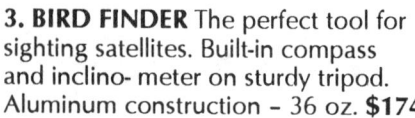

7. ANGLE FINDER (SMALL) Accurate to 0.5°. Easy, accurate reading from 0° to 90°. **$20**

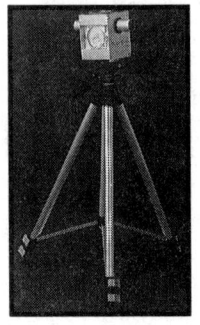

4. SILVA COMPASS/INCLINOMETER #15CL Compass & inclinometer in one, with sighting mirror for 1° accuracy. Has rotating capsule, sight line and adjusting screw for declination readings. Azimuth and quadrant scale. **$59**

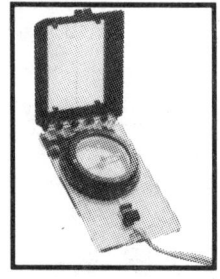

8. SATELLITE TUNING METER Bulz-I meter operates in the 70 to 1.4 GHz range. Useful for peaking dishes. For single and block downconversion receivers. **$139** With audio output **$159**

4A. TWIN SUNTO COMPASS/INCLINOMETER Sight through or prism views for measurements of both azimuth and elevation angles. Jewel bearing with accuracy to 0.25°. Threaded tripod socket. Clinometer scales in degrees or percentages. **$155**

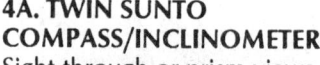

9. SCOTCH LOK CONNECTORS 100 connectors per box. Self-stripping, insulated and moisture resistant with either 2-wire or 3-wire conductor. **$28**

10. INCLINOMETER Sight through to see slope, percent or degree scales for measuring elevation angles with precision and accuracy. Cross hairs extended by optics for ease of surveying or installing satellite dishes. **$105**

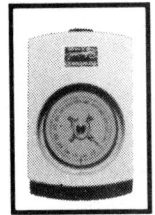

11. COMPASS View-through, liquid-filled instrument with sapphire bearings and rapid settling time. Precision designed for professional use. **$95**

12. COMPASS – SILVA 80 Employs glass optics and is liquid dampened. Scale has sapphire bearing yielding an accuracy of 0.5 degrees. Water resistant. Has internal light. **$118**

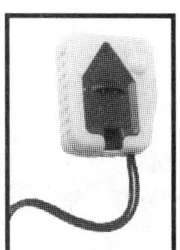

13. KWIK-STRIP Coaxial cable stripper. Cuts and strips RG-58/U, RG-59/U or RG-6 coaxial cable. Replaceable precision ground cutter blades **$24** Extra blades **$3**

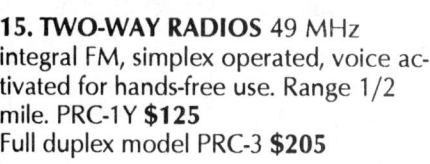

15. TWO-WAY RADIOS 49 MHz integral FM, simplex operated, voice activated for hands-free use. Range 1/2 mile. PRC-1Y **$125** Full duplex model PRC-3 **$205**

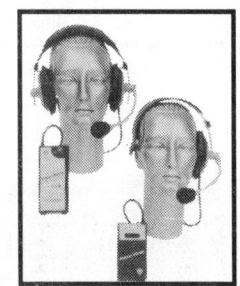

16. FOCUS FINDER CENTERING TOOL 1- or 2-piece ball fits either C-band or Ku-band feeds. Ball inserts into feed to easily measure focal distance. Telescoping tube to 60". For C- and Ku-band **$54**

17. LASER FOCUS FINDER. Beam of red laser light targets center of dish. Small, portable tool. Fits both C-band and Ku-band feeds. **$199**

18. DIGITAL INCLINOMETER 360° Digital read-out. 0.1° accuracy. Reads degrees of slope, % of slope, inches per foot (pitch) & electronically simulated analog bubble display from distance of 10-feet. Large LCD display. Length 16cm. **$124** With magnetic base **$130** 24, 48 and 72 inch rails. **$39, $65** and **$95,** respectively.

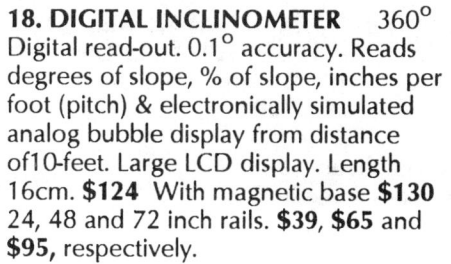

18b. SATLOOK SPECTRUM ANALYZER This unit features a built-in satellite receiver (900 to 1750 MHz), a 4.5" b/w monitor, an audible peaking tone and rechargeable internal battery. Switching on LNB voltage, 13 or 18 Vdc. Ideal for DSS With carrying case. **$799**

19. DATA SCOPE DIGITAL COMPASS/RANGE FINDER Monocular, fluxgate compass and electronic range finder in one! Compass: 0.5° accuracy, stores 9 bearings in memory; calculates change between bearings. Range Finder: determines range from known height & relative size. Includes chronometer. **$479**

20. DISH CRANE For lifting satellite dishes. Model for 10' dishes. **$360** Model for 12' dishes. **$460**

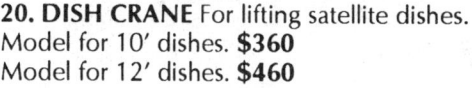

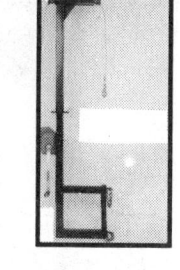

21. E-Z TRENCHER Cuts 7-inch deep, 100-foot trench in approximately 5 minutes.
Model J-100 cuts 1.5" x 7" **$1155**
Model I-2000 cuts 2.5" x 13" **$2600**
Model TS-1000 cuts 4" x 7" **$1176**

22a. WAYFINDER DIGITAL NUMERIC COMPASS. LCD display, water resistant digital compass with backlighting. Easily calibrated to zero out effect of nearby metal. For boating, provides heading average in rough conditions. **$135**

417

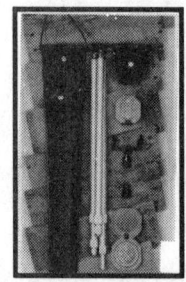

25. BRUNTON POCKET TRANSIT with INCLINOMETER. HIGHTECH VERSION. Dial graduation of 1°. Water proof. 360° rotation suspension. **$204**
Tri-pod **$99**
Ball & socket head **$39**
Case for complete set **$88**

26. PORTABLE ACTUATOR. Self-contained, operated by rechargeable batteries. With power supply to operate LNB and Bultz meter. **$189**

28. SERVO MOTOR TESTER. This battery-powered, accurate pulse generator tests servo motor movement. Also used to check receivers/cables for LNB power and presence of the necessary pulses to drive servo motors. **$93**

29. PORTABLE RECEIVER/TEST SET. C/Ku-band systems. 950-1750 MHz input. Digital read-out. Baseband output for decoders. Polarity test. Video and audio fine tuning. Built-in 3" color monitor and actuator. Operates from rechargeable batteries. 24 pounds **$1310**

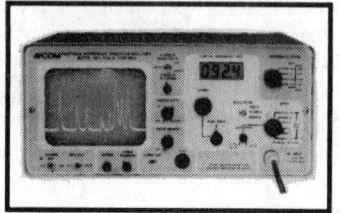

30. AVCOM PORTABLE SPECTRUM ANALYZERS PSA-65A has a 0.1 MHz accuracy at 0 span with −95 dBm sensitivity. Line or battery operated, LCD digital frequency display.

PSA-65A. Spans 2 to 1000 MHz. **$2945**
PSA-37D Spans 10 to 1750 MHz and 3.7 to 4.2 GHz. Powers LNAs and LNBs. **$2545**
Carrying Case $109

31. AVCOM PTR-25 PORTABLE TEST RECEIVER C/Ku-band, battery or ac power operated, audio and video signal strength indicators and polarotor controls. **$1445**
Carrying Case $109

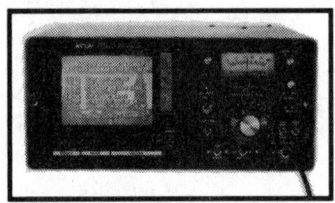

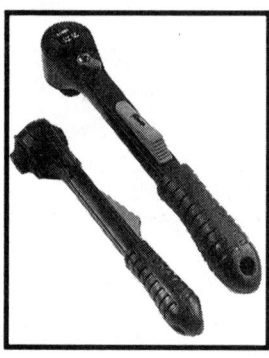

33. JAWS POWER Continuously adjustable, patented ratchet with contoured sure-grip handle and inverted jaws gear system. Small size adjusts from 8 mm (under 5/16") to over 14 mm (17/32"); large size adjusts from under 11 mm (7/16") to over 19 mm (3/4").
Large size $42
Small size $36

ORDER INSTRUCTIONS for SCPC RECEIVER and TOOLS

TOOLS MUST BE PURCHASED ONLY FROM OUR BOULDER, COLORADO OFFICE

1.	COAXIAL CABLE CUTTER & STRIPPER		$26.00
2.	SILVA POLARIS COMPASS		17
3.	BIRDFINDER		174
4.	SILVA COMPASS/INCLINOMETER – #15CL		59
4A.	TWIN SUNTO COMPASS/INCLINOMETER		155
5.	DIGITAL COMPASS		167
6.	ANGLE FINDER (Large)		23
7.	ANGLE FINDER (Small)		20
8.	SATELLITE TUNING METER		139
	with audio tone		159
9.	SKOTCH LOKS		28
10.	INCLINOMETER		105
11.	SIGHTING COMPASS		95
12.	SILVA COMPASS – 80		118
13.	KWIK-STRIP COAX STRIPPER		24
	extra blades		3
15.	TWO-WAY RADIOS		
	PRC-1Y		125
	PRC-3X		205
16.	FOCUS FINDER TOOL (C/Ku-band)		54
17.	LASER FOCUS FINDER		199
18.	DIGITAL INCLINOMETER		124
	with magnetic base		130
	24" rail		39
	48" rail		65
	72" rail		95
18b.	SATLOOK SPECTRUM ANALYZER		799
19.	DATA SCOPE		479
20.	DISH CRANE (10 ft. antennas)		360
	(12 ft. antennas)		460
21.	E-Z TRENCHER		
	MODEL J-100		1155
	Model J-2000		2600
	Model TS-1000		1176
22a.	WAYFINDER DIGITAL COMPASS		135
25.	BRUNTON TRANSIT–HIGH TECH VERSION		204
	Tri-pod		99
	Ball & Socket Head		39
	Case		88
26.	PORTABLE ACTUATOR		189
28.	SERVO MOTOR TESTER		93
29.	PORTABLE TEST RECEIVER		1310
30.	AVCOM PSA-65A		2945
	PSA-37D (900 to 1450 MHz)		2545
	Case for each		109
31.	AVCOM PTR-25A TEST RECEIVER		1445
	Case		109
32.	PANASONIC		
	EY6040 Portable drill		75
	EY6207 Portable drill		339
	EY6220 Drive & driver		146
	EY6990 Hammer drill & driver		275
	Case for each		109
33.	JAWS POWER		
	Large		42
	Small		36

SHIPPING CHARGES for TOOLS

Exact air or sea shipping charges will be added to credit card orders, the preferred method of payment. As a high estimate, add 20% for overseas air, 10% for overseas sea, 10% for US/Canada air, 5% for US/Canada sea.

Total Cost for this Page $_____
Shipping $_____
Total for Tools $_____

TRANSFER TOTAL TO OPPOSITE PAGE

ORDERING INSTRUCTIONS for BOOKS

Baylin Publications
1905 Mariposa
Boulder, CO 80302, U.S.A
Telephone: 303-449-4551
FAX: 303-939-8720

Distributed from England by:
Swift Television Publications
17 Pittsfield
Cricklade, Wiltshire SN6 6AN, U.K.
Telephone: 44 (01) 793-750620
FAX: 44 (01) 793-752399

ITEM	PRICE $	PRICE £	NUMBER	TOTAL
World Satellite TV and Scrambling Methods – 3rd edition	$40	£29		
Ku-Band Satellite TV – 4th edition	30	25		
Satellite Toolbox Software (5-1/4" or 3-1/2" diskette)	95	59		
– ITU Region 1, 2 and 3 Databases	See page 11 for prices			
Update to Satellite Toolbox – Satellites/Programming Databases	20			
Wireless Cable and SMATV	50	35		
Satellite & Cable TV Scrambling – 2nd edition	20	21		
Home Satellite TV Manual, English – 3rd revised edition	30	25		
Install, Aim & Repair Your Satellite TV System – 2nd edition	10	12		
Satellite TV Installation Video – English ❑VHS only	40	27PAL		
Home Satellite TV Manual ❑Arabic ❑Spanish ❑Portuguese	30	25		
Satellite TV Installation Video – Spanish ❑BETA ❑ VHS	40			
Hidden Signals on Satellite TV – 3rd edition	20	23		
The "How To" Book of Satellite Communications	25	22		
1995 World Satellite Yearly	90	59		
What's On Satellite? (3 issues)	95	Shipping: US $10; other $26		
1995 International Satellite Directory	260			
Future Talk	20			
Repairing Satellite Equipment (The Insider's Notebook)	80			
European Scrambling Systems – IV	55	32		
Miniature Satellite Dishes – The New Digital TV	20			
The Satellite Book	50	32		
	Total for this page			
	Merchandise Order Total from previous page			
	Sales Tax - Colorado Residents only (3.7%)			
	Shipping (see below for detailed costs)			
	Grand Total			

Method of Payment:
Check Enclosed ❑
Cash ❑
Credit Card ❑
COD ❑

Shipping Method:
Air Mail ❑
Surface ❑
UPS Blue ❑
UPS Red ❑

Visa, Mastercard & Amex – U.S. Office
Visa, Mastercard, Eurocard & Access – U.K. Office

SHIPPING INSTRUCTIONS

to UNITED STATES: Add $4.00 for each item ordered. UPS regular, 2nd day or overnight may be requested at additional charge. UPS COD orders are an additional $4.50. MasterCard, Visa and Amex cards accepted.

to CANADA: Add $5.00 per item ordered. Please remit funds in U.S.$ or by credit card. Orders are shipped via book rate mail. UPS shipping can be requested, but no COD is available to Canada.

WORLDWIDE (from our US office): Remit payment in U.S. funds as checks drawn on U.S. banks, money orders, Master, Visa or Amex credit cards, or cash. Shipments are by surface mail unless additional air mail charges can be added onto credit card billing.

from UK OFFICE: Add £2.50 postage per book within UK. Add £5 per book to Europe; beyond Europe add 30%. Optional insurance £4 extra. Remit payment in UK £ Sterling drawn on UK bank, Eurocheque, Postal Order, cash, Visa, Mastercard, Access or Eurocard.

ORDER TODAY
(Quantity Discounts Available)

Name:_____

Company:_____

Address:_____

_____

_____

Country:_____

Telephone:_____

FAX:_____

Credit Card No.:_____

Expiration Date:_____

Signature:_____